THE GENE KEYS

Embracing your higher purpose

Richard Rudd

WATKINS

Sharing Wisdom
Since 1893

This edition published in the UK and USA 2015 by
Watkins, an imprint of Watkins Media Limited
Unit 11, Shepperton House, 89-93 Shepperton Road, London N1 3DF

enquiries@watkinspublishing.com

First published 2009 by Gene Keys Publishing
Second Edition 2011
Third Edition 2015

Book Design by Lynda Rae
Aurora Design Studio
www.auroradesignstudio.com

17 18 19 20

Printed and bound in Turkey

A CIP record for this book is available from the British Library

ISBN: 978-1-78028-542-9

www.watkinspublishing.com

To all who are reading or hearing these words:
May your heart burst open in unconditional Love
May your mind be illuminated by infinite Peace
May your body be flooded by the light of your Essence.

May all whom you touch in this life
Through your thoughts, your words and your deeds
Be transformed by the Radiance of your Presence.

CONTENTS

ACKNOWLEDGMENTS

Writing this book has been a great adventure for me. Right from its beginning, five years ago, it has led me a merry dance. I know that many writers experience this same feeling of their book having a soul all of its own. The Gene Keys have overseen the creation of a rich new landscape all around me, weaving together the many multi-coloured strands of my life into a vast magic carpet of pure possibility.

Many people, both seen and unseen, have made this journey possible for me.

Sheila Buchanan and Neil Taylor have walked the path through the Gene Keys with me from the beginning and I will always feel indebted beyond words to the unending faith, love and wisdom they have both for me and for this work. That this book exists at all is due in no small part to the altruistic spirit in which they have supported me down the years.

More recently, my life has also been blessed with the presence of Teresa Collins and Marshall Lefferts, both of whom have imbibed the Gene Keys deeply into their being, thus empowering me to discover new levels of synthesis and integration in the living body of the text. Both Teresa and Marshall have given their love, time and copious gifts unconditionally to the Gene Keys and continue to do a great service to the work and our growing community. Their combined skills have overseen all aspects of the publication of this book. Once again, my gratitude towards them both is beyond words.

There have of course been many teachers who have influenced me down the years and many of their hearts beat within the living veins of this teaching. Above all, Omraam Mikhael Aivanhov has been a constant internal reference for me, guiding me from a plane beyond our own. It is his prophetic understanding of the notion of the Great White Brotherhood and the coming planetary synarchy that permeates much of the wisdom behind the Gene Keys. I feel personally enriched to be able to honour and bow to the purity and inner light of this great *Rishi* who has guided the upper reaches of my awareness for many years now.

On a more practical plane, the Gene Keys owe a great debt to Ra Uru Hu, the founder of the Human Design System. It was Ra who opened my eyes to the real nature of the I Ching, and it was Ra who also taught me how to read the codes hidden inside it. I will always feel a deep love and gratitude towards this man who guided me towards my higher nature, and whose genius ultimately paved the way towards the Gene Keys themselves.

Still on a practical level I am also exceedingly indebted to Barbara McKinley for her excellent editing and for reorganising the text and even adding to its fluidity and natural cadence. She is a woman of rare ability and great generosity of spirit. Also many thanks to Lynda Rae for laying out the book and Jackie Morris for her beautiful dragonfly. A special thanks too to Melanie Eclare and Tom Petherick (a pair of stars sent from heaven) for their part in helping to bring this book and the Gene Keys into the world.

The journey that has led to the existence of this book has been one of many twists and turns, and I would like to thank some of the heroes and heroines I have met along the way whose destinies and hearts are woven at some level into its story.

My gratitude to the following: Werner Pitzal for his amazing brotherly love, Linda Lowrey for her trust and devotion, Peter Maxwell Evans for his totality, Marina Efraimoglou for her warmth and generosity, Chetan Parkyn for his endless enthusiasm, Sally Searle for her empathic friendship and Shofen Lee for her consistently pure heart.

The profound love and recognition I have felt from all these people has greatly accelerated my own ability to dive into the higher frequencies of the transmission and pull down its most exquisite jewels and insights. Again, I feel gratitude beyond words.

My penultimate acknowledgment is towards all the students I have met and come to know down the years. Many of them have become my trusted friends, allies and in some cases co-teachers. None of this would be possible without the love, support and encouragement of the core community that has formed around these teachings. I know that great things lie ahead for us all, and I am excited to explore our deeper communion together as a single unified field of consciousness. I drink to you all!

Finally, I must honour my family, my parents and my beautiful children, who are a continual inspiration and delight to me. Above all I bow to Marian my beloved wife. Her strength, radiance and pure spirit have enabled me to bring a new higher teaching down to earth. So often it is we men who are afforded all the praise for our creations. Even though it was I who spoke the actual words, it is Marian who has provided the space for this magic to occur, and for that I will always hold her more precious within my heart than words can say. Both as a muse and a mother, and as a friend and wife, she has grounded my dreams and been the constant earth to my star-drawn heaven.

FOREWORD

Welcome to the Gene Keys.

This book is an invitation to begin a new journey in your life.

Regardless of outer circumstances, every single human being has something beautiful hidden inside them. The sole purpose of the Gene Keys is to bring that beauty forth — to unveil your *incandescence*, the eternal spark of genius that sets you apart from everyone else.

Recent breakthroughs in biology point towards an amazing truth — your DNA, the coiled code that has made you who you are today, is not in control of your destiny. Rather, it is your general attitude to life that tells your DNA what kind of person you want to become. This means that every thought, feeling, word and action that you make in life is imprinted in every single cell of your body. Negative thoughts and emotions cause your DNA to contract, whereas positive thoughts and emotions cause it to expand and relax. This process is going on all the time, from the moment you come into the world to the moment you leave.

You alone are the architect of your evolution.

When you fully embrace the implications of this truth, then your journey has begun. Even if you read no further, this truth can transform your life. It is reflected in the lives of all the great saints and spiritual teachers throughout history — that inside each of us lies vast potential and beauty, and it is limited only by the way we see ourselves and the world around us.

Our planet is in the midst of a vast transition in which humanity is playing the central role. A great quantum leap is in the air. The Gene Keys offer us a vision of a very different world than the one we see today. They show us a world where human beings are governed by higher principles such as love, forgiveness and freedom. Such a world is not a dream, it is the next stage of our natural evolution and it depends upon each of us unlocking the higher purpose that lies hidden in our DNA.

I hope as you enter and explore the wonderful labyrinth of the Gene Keys that they will ignite the spark of your highest potential, and that you will embrace the beauty of your personal dream and allow it to grow inside you. And as your higher purpose begins to impact those you love and the world around you, may you join your genius with all of us who dare to dream of a higher and better world, and together let us make that dream a reality.

Richard Rudd

HOW TO USE THIS BOOK

The Gene Keys teachings are designed as an open system that you can explore in a wide variety of ways. Unlike many teachings, they are a self-teaching transmission that awakens within you, rather than a structure imposed from the outside through a set discipline or teacher. In this sense, the Gene Keys are an adventure that you can tailor to your own life, moving at your own pace and trusting in your intuition and imagination.

CONTEMPLATION

The great potential of the Gene Keys as a teaching is to awaken a powerful new creative impulse inside you, and as you follow this impulse you begin to witness the emergence of your genius. The central technique that makes this possible is Contemplation. Contemplation is something of a forgotten spiritual path. Unlike meditation, it does not completely bypass the mind, rather it uses the mind in a playful way to open new inner pathways inside our brain and body. It is through sustained gentle contemplation on the Gene Keys that we can affect subtle changes in our biochemistry.

The Gene Keys are designed to be contemplated and digested over time. Each Key contains a unique message, and each message takes time to absorb before you feel a change occurring in your life. Contemplation is far more than simply thinking about something. It is the direct imbibing of a universal truth at a physical, emotional and mental level. Therefore it is a good idea to begin your voyage into the Gene Keys with a sense of relaxedness and patience. To enter on a path of contemplation is to slow down inwardly in order that you begin to see things around you more clearly.

A CODEBOOK OF CONSCIOUSNESS

The Gene Keys are a new codebook of consciousness. To apply them directly to your life, you need to know which codes apply specifically to you and how. All biological codes are imprinted at specific times and those times can tell us a great deal about you.

When you read the Introduction to the Gene Keys, you will learn about your Hologenetic Profile. Your Profile is calculated from the time, date and place of your birth and can be accessed free online at: **www.genekeys.com**. Once you have your Hologenetic Profile, your journey of contemplation begins in earnest, as it highlights specific Gene Keys that have a powerful bearing on your purpose, your relationships and your prosperity.

TREADING THE GOLDEN PATH

Your Hologenetic Profile invites you into a deeply personal voyage through the Gene Keys. Certain sequences of Gene Keys imprinted at your birth unlock doorways of awakening inside you, and as you contemplate them and apply their teachings to your everyday life, you may feel a new spirit coming alive in you.

There are three sequences:

> The Activation Sequence – Discovering your genius through the four Prime Gifts
> The Venus Sequence – Opening your heart in relationships
> The Pearl Sequence – Attaining prosperity through service

Taken as a single integrated journey, these sequences are collectively referred to as the Golden Path.

As you contemplate the message in the Gene Keys you can pay special attention to those Gene Keys that comprise the sequences of your Golden Path. A full understanding of each sequence and its bearing on your life can be found on the website www.genekeys.com, through the Golden Path Program. This program is a step-by-step journey of deep contemplation into the forces that make you who you are. As you tread your Golden Path of contemplation through your sequences and their Gene Keys, considering and applying them in the light of your life, you will discover that you are treading a path of very powerful transformation.

THE GENE KEYS AS A CREATIVE COMPANION

There are many other ways to use the Gene Keys. You might like to use the book like an oracle, in the spirit of the original I Ching. You can open it at random in response to a question you have, or a challenge you are facing. At such times, the Gene Key you find yourself reading often has an uncanny knack of highlighting the hidden essence of the issue.

At the back of the book you will find a list of words known as 'the spectrum of consciousness'. This lists the words for each of the 64 Gene Keys at each level of frequency. If you are experiencing a certain Shadow state or are on the receiving end of a certain negative behaviour, you can scan down the various words for the Shadows until you find the Gene Key that matches that state. When you read that specific Gene Key it can help you to see the hidden higher consciousness behind that state. This can greatly enhance your compassion, either for others or for yourself.

However you choose to use the Gene Keys book, it is intended as an ongoing creative companion, rather than a book that you read and then discard. Because the Gene Keys are an open system, you are invited to invent new ways of using them or adapting them in your life. The most important thing is to use your imagination and enjoy your journey.

INTRODUCTION

PART 1:
UNLOCKING THE HIGHER PURPOSE
HIDDEN IN YOUR DNA

ST. BENEDICT'S EYES

It was a perfect autumn morning. The golden sun was just coming up over the Sibruini Mountains, lightly brushing the hoary frost from the earth's sleepy eyes. One of the nuns let me in through a side gate and in my halting Italian I asked her the way to the sacred grotto. Kindly and without any words, she led me through a maze of monastic passages and down countless sandal-worn steps until finally I stood in that most magical of places.

For the next two hours I lay with my head on the same stone pillow where a humble hermit once spent his life in a daily vigil, waiting and praying for a vision of God. After three years, his vision was granted in a thunderbolt flash from the heavens, he emerged from his inner sanctuary and in time founded the greatest and most successful Christian monastic order in history. His name was St. Benedict.

I tell this story because as I lay upon that cold stone slab, I too experienced a vision, or at least the reflection of a vision. Suddenly and without warning I saw a pair of eyes looking at me. I have no recollection whatsoever of a face around the eyes, but as long as I live I shall never forget those eyes. They were the eyes of one who has seen the Truth, brimming over with the profoundest love and knowing. A single string of familiar words accompanied the vision and rang out like a mantra inside my head:

"Mine eyes have seen the glory of the coming of the Lord..."

These were, I knew, St. Benedict's eyes.

Many new things begin with a vision. The thing I have learned about all true visions is that they are not limited to a single event. Once the initial download has occurred, the vision begins its real purpose — to transform you into the heightened frequency of the original experience. This book sprang up around a vision and it conveys, in as much as words can, the core of that vision. The vision I personally received came long before the experience of seeing St. Benedict's eyes. It came to me early in my life and I have carried it secretly inside me ever since. I might better say that the vision came *out* of me rather than *to* me, because as this book will testify, our destinies are written inside our very DNA.

Like St. Benedict's vision, my own was a direct experience of the perfection of all creation. I too saw the future in my vision, but I saw it as already having occurred, which gave me the same conviction that I glimpsed in the hermit's eyes. This book then affords a view through the universal eye — the eye that has already seen the beauty, wonder and certainty of the future of humanity. We are entering a new Solar Age and, like the sun, the Gene Keys offer nothing but optimism.

WHAT IS YOUR HIGHER PURPOSE?

You are a living genius. Every human being is born a genius. I am not just saying you have the *capacity* to be a genius, but that you *are* one, right now. Your higher purpose in life is to share your particular genius with the world. But what is genius? The original root of the word refers to a kind of overseeing spirit or guardian who looks after a person. It also has a clear connection with the word *gene*, which is one of the reasons it has come to be associated with certain *special* people with a genetic hereditary gift of intelligence or intellect. Today when we think of genius we generally think of it as intellectual prowess, as for example in the case of Einstein.

I would like you to understand genius in a new way. To begin with, you don't have to be intellectual to be a genius. Genius describes you living at your zenith. It is you living your life without holding anything back, owning and therefore transcending your fears. To be a genius is to have the courage to live life with an open heart, as a deep romance. As you journey through the Gene Keys you will be traveling through all the great human lives. Although their names may not be mentioned directly, you will recognise what it takes to live a great life. A great life does not necessarily mean a famous life. Out of all the people you know, think of a person you truly admire — not for their deeds, but for their character, their patience, their tireless optimism, their courage. Genius can thrive in the most mundane of environments.

Your genius simply makes you a truly joyous human being. That is your higher purpose — to be radiant for no reason other than being alive. Your genius can only emerge out of that inner radiance. If it doesn't make you truly joyous, then it isn't your genius. Your genius is the soil and your higher purpose grows out of that soil — whether it turns out to be a humble herb, a delectable fruit or a great oak tree. The Gene Keys are the gardening manual that will guide you through the growing process, but the seed is already there inside you, waiting inside your DNA.

You may also notice from the name of this book that the process of living your higher purpose has much to do with the notion of keys and locks. This is because your DNA is an actual code, and that code can only be unlocked when you have the right keys. We will learn more about how this works later on, but for now it is important to realise that you have in your hands a codebook whose sole purpose is to guide you into the field of your own genius. The universal keys are in your hands, but only you can discover the right keys and the correct order in which they fit your specific genetic code. It is a wonderful puzzle for each of us, and only our hearts can show us the way. The process of unlocking your higher purpose is one of turning the authority of your mind over to your heart. This alone will transform your life.

That is your higher purpose — to be radiant for no reason other than being alive.

A VOYAGE OF LETTING GO

The Gene Keys are presented to you as a voyage, a voyage that will change your life forever. For me, the writing of this book has been an act of inner dedication and transformation. The transmission contained in each of the 64 Gene Keys has unlocked a new interpretation of my own genetic code at a cellular level, which has revealed my true higher purpose to me. Greater possibilities have wakened inside me, and a new and higher frequency continues to draw me out and beyond the negative patterns that we all carry through life. This has not always been easy. Much of the language and understanding of the Gene Keys is rooted in our unconscious fears. As you will discover for yourself, such fears are so tightly woven into your DNA that it takes real courage to confront them directly. However, fear is the raw material of the higher states, and it must be passed through.

As the first recipient of the Gene Keys, I have learned a few things that I wish to pass on to you, my fellow intrepid voyagers. The first is that the Gene Keys represent a living energetic field that exists at all times inside you. They are a wild wisdom rather than a systematic, logical process. This has been personally very challenging for my mind because I have discovered that there is no set pathway through your inner being. You have to work out these teachings for yourself, inside yourself. There is no guru or guide to show you the way. There is no technique unless you invent your own. There are simply gentle pointers which I will continue to share with you throughout this introduction. For me, the Gene Keys are more about dismantling concepts rather than adding any new ones. At the end of the day, the transformation happens of its own accord, simply because you are ready for it.

The Gene Keys are more about dismantling concepts rather than adding any new ones.

My personal voyage through the Gene Keys continues to open up new and daring panoramas within me. Above all, the Gene Keys have given me inner freedom. At times I have had to move on from many old systems, teachers and even friends because their views were neither compatible nor comfortable with such vast inner vistas of freedom. My greatest breakthrough with the Gene Keys took place as I came to write the 55th Gene Key, after returning from the USA where I had been teaching. This Gene Key is the activation code for freedom itself, so perhaps I should have been more prepared! However, as this Gene Key opened up inside me I was shocked to discover how deep was my terror of true freedom. I now realise that this is one of the greatest of all human fears, especially at this time in history — the fear of freedom.

When you read the 55th Gene Key, you will have a taste of what is to come for humanity. Indeed, the time we are currently living through is referred to throughout this book as the *Great Change*. It is a change that is taking place at a molecular level within humanity, and it is also impacting all natural systems and all creatures. Everywhere you look today — at the environment, our political and social structures, the world economy, religion, science and technology — you will see a world preparing itself for a leap in consciousness. Such times are inherently unstable and a deep collective fear of the coming change is moving like a spectre through our world. It is this fear that the 55th Gene Key directly challenges.

As I dove deeper and deeper into this fear and encountered the living transmission field of the 55th Gene Key within me, a vast and unbridled sense of inner freedom swept through my body like a tornado. I touched the simplicity of the freedom that comes when your mind is relieved of control over your life. The fear simply evaporated. Another far more powerful presence awoke inside my solar plexus and stepped into the driving seat. A new awareness dawned within me opening its eyes like a newborn baby in my heart. The rush was immense. Released from inside my DNA, light was pouring out of my body, connecting me to every conceivable aspect of the universe.

I learned then the great inner secret of DNA itself.

Your DNA is a wormhole. It contains a code that, when activated, opens up to the core of the holographic universe. As such, the DNA molecule is really a transducer of light. The more open the wormhole is, the more light pours through it. Like a torus, it both draws light towards itself and emanates it outward. Eventually, so much light will radiate through you that the wormhole itself collapses. The resulting supernova reveals to you your true universal nature as one with all creation.

This is the highest role of the 64 Gene Keys — to illuminate your path towards this final glorious flowering — union with your own Divinity. Then perhaps your eyes too will blaze with the fire of pure knowing, deep compassion and fearless freedom, as you emerge triumphant like St. Benedict from the cave of your old self.

PART 2:
THE 64 GENE KEYS —
A BOOK THAT TALKS TO YOUR DNA

Who, what and wherever you are,
if you are not continually transcending, then you are dying.
— The 3rd Gene Key

THE AGE OF SYNTHESIS

As the Great Change gradually dawns within humanity, much that we now take for granted in the world will be transformed. One of the greatest shifts currently taking place in the human sphere is in the role of science. For hundreds of years now science has been founded upon the left-brain approach of observing nature objectively and making logical assumptions based upon empirical evidence. However, given that the new human being, *homo sanctus,* will no longer operate primarily from mental awareness, a whole new way of seeing the world will emerge. In fact, it will no longer be possible to *observe* life without *being* life. To a fixed modern logical mind this is not an easy thing to understand. A fundamental change is coming to our brain structure and even today we see its early manifestations. We are entering the Age of Synthesis.

True synthesis can only come about when the left and right hemispheres of the human brain are in balance. This means that a new kind of thinking is emerging in humanity. It is really not thinking at all, but *knowing.* For example, as you read the words in this book, there may be insights within it that you simply know to be true right to the core of your being. This kind of intuitive knowing becomes stronger and more consistent as your life becomes more harmonious. What is occurring is that you are tapping into a fundamental holographic pattern found throughout the universe. The same geometry found in your genes is also found in vast, wheeling galaxies. As the Great Change moves through your DNA, it begins to realign every aspect of your life, bringing you gradually into harmony with these omnipresent universal patterns.

The 64 Gene Keys herald this new approach to truth, since they are the core archetypes found in every conceivable aspect of our universe. The 64-bit matrix is integral to physics, biology, music, geometry, architecture, computer programming and most fields of human research and endeavour. They form the foundational tetrahedral structure underlying space-time itself. It is therefore no wonder that they continue to be discovered at the core of all natural systems. As we shall see, our very DNA is based on this same 64-fold geometry, making us a holographic microcosm of the entire cosmos. Many of the great ancient civilisations and wisdom traditions, including the Vedic, Egyptian, Mayan and Chinese, encoded this mathematical structure into their art, cosmology and science.

THE 64
GENE
KEYS — A
BOOK THAT
TALKS TO
YOUR DNA

Wherever we look, from the structure of our cells to the rhythm and movement of the celestial bodies, we see the same fractal patterns, endlessly repeating themselves in ever more unique shapes and forms.

THE GENE I CHING

One of the best-known systems based on this grid of 64 is the Chinese I Ching — the *Book of Changes*. This extraordinary system, whose mysterious roots date back thousands of years, is one of the major inspirations for the Gene Keys. Generations of sages and lay people have used the I Ching as an oracular tool to help them make clear decisions in harmony with nature.

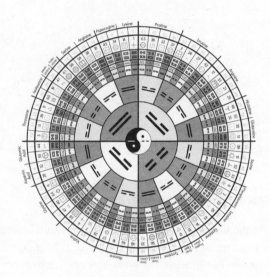

More than this however, the I Ching provides an extensive digital map of the energy dynamics within all living systems. This is especially fascinating because of its mathematical similarity to the genetic code. Many scientists, metaphysicians and mystics have begun to probe this striking relationship between DNA and the I Ching.

Without diving into a deep treatise on the subject, here is a simple version of the relationship between the I Ching and the genetic code.

Your DNA is made up of two strands of nucleotides, one strand being a perfect reflection of the other. This binary pattern is also the foundation of the *Yin* and *Yang* of the I Ching. Your genetic code is also made up of four *bases*, which are arranged in groupings of three. Each of these chemical groupings relates to an amino acid and forms what is know as a *codon*. There are 64 of these codons in your genetic code. Similarly, in the I Ching there are only four basic permutations of yin and yang, which are also arranged in groups of threes known as *trigrams*. In the same way that the two strands of your DNA reflect each other, each trigram has a partner. Together these two symbols create the *hexagram*; the basis of the I Ching. Just as there are 64 codons in DNA, so there are 64 hexagrams in the I Ching.

This precise mathematical correlation between the I Ching and the genetic code allows us to create a new holistic language that resonates within the living cells of your body. In effect, the 64 Gene Keys are a Gene I Ching — a book that talks directly to your DNA.

A GENETIC DOWNLOAD OF FREEDOM

Now that we have covered some simple background information on the Gene Keys we can begin to look at how they actually work.

Our very DNA is based on this same 64-fold geometry, making us a holographic microcosm of the entire cosmos.

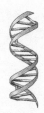

I would like to invite you to approach the Gene Keys in a completely original way. First of all, I would like you to let go of all the normal rules you have learned for reading a book. This is not a normal book. It is a genetic download specifically designed to penetrate the building blocks of your everyday reality — your DNA. Secondly, as you enter into this process of digesting the Gene Keys, I would like you to begin imagining what your life would look like in your wildest dreams. It doesn't matter so much what specifics your mind fabricates. What is important is that you recapture inside yourself a feeling of absolute inner freedom.

The Gene Keys process is about you giving yourself freedom, and this freedom begins in your imagination. You have to open yourself up to the highest possibilities of your own nature. As this process continues, it will most likely put you in touch with some very deep fears. The good news is that you no longer need fear these fears. We now know that they are inherited ancestral fears carried in the DNA of the bloodlines of all human beings. We understand that they had to be there for our survival as a species. The Gene Keys offer you the opportunity to face and eradicate each of the specific fears that stand in the way of your freedom.

Every day a scientist somewhere in the world discovers something new and amazing about DNA. Genetics is one of the hottest and newest frontiers in all of science. Even so, some of the greatest scientific insights in this field may lie way ahead in our future. But you do not have to wait for science to prove anything to your mind. You already have direct access to the laboratory of your DNA just because you are alive. You will discover that your DNA is simply waiting for you to input the right code. Once it has received your instructions, it will run the new program and build you a new body, a new life and a new reality. The evidence is lying within you.

Let's take a little tour around your laboratory and have a look at some of the equipment nature has given you. Take a moment to look at the palm of your hand and observe your own skin. It is made up of millions of tiny skin cells. Let's take a look inside just one of those cells.

The Gene Keys process is about you giving yourself freedom, and this freedom begins in your imagination.

The basic elements of a cell are threefold — there is an outer membrane, an inner mantle (cytoplasm) containing the working machinery of the cell, and a nucleus containing the cellular DNA and its instructions. You have about 60 trillion of these cells in your body, all with different roles and responsibilities. But most importantly, every single cell in your body right now is doing two essential things:

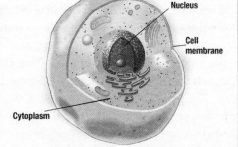

It is listening, and it is responding.

Each cell is listening to your environment through its countless molecular antennae embedded in the cellular membrane. This cellular skin interprets the environmental signals and relays the corresponding instructions to the DNA within the nucleus of the cell. Your DNA then responds by activating the necessary machinery within the cell.

THE 64
GENE
KEYS — A
BOOK THAT
TALKS TO
YOUR DNA

It is rather like the captain of a steamship relaying information from his lookouts down to the engine room. If the lookouts spot an obstacle, they alert the captain and he orders the workers in the engine room to either open or close the furnaces that drive the turbines or shift the gears that engage the propeller. In your body it is the same: the molecular switches in your cells tell the DNA which genes to switch on and which genes to switch off. This process goes on in all 60 trillion cells all the time, day and night, as long as you are alive. And you are designed to unlock the awesome molecular power stored within your body.

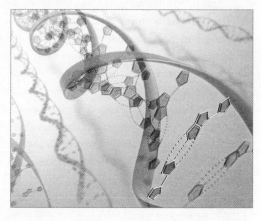

Now let's have a look inside the engine room — the nucleus of the cell. Here we can see the familiar double helix shape of DNA. What most people don't realise is that as a salt, DNA is a natural conductor of electricity. It is extremely sensitive to electromagnetic waves. Even a slight shift in your mood will create enough of an environmental signal to trigger a response from your DNA. Likewise a negative or a positive thought will generate a subtle electromagnetic current throughout your body that will stir your DNA into some form of biological response. Most of us are completely unaware of how our moods, thoughts, beliefs and general attitude literally mould our bodies.

Because of the heightened sensitivity of your DNA, everything in your life, from the food you eat to the people you live with, is co-creating your body via your attitude. Your attitude determines the nature of the electromagnetic signals that reach your DNA. For example, if you are having a bad day and find yourself in a negative frame of mind, this attitude will generate a low frequency impulse throughout your body. Your DNA will respond to this by shutting down certain hormonal pathways in your brain, and you will feel sad, depressed or frustrated. On the other hand, if you are having a bad day and are able to break out of your negative mindset and laugh at yourself, a high frequency electrical signal will reach your DNA and you will feel lighter and more joyful. Your DNA will respond by activating certain hormonal signals that will lead to your day feeling much brighter.

The process of programming your DNA through attitude is the foundation of the well-known placebo effect and is also the core of an important new branch of genetics known as epigenetics. Epigenetics is the study of how the environment affects your genes. This exciting new field of biology is far more holistic in nature than the old models we all learned at school. With epigenetics, when we extend the idea of environmental impact to the electromagnetic world of quantum physics, then it must also include human attitude. At a quantum level, your environment is your attitude.

What all of this means is that you can never be a victim of your DNA. Neither can you be a victim of fate. You can only be a victim of your attitude.

Even a slight shift in your mood will create enough of an environmental signal to trigger a response from your DNA.

Every thought you think, every feeling you have, every word you utter and every action you take directly programs your genes and therefore your reality. Consequently, at a quantum level, you create the environment that programs your genes. In the following section, we will learn more about how this works and how you can create the optimal environment to unlock the highest potential in your DNA.

This is the great secret that the Gene Keys hold — the secret of freedom. As you discover it for yourself, your life will be transformed before your very eyes.

PART 3:
THE LANGUAGE OF LIGHT

In the beginner's mind there are infinite possibilities. In the expert's there are few.
— Suzuki

THE ANCIENT ART OF CONTEMPLATION

If I had to say in a single line what the Gene Keys are, I would say they are a universal language made up of 64 genetic archetypes. If I had to say what the Gene Keys do, I would say they allow you to completely re-envision yourself and recreate your life at a level limited only by your own imagination.

The Gene Keys are also a transmission. In Buddhism there is a wonderful word known as *dharma*. It is one of those words pregnant with many dimensions of meaning. It points towards the existence of a higher truth or universal law pervading the universe. Because the realisation of the dharma is beyond words, its transmission can only be received through silence and deep meditative absorption. The Gene Keys are just such a transmission. As archetypes they each contain a fractal aspect of the same universal Truth. As genetic archetypes they allow you to resonate that Truth deep within each cell of your body.

This brings us to a very important point that you must know before you enter the living dharma-field of the Gene Keys. Because the Gene Keys are a transmission beyond words, they will not yield their secrets to an intellectual, grasping mind. The more you chase after them with your mind the more frustrated you may become. As archetypes, the Gene Keys are designed to be contemplated, and contemplation demands relaxation and patience. Contemplation is one of the greatest and least understood of the ancient mystical arts and paths.

THE THREE CLASSIC PATHS TO TRUTH

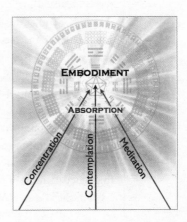

There are three classic paths that lead towards higher consciousness; they are meditation, concentration and contemplation. Even though each path is markedly different, they all lead to the same final goal — absorption, which we will hear more about later. Meditation is the great passive art in which all forms, thoughts and feelings are simply witnessed exactly as they are. Over time, this continued witnessing gives way to the natural arising of great inner clarity, which culminates in an enlightened view of the nature of reality. Meditation is rooted in the holistic right hemisphere of the brain. Concentration, on the other hand, is the path of effort.

*True
intelligence
is activated
through the
patience and
softness of the
heart, and only
later confirmed
by the mind.*

Through concentration, you strive with your heart, mind and soul to reunite your inner being with your true higher nature. Most mystical systems and all types of yoga are based on the path of concentration. Concentration is rooted in the left hemisphere of the brain. Through a gradual process of continual refinement, this path too arrives at an enlightened view of the nature of reality.

Right in the middle of these two paths is contemplation. Contemplation utilises aspects of both meditation and concentration. It is rooted in the corpus callosum, that part of the brain that unites both left and right hemispheres. Contemplation involves a kind of cellular digestion. You take the object of your contemplation and you concentrate your whole being on it but without any effort or tension. In a certain sense you play with it with your mind, your feelings and your intuition. It is as though you are holding the velvet case of a diamond ring in your hands and continuously running your fingers gently over it, enjoying the sensation and the mystery of not knowing what lies inside. Then, at some unknown moment, your fingers suddenly and unexpectedly discover the tiny hidden catch embedded in the deep folds of velvet. All of a sudden the case opens and the treasure is revealed.

This is how the Gene Keys are best approached — in a spirit of deep relaxation. True intelligence is activated through the patience and softness of the heart, and only later confirmed by the mind. Here you need to be a lover of mystery with a beginner's mind, rather than arriving as an expert determined to solve a riddle.

THE SPECTRUM OF CONSCIOUSNESS —
A NEURO-LINGUISTIC ALPHABET

As you enter the Gene Keys you are stepping into a world of words. The words themselves are simply pointers and codes that invite you to move into a state that lies beyond words. All words resonate within the chambers of your body. They carry frequencies in and out of your being. If for example you take the word *conflict* and sound it silently within your mind, it creates an electromagnetic pulse that is heard throughout your body. If you then imagine the feeling that this word engenders, you send an even more powerful signal into the inner recesses of your physiology. Remember, your DNA is so incredibly sensitive that it hears everything and responds accordingly.

At the end of this book you will find a table of words known as the *Spectrum of Consciousness*. These are the specific word codes that relate to each of the 64 Gene Keys. You will learn that each Gene Key spans this spectrum and is divided into three levels or frequency bands known as the Shadow, the Gift and the Siddhi.

The Spectrum of Consciousness is your personal genetic programming language. It is really a neuro-linguistic alphabet, which is to say that as you apply these words and their meanings to your own life, they will deprogram and reprogram your genes with healthy high frequency electromagnetic signals. The goal of the Gene Keys is first to deprogram your DNA of all its low frequency patterns (the Shadows), then reprogram your cells with the higher frequency patterns of your genius (the Gifts and Siddhis).

For the sake of clarity, let's summarise everything we have learned thus far before we move deeper into the language of the Spectrum of Consciousness and learn how to activate the Gene Keys themselves.

You have seen how your genes hold the blueprints of your nature, while it is the environment that determines how they are activated via the cellular membrane. You also know that the concept of environment must be extended to include your thoughts/feelings/words, all of which generate subtle electromagnetic signals that have a profound effect on your DNA. You have been introduced to the Gene Keys, a highly specific language whose purpose is to communicate directly with your genes in order to adapt their functioning and affect a transformation throughout your whole being. Finally, you have learned that the primary means of working with the Gene Keys is through the art of contemplation, a playful but sustained method of ingesting the truths that they contain.

LOCKS, KEYS AND CODES – THE PATH FROM THE SHADOW TO THE SIDDHI

THE 64 SHADOWS – YOUR PASSAGE THROUGH THE UNDERWORLD

As you begin to enter this new language of the Gene Keys you will be traveling through a world of vibrations. All life is simply vibration and, as we have seen, your DNA creates your life based upon the frequency of the vibration it receives. Fear generates a low frequency energy field, whereas love generates a high frequency energy field. Different frequency bands activate different codes within your DNA. For example, your body contains a code that when unlocked will manifest a feeling of peacefulness so profound it will actually silence the thoughts passing through your mind. However, your genes can only create this state when you send extremely high frequency energy to your cells. This is why such higher states are referred to as being *hidden* in your DNA. They are hidden by the lower frequencies, known as the Shadow frequencies.

The 64 Shadow frequencies are states of consciousness that many people consider *normal* in human beings. In some cases, we are even told these attributes are healthy, which is definitely not the case. The 64 Shadow frequencies form a collective energy field that is generated by ancient genetic memories from the time when we were part of the animal kingdom. The primary focus of the 64 Shadows is individual survival based on fear, and as such they always stimulate the oldest parts of the brain and its related physiology.

When you work with the Shadow Gene Keys, you are working with real physiological fears lodged deep in your unconscious.

Despite the incredible evolution of the human brain over hundreds of thousands of years, the group consciousness of humanity is still powerfully influenced by these ancient fear-based codes. No matter how positive you try to be in life, if you do not become fully aware of your own Shadow frequency patterns — your so-called *dark side* — you will never be able to unlock the higher frequencies.

This is the real work of the Gene Keys — to give you an inner language that allows you to face the unconscious fears at work inside you. Just by dint of being born, you have inherited some ancestral memories and fears that actually have no root in your current lifetime. They come from the collective human gene pool. When you work with the Shadow Gene Keys, you are working with real physiological fears lodged deep in your unconscious, as well as the collective unconscious. As your contemplation of the Shadow frequencies stirs up these ancient codes in your DNA, you will begin to see how much of your life and the world around you is governed by these frequencies rooted in fear. By bringing awareness to these hidden aspects of your unconscious on a daily basis, you will gradually and effectively disarm them.

THE AUTHENTIC YOU

In regard to the 64 Shadows, you will also notice that each one operates either through a repressed expression or a reactive expression. Depending on your character, your culture and the nature of your childhood conditioning, you are more likely to act upon one more than the other of these energetic patterns. The repressed nature manifests as a more introverted psychological pattern rooted in fear, whereas the reactive nature is more extroverted and manifests as anger, which is really fear turned outwards. It is also common for many human beings to move and swing between both of these poles, so it is important to consider both.

As you read and contemplate the Gene Keys and particularly their Shadows, keep in mind that the foundation of this journey is about becoming authentic. When you know and understand your own Shadows you become an authentic human being — one whose frequency automatically evolves to higher and higher levels through the power of self-acceptance. The beginning of your voyage into the Gene Keys can be a most challenging time, as the keys unlock the very codes that you have been avoiding for so long. It is quite a revelation to realise that you have locked yourself out of the higher frequencies because you are caught in these repressed and reactive Shadow patterns. But take heart from every insight and breakthrough that comes your way. Each time you make a leap, the frequency of your DNA leaps with you. This inner process takes time and requires as much patience as courage, but as you stay with the process, so you begin to taste the higher frequencies pulsing through your blood. This is an indescribable gift because it signifies the rebirth of the real you — the authentic you.

THE 64 GIFTS – THE OPENING OF YOUR HEART

The higher the frequency of your DNA, the more sensitive you become to the energy fields around you.

Every Shadow contains a Gift. This is the heart of the transmission of the Gene Keys. It is one of the amazing twists woven into the human story. The basis for all our myths, fables, novels, movies and stories is that our suffering contains the seed of our eventual transcendence. When you accept and embrace your Shadows, they suddenly reveal their true nature and a new creative impulse is released through you. Inside your DNA a subtle but potent mutation takes place. In genetics, a mutation refers to a change in the way the genetic code is copied within the cell, which leads to a change in the way proteins are made, which in turn leads to a change in your biochemistry. As your Shadows reveal your hidden Gifts, the whole tempo of your life changes. Your blood chemistry changes, your biorhythms change, your moods stabilise, your eating patterns shift and your general attitude to life becomes uplifting and optimistic. All of these changes occur naturally and in their own time, as you continue to contemplate the Gene Keys.

Once you have begun the process of transforming your Shadows, the view you have of yourself and your entire belief system undergoes a quantum leap. As you begin to live more and more at the Gift frequencies, you even find yourself opening up to the possibility of attaining the very highest states known as the 64 Siddhis. The higher the frequency of your DNA, the more sensitive you become to the energy fields around you. This can be a challenge for you since much of the world operates at the Shadow frequencies. However, the nature of the Gift frequency is to open your heart, especially towards the Shadows. Once you know the Shadows inside yourself, evolution will begin to use you to raise the frequency of others, which will undoubtedly draw you into some form of service.

AWAKENING THE POWER OF YOUR AURA

The human body emits a range of subtle bio-electric energy signatures, which taken together form a powerful electromagnetic field known as the aura. Over a hundred different cultures have given names to this phenomenon, so it is in no way a new revelation. What is new is the understanding that the DNA molecule's primary role in the body (even before protein synthesis) is, in fact, electromagnetic reception and transmission. By converting frequency into chemistry, your DNA generates the overall vitality and quality of your aura. As the Gift frequency grows in your life, so your aura grows. Your aura is directly linked with your health and with your ability to generate and transmit light waves out into your environment. Within your body, higher frequencies open your heart and flood you with periodic rushes of love and even ecstasy. Such waves reach right into the auric fields of others where they can unlock blocked pathways and give you access to powerful healing energies.

Every one of the 64 Gifts unlocks a particular genius that then finds expression through the radiance of your aura.

In recent years the human aura has also been likened to an attractor field. Through its frequency, it attracts similar frequencies towards itself, which is the foundation of all relationship attraction. More than this however, as your aura expands, it begins to bring you into a deeper harmony with the greater rhythms of the cosmos, and this in turn produces potent manifestations in your life. Your aura engages the power of synchronicity, the universal law of good fortune. This draws prosperity into your life at all levels. It unites you with your true allies in life through their natural resonance with your expanding energy field. Every one of the 64 Gifts unlocks a particular genius that then finds expression through the radiance of your aura.

THE 64 SIDDHIS – AN ENCYCLOPEDIA OF ENLIGHTENMENT

As your contemplation of the Gene Keys continues to deepen and activate the Gifts in your life, your frequency becomes gradually higher and more refined. At a certain point in this process, contemplation spontaneously gives way to absorption. Absorption is a very high frequency state of consciousness in which your DNA begins to trigger your endocrine system to secrete certain rarefied hormones on a continual basis. These hormones, which include pinoline, harmine and melatonin, are associated with higher brain functioning and involve states of spiritual illumination and transcendence. Absorption can only occur when your aura is generating a frequency high enough to continually feed off its own electromagnetic field.

At such a stage, it is no longer possible to be drawn back into the lower frequencies for more than brief periods of time.

Both the Gift frequency and the Siddhi frequency involve quantum leaps in consciousness. The group consciousness of humanity stands poised to make the huge transition out of the Shadow frequency and into a newly stabilised awareness in the Gift frequency. This is what will catalyse a global genetic mutation. At the same time, there is another far smaller group of human beings poised to make the transition from the Gift to the Siddhi frequency. The word *siddhi* comes from the Sanskrit meaning *divine gift*. There exists much ancient lore surrounding the siddhis (there is even an established tradition that states there are exactly 64 of them!). In the context of the Gene Keys, the 64 Siddhis are the biological expressions of the ultimate human state of enlightenment. They are quite literally an encyclopedia of the many expressions of spiritual realisation.

YOUR DNA IS A SUPERCONDUCTOR

Before a human being makes the quantum leap to the rarefied and wordless realms of the Siddhis, some unusual phenomena must occur. A vast amount of your DNA (over 90%) appears to geneticists to serve no purpose whatsoever. It has therefore earned the unglamorous name *junk DNA*. This is a huge misinterpretation of the true role of this type of DNA. It is the junk DNA that holds the entire collective memory patterns of your past — not just your past as a human, but also your distant past as an animal, a reptile and going even further back, a plant and a bacterium. Before you can attain a siddhic state, all this genetic memory has to be purged from your DNA. This also means that as your frequency becomes higher and higher, you have to process deeper and deeper Shadow patterns that come from our collective ancestral past.

Known in the Indian yogic tradition as *sanskaras*, these ancient Shadow frequencies are literally wound around all human DNA. The only thing that can unwind them is light itself. Research into DNA has demonstrated that one of its more unusual electromagnetic properties is its ability to attract photons (elementary light particles), causing them to spiral along the double helix. It is this ability of DNA to weave light around itself that reveals its true hidden role within your body — to act as a superconductor whose sole purpose is to exponentially increase the frequency passing in and out of your body. This in turn leads to a complete transmutation of the fabric of your being. This very rare phenomenon is theoretically known as a *transposition burst* and involves the sudden and synchronised movement of thousands of DNA elements inside your body to new and different genetic locations. It signifies the extraordinary birth of a completely new kind of human being — *homo sanctus* — the sacred human.

It is this ability of DNA to weave light around itself that reveals its true hidden role within your body — to act as a superconductor whose sole purpose is to exponentially increase the frequency passing in and out of your body.

When you consider this passage from the Shadow to the Siddhi, perhaps you can see what an awesome evolutionary voyage the Gene Keys represent. They are the original inner language of light, and even though woven from words, those words are really nothing more than the frequencies of light itself. The Gene Keys are simply the messengers that you send into the living structure of your DNA, and their role is to instruct that DNA to build you a higher body fully equipped to handle a far more joyous life. Even the deepest fears in your body are nothing more than low frequency patterns that obscure the pure and radiant light that lies at the heart of your DNA.

The voyage through the Gene Keys is a great adventure into the final frontier of inner space. It will call upon the warrior spirit inside you to face the personal and collective demons that lurk within your cellular DNA. Nonetheless, armed with the art of playful yet focused contemplation and a healthy dose of patience and courage, you will make it through to the highest levels of human potential. Always remember, this is not a complicated journey. Your mind may try and make it complicated, but it is really as natural as breathing. Your contemplation of the Gene Keys can go on while you are about your everyday life — working, cleaning the dishes, looking after the children, relaxing, sleeping and even dreaming. The pattern of your Gene Keys will automatically manifest, so all you have to do is listen and learn. The life playing out before you is a mirror of what is occurring inside your DNA. If you are contemplating it, it will appear as if by magic, right before your eyes!

NAVIGATION TOOLS

Before beginning any voyage into unknown territory it is always a good idea to prepare. The language you will find in the Gene Keys uses specific terms drawn from a wide variety of fields. As you move through this labyrinth of new ideas, concepts and feelings you may need to adjust yourself to this new terminology. When you look at each Gene Key, you will notice that it is related to an aspect of your physiology and an amino acid, is part of a genetic family called a *Codon Ring* and has a *programming partner*. Each of these aspects is a portal for your further contemplation. In the back of this book you will find a resource called the *Glossary of Personal Empowerment*. This is a profound guide to all the major terms and concepts in the Gene Keys, and it also describes their practical application to your life as well as their meanings. It is important to remember that each of the terms used within this book have specific meanings within the context of the overall synthesis of the Gene Keys. These meanings may differ from more traditional interpretations of the words. You are recommended to use the Glossary as an ongoing font of meaning and inspiration. As you will discover, it is far more than just a Glossary. When contemplated over time, every term contains an empowerment designed to raise the frequency of your DNA.

PROGRAMMING PARTNERS

One of the biggest insights into the Gene Keys comes from a close understanding of the pairing of Shadows, Gifts and Siddhis. As the ancient Chinese sages developed the I Ching, their *Book of Changes*, they discovered many different ways of adapting and amplifying their revelation. A major breakthrough came when the 64 archetypes, known as hexagrams, were arranged in a circle as opposed to being seen as a linear sequence from 1 to 64. This circular arrangement of the Gene Keys allows you to see how each Gene Key operates as one half of a binary programming field.

When you contemplate the Gene Keys alongside their programming partners, you can see how these genetic pairings create biofeedback loops within your body, mind and emotions either blocking or releasing the higher frequencies.

THE 21 CODON RINGS

In genetics, the role of the genes in your cells is to synthesise and blend various amino acids in order to make proteins, which are the building blocks of your life. The major amino acids are also grouped into genetic families, which are known in this book as the 21 Codon Rings. Each of these Codon Rings has a specific name and they act as collective programming bodies within the greater body of humanity itself. Much of our human mythology emerged from these chemical groupings and each ring contains great mysteries. They have direct connections to many of the original sacred alphabets, such as the Hebrew alphabet, as well as a relationship to the rich symbolism of the tarot. Their deeper significance lies outside the domain of this book, but as you consider the Gene Keys through these Codon Rings, you may also unlock their secrets from within your own DNA. Above all, this is your voyage, and how daring you are is completely up to you!

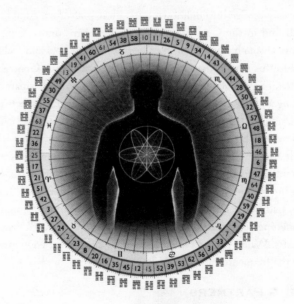

PART 4:
ANGLES OF APPROACH TO THE GENE KEYS

WALKING YOUR OWN PATH

There is a wonderful expression by St. Augustine, *solvitur ambulando*, which means *by walking it is solved*. You can take this as the creed of the Gene Keys. Whatever question you have about any aspect of your life, the Gene Keys will point the way to an answer, simply because they point the way inward, which is where you will always find the answers. All you have to do is keep walking, continuing in your quest. The original Chinese I Ching is an oracle with the capacity to guide any decision you make in life. The Gene Keys take this much further by guiding you into the actual living I Ching inside your genes. Once this book has done its job, you will never again have to look outside yourself for truth.

Throughout this introduction, this book has been represented to you as a voyage — an adventure into a great mystery — the mystery of who you are and why you are here. As you will see, there are many different pathways you can take into the Gene Keys. As well as this book, there are a host of different tools and systems that utilise the Gene Keys in myriad different ways. Rather than feeling overwhelmed by all these different paths, I encourage you to pace yourself and find the way that feels most comfortable to you. As you may now realise, the Gene Keys are a living transmission rather than just another information system. Therefore they will only yield their secrets to a relaxed and contemplative attitude. The quality of the first steps you take determines the shape of the journey to come. So know that you are welcome here. There is plenty of time for you to explore the Gene Keys at whatever level you wish.

ANALOGUE AND DIGITAL

At the broadest level, there are two ways you can approach the Gene Keys — the analogue way, which embraces the holistic view, and the digital way, which dives into the detail and the parts. Life is made up of both, and it is beneficial to maintain a healthy balance between the two. As we have seen, our DNA itself is a digital binary code laid down in patterns and sequences that can be interpreted in a logical, binary way. The analogue way is completely different. It is mysterious, playful, spontaneous and intuitive. The higher frequencies can only be unlocked by the analogue way, but the digital way is how we ground those frequencies and experiences into our mental understanding. **It is the combining of both analogue and digital that results in contemplation — the primary pathway into the Gene Keys.**

ANALOGUE SEQUENCES – ALLOWING THE MAGIC TO OCCUR

In light of the above, you are invited to create your own analogue pathway into the Gene Keys, while your intellectual understanding travels underneath. Your natural genius can only flower through your heart.

The intellect can be the great enemy of your natural genius. However, when you allow the Gene Keys to sing inside your heart, your intellect finds its natural place, serving your heart rather than trying to lead it. I therefore recommend that you travel through the Gene Keys in a non-sequential way, as this will distract your intellect and unlock your own unique passages and pathways through the matrix. This is the kind of book that needs to be allowed to fall open randomly to different pages at different times, in the spirit of the magic of the original I Ching.

Through playing in the field of the Gene Keys in this analogue way, you will learn the true mystery of sequences. We tend to see sequences as digital (e.g. from 1 to 10), but inside your DNA there are geometric sequences that follow no clear logical pattern. Perhaps in your journey through the Gene Keys you will return to one specific Gene Key over and over again until you learn that lesson. Perhaps certain Gene Keys will be held back until you are ready to activate them inside. Allow yourself to dip and dive through the text and find your own unique sequence. Remember that your voyage through the Gene Keys is also taking place in your physical biochemistry. Every Gene Key is linked through the DNA directly to your endocrine system, which influences everything inside your physical body, from your breathing pattern to your heart rate. Your attitude talks to your genes, your genes talk to your glands and your glands reorganise your life in line with a higher harmony.

There will undoubtedly be times in your process in which the light inside your DNA seems very dim and far away. Your attitude during such times is of utmost importance. You may have arrived at a turning point in your own sequence, and you must honour the feelings that come up inside you. Your sequence can never be wrong! Understand that your DNA is throwing up a Shadow pattern so that you can erase it from your genetic memory forever. In every person's sequence there are twists and turns that lead to revelations, insights and break-throughs. Each revelation you have with this book leaves you different. Every breakthrough causes your DNA to mutate to a higher frequency, which in turn makes you feel and behave differently from before. The real magic comes alive as the inner sequence is reflected in your outer life. Your attractor field changes and a new outer sequence of events takes place as the cosmos begins to work for you instead of against you.

This is the kind of book that needs to be allowed to fall open randomly to different pages at different times.

Above all, know that your DNA is a store of miracles waiting to be tapped. Allow yourself to let go of all the normal rules you may have for reading a book. Let the pages fly and enjoy the flow of analogue magic as it unlocks those miracles!

DIGITAL SEQUENCES – STEPPING INTO THE HOLOGRAM

Once you have imbibed the spirit and magic of the analogue way, your voyage with the Gene Keys will begin in earnest. You can then take a dive into digital Disneyland! As said earlier in the introduction, our current scientific understanding of the universe is changing rapidly and radically, and the picture that is emerging is nothing short of mind shattering. Quantum physics shows us that everything we see in the universe appears to be reflected in everything else in the universe. The mind itself, enmeshed as it is with the very stuff of the cosmos, can no longer be relied upon for objective reasoning.

Consciousness studies, a field avoided by science for generations, is now becoming one of the hottest new areas for scientific exploration.

Accordingly, as you play with the Gene Keys, you should know what you are getting yourself into. You are stepping into the heart of the holographic universe and entering the programming matrix for all life. Every frequency shift you experience inside your DNA will affect every single atom in our universe. As you evolve, so all beings will evolve with you.

In the digital universe, the holistic fabric of all life is subdivided into its infinite fractal aspects. All these diverse elements, relationships and systems can then be dismantled and viewed from a microcosmic perspective bringing increased understanding for us as individuals. In other words, if you really want to get to the bottom of something, you have to take it apart! This is the foundation of the digital way.

YOUR HOLOGENETIC PROFILE

Every human being is born with a unique sacred geometry embedded deeply within their being. This geometry can be plotted through the precise timing and placement of your birth in our constantly shifting universe. This same structure is also holographically encoded in your DNA, forming a distinct personal profile of genetic patterns and digital sequences, each of which relates to a different aspect of your life. This is your Hologenetic Profile, the original blueprint that tells us who you are, how you operate, and above all why you are here.

Your Hologenetic Profile is a personalised map of the various genetic sequences that will unlock or awaken different aspects of your genius. There are sequences governing your life purpose, your relationship patterns, your financial prosperity, your family dynamics, your developmental cycles as a child, your health and healing and your spiritual awakening. Each sequence has its unique application in the world and comes with its own set of tools and teachings. One example is the relationship sequence, known as the Venus Sequence. This system accurately pinpoints specific mental issues and emotional blocks lying within an individual's DNA. Such blockages can cause common difficulties in human relationships and health. Through simple techniques of pattern recognition using the Gene Keys, any individual can

Your Hologenetic Profile is a personalised map of the various genetic sequences that will unlock or awaken different aspects of your genius.

be guided to see their self-destructive tendencies and transmute them into far more beneficial patterns. The Venus Sequence has been shown repeatedly to have huge transformative effects on countless relationships and individuals.

*The Synarchy
points to a
higher form of
organisational
intelligence that
operates across
all human gene
pools.*

You may choose to enter the Gene Keys through one of their many applications, like the Venus Sequence, or you can simply contemplate this book as a whole. Either way, it is important to know that when you raise the frequency in one area of your life, you will also be raising it in all the others.

CONTEMPLATION, ABSORPTION AND EMBODIMENT – THE CORE PATHWAY

Your voyage into the Gene keys begins with contemplation. Even as you have been reading this introduction, your contemplation has begun. The truths inherent in these teachings are already entering your aura where they must be allowed to percolate. The art of contemplation is about patience and digestion. If you have an open mind, then the transmission of the Gene Keys will move deeper into your feeling body, the *astral* body. Depending on your own readiness to receive the higher frequencies, your DNA will respond when the transmission reaches down into your physical body. In order for this to occur, it must first pass through your mind and your feelings. At any point the transmission may come up against a blockage or Shadow pattern. In your mind this may appear as a judgement, opinion or belief. Through your emotions it may show up as a repressed unconscious memory or a constrictive emotional pattern such as guilt, shame or fear.

Such Shadow patterns lie hidden inside all human DNA and contemplation on the Gene Keys will bring them naturally to light. As you journey into the Gene Keys, keep your inner ears open at all times and listen to your natural responses, gut feelings and thoughts. Anything that causes you discomfort is of primary importance and I recommend you give it your full attention. This is a deeply personal voyage you are on, and as you allow the higher frequencies to resonate inside you, through thinking about them, through feeling your way inside them and through practicing them, they will find their way into your waiting DNA. You will know when a high frequency pattern in your DNA has been unlocked because your body, mind and heart will feel it as a rush of illumination. Such moments are to be treasured, and as your contemplation deepens, they will come more often.

Absorption occurs when the body opens up at a cellular level to receive the nutrients from the higher planes of reality. It requires deep emotional maturity and great mental clarity. In the state of absorption your aura swells as your DNA begins to suck light into the cells of your body. At this stage even your physical body is contemplating its highest nature and you begin to taste the Siddhis, the highest manifestations of your genius. As the cells of your physical body become more and more accustomed to these higher frequencies, you begin to embody the heart of the transmission that the Gene Keys have been pointing you towards. This final stage of embodiment is the culmination of your contemplation. As you enter into this beautiful state, you find that you no longer need techniques or tools.

EMBODIMENT – MAKING THE LEAP BEYOND LIGHT-SPEED

Embodiment brings an end to all words and explanations. Even though you may still use words, you have now entered into the language of light itself, which is represented by the

64 Siddhis. You will see at this stage how the Gene Keys are simply messengers that relay information in and out of the cell in order that this miracle beyond words can occur. As you recognise yourself to be a part of a higher evolution, your life's work and your inner purpose finally come into alignment. You are part of the grand awakening. You are a DNA molecule within the body of humanity, and you are fully awake. Your only role is to begin programming all the cells around you with the new higher frequencies. As you embody the transmission of awakening, you will find your own voice and you will adapt your language for those who are drawn to your radiance. You will realise that it is not the language that matters — it is the frequency of the message and the messenger.

When you look at the children of the world today you can easily see the disenchantment in their eyes as their natural state of innocence meets the modern world we have created. What these children need most of all now is for us to reclaim our own innocence and bring our romantic genius back into the world. The embodiment of the truths inherent in the Gene Keys is a transmission of pure romance — it demonstrates that magic is alive inside each of us, that anything is possible and that miracles are inevitable. The future world will be a world created by adults who can see through the eyes of children. The coming genetic mutation from *homo sapiens* into *homo sanctus* will bring wave after wave of enlightened children into the world who will seek only that which is built on natural cosmic laws, not the old systems which were rooted in fear and competition. As the 55th Gene Key testifies, it is truly an extraordinary time to be alive.

These changes will only occur when you who are reading this can physically embody the living transmission of light. You will have to come to a point of profound relaxation inside your being in order for this to occur, because embodiment is rooted in relaxation. As we have said, contemplation is the core pathway. It requires a relaxed and reverent attitude. We must at all times remember that Shadows are simply Shadows, and every one contains a Gift. When you change your attitude, you will see that all the Shadows are cast by the same light. This is a journey of self-forgiveness. You must be compassionate first and foremost with yourself. As you forgive yourself the Shadows, you will be giving yourself the Gifts — the 64 Gifts. And as you now know — inside every one of the Gene Keys lies something even more precious — the keys to unlock the higher purpose hidden in your DNA.

I hope that you enjoy your voyage through the Gene Keys and their many applications. They are offered to you in the spirit of a feast for your contemplative delight. Because they are also designed to attract others who are ready to embody this new language, I hope that you will continue your voyage beyond this book and explore them as a co-creator, celebrating your own radiance and genius in the ever-expanding field of their mystery.

In the state of absorption your aura swells as your DNA begins to suck light into the cells of your body.

1st Gene Key

From Entropy to Syntropy

SIDDHI
BEAUTY

GIFT
FRESHNESS

SHADOW
ENTROPY

Programming Partner: 2nd Gene Key

Codon Ring: The Ring of Fire (1, 14)

Physiology: Liver

Amino Acid: Lysine

THE 1ˢᵀ SHADOW – ENTROPY

THE DANCE OF SHIVA

Once upon a time, perhaps even an aeon ago, in a land now become a myth, a young man sat in deep reverie beside a great river. Although this river no longer exists today, legend tells us that its descendant may well be the great Yangtze River which flows through the heart of the land now known as China. Our young man gazed out upon the sinuous, soft waves as they lapped the shores by his feet, when all of a sudden a small turtle emerged from the great green mother lode, and drawing itself proudly out of the water, sat down beside him to share his gentle contemplation.

For some considerable time neither party spoke, until finally our young man, evidently on the cusp of some great universal epiphany, exclaimed to the little turtle in a great and wondrous sigh: "Ahh little one, what's it all about then...?"

Rather to his surprise, the turtle revolved in a half circle and nonchalantly turned its back on the young man, whilst continuing to sun itself lazily and silently.

The young man gazed intently at the little creature's back and the intricate interlocking patterns and plates of its shell drying in the spring sunshine. As he gazed, a strange thing began to occur to him — the more he looked, the more he understood the nature of his question. And so it was that he gave himself up to the moment and gazed with all his heart at the little turtle's back. Slowly, almost imperceptibly, everything began to evanesce and disappear — first the turtle, then the cosmos, and finally the young man himself. It is said that when the young man regained consciousness several hours later the turtle was gone.

Ever since that day humanity has had a means of understanding every aspect within the universe. It was discovered in the interlacing patterns on a humble turtle's back — and in time it became arguably the most profound knowledge ever discovered by a human being. It became The I Ching.

The Chinese I Ching is one of the greatest spiritual books of all time. Written down thousands of years ago by a legendary Chinese Emperor called Fu Hsi, it encapsulates a binary code that charts all the seasons and cycles of life. In the book, these life processes were measured by a simple code of six lines, either male or female, and mixed together in a total of 64 combinations. The very first archetype that was represented, known as the 1st Hexagram, consisted of six male lines all in a column. This archetype was seen as the primary code for all creative life in the universe. Its opposite, the 2nd Hexagram, consisted of six female lines in a column, and it was seen as the primary code for directing all creative life in the universe. Here, right at the beginning of your journey into your own DNA, hides the greatest secret of all — the secret dynamics of duality. The other thing that we discover right here at the beginning is that in the I Ching and therefore in life, it is always the women who are in charge! As you enter more and more deeply into the mysteries of the Gene Keys, you will gradually begin to understand what is truly meant by this statement.

The I Ching is a mathematical mirror of the genetic code, and one can actually reduce all 64 archetypes down to four essential principles. These are captured at the beginning and the end of the I Ching sequence itself — the 1st and 2nd Gene Keys and the 63rd and 64th Gene Keys. These two pairs are rather like the prologue and epilogue to the great book itself. Upon these four principles or pillars life is built.

The lower frequency of the 1st Gene Key — the 1st Shadow — is captured perfectly by the word *entropy*. A simple definition of entropy is:

"A measure of the disorder or unavailability of energy within a closed system.
More entropy means less energy available for doing work."

Modern physics and the laws of thermodynamics are based upon the basic perceived law of entropy. According to what we can see through the mind, the universe appears to have a single direction — it moves from order towards chaos. This 1st Shadow keeps the entire planet living at a low level of frequency — it is like a blanket thrown across our civilization. According to our mind, we cannot do anything about entropy. That is our chief problem. Human beings do not generally accept themselves, and when you convert entropy into human feeling it becomes a kind of deep numbness or sense of gloom. Entropy is in effect the opposite of love.

Numbness is actually an extremely fertile state of awareness.

If we learn anything from these first two primary Gene Keys, it is about the nature of duality itself — which is that life cannot exist without polarity. The entire notion of the Spectrum of Consciousness at the heart of the Gene Keys revelation depends upon this polarity of the Shadow at one end of the spectrum and the Siddhi at the other. Each gives birth to the other. Entropy is the black hole to creativity's white hole, and this first Gene Key is about creativity. The secret to harnessing creativity actually lies in the 1st Shadow. In fact, the secret to every Gene Key in this book lies in harnessing and accepting the energy latent within each of the 64 Shadows. In grasping this fact right here in these two primary Gene Keys, you truly prepare yourself for the voyage to come.

What is the meaning then of entropy in your life? As was just mentioned, entropy manifests in human beings as numbness, and this numbness is actually an extremely fertile state of awareness. First of all, you have to understand that it is a chemical state, and secondly, it comes on suddenly and, accepted and fully embraced, it passes just as suddenly. Entropy and creativity are an eternal dance played out within the universe we inhabit. Many mythologies have captured this dance — the dancing figure of Lord Shiva from the Hindu pantheon is just one of these figures whose dance both destroys and creates all that exists.

We do not know exactly why we humans feel gloomy on certain days and happy on other days, despite our thousands of theories. Just as there is external weather, there is also internal weather, and it differs in every human being. Yet it is the very unpredictable nature of these internal weather patterns that so many of us struggle with. When you feel the movement of creation inside you, you are happy. When you feel entropy inside you, you are no longer happy. This continual interplay of energy in your life makes you want to always maintain the happy side and escape the gloomy side. Herein lies the greatest flaw in your nature and the distortion of the true energy of entropy into depression.

The 1st Shadow appears in your life whenever you feel flat or sad or low. This is a chemical process that the body enters into, and if you try to comprehend it, find a reason for it, or worst of all try to fix it, then the natural process will not complete itself cleanly. If you resist this state mentally, one of the great dangers is that you will interfere with its process and *fix* it inside yourself as depression, because this 1st Shadow triggers the chemical processes that can lead to depression. Most depressive states are the result of resisting certain Shadow frequencies within our genetic makeup. Depending on your individual genetic predispositions, you may be more or less prone to feeling low than others. Generally speaking, the more creative a person you are, the more deeply you are affected by this kind of melancholic chemistry.

The state of entropy is rather like a vacuum state. Your system is recharging, so the energy within you withdraws into a kind of stasis. The resulting feelings or lack of feeling and/or enthusiasm provide a delicate environment for something quite special to occur, if you are patient enough to allow it. This something is the creative process. In other words, your low energy means that something intangible is gestating inside you even though you cannot yet see it. Only when the state mutates to its expressive stage will you see what the process is about. These low times in your life are therefore very special times, and they generally require aloneness and withdrawal in order for the seeds sprouting inside you to germinate. The worst enemy at such times is interference from your own mind (or indeed someone else's mind) wondering what is wrong with you. The programming partner of this 1st Shadow is the 2nd Shadow of Dislocation, which can only further worsen your state of mental turmoil if you try to intellectually understand what is happening to you. The 2nd Shadow tends to fuel the fire of your anxiety, since it gives you the feeling that everything is out of sync with the whole, even though this is not the case.

Because of this 1st Shadow, humanity as a species is not nearly as creative as it could be. The reason for this is the weight of the collective denial of the natural phases of entropy that individuals experience. This Shadow can only be resolved and accepted at an individual level, which takes a great leap of courage, patience and trust. When you suddenly find yourself blind and lost, the best thing to do is to stay still and allow it to pass gently through your system, giving it the minimum mental attention possible. This deep acceptance of entropy as natural in your life will eventually allow you to unlock its true potential and in time transcend it entirely.

REPRESSIVE NATURE – DEPRESSIVE

The introverted nature of this Shadow inevitably leads to depression. States of depression can be caused by the *freezing* of a low frequency emotional state due to mental collapse rooted in fear. Once fear takes over the physical system, the entropy goes on pulling more and more energy away from the surface of one's life. Such states can occur at different levels — some can be permanent, others sporadic. Some can leave one bed-ridden, others can simply take away the lustre in one's eyes. Once a depressive state has been fixed, it can only be broken by the individual, and without help. The individual must face down the very fear that caused the depression and shift the frequency of their attitude on all levels.

REACTIVE NATURE – FRENETIC

The reactive side of this Shadow manifests as a frenetic urge to escape the way one is feeling at all costs. Instead of moving in harmony with entropy by closing their doors and being alone, these people immediately increase their activity and contact with others. They become frenetic in their bid to suppress what is going on inside them, and can become engaged in wild schemes or locked in monotonous patterns that quickly undermine their health. Such people put themselves in great danger because they are moving in the opposite direction from the chemistry of their body. Their urge to escape their feelings opens them to all kinds of illnesses in their body that would otherwise never have troubled them.

THE 1ˢᵀ GIFT – FRESHNESS

THE BEAUTY OF MELANCHOLY

In most of the ancient creation myths, one of the first manifestations of life is seen as light. In the Bible in Genesis, this is forever implanted in the western psyche through the clarion call: "Let there be Light!" This 1st Gift is rooted in this notion of light as the manifestation of creative energy in the universe. One of the other common creation myths is based upon sound, captured in the above words, spoken out loud by God in the Bible. Here in this 1st Gift we have the bringing together of these two essential principles of light and sound. The third great symbol of creativity unites these two principles through the symbol of fire. Fire is perhaps the greatest creative archetype since it not only consumes but also transforms. Whilst the 2nd Gene Key indicates your true direction in life, its programming partner the 1st Gene Key provides the actual energetic thrust to get you there.

Every time an individual moves cleanly through their low frequency chemical process, they re-enact this creation myth — out of the darkness suddenly light emerges — and as if by magic, the low energy field switches and is experienced as joy. This joy comes as unexpectedly as the sadness does, but with the joy also comes the need to express it, and especially to express it through your voice or your art. This 1st Gift is called the Gift of Freshness because whatever emerges out of its numbing chemistry field is absolutely new. Every word for every Gene Key is highly specific, and the word *freshness* is different for example from the word *newness*. Freshness conveys aliveness, as though it refers to something burning with an inner fire. This is exactly how people empowered by the 1st Gift express themselves — as though surrounded by a halo of something brought with them from another world.

This 1st Gift can perform wonders within small organizational groupings. As you raise your frequency through this Gift, you become a person likely to be singled out by others as a natural born leader. At the same time, you will find yourself reluctant to assume any leadership role since you are not in the least bit interested in being followed by others! All the 1st Gift of Freshness really wants is to express itself fully through you so that you can enjoy watching the impact you have on others. As this Gift awakens inside your DNA, you will find that you automatically inject life and light into any group of people you are a part of.

For these reasons, it is the domain of the family, compact team or intimate group where the 1st Gift is designed to excel. Freshness is a Gift that needs the right environment in order to bloom — it needs open-minded people who give you centre stage exactly when you need it. After the fresh energy has been released and your creative impact has been felt, you will usually need to retreat as quickly as possible in order not to spoil the power of your release. Your secret lies in the knowledge that fresh flowers soon wilt — for just as your light can infiltrate any group bringing inspiration and joy, so can your melancholy equally draw energy away from the very same group.

The Gift of Freshness relies on one immortal truth — creativity can never be controlled. It simply comes when it comes, and when it is absent there is nothing for you to do but wait and relax. Through the Codon Ring known as *The Ring of Fire*, this 1st Gene Key is chemically bonded to the 14th Gene Key whose Shadow is Compromise. When your creative fire is burning, everyone wants to gather around you and partake of your warmth and inspiration, but when your fire dies to a mere flicker, it is as though you become unnoticeable. If you then try to rekindle your creativity through force of will, you will only end up making huge compromises, both with yourself and with others. For you, life is either fully engaged or completely at rest.

If your life is marked by such a creative pulse, you will probably have a strong genetic activation within this 1st Gene Key. As such you are here to dissolve the 1st Shadow by being a living example of the unpredictable power of the creative process. Your true power lies in your ability to be alone with yourself and to trust in the power of your own uniqueness and timing. Out of every dark hole into which you dive comes forth a sally of startling and profound creativity. The genius of freshness is to bring something to the world that no other has ever seen before and that no one else could replicate.

That the very first Gene Key in the Book of Life inside you is dedicated to creativity speaks volumes about the human species as a whole. We are designed to overcome the Shadow states inside us so that our true genius can emerge and we can add our spirit to the world. It is through individual creativity that all diseases and negative patterns will eventually leave this planet. This is the true meaning of Freshness — to be a clear vessel for the creative process so that evolution can move through you and find its way towards a permanent state of eventual love, beauty and unity.

The genius of freshness is to bring something to the world that no other has ever seen before.

THE 1ST SIDDHI — BEAUTY

THE PROMETHEAN FIRE

As we have seen, the 1st Gift is rooted in the power of light and fire. Here at the siddhic frequency this light is all that exists, and as it shines through human awareness it becomes what we call beauty. Beauty is the reason for life, and life is the reason for beauty. As one of the four great pillars of the cosmos, whenever this Siddhi blossoms within a human, that person's life becomes a symbol of a directional shift within humanity as a whole. Therefore, it has a great importance within the genetic matrix, even at the Gift frequency.

These four cornerstone Gene Keys mentioned earlier (the 1st, 2nd, 63rd and 64th) at their highest level represent a divine archetypal foundation that has been sensed by many different cultures and is embodied in the mystical literature of many diverse pantheons. These four principle energies are known to Cabbalists as the *Hayoth Ha Kadosh* or Four Holy Creatures and are also embodied in the *tetragrammaton* — the so-called mystical name of God. The ancient Gnostic traditions worshiped them as the four elements, Native Americans knew them as the *four directions* and the Egyptians carved them into the figure of the sphinx. The Chinese I Ching, which is a keystone of the 64 Gene Keys, has at its foundation the *four bigrams* — universal principles that underpin the entire foundation of life in our cosmos. In our genetics, the principle of these four archetypes is reflected in the *four bases* — the four key letters out of which all genetic language is created.

In light of all these correlations, we can see the great importance of this 1st Siddhi to our species as a whole. In human beings, the 1st Siddhi is Beauty, and as with all Siddhis it is also reliant upon its programming partner the 2nd Siddhi of Unity. Beauty lies in the unity of all. True beauty, when manifested and realised in a human being, presupposes a state of union with the totality. This union or *oneness* is not like anything we can imagine. Its beauty lies in the fact that it is simply so natural that it cannot be expressed. Every time it has been expressed, it has been misunderstood. The only way to understand the siddhic state is to die into it, and that means you must die into beauty. The moment we humans look at something or feel something and proclaim it as beautiful, we have divided from it and left the state of unity. This is not true beauty, although obviously it has its reflections on the lower planes.

To be awakened in this way is to become a great inspiration to humanity.

True beauty is emptiness. There is no one to comprehend it and nothing to feel it. It simply does not exist — and this is its paradox. It is not true beauty if it can be spoken of, copied or shared. It is unique and inexplicable. Beauty is the expression of unity in human form. At a lower frequency we can only comprehend beauty through light or sound — whether that be a beautiful face, a sunrise, or a piece of music. Our notion of beauty is always rooted in the presence of something rather than in its absence. True beauty has a greater connection to absence — to darkness or silence, but even these words cannot approach it being rooted as they are in the world of duality, language and opposites.

When the 1ˢᵗ Siddhi dawns inside you, everything in this cosmos is experienced as beautiful and fresh. Even the Shadow state is beautiful. In beauty, nothing lies outside the unity of everything. Everything is experienced as a unique creative expression of this unity. Everything shines and drips with the essence of its uniqueness and yet at the same time, it all shares a single source. When beauty catches fire inside you in this way, you cannot really become a teacher of others because you now no longer have anything to teach. How are you possibly to teach the experience of beauty? It is a fire that can only be embraced. Thus the only fate left to you is to live your life as an example of what humanity will one day become, and your expression of that beauty in form will become a testament to our common future. To be awakened in this way is to become a great inspiration to humanity. In the past, many such people have suffered greatly at the hands of the masses. To be so dangerously unique and beautiful is to draw fire from the forces of jealousy and denial.

The Siddhi of Beauty is incomprehensible to *normal* consciousness as it can only live out its uniqueness without compromise, no matter what happens to it. Thus, it endures in our mythologies and often becomes attributed only to our deities. It is the creative Promethean fire of the gods that is stolen by those few who have the courage to be obliterated by it. It is the primal masculine creative force within our universe and is embodied in the great lingam or penis of Lord Shiva, as well as all our great cultural symbols of male fertility. Ironically, beauty becomes enshrined as something outside of us — of something we should aspire to, of something we can never really attain whilst alive. In truth, beauty is our nature. It is right here, right now, inside each and every one of us. It is paradoxically found both in our absolute ordinariness and in our utter extraordinariness.

You will recall that the foundation of this 1ˢᵗ Gene Key at the Shadow frequency is the field of entropy, in which energy ceaselessly moves from order to chaos. Well, at the siddhic level of frequency, where frequency itself is obliterated and transcended, we find ourselves in an eternal field of Syntropy. Syntropy refers to the movement of energy within infinite dimensions, all imbued with consciousness and all united by order and love. When the inner divine fire is released through your DNA, you become capable of miraculous power. The Codon Ring of Fire unites the two Siddhis of Beauty and Bounteousness (the 1ˢᵗ and 14ᵗʰ Gene Keys). As such it represents the energetic germ at the heart of creation and the purpose of existence itself — to create that which is endlessly beautiful and to imbue it with the awareness that it is endlessly beautiful.

SIDDHI
UNITY

GIFT
ORIENTATION

SHADOW
DISLOCATION

▦ 2nd Gene Key

Returning to the One

Programming Partner: 1st Gene Key
Codon Ring: The Ring of Water (2, 8)

Physiology: Sternum
Amino Acid: Phenylalanine

THE 2ND SHADOW – DISLOCATION

AS YOU SEE IT, SO IT CHANGES

As the most archetypically feminine of all the 64 Gene Keys, the 2nd Gene Key and the journey it represents contain a beautifully simple distillation of cosmic wisdom. If you wanted to describe the nature of humanity and the reason for our existence to an alien being, you would have to look no further than this 2nd Gene Key. As the 1st Gene Key tells the story of the great masculine principle of energy and light, the 2nd Gene Key grounds this story into the world of form. Even at the Shadow level of consciousness, this Gene Key teaches us that there is a purpose to everything in existence. At no point in evolution has anything ever occurred that was not a part of a huge interconnected plan. This is the truth that resides within the feminine principle — it is the binding force for all the seemingly disparate cells and events within existence and in this sense it represents the great motherly embrace that pulls us all into a single unity. Your personal resonance to this great truth of our unity determines the overall frequency passing through your DNA. Whether you believe it, deny it, wish for it or embody it, the truth does not change. There is a force within the universe that choreographs everything, and it is found right inside you.

The 2nd Shadow is the Shadow of Dislocation, which is an interesting word. It implies both the sense of being lost in time and space and of dismemberment. However, when you feel lonely, cut off, afraid or dejected, you have never actually lost your way — it just feels that way. The Shadow state is really only a human perspective rooted in your biological functioning. There is not a single moment in your life that is not perfectly in harmony with all creation. Neither is there the slightest possibility that you can make a wrong decision or take a wrong turn in life. It is all simply nuances in your biology.

Your biology determines your perception, and your perception is the measuring stick for your evolutionary frequency. Evolutionary frequency refers to the current state of advancement of your awareness. Human awareness is following an evolutionary curve. It began as a primitive form of instinct rooted in our animal origins, then it took a great leap at a certain point in history as the brain began to evolve and we moved into this current phase as thinkers. We have now attained our zenith as thinkers, and are preparing for another great leap — the leap into a new biological awareness rooted in the nerve ganglia of the solar plexus. You have to understand where you have come from if you are to see where you are going, and where you are going is into the awareness of the unity and oneness of all beings. It is such a paradox: You have never left this state of oneness, and yet the operating system within human biology does not currently allow you to feel this continual sense of connection.

It is tempting to look at so-called primitive human awareness — that of the aboriginal tribal groups still left across our planet — and dream of the return to that primal awareness that existed long before the human brain developed so rapidly. Most aboriginal cultures do not live with a sense of separation from life itself, so modern human beings are often left with the sense that we have somehow gone adrift. We tend to think that there is something wrong with the direction we have taken — exemplified by the massive technological revolution that is sweeping us before it. But the rapid development of the human brain is a vital bridge to an even greater leap in awareness. Even so, the mind is our great blind spot. In evolving the way it has, it prevents us from inhabiting that older instinctive awareness, and so across the world a great fear has taken root — the fear that humanity might actually destroy itself.

The fear that we are going in a bad direction comes from the 2nd Shadow. It is because we now see ourselves as separate from nature that we feel this great collective fear. Our perception dislocates us from the truth. This collective fear pervades our individual lives as well. The 2nd Shadow, along with its programming partner the 1st Shadow of Entropy, pulls us away from living in a state of trust and connectedness and also reinforces this sense of isolation through our actions. Actions that come out of trust have very different results from actions that come out of fear. The former creates more energy for everyone and the latter takes energy away from everyone. In your personal life, if you allow it, the 2nd Shadow will affect every decision you make, pulling you into an interference frequency. This means that you don't appear to be in synchronicity with life, so you miss opportunities that would serve you and end up in repetitive patterns that can be very draining for all concerned.

However, the 2nd Shadow of Dislocation is an integral part of the script for life. It actually allows you to experience being out of the universal flow even though this experience is an illusion. Ultimately, even the level of frequency that courts chaos is a part of the overall fabric of existence. The 2nd Shadow allows you to witness your own helplessness as you seek to escape the feeling of dislocation and loneliness. As soon as you enter deeply into this shadow world with your full and honest awareness, it magically appears to change. This shift in your inner honesty precipitates a leap out of the mind and into a newer and far more expansive awareness. One needs to really grasp this insight — *you do not change your reality through doing anything.* There is a perceptual shift pre-wired into the DNA of every human being, and when it is activated it occurs in spite of you, as an aspect of your biological evolution. At a certain point your new awareness simply begins to open up. It does so gradually at first, but over time it coincides with a remarkable improvement in the quality of your life. As you see it, so it changes.

In the earliest versions of the I Ching that have come down to us, there appear to be quite a few anomalies with the more modern translations and sequences. One of the most interest-

It is because we now see ourselves as separate from nature that we feel this great collective fear.

ing of these concerns is the ordering of the 1st and 2nd hexagrams. There is strong evidence that the oldest versions began with this 2nd hexagram, representing complete yin. The patriarchal systems that translated the original text must have swapped the prime yin for the prime yang! Esoterically and mythically it makes far more sense to begin with the feminine rather than the masculine. Again, it depends on your frequency. At the Shadow frequency the masculine always comes first, which becomes the way of distrust, separation and force. The feminine approach however is based on unity, surrender and trust, which are hallmarks of all higher frequencies. Furthermore, one of the translations for this hexagram, known as Kun, is the word *Field*. This is a beautifully apt word to describe this 2nd Gene Key, because it represents the universal field in which we live. To move in harmony with this field is to be oriented, and to move out of harmony with the field is to be disoriented and dislocated.

REPRESSIVE NATURE – LOST

The two pattern nuances of the 2nd Shadow — that of being lost or regimented — describe the state of the great majority of mankind. In the case of the repressed nature, being lost describes the state of being out of alignment with one's true universal destiny. The repressive nature aptly describes the path of materialism and selfishness which follows its own path without regard to the greater environment. Our true destiny is to awaken out of our selfishness to our universality. Those with no sense of the spiritual dimension to life inhabit this state of being lost and the by-product of being lost is misery and suffering. Without the experience of a direct connection to the greater cosmic forces, one has nothing within to bolster one in dealing with life's trials. Life itself seems to have no purpose. Without a cosmic connection within one's being, one will always flounder in the world.

REACTIVE NATURE – REGIMENTED

The reverse pole of this Shadow is the enforcement of an external rhythm or structure over the top of life. This stems from deep anger that has never been resolved. Regimentation actually describes many of the world's great religions, which put themselves between an individual and their direct experience of the Divine. Regimentation can also describe science, which attempts to organise life into some form of meaningful and logical framework. Anything that tries to control life and pin it down emerges from the reactive side of the 2nd Shadow. None of this is to say that religion and science are wrong, except when they serve to obscure the hidden harmony that lies within an individual. The true meaning and purpose of life is only to be found within one's heart — in the mystical experience of moving in perfect synchronisation with all that is.

THE 2ND GIFT – ORIENTATION

MINERAL MAGNETISM

The whole process described in the 2nd Shadow involves orientation, the 2nd Gift. At the Shadow frequency you experience *dis-orientation*, whereas as you begin to reach into the Gift dimension, you become *re-oriented*. Again, as was discussed in the 2nd Shadow, there is no question of any doing on your part as this shift in your process engages even though it may feel to you as though you are *doing* quite a lot. Perhaps you started going to a therapist who helped you to see aspects of your Shadow, which in turn began to change your decisions and therefore your life. Perhaps you discovered a great mystical system or teacher who catalyzed this process. Perhaps it happened as a result of a personal crisis or perhaps it just spontaneously occurred as a complete surprise to you. The point is that all human life follows the same archetypal pattern laid down inside your DNA. Evolution itself is drawing you inexorably along the path towards awareness of your unity and oneness.

The Gift of Orientation has two faces. It either comes as a shift in your awareness that then affects your actions, or it comes as a shift in your actions which then catalyses a shift in your awareness. Whichever tone your experience takes there are a number of indicators that precede your permanently residing at the Gift level of frequency. One of the key experiences that people have is an increased feeling and occurrence of synchronicity. Synchronicity is the direct manifestation of the Gift of Orientation — it allows you to peep through the keyhole of existence and place yourself in a wider perceptual context. Synchronicities cannot be forcibly created but flow out of the feminine nature of the 2nd Gift — in other words, they happen when you are not looking. As your awareness begins to operate at its highest biological level — through the solar plexus system — so you fall into an easier rhythm of life. You no longer feel dislocated but begin to experience life in a more and more magical dimension.

One of the other indicators of your rising awareness is to be found in the 8th Gift of Style. Through the chemical family known as *The Ring of Water* this 2nd Gift shares a strong genetic link with the 8th Gift, which presents as a new and original way of living. As you become oriented into a wider context, your real face begins to show itself in the world. Instead of being weighed down by your changing moods and commitments, you use them as a means of developing a new flair. You also begin to override the perceptions and projections of others. In short, you begin to enjoy life so much that this rising energy spills over into your life as panache — a unique sense of individual style that is inimitable and always fresh, if even at times a little dangerous. Your panache or style is the sign of another paradox — that as you come more into contact with your unity with all creatures, you also witness an enhancement in your own uniqueness, particularly through your creative process.

The 2nd Gift has a special role within your DNA in that it creates a kind of attractor field around you. It not only unifies the microcosm with the macrocosm, but it also unites matter and spirit. A secret resides in the chemicals and amino acids for which this Gene Key codes. In every human being there are certain minerals that have magnetic properties, and this 2nd Gift concerns the chemical composition and purpose of these minerals. These minerals in our bodies, and particularly those in our endocrine glands, appear to tell us how to live life in or out of harmony. For example, in the tissues of the pineal gland, biologists have discovered a ferrous chemical known as magnetite. This mineral has been proposed as a key in linking electromagnetic activity to cellular function. The fact that it has also been found in most animals suggests that all creatures have a built-in magnetic guidance system that keeps them in alignment with

The more you let go into the feminine yielding quality of this Gift, the more universal power floods through you.

wider rhythms. It is through this magnetism that all life is linked, from the spin of the atom to the wheeling of the great galaxies.

As the frequency through the 2nd Gift rises, you live more harmonically and the electromagnetic power of your aura increases. The more you let go into the feminine yielding quality of this Gift, the more universal power floods through you. Your timing becomes more and more fine tuned, so much so that if you find yourself out of harmony the magnetic mineral transducers in your body immediately convey this to your brain. It is through the 2nd Gift that you can see the hidden agenda of life — to bring all beings into awareness of their unity. The magnetic attractor field around a person who is centred in the 2nd Gift exerts a powerful effect on all those around it, meaning that harmony is catching. This is why people who are deeply surrendered and attuned to life's processes can be so empowering. They can intuitively sense how surrendered or resistant others are to the great truth of their universality. Over time, such people bring others into their own personal harmony through the magnetic power of their aura.

THE 2ND SIDDHI – UNITY

DIVINE LOGIC

The 2nd Siddhi describes the experience of enlightenment or awakening. It is the cornerstone of all the siddhic states, being as it is the essence of the Divine Feminine. The Divine Feminine pole has a great mystery within it, because in a certain sense it cannot be said to be a pole at all. The masculine pole is very simple and straightforward, but the feminine is beyond any sense of reason and understanding. The masculine principle is really an externalisation of the feminine rather than a duality. At the siddhic level there is no such thing as duality. Duality is destroyed by a strange kind of Divine logic. At this heightened level of awareness, mathematics works differently than on the mental level. One plus one doesn't equal two, but always makes three. The only numbers that really exist at the siddhic level are one and three. One is one — it is consciousness resting in its own nature. It is the ultimate state of yin or femininity, or as this 2nd hexagram is often known — it is the *receptive*.

Words such as *receptive*, *feminine*, *yielding* or *mother* can be misleading as we tend to understand them as poles with opposites. However, they hint at what lies beyond them, so we need to consider them in a different way if we are to grasp what the 2nd Siddhi really means. This is something that can only be grasped intuitively. Thus when the One externalises itself as a manifestation in form, it has not created a duality, but a trinity. Every duality is really a relationship, and every relationship is really a three — there is a man, a woman and instantly there is also a couple — the relationship itself. In Divine mathematics, the number two is always an illusion — it cannot logically exist. If you can say anything at all about the number two you might say it is a bridge — a dynamic process that is instantaneously transmuted before it is even born.

These are concepts that cannot be approached with ordinary logic. Just like the quantum particles in physics that avoid our definition because they appear to be linked to our very perceptual apparatus, oneness cannot be comprehended, only lived.

Oneness cannot be comprehended, only lived.

Enlightenment is not an experience! This is a sentence to meditate upon like a Zen koan. If you see oneness as an experience to be attained or that may one day happen to you, then you are caught within that straight line between two points. The third thing is transcendence. It does not occur to you — rather it negates you. Ironically, transcendence does not take you away from life, as its name might suggest — it places you right in the heart of life, where you have always been. It unifies all opposites, ends all riddles, leaves all mysteries just as they are and brings a sense of trust that cannot be described. One cannot even really use a word like *trust* to describe the Siddhi of Unity because trust suggests duality again — that there is somehow a *truster* and a *trustee*. This is the wonderful dilemma of the siddhic state.

As do all the Siddhis, the 2nd Siddhi carries a specific mythology when it manifests through an individual in the world of form. As one of the primary pillars of our DNA, the life of one who lives within this Siddhi is of great relevance to our evolutionary history. These beings exert an enormous magnetic influence on our entire planet. Although it may sound like science fiction, the expression of the 2nd Siddhi in a human being actually changes the direction of the earth as it moves through space. The 2nd Siddhi can only therefore be born in the world of humanity if our whole species makes a leap in consciousness. In order for such a leap to be made, we have to wait for a certain set of geometric coordinates to line up in the cosmos. These are the alignments that astrologers are always seeking to find and understand in the heavens.

The 2nd Siddhi is mythically represented by the star of Bethlehem in the Christ myth. Other cultures also have stories of great beings connected to the appearance or alignment of stars and comets in the heavens. This Siddhi therefore tells us something about our ultimate state of consciousness; it is linked not only to when we are but also to where we are. The earth itself is on a trajectory through the galaxy, and at certain points in our time frame it moves into alignment with other geometric aspects of the cosmos. The ancient Mayans for example, believed that in the year 2012 the earth would be in direct alignment with the hub of our galaxy, which for them signified the birth of an age of heightened consciousness. At such junctures in time and space, the 2nd Siddhi may well incarnate again on our planet, and in the case of 2012, it is most likely to do so through a whole generation rather than through a single individual.

According to the planetary genetic time clock that is derived from the I Ching, we will experience another great axis point in the year 2027 as the precession of the equinoxes transits into the 55th Gene Key, opening up the potential for a genetic shift in humanity. Both dates — 2012 and 2027 — are hugely significant within the time frame of the publication of this book, but many axis points will follow in the near and far future. The 2nd Siddhi is the original nature of consciousness itself, and it manifests as a highly beautiful plan unfolding in time and space, swept along by the currents of evolution. All such mythic journeys echo the journey of the earth and of our universe, and as they all begin by leaving the warm comfort of the mother and the home, so they must one day return again to that same embrace. This indeed is our final destiny as a species — to realise our state of oneness and unity with all that is.

☷ 3rd Gene Key

Through the Eyes of a Child

SIDDHI
INNOCENCE

GIFT
INNOVATION

SHADOW
CHAOS

Programming Partner: 50th Gene Key

Codon Ring: The Ring of Life and Death
(3, 20, 23, 24, 27, 42)

Physiology: Navel

Amino Acid: Leucine

THE 3ᴿᴰ SHADOW – CHAOS

FROM CHAOS TO COSMOS

The 3ʳᵈ Shadow lies at the core of all our beliefs that the human individual is basically powerless in comparison to nature or the infinite. This is the programming domain of both religion and science, the two cornerstones of human belief. On the one hand we have religion, which separates humanity from nature by the interposition of a God or gods, thereby creating a division in our entire unconscious reality. This reality must then be based upon the worship of either our own or someone else's projection of what God is. Such a situation in turn creates the notion of free will and its judgment by the presiding deity. On the other hand we have science, which sees our nature as predetermined by our genes, which come pre-programmed only for survival, thus leaving all human beings victims of the vagaries of chance. In either scenario the individual comes off badly — we are either shown divinity only to be denied it, or we are shown freedom and then placed in a world of merciless competition in which only the lucky or strong can thrive.

Perhaps more than any other, the 3ʳᵈ Shadow captures the essence of the role of human DNA through an understanding of the structure of the single cell, the prime unit of life on our planet. Through an understanding of the single cell we will see how both of the above belief systems have grown out of a very natural organic progression that has brought us to exactly where we stand now — at a great evolutionary fork. As a planetary consciousness, we now stand before a chasm, and humanity's response to this threshold must either lead to chaos or it must lead to cosmos. Never before in history has the future of our entire planet and all its organisms hung in the balance of human behaviour. It is interesting to note that the words *chasm* and *chaos* are derived from the same root, since this 3ʳᵈ Gene Key governs what we humans understand as chaos. It is also intriguing to note that the original meaning of the word *chaos* is closer to *primordial space*. It was only due to a misunderstanding based on fear that the word came to be equated with disorder.

In order to get to the bottom of the profound mystery within the 3ʳᵈ Shadow, it is perhaps helpful to overview the direction of modern scientific thinking in response to our current evolutionary fork.

Contemporary mainstream scientific thinking is still based mostly upon two giants — Newton in physics and Darwin in biology. Despite what we hear, the scientific world is still reeling from the aftershock of Einstein and what he discovered. Einstein's breakthrough opened physics and all of science to vast dimensions that can barely be comprehended. All an honest physicist can say for sure any more is that the very bedrock of scientific thinking has been so severely shaken that no scientific premise can ever be taken for granted again. Einstein's quantum world has still not filtered through to mainstream biology except for a few brave pioneers who are willing to put their reputations on the line. Darwin's dogma of genetic determinism prevails and still remains at the root of all modern medical practice. However, a new quantum biology is indeed in the process of forming at the frontiers of modern science and it holds many fascinating possibilities.

At the core of the new biology lies a completely new understanding of the single cell. According to the mainstream view, the brain of the cell resides within its nucleus where the genetic instructions for life (DNA) also reside. The thinking goes that if the brain is in the nucleus and the nucleus contains the instructions, then the instructions control the cell and therefore us. However, quantum biology has discovered something very remarkable that directly challenges this. The brain of the cell does not reside within the nucleus after all, but in the cellular membrane, which provides an interface with the environment. In a nutshell, this means that life is designed to be cooperative rather than competitive. This new biological view makes enormous sense when considered alongside quantum physics, which holds that all of life is interrelated and holistic rather than separatist. In the old view, we humans are the victims of our selfish genes; in the new view there are no victims, only a huge interconnected and interdependent cosmos.

In the old view, we humans are the victims of our selfish genes; in the new view there are no victims, only a huge interconnected and interdependent cosmos.

The original Chinese name for this 3rd hexagram of the I Ching is an unusual expression that is traditionally translated as *Difficulty at the Beginning*. This is a profound insight because chaos is about beginnings. Modern scientific chaos theory is founded upon tiny variations in *initial conditions* within a system. In terms of the evolution of life however, the single cell's prime directive is indeed to survive at all costs, which is what forms the consciousness of this 3rd Shadow. The greatest challenge in evolution is always at the beginning because every cell has to learn to fend for itself. Only those that become strong will survive. Likewise, human beings are locked into single-celled consciousness at the frequency of the 3rd Shadow. This is the same consciousness that dominates across our planet, and its by-product is what we call chaos. However, chaos is really only a perspective that conceals a hidden pattern that inevitably leads to order. If life truly were that selfish, it would never have made it off the starting block.

According to the law of fractals, the same law that governs single cells also governs individual human beings, which means that early in our evolution we were programmed to survive. Our survival instincts are rooted in the primitive aspects of our brains, which became dominant as we learned to evolve from our apelike ancestors. Most of our modern beliefs and ideologies are based on these ancient aspects of our awareness that are rooted in fear, and we remain walled off within our individual cells. Our mainstream scientific thinking is still based on the divisive worldview that there exists no organising force in the universe other than chance. Likewise our religious thinking is locked into the same single cell consciousness and so we have divided the world into an inside and an outside — an *us* and then God.

What then does all of this really mean, and how does it affect you in your daily life? One answer lies in the polarity of this Shadow, the 50th Shadow of Corruption. It means that you are only trapped by your own thinking — *it is you who corrupts the data within your DNA by only allowing in a frequency that activates its most primitive components.* The wider you open yourself to the chasm that lies ahead of you, the more you will realise that there never was any chaos. It was all simply the difficulty at the beginning. It is this 3rd Shadow that makes human beings afraid to change, despite the fact that we are wired to change. In order to evolve, a human being must embrace chaos rather than try to protect against it. The 3rd Shadow causes human beings to distrust life itself and therefore adopt the old survival strategy known as *dog eat dog*. Incredibly, when you trust in chaos and allow your environment to mutate you, rather than trying to control it and stay the same, the greatest magic is revealed to you — that in chaos there is and always has been a vast underlying transformative order.

REPRESSIVE NATURE – ANAL

Our deep-seated fear that life is simply based on chance leads to one of the first great human control mechanisms — that of anal retention. Despite the discoveries made by Freud regarding this phenomenon, it has been known about for millennia by many systems of yoga and meditation. Fear activates a very subtle restriction of the muscles around the anus, which in turn affects our entire breathing pattern. As our breathing becomes shallower, the fear manifests through our need to maintain control over our lives. Almost all human beings suffer from some level of anal retention, ranging from mild to acute. It is only when we begin to touch the core of the primal fear within us that our system begins to let go of this deep-seated tensing of our being. The more we feel and allow the fear, the more we let go into the embrace of the cosmos, and the more we realise how deeply we are held and protected by it.

REACTIVE NATURE – DISORDERED

The adverse reaction to our fear of life is to express it outwardly as rage. The result of this is the creation of the very thing that terrifies us — chaos and disorder. These natures have absolutely no predictable direction, rhythm or purpose. They bring into the world the very vibration that terrifies the repressed nature — sometimes in the form of aggression, sometimes anarchy and always destruction. Again, there are varying degrees of this disease, from the mild to the acute. Every time we stop trusting in life and believe in our fears, we begin to co-create the vibration of chaos. Whenever that fear turns into anger, even in the most mundane circumstances, it carries a destructive force that will inevitably come back to haunt us. In this sense, all anger that is not owned fully strengthens the frequency of the 3rd Shadow in the world.

In order to evolve, a human being must embrace chaos rather than try to protect against it.

THE 3ʳᴰ GIFT – INNOVATION

THE END OF ISLAND MENTALITY

What is so wonderful about the 3ʳᵈ Shadow is that despite its narrow, fear-based, mono-celled outlook on life, it also holds the secret to our future. All one has to do is look back at our own evolution to see where it must inevitably lead — to the 3ʳᵈ Gift, the Gift of Innovation. As single-celled organisms multiplied all across our planet, evolution prepared itself for a quantum leap — the great transition into multi-cellular life. If it were true that deep within the nucleus of every cell a selfish drive were paramount, then it is highly unlikely that any two cells could ever cooperate, since their very design would drive them to compete. The genius of the cell and its true brain must therefore lie where the quantum biologists say it lies — within the cell's membrane. We know that the membrane allows the single cell to respond to its environment. But it goes one stage further than this — it must allow its DNA to be *influenced* by its environment. This is the foundational principle of quantum biology, and it is regarded as nothing short of heresy in the face of current mainstream biological dogma.

Who, what and wherever you are, if you are not continually transcending, then you are dying.

What we can now see is that life does indeed work in this way. Millions of years ago, single-celled consciousness made a quantum leap into multi-celled consciousness, which means that the survival-based programming within DNA must have either mutated or adjusted itself in order for the cell to be assimilated into a greater organism. When we apply the same metaphor to humanity, what we see is the empowerment of the human individual through the 3ʳᵈ Gift of Innovation. Innovation is built into life. In other words, life itself is designed to transcend its own initial programming (Difficulty at the Beginning) and discover new and higher forms of consciousness. Beyond selfishness and chaos lies cooperation and innovation. Innovation only occurs when you truly begin to think for yourself. This may sound like an everyday occurrence but in fact it is relatively uncommon. In order to be truly innovative you have to attain a very high frequency that allows you to see beyond the collective worldview.

The Gift of Innovation is far more exciting than simply being creative. It implies that you have permanently escaped the fear-based perspective of the Shadow frequency. Innovation thrives on optimism, although this is not the kind of optimism that is built upon hope. True optimism is the dynamic energy at the heart of creation. Through the 3ʳᵈ Gift, you begin to tap into the higher aspects within your own DNA. It is not that the DNA itself changes, but that the frequency passing through it activates its hidden programming. This is how single-celled life gave way to cooperative life, and it is how human selfishness must and will give way to collective consciousness. The whole is far more powerful than its individual human components — that is the rub.

Innovation implies cooperation by its very nature. In order to mutate life to a higher order, you must integrate and synthesise. The path of innovation broadly means to improve something through the introduction of a new element or elements. Those who work with the 3ʳᵈ Gift are life's great synthesisers because they understand the primal law wound into the substructure of all life — that unity equals efficiency.

This is the integral message encoded in the codon group known as the Ring of Life and Death. It is all based upon change, which is what the Chinese words *I Ching* mean — *The Book of Changes*. Life is continually mutating, and as it mutates it transcends and includes those levels and views it has just transcended. Who, what and wherever you are, if you are not continually transcending, then you are dying.

The 3rd Gift also teaches us something else wonderful about synthesis which has to do with another often-overlooked human gift — the gift of play. If you truly want to see the genius of innovation, you only have to watch a young child at play. When seen through the eyes of the 3rd Shadow, children appear to create nothing but chaos, but to the higher eyes of the 3rd Gift, a child is the living expression of genius. If we can recall the image of the cellular membrane allowing in certain frequencies from the environment which then impact on our DNA, this is exactly what a young child mirrors back at us. The child both mutates its environment (as any parent knows!) but is also mutated by its environment. The Gift of Innovation requires then that adults also allow themselves to be shaped by their environment. This means that you must be as open-hearted and open-minded as a child. All preconceived notions, dogmas or beliefs must be discarded when they no longer serve the developing synthesis. Innovation also requires a deep sense of inner trust. As you continue to work with this Gift, you continually update and change your position. Even though you may not yet see how everything fits together, you feel the underlying unifying spirit, and above all else you have enormous fun.

Through the Gift of Innovation, humanity must now improvise in order to survive the coming centuries. Just like those early single-celled organisms, we must mutate to become a complex multi-celled organism. Just as those early organisms developed a nervous system culminating in a brain, we developed our governments and our modern global communication-rich culture. However, our greatest innovation is yet to come which is literally to relocate our brain from the cranium to the far more advanced solar plexus system. Just as the cell's true brain is in its membrane rather than its nucleus, so our true brain is through our emotional system. Like the cellular membrane, the solar plexus system determines what frequencies are allowed in and out of the body. Thus as the individual mutates to receive higher frequencies, these frequencies unlock the higher organising principles of collective life embedded within our DNA. It is not we who make this decision, but life. These codes are already waiting inside us, threaded along those mazy pathways deep within our DNA. This is the sacred secret: life is designed to keep on innovating, and the old human with its single-celled island mentality has had its day.

THE 3RD SIDDHI – INNOCENCE

ALL PLAY AND NO WORK

In discussing the 3rd Shadow and its Gift, we have travelled into some fairly complex territory in order to understand the true depth and relevance of this 3rd Gene Key. Actually, it is really one of the simplest of all to understand once you move out of mental territory. At the absolute zenith of its frequency the 3rd Siddhi is about innocence.

This Siddhi reminds us that all of life, and that includes human beings, is essentially inno-cent. Our most common misconception as human beings is that we are not a part of life. Our mind seems to tell us that we can control life, but in truth this is an illusion born of the brain. Human beings are an instrument *of* life, an experiment *in* life, but we can never be masters of life as long as we are only a part of life.

Human beings remain a brain-centred species today. We have seen how deeply we feel the need for a centre or a nucleus to be in control of life. This need is expressed within every level of our thinking and at every level of our society. It is epitomised in our need to find a God. The thought that no one might be in control terrifies us! What would happen if the world had no governments? What might happen if we had no religion, education, police, army or money? Because we do not trust life, the only answer we can give to this question is chaos. The truth is that we do not know what would happen, but the child inside us longs to find out. That is what this 3rd Siddhi is all about — it is about no barriers, no laws and no work — just play!

As a species, humanity has hugely misjudged itself. We represent the cutting edge of con-sciousness on earth at this moment in time. One day, when the human we see now is a relic of the past, we will look back at this time through the 3rd Siddhi and wonder to ourselves why we became so serious during this epoch in our evolution. From the viewpoint of the 3rd Siddhi we are eminently laughable, in the warmest possible way. That human awareness had to go through a phase in which it believed itself to be the centre of the universe is an enormous joke to the 3rd Siddhi. However, to a child, play *is* absolutely serious. The toy in the child's hands *does* become the centre of their universe, just as our individual lives become the centre of our universe. We are innocently unaware of the magnitude of our great good fortune to be alive and playing out this epic exploration of consciousness with all of its toys in the world of form.

One day inevitably the child will grow up. Humanity will evolve towards its awesome final destiny of global and celestial harmony. But one thing is certain: consciousness will never grow up. It will keep on exploring, playing and experimenting forever, all because its nature is inno-cence. Life is innocent, we are life, and therefore we are innocent. That is the equation for the next millennium. As the 3rd Siddhi comes into the world, which it will do very shortly, a great remembering of our innocence will emerge in humanity because the first vessels of the new human destiny (spoken of in the 55th Gene Key) will all be children. The strange thing about these children is that the consciousness within them will never *grow up* and become serious. Their awareness will function in an entirely new and different way from our current brain-centred awareness. Their brain will be the environ-ment itself — it will be inside every other human being, creature, plant, stone and star. We cannot yet imagine that kind of cosmic consciousness, but it will soon come pouring into the world like a great breaking wave.

One thing is certain: consciousness will never grow up. It will keep on exploring, playing and experimenting forever.

Humanity will change from within, as life always has. It is shaped by its own environment. If it becomes a threat to that environment then life relays that message back into its cellular structure, causing it to mutate a more efficient program. This is not a decision that can be made by humanity but by the whole organism.

There are no losers in this game because every cell innocently contributes its own unique information to the whole. In the process of mutation, new life codes are activated and any cells that do not mutate will simply die off. This is a totally natural and organic process that is already beginning to occur to humanity. We fear it because we do not trust our own innocence. The great human fear is of losing our own uniqueness. If we lost that and became an amalgamated soup, it would indeed be a step backwards in evolution. However, evolution does not travel backwards — it only knows progress. By giving up our own individual identity, we actually create another greater individual identity that is both collective and individual at the same time.

The human body itself is the prime example of the direction that humanity is moving. Our body evolved from single-celled amoebas responding to their environment. Not a single cell within our body can afford to be selfish or we would die. Life spontaneously discovered in its innocence that the easiest way to keep evolving was to integrate all those competing individual cells into a single body. What incredible things life discovers when it is left alone to play! This is the central message contained within the Ring of Life and Death — that the chaos of play is to be deeply trusted and revered. Play is the expression of genius and genius always finds a new solution to the challenges it meets along the way. We are all really children — children of the cosmos — and our only real job is to let go of our seriousness and find the delight in every exquisite jewel that life places before us.

From reading this 3rd Gene Key and its various levels of frequency you can perhaps feel the wonder and the scale of the experiment you are involved in. As Einstein said: "God doesn't play dice." The human brain was not designed to grasp insights as profound as these. The certainty of order within the universe is a feeling that can only be captured by a more advanced system of awareness in human beings that our mystics have called by many names. Above all it is equated with the human heart and the feeling of love and unity that lies between all creatures. Your true home is not located within your body, nor is there somewhere in the universe a single centre through which everything is orchestrated. The centre you are looking for is the feeling of love itself, which is the omnicentric manifestation of your eternal state of innocence.

4th Gene Key

A Universal Panacea

SIDDHI
FORGIVENESS

GIFT
UNDERSTANDING

SHADOW
INTOLERANCE

Programming Partner: 49th Gene Key

Codon Ring: The Ring of Union
(4, 7, 29, 59)

Physiology: Neocortex

Amino Acid: Valine

THE 4ᵀᴴ SHADOW – INTOLERANCE

THE FOLLY OF YOUTH

When the ancient Chinese named this 4th hexagram of the I Ching, they gave it the wonderful name *Youthful Folly*, and in doing so, they showed a profound understanding of its lower nature. The 4th Shadow, the Shadow of Intolerance, is rooted in the mind's habit of becoming tangled up with human emotions. Intolerance is best understood through its relationship to its programming partner — the 49th Shadow of Reaction. Because human beings are governed by their emotions, the general emotional state of humanity is unstable and chaotic. We tend to react to our whims instead of attuning to the quiet clear guidance that lies within each of us. In reacting to our own and others' emotions, we decide that what we feel must be the truth, and our mind agrees with us.

The 4th Shadow is essentially a misuse of one of the great gifts of the human mind — logic. The power of this 4th Gene Key is the power to read and solve logical patterns, and as we shall see in the case of the 4th Gift, this ability leads you to a universal understanding of the rhythmic patterns and tendencies of all life. However, at a low frequency based on emotional reaction or over-reaction, this Shadow uses the power of distorted logic to support and uphold its volatile nature. In other words, if the 4th Shadow has a bad day and decides it doesn't like someone, it will find a whole list of logical reasons to support its dislike. This is what intolerance really is — a slanted perspective of logic. At the Shadow frequency, the mind is actually given the authority to make important life decisions, and this is often disastrous because its true role is not to decide anything. Its true role is to understand and communicate.

The aptness of the name *Youthful Folly* becomes apparent when you see what happens as this Shadow reaches an extreme emotional state. At the Shadow frequency, you identify absolutely with your emotional state, which defines the way you live your life. Any unresolved emotional pattern is taken up by your mind and built into a highly intricate logical framework that masquerades as the absolute truth. Through the medium of the 4th Shadow, opinions, judgments and resentments are transmuted into convictions and certainties. In this way people blind themselves with their own logic and can become deluded and sometimes dangerous. Intolerance is based on a subjective distortion of logic, which only measures those patterns that a person wants to see, rather than seeing all patterns from both sides of an argument. The power of the 4th Gift of Understanding is founded upon this ability to objectively assess all aspects of a viewpoint, thus avoiding the pitfall of taking sides.

The human logical mind is actually not designed to take sides. The essence of logic is its foundation upon objectivity, but in the hands of fear, objectivity dies and logic becomes subjective, even at a collective level. You may wonder how something can be both collective and subjective. Racism and prejudice are examples of genetic or ancestral fears that manifest through certain population groups who then reinforce these fears through subjective logical argumentation. Even science is rarely truly objective, unless it remains open to all counter-arguments. In the case of science, a counter-argument might come from religion, which challenges logic as the only means of acquiring truth. Only when science is skeptical of its very own nature can it really be regarded as truly objective. The 4th Shadow forms a very subtle undercurrent to all human mental structures, from the scientific to the spiritual — it cannot help taking sides.

The underlying nature of all the Shadow states is fear; in the case of the 4th Shadow, fear is projected onto others and then reinforced through taking a defensive (and sometimes offensive) mental position. This is how intolerance is created, and it is sometimes extremely subtle. Intolerance bases its position on opinion rather than fact. If you took the time to examine the other side of the argument, you would immediately understand that your opinion is rooted in a deep-seated emotional fear of something. The great problem with logic is that it can only disprove itself, which does not make human beings feel at all secure. Therefore, most people choose one single side of an argument because it makes them feel a certain mental solidity. Ironically, however, mental assuredness cannot make the body feel safe. The body can only feel safe when it surrenders to the moment without wanting something else.

The 4th Shadow is endlessly restless in its need to examine patterns and resolve questions. One answer is simply replaced by another question. The role of this Gene Key is to understand, but understanding cannot come through the mind itself. This is the essence of the 4th Shadow's dilemma, and it keeps many people from realising its Gift. Understanding, as we shall see, only comes when you realise that your mind can never truly understand anything! Before you come to this huge inner realisation, you will live your life under the influence of the 4th Shadow and its persistent promise that it will one day arrive at an answer that brings you lasting peace. This is youthful folly, because only after much anguish and experience do you come to realise that there are no intellectual answers that bring about such peace. There are only two options for the 4th Shadow — you either settle for a one-sided opinion and deny the other side, or you become lost in the fruitless quest to bring an end to the feeling of uncertainty that lies deep within you. Unless you make the leap into the true understanding of the 4th Gift, you have no choice but to be caught in eternal misunderstanding and intolerance.

REPRESSIVE NATURE – APATHETIC

When the mental dynamism of this 4th Gene Key is frozen by the unconscious fear of a repressive nature, the result is an apathetic mind. An apathetic mind is a collapsed mind that is no longer bright or intelligent but has given up on understanding anything and sunk into a kind of mental lethargy. These people believe themselves to be less intelligent than others, when in fact they are really paralysed by an unconscious fear. Their fear is that they will have to assume responsibility for themselves, their decisions and their actions. Instead they choose to have no opinions about anything. Such people can pretend to be quite enlightened and very open, but there is a vital energy lacking inside them. Thus they can have problems in motivation as well as with their health. To escape their apathy, they simply have to start thinking again, but without letting their thinking rule their lives.

REACTIVE NATURE – NIT-PICKING

In the reactive nature, thinking *does* rule a person's life. The reactive nature projects out its eternal need for answers to questions, and it does so in the belief that these answers will bring them a sense of security. When they discover that this is not so, they become angry and blame someone — often the person or system that they supposed would give them all the answers! These people cannot let go of their need for some feeling of resolution so they make their mind the authority for bringing about this feeling, only to be endlessly disappointed. These people hone in on the most irrelevant details, unconsciously looking for a vent for their frustration. When they find such a detail, it affords them the opportunity to criticise or complain and thus release some of their pent up anger and tension. These people, above all, need to find a way to let go of the hope that their mind can ever bring them solace. When they do this, they can finally stop projecting their eternal disappointment onto others and begin to find a new awareness arising in them, outside of their mind.

THE 4TH GIFT – UNDERSTANDING

QUANTUM KOANS

If you are someone with powerful intellectual capacities, the 4th Gift represents a wonderful breath of fresh air for you. At the same time, it requires a huge quantum leap in your whole being. The Gift of Understanding has nothing whatsoever to do with knowledge. Knowledge is what your mind thinks it needs in order to take away its permanent feeling of unease. But knowledge can never bring a sense of peacefulness. At most it can give you the hope of that peace, although ironically it is this very hope that sustains your intellectual quest and keeps you within the confines of the Shadow frequency. Only true understanding can bring peace along with it because true understanding lies outside the domain of the mind. Understanding is of the whole being, and it does not and cannot require agreement from the cognitive capacities within your brain.

True understanding lies outside the domain of the mind.

If you allow the 4th Gene Key to run its natural course without giving it the responsibility to make decisions, it actually does something quite magical — it propels your awareness out of the mind. The very desperation of the mind to come to understanding through knowledge constantly thwarts itself by looking at life from every conceivable angle. At a certain point, all this pent up energy explodes into a quantum leap out of the mind. This is precisely how the concept of the Zen koan operates. A koan is a paradox given to the mind to solve, and at the precise moment when the mind has finally realised that its own logic can never solve the koan understanding dawns. This quantum leap is true understanding, which is a feeling of knowing that floods your whole body and radiates from the solar plexus area.

The Gift of Understanding is the only answer that will satisfy a person's dissatisfaction, and it must come about through the exhaustion of your mind rather than any other way. When you look logically at all angles of any concept, you begin to realise that nothing can ultimately be proved through logic, because logic can always be used to prove the opposite. When you finally see this, your whole being lightens because you realise once and for all that the mind is of no use for resolving anything of real importance. This in turn releases the mind to do what it loves best — research and communicate and play.

When the 4th Gift is freed from having to solve your existence, it finally comes into its real genius — to play with the patterns of existence and arrange them in new and original ways. When you have the feeling of visceral understanding deep in your belly, your mind is no longer hampered by the need to defend your own viewpoint. In fact, you realise that all logical formulae can be manipulated to prove or disprove anything. The higher frequency of such understanding also brings with it the urge to be of service to the world, and you can use the mental alacrity of the 4th Gift to follow the dictates of your higher self. This newfound genius at seeing the underlying patterns of life also affords you direct access into another aspect of this 4th Gift — the ability to understand people.

Through seeing all sides of any mental construct, the Gift of Understanding sweeps aside the possibility of intolerance and uses its gifts to create new roles or systems to bring positive change into the world. The programming partner of the 4th Gift is the 49th Gift of Revolution, and this energy always accompanies true understanding. The very nature of understanding is to bring about improvement for society in general, since the dynamic energy within this 4th Gift is still experienced as a certain restlessness. Whereas at the Shadow level, this was the restlessness to resolve your own insecurity, at the Gift frequency this becomes the restlessness to resolve the insecurity within society in general. Thus understanding always carries within it the seed intention to solve the problems of intolerance and division in the world.

In the coming genetic shift triggered through the 55th Gene Key and the 49th Gene Key the role of this 4th Gift will undergo some very important genetic changes, which will gradually sweep through humanity. The involution of powerful siddhic energy from the highest aspect of this archetype will bring about a minor but extremely important genetic mutation in this 4th Gene Key and its associated amino acid valine. This mutation will essentially phase out the 4th Shadow of Intolerance.

The children that come into the world carrying this new sequence of mutations will not be emotionally polarised; their mental system will not run riot throughout their lives. The 4th Gift of Understanding will govern the way in which their minds operate from birth. They will bring about a social revolution on a global level, and this revolution will be based upon a logical understanding of the folly of the existing systems and structures. New formulae will come into the world through this 4th Gift that will undoubtedly lead to technological breakthroughs that solve longstanding problems rather than create new ones.

THE 4TH SIDDHI – FORGIVENESS

MERCILESS FORGIVENESS

Not only will the coming shift bring about a social revolution, but it will also bring an end to one of the great searches of modern man — the search for knowledge. Through the rupturing of the 4th Shadow, understanding will take the place of knowledge and much of the thrust of our modern world will die down. We will no longer need to logically make sense of the paradoxes of existence because our new centre of awareness will give us a physical and energetic understanding of existence. Thus the role of logic in our world will alter. It will no longer be used to defend our prejudices and fears and it will no longer be used for purely personal benefit. Logic, at its highest frequency, is a means to orchestrate the most efficient society possible. True efficiency is based upon a higher holistic understanding of living systems. Once our understanding shows us how connected we all are to each other, we will see for ourselves that selfishness is the most inefficient frequency of all.

The 4th Gift forms the launch pad for an even more refined frequency — the Siddhi of Forgiveness. Forgiveness is born out of understanding, but it occurs when a being makes a leap *beyond* understanding. Forgiveness is a stage further on from social revolution. Just because a person has understanding and good intention, does not mean that they can orchestrate a perfect society. History has shown that revolutions never change the world — they just change society, and usually only briefly. The highest possibility of the 49th Gene Key is the Siddhi of Rebirth, and this is the Siddhi that always awakens alongside Forgiveness. As we have seen, understanding leads to the urge to serve the totality through instigating some kind of social reform. Forgiveness, however, is a pure siddhic state and as such it has no sense of restlessness at all. All siddhic states are the end of the line — they represent the absolute transcendence of our genetics and the end of being human.

Forgiveness is the thunderbolt that is released when a being attains Christ consciousness.

Forgiveness is the thunderbolt that is released when a being attains Christ consciousness. It is like a kind of cosmic warmth that melts the borders and boundaries within the world of form. Forgiveness allows the Truth behind all form to be seen. Further than that, it allows one to see through and thus become one with Truth.

There exists a great mystery about the power of forgiveness concerning time. Forgiveness represents an involving force rather than an evolving force because it literally comes from the future towards the past. It is a Divine quality that descends, like Christ, into the world of form. In descending into human form, forgiveness lays its hand upon all humanity and works its way back through time, burrowing into our collective past, releasing and freeing energy that has lain trapped and stagnant for aeons. Forgiveness moves down the ancestral bloodlines of all humanity in this way, dissolving genetic blocks and lifting karmic curses wherever it travels. This is why the Siddhi of Forgiveness is often credited with being capable of inducing miracles, because it can release a karmic debt that has stagnated for generations. As such debts are released, those in whom they live can move through incredible transmutations. These mysteries are explored in more depth in the 22nd Gene Key through a transmission known as *The Seven Seals*.

The 4th Siddhi is a primary agent of Divine Grace, that is to say it does not adhere to human laws. It concerns the resolution of old debts, at all levels. At a purely individual level, the whole process of human incarnation is based around the notion of karmic debt. Until you have paid off all your debts, particularly through your relationships, you cannot escape the game of incarnation and reincarnation. Because it is also a part of the Ring of Union, the 4th Siddhi's ultimate role is to bring humanity into a collective union through the resolution of karmic debt, individually, racially and mythically. In the material realm we will see this manifesting one day through nations forgiving each other their financial debts. Forgiveness as such is really a collective phenomenon, which is why we humans have never been able to control it or fake it. It falls upon you like a surprise, and something inside you opens up that was previously choked. It really is a miracle.

As more beings bring this Siddhi into the world, they play their part in the releasing of humanity's collective karma. Such beings hold onto nothing in life because they have moved beyond understanding into pure Truth. The forgiveness that we know of now is but a tiny shadow in comparison to the *beyondness* of this 4th Siddhi. Pure Forgiveness is a universal panacea that radiates in every direction throughout time and space. It is the final answer to end all questions, and when the first atoms of forgiveness finally travel back to the very beginning of time, as they already have, the world we know will begin to dissolve. When all is forgiven, then forgiveness itself no longer exists, only Truth. The final destiny of the 4th Siddhi is to rupture the connection between the past and the future, between the black and the white, between the yin and the yang — to finally bring an end to the logical fabric of spacetime itself. True forgiveness is merciless because it returns everything to its source and is a force of pure annihilation. The ultimate goal of forgiveness is in fact to bring an end to the world of form itself.

≣ 5th Gene Key

The Ending of Time

SIDDHI
TIMELESSNESS

GIFT
PATIENCE

SHADOW
IMPATIENCE

Programming Partner: 35th Gene Key
Codon Ring: The Ring of Light
(5, 9, 11, 26)

Physiology: Sacral Plexus
Amino Acid: Threonine

THE 5ᵀᴴ SHADOW – IMPATIENCE

THE NEW GENETIC CODE

It is the 5th Gene Key that really forms the backbone of the 64 Gene Keys. Containing all the codes and patterns of life, the 5th Gene Key represents the great digital library of consciousness in form. These codes lie coiled and concealed within every single living cell, wound into the famous helical patterns of your DNA. The 5th Gene Key is also one of those Gene Keys that is found in all life forms, since it alone maintains the very rhythmical patterns that allow an organism to stabilise within its particular environment. Furthermore, the 5th Gene Key is a great mystical chess piece within the genome, since it also unites all these separate organisms into one great universal rhythm — the pulse of life.

Because the 5th Gene Key binds all living forms through these universal patterns, at its lower frequencies it tends to display a deep distrust of life. This manifests in human beings through the 5th Shadow of Impatience. We know that all human beings carry inside themselves a deeply ingrained fear of death. What you may not realise is that there are many layers around this fear. At the level of your personality there is your outer fear, the prime fear pattern you absorbed through the events of your childhood. This personal fear in turn gives way to the great collective fears; for example, the fear of change. At the deepest level however, at the very precipice of awareness, lie the most ancient human fears, and this 5th Shadow represents them. These ancient collective fears stem from one prime source — the fear that there is no underlying order to the universe. If the frequency of your genetics is tuned into this fear, no matter what you do to try to bring a sense of stability to your life, your body itself will never feel safe. This is in fact the normal state of consciousness of the mass of humanity.

Within the 5th Gene Key lies the great secret of the timing of life. This Gene Key is about trusting or not trusting universal rhythm and natural timing. It sets the rhythms of the seasons, it fixes the timing of cell growth and decay in every living creature and it governs all patterns of animal and human migration. As we have seen, all distrust concerning the timing of life manifests in human nature through the 5th Shadow of

Impatience. This impatience is one of the greatest causes of disease on the planet, for it will undermine your health and well-being and cut you off from the very heartbeat of life. In a certain light, impatience can be viewed by human beings as a positive trait, since it can goad you into action instead of remaining complacent. This is a mistaken impression since impatience is rooted in agitation and all action arising from agitation is out of harmony with the whole. There is a vast difference between acting out of impatience and acting out of resolve.

Impatience is not a natural feature of human character; rather it is the result of a loss of your natural rhythm at a biological level. If you are feeling impatient, your breathing has become shallow and your nervous system is over activated. Your core feeling is that all is not as it should be. Of course this feeling is absolutely untrue. Everything is always exactly as it should be. What has happened is that you have fallen out of your natural state — the state of trust. Impatience is always rooted in the mind and as such is unique to human beings because of the unusual nature of our neocortex which processes information in such a way that we perceive everything as happening in time — with a past, a present and a future. The only way to escape impatience is to escape the mind and its realm of time, which is precisely what happens at the higher frequencies of this Gene Key.

There is so much you can learn about the 64 Shadow archetypes by looking at them through their harmonic pairings. Psychologically, you can find the root of all human issues and complexes within the 32 Shadow pairings. Each of the Shadow pairings is literally set up to create obsessive patterns. The programming partner of the 5th Shadow of Impatience is the 35th Shadow of Hunger. These two states literally feed on each other — the hunger to escape the feeling of insecurity leads to impatience and vice versa, and it is all driven by the fear of time running out. In the modern world, you can really feel this deep-seated fear rippling through humanity. We have forgotten that life itself sets the pattern of evolution, and that we are merely the agents of those patterns. From time to time, both in our personal lives and at a collective level, natural mutations occur when the stability of life is overturned. If you can enter deeply into the spirit of such times you will come to realise that nothing is really out of balance and that all will reveal itself in time.

Because the 5th Gene Key connects all life forms into a wider pattern, you have to learn to realise that nothing happens by chance. Everything is connected to everything else, so when you find yourself moving through a difficult period it means that all life is moving through a difficult period. The themes that punctuate your daily life are universal themes that are lived out by all humans and all creatures simultaneously. If you look at the state of the mass consciousness on our planet today you will see a universal mutation underway. Our entire species is undergoing a deep genetic quantum leap. As you come to terms with your personal fears, humanity comes to terms with its collective fears. The very fact that you are reading this right now is telling you that life is examining itself in enormous depth, and in doing so it is discovering all manner of anomalies or glitches in its programming matrix. As each glitch is observed, it is deleted, and as each fear is accepted, it diminishes. Beneath the old genetic code a new one is thus revealing itself.

When you find yourself moving through a difficult period it means that all life is moving through a difficult period.

In summary, this 5th Shadow of Impatience is nothing more than a low frequency human response to a particular set of environmental conditions. Like all the Shadows, it is simply a matter of perception and attitude. Time always moves according to your perception or mood. If your breathing is calm, rhythmical and deep then time appears to dissolve. At the highest frequencies of this 5th Gene Key, you do not even notice time. The more you become aware of your own impatience and restlessness, the deeper you begin to sink into your true centre and the less you are concerned with time and timing. This is the birth of a higher frequency, and that frequency is what we call patience. This 5th Gene Key contains the most beautiful of equations — acceptance equals patience because the deeper you enter into your fears the more patient you become.

REPRESSIVE NATURE – PESSIMISTIC

It is interesting to consider that pessimism is really rooted in impatience. One of the key features of all repressive natures is collapse. If you have a repressive nature, then you have at some level moved through some kind of energetic collapse. When impatience drives a person to give up on life, it manifests as pessimism. Pessimism is nothing but the vestige of a complete loss of rhythm in one's life. It is an expression of a deep-seated fear that nothing can or will ever get better. Pessimism pulls one into a downward spiral, feeding on itself over and over until it finally ends in crisis or some kind of psychological and/or nervous collapse. If a person has been swept into pessimism there is nothing proactive they can do to break free. As their friend or an observer, you must trust that life will create a crisis for them, because eventually it will. Only such a breakdown or breakthrough has enough energy to snap them out of their pattern.

REACTIVE NATURE – PUSHY

The anger-based version of pessimism is pushiness. This is simply a different type of nervous system, and it reacts to impatience in an extroverted manner rather than collapsing inwards. Pushy people are constantly trying to force life's flow. They push others around, becoming tetchy and aggravated, and can suddenly lash out for no apparent reason. These people also tend to manifest very inharmonious situations where natural timing is out of synchronisation and nothing seems easy, as though life were deliberately blocking one's path. Even so, these kinds of natures stubbornly push on, making the situation worse and worse until something or someone has to snap in order to release the pressure. You can see how resistance to the natural patterns of life eventually brings one to a turning point, both in the repressed or reactive natures. At such turning points, the only decision is really — do I want to begin living again? Whatever a person decides, it must be respected as a part of the wider pattern of life.

THE
ENDING
OF TIME

THE 5ᵀᴴ GIFT – PATIENCE

THE LIBRARY OF LIGHT

The antidote to impatience is patience. This may sound like a very obvious truth, but it is actually rather profound. The ironic thing about patience, though, is that it takes patience to learn patience! And patience can indeed be learned. It is a self-proving Gift. In other words, the more patient you become, the more you learn that patience always pays off, and thus the easier and more natural it becomes to wait. However, patience is not the same thing as waiting. You can wait both patiently or impatiently. Patience is the natural ground of your being, whereas impatience arises out of fear and conditioning.

Patience is about Trust. These two words have very similar meanings. If you trust in life, you will trust life in every moment, even the challenging moments, and in so doing you will always remain in the flow. As you live your life you may notice many rhythms all around you and it is the underlying presence of these rhythms that makes you feel stable. The most obvious annual rhythm can be seen in the shifting of the four seasons. If you are inwardly quiet, the passage of the seasons will make a very powerful impression within your psyche. It tells us some very deep truths — that spring always follows winter, for example. In your daily life you will experience periods of winter where resources may be scarce, where you may feel lost for a while, or when you simply feel melancholy for no apparent reason. Such stages are built into life, and if you are patient they change of their own accord and they always reveal some magic.

As a vital element of the chemical family known as the Ring of Light this 5ᵗʰ Gene Key plays a profound role within your genetics. Coding for an amino acid known as threonine, it determines the blueprint for how your individual cells trap light and convert it into energy. Through the magnetism of your living aura, you draw in or limit the frequencies of light deep into the cellular structure of your body. If you live your life out of fear, magnetically speaking you limit the amount of light that *touches* your DNA. The more open-hearted you are, the more you magnetise higher frequencies into that very same DNA. It is only at certain frequencies that particular codes can be activated. When you are in love, for example, you will always experience higher codes within your DNA awakening. Time will pass differently, almost as if you were in a timeless bubble. Anyone who has been in love can recognise or remember this kind of experience.

It is patience that has always been the true gauge and measure of the greatness of a person's soul.

The 5ᵗʰ Gift of Patience is a far greater gift than it may sound. Patience, when really experienced deep within the body, leads to the opening of your heart and the consequent turning on of a higher code within your DNA. Your DNA is a library of consciousness that is completely dependent on the frequencies of light. The body is designed to awaken to these higher frequencies, which is why humanity really only seeks one thing — to live in the state of love. The more you are able to settle into a deep trust and patience with the rhythm of your own life, the more your heart will open and the softer and more yielding you will become in your attitude to everything and everyone that comes your way.

You will begin to discover a higher functioning of your DNA — a code within a code. This code is not dependent on anything or anyone outside you. The state of open heartedness or of being in love is a completely natural human state. There are people in the world who live permanently in this state.

The gifts that come to the patient person are not only inner quietness, but also integration. When you wait and allow life to reveal its natural rhythms, you also come to realise that life knows best. Hindsight shows you repeatedly that life is constructed of beautiful and perfect patterns woven into wonderful tapestries and that your individual life follows such cosmic patterns. If you know how to wait calmly, you will realise that you are always a part of these wider patterns, and that they actually support you at all times — even the most challenging times. Above all, the Gift of Patience allows you to hear life's music. It tunes you into the subtle metronome behind your life. It allows you to breathe deeply into your belly and never feel trapped by your outer life situation. Patience smoothes rough edges, keeps your heart and mind open and makes life seem simple and easy. When you become impatient, even for a moment, you have stopped listening and trusting in the greater life. For these reasons, it is patience that has always been the true gauge and measure of the greatness of a person's soul.

THE 5TH SIDDHI – TIMELESSNESS

HITTING THE SPEED OF BEING

Since the discovery of DNA and the mapping of the human genome, a huge amount of energy has gone into trying to identify and locate genes that seem to have specific links to certain diseases and illnesses. Now that we humans have cracked the code of life, our goal is to try and use it to make our lives more secure, which is understandable. However, what the Shadow frequency fails to realise is that just because you can read a code does not mean that you are bound by that code. For example, knowing a child's genetic predisposition to heart disease may well end up being a double-edged sword because the very knowing increases the possibilities of its occurring. If the parents respond through fear, then the frequency is indeed strengthened. At the Gift level of frequency, you have just learned that higher manifestations of your DNA also exist. To live an open-hearted life is to operate at peak health and therefore to override any such genetic predispositions. However, there exists an even higher truth hiding here at the siddhic frequency — it is that all codes can eventually be transcended and are in fact *designed* to be transcended.

This truth is momentous. What the Shadow frequency views as a problem may actually be a higher genetic mutation waiting to be unlocked. Here at the highest siddhic frequencies, codes can no longer be unlocked digitally, which means to say that the mind cannot comprehend them. Only the human spirit can unlock the deepest secrets in your DNA. It is here in the 5th Siddhi that the great secret known as enlightenment lies waiting. The highest code can only be awakened by the flood of high frequencies contained within the spectrum of light. This is why the ancients named the phenomenon of awakening *enlightenment.*

The more you open yourself to higher frequencies of light, the more heightened the vibration becomes within your cellular DNA. Awakening usually occurs in a rhythmic or logarithmic manner. That is to say that the body is periodically flooded by higher siddhic frequencies prior to full-blown enlightenment. The distance in time between such events decreases by approximately one half every time it occurs. The body is readying itself for an experience that will alter its own functioning permanently.

Every time you experience such an inflow of siddhic high frequency light, you move through an intense period of mutation in which any glitches in your genetic code are erased. This translates into experience as deep unconscious fears rising up to the surface of your awareness through the medium of your body. After such periods, your whole system must literally be rebooted. These times in your life can be very challenging and you may well experience some disorientation. As it nears the siddhic threshold, the frequency of your body will begin to approach the speed of light. It is only when your frequency hits the speed of light that time finally dissolves and pure being is experienced. This is the higher purpose hidden in all human DNA — to hit the speed of Being.

It is only when your frequency hits the speed of light that time finally dissolves and pure being is experienced.

At the level of the 5th Siddhi, patience itself is completely transcended and something wonderful happens — you as a separate entity cease to exist! Impatience and Patience are two ends of the same spectrum and as such they are rooted in the existence of time. There is a fascinating lesson to be learned from the Spectrum of Consciousness, and patience is the only key that allows you to see this lesson. All beings in life are waiting. Life itself can be seen as a waiting for death. We each wait for the future to reach into us, and as it does, so we see what life has planned for us. Consciousness dances through our vehicles in unique archetypal patterns. Sometimes we may find ourselves struggling in a Shadow state, whilst at others we feel a deep sense of peace. Our lives are like the candle flame flickering in the winds of duality — impatience to patience — fear to trust and back again endlessly.

A siddhic state is different from this. In the siddhic state, you have jumped out of the game. You are not identified with the candle flame, even though it may go on flickering. You cannot talk in terms of patience or impatience because you cannot talk in terms of time or identity. This is the meaning of the Siddhi of Timelessness. Timelessness is the nature of consciousness itself. It is unborn and undying. Timelessness does not care whether the body is exhibiting patience or impatience. It simply is. Timelessness is thus the death of all fighting. Patience and impatience are not absolute states. They are the poles within life, experienced through human awareness. Only timelessness is an absolute state, a siddhic state.

The nature of the 64 Siddhis can be seen very clearly through this 5th Siddhi. These are not glamorous states, even though the words may sound captivating. Most of the Siddhis are actually very ordinary. They occur when the set of patterns that you call yourself finally experiences itself as nothing more than a set of patterns! At this point, consciousness sees itself, and the awareness operating through you is never quite the same again, even though the patterns of your nature continue to run until you die.

This great change occurs when the smaller self stops trying to interfere with the flickering of the candle flame of awareness from one state to another. In other words, when you stop trying to leave the Shadow frequency, only then can the paradox happen — when there is no more fight left in you.

This 5th Siddhi also shows up the problems of language. In this book, we talk about states in terms of frequency bands. A higher Gift state is spoken of as having a higher frequency, and a Shadow state as having a lower frequency. This language feeds the human evolutionary urge to achieve higher states. Because of this language, it may be assumed that a higher frequency is therefore your goal. Ironically, the siddhic states cannot really be spoken of in terms of frequency at all, since frequency depends on oscillations of wave patterns within time. How then, can we discuss timelessness, or indeed any of the other 63 Siddhis? The answer is that we cannot, but we can still play within the language. This is the hidden lesson of the Spectrum of Consciousness — that it is actually a fallacy because you cannot measure or divide consciousness.

You may well be wondering by now whether there is any point in going on in this exploration of the 64 Gene Keys! Well, there is and there isn't! One thing is true to say — the more deeply your mind grasps the fact that you cannot *do* anything in life, the more you must begin to surrender both to your smaller, genetic nature and your greater, galactic nature. In the end, only timelessness will tell.

☷ **6th** Gene Key

The Path to Peace

SIDDHI
PEACE

GIFT
DIPLOMACY

SHADOW
CONFLICT

Programming Partner: 36th Gene Key
Codon Ring: The Ring of Alchemy
(6, 40, 47, 64)

Physiology: Mesenteric Plexus (Lumbar Ganglia)
Amino Acid: Glycine

THE 6ᵀᴴ SHADOW – CONFLICT

THE BATTLE OF THE SEXES

The 6th Shadow of Conflict is the single most influential Gene Key in regard to the issue of human communication. At its highest potential, the 6th Gene Key is the archetype of peace on earth, whilst at its lowest potential it is the root cause of all human conflict. This conflict stems from the human emotional system and our inability to handle the voltage of extreme emotional states. Conflict breaks out whenever two or more people *agree* to identify with their emotional state. As long as you surrender your will to the emotional system, then you will be trapped by its volatile nature.

Within the human body, the 6th Shadow relates to the pH level of your blood. Its job is to maintain an optimum balance of acidity and alkalinity so that your cells can thrive. As a metaphor on a wider scale, we can see that the 6th Shadow is about the loss of this balance in the world at large. In particular it is about the imbalance between male and female and over time has given rise to the notion of the *battle of the sexes*. This battle or conflict is not just about men and women — it is about the balance of all polarities — religion and science, east and west, rich and poor. The world itself has its own kind of pH level, and wherever it is imbalanced, conflict ensues. In the same way that an overbalance of acidic body tissue becomes an environment for viruses and cancer to thrive, so do social imbalances result in upheaval, corruption and at their worst, war.

The 6th Shadow can be interpreted individually through relationships or collectively through communities. At an individual level, this Shadow manifests through your emotional state. If you have ever been emotionally repressed through shame or guilt or abuse, the entire culture within your being has been disturbed. Likewise, if you are utterly ruled by your emotions, there can be no sense of harmony inside you. It is well known how much your emotional state influences your biological health. If you are stressed emotionally your body will suffer. Emotional problems are the greatest cause of illness on our planet, and the 36th Shadow of Turbulence, the programming partner to this 6th Shadow, reinforces this fact.

*Conflict
breaks out
whenever
two or more
people agree
to identify
with their
emotional
state.*

The 36th Shadow conditions you to be nervous when you feel uncertain or insecure about anything in life. It is this nervousness that forms the background frequency of our whole planet. The biofeedback loop between these two shadows is rooted in nervousness and defensiveness. The 36th Shadow makes you nervous, feeding the 6th Shadow, which responds through making you behave defensively. Likewise, your own defensiveness makes other people behave in a nervous manner around you.

Human beings are unconsciously addicted to conflict. We long for peace individually and globally, but our collective low frequency ensures that we keep reinforcing the patterns of conflict. Nowhere is this seen more clearly than in our relationships. The conflict between male and female is part of the oldest wound there is, and it is hot-wired into our genetics. You are genetically sub-programmed to defend yourself from the opposite sex, and until your frequency rises above the emotional gravitational pull of the opposite sex, you can never really know peace. This is the deep irony of the 6th Shadow — to bring an end to conflict, you have to give up your attraction to the opposite sex. Sex and war are deeply interrelated. This is a profound and perhaps disturbing truth for many people, and especially so because there is nothing we can do about it. Human sexuality can only be transcended through physiological mutation, which is dependent on the extent of your connection to the higher planes of reality.

When the 6th Shadow is viewed at a social level, it dictates the relationships between different racial groups. Here the genetic defence reflex becomes truly dangerous, since we are sub-programmed to distrust other gene pools and cultures. It is this 6th Shadow that first gave birth to the idea of national borders and boundaries, and it is the 6th Shadow that is responsible for war. War has been a by-product of our genetic makeup for millennia. However, the 6th Shadow is not as frightening as it sounds when viewed in evolutionary terms. It is an ancient part of our genetics, and it has been a necessary part of our evolution. It ensured that racial diversity existed in the first place, allowing different gene pools to expand and thrive without too much inbreeding. Only relatively recently have all the gene pools begun to merge with each other, which suggests that the human race has arrived at a very important genetic axis point.

The 6th Shadow is about maintaining boundaries and borders. It is about who is included and who is excluded, and its entire basis is defence. The 6th Shadow causes you to believe that you have to defend yourself from danger. You can see this at both an individual level and a national level. Your emotional defence strategies are primarily laid down during your second seven-year cycle — from the age of 7 until 14, as you navigate puberty. An enormous amount of life energy is tied up in protecting ourselves from the volatile emotional situations we felt as children. Unless we enter some deep process of de-conditioning, we will continue to carry these defences throughout our adult lives. In the same way, on a national level, we invest the majority of our energy in defence. The aggregate world defence budget is equivalent to well over one thousand billion dollars. One can only imagine how different the world might be were even one tenth of this sum used in a creative way.

The 6th Shadow is the reason we cannot create world peace. Until such time as human beings move through an emotional revolution at an individual level and find peace in our relationships, this Shadow cannot be transcended.

However, as we shall see, it is also written into our human story that revolutions do occur, and in fact one of the biggest is beginning right now.

REPRESSIVE NATURE – OVER-ATTENTIVE

The repressive nature of the 6th Shadow is concerned with maintaining peace between people, regardless of the cost. This pattern is based entirely upon fear and means that these people will totally compromise themselves in order to maintain control of their emotional environment. This is the defence pattern of over-attentiveness, and these are the *people pleasers*. The repressive nature will sweep conflict under the carpet by adopting any behaviour that will keep it at bay. The problem here is that unless conflict is dealt with transparently it tends to eventually explode. The over-attentive nature is at its core deeply false, which means that it will invoke unconscious distrust in others. Over-attentive natures also tend to draw tactless natures to them, resulting in some spectacularly dysfunctional relationships and family dynamics. When these people finally develop the courage to face conflict, they realise it is never as bad as they feared.

REACTIVE NATURE – TACTLESS

The other side of this 6th Shadow is unable to contain its emotions at all. Lack of emotional tact and timing betrays a low frequency nature that inevitably results in a backlash. These people are always finding themselves in difficult emotional situations, which they make worse by giving vent to their anger, as well as projecting it onto others. They are unable to take responsibility for their own volatile feelings. Their most common defence is to lash out and then storm away. The dilemma of the tactless nature is that it is deaf to its own lack of tact. These people assume that the problem is always in the other person which, unfortunately, earns them no friends and makes it very difficult for others to get close to them. The only kind of person who will remain with such a pattern is the over-attentive nature described above. The secret to breaking this pattern is for reactive natures to take full responsibility for their emotions, instead of remaining stuck in a *teenager* mentality.

THE 6TH GIFT – DIPLOMACY

DROPPING YOUR DEFENCES

At a heightened frequency, the 6th Gift pulls rapidly away from the world of conflict and argumentation. This is the Gift of Diplomacy — the ability to adjust your own behaviour to create a harmonious exchange with others. This gift is a by-product of opening your heart to another person. As you become clearer about your own emotional conditioning, you begin to feel more peaceful. Because this Gift is so deeply connected with the pH balance in your physical body, it also has the effect of stabilising the emotional aura in your environment. In other words, when one is able to see and own their unconscious projections, they actually rupture the collective Shadow patterns wherever they go. If one person is not playing the reaction/blame game, the other is forced to come to terms with his or her own demons.

You can see how the balance of pH in the body can be further understood through this Gift. For a relationship to be healthy there must be a balance of yin and yang, of receiving and giving, listening and expressing. Conflict ensues when this balance is lost. The 6th Gift has the effect of instantly applying the required amount of give and take in order to maintain peacefulness. As an example, if one person in a relationship becomes aggressive, the diplomatic countermeasure is to absorb the aggression and then pass the energy back without adding anything to it. This may be done in any number of ways, but tactful honesty is one of the most common. Honesty carries an extraordinary power and is one of the keys to Diplomacy. The other key is timing. You have to be honest in the right way and at the right moment.

People displaying the 6th Gift are always attuned to the right timing of how to act and what to say through the strong resonance field of their aura. These people can physically feel into another person's aura. They can do this because it is a specific genetic gift and also because they are emotionally mature. Emotional maturity means that your awareness operates even during the most profound emotional states. As you become more and more aware of your own emotional patterns, the frequency passing through your emotional system is released and quickened. This makes you much more sensitive at an energetic level, like an early warning system for conflict. When you are operating out of this 6th Gift you can sense conflict in others before it even arises. This enables you to temper your own actions and/or words to disperse the conflict.

Emotional maturity means that your awareness operates even during the most profound emotional states.

The Gift of Diplomacy is far more than the ability to speak the right words, however. That is a mere surface skill that can be mastered by anyone even at a low frequency. True diplomacy is an energetic gift operating through a person's aura. Because this 6th Gene Key is so embedded within human sexuality, it is about the penetration of borders and boundaries. Because of the 6th Shadow, there is tremendous friction between the opposite sexes. Both sides are so busy defending their individuality that there is very little real love or connection. But this friction is maintained only as long as each party hangs on to its defences. As the 6th Gift enters into human relationships, the emotional barriers begin to break down between people. The 6th Gift catalyses the process of dissolving the natural friction between opposites and in doing so, it allows a far greater exchange of energy between them. This is exactly what the experience of falling in love is.

As part of the chemical genetic family known as the Ring of Alchemy, the 6th Gift is playing a pivotal role in the transformation of the human species. This codon ring is formed by four Gene Keys — the 6th, 40th, 47th and 64th — whose themes are Diplomacy, Resolve, Transmutation and Imagination respectively. This is a very powerful genetic grouping. The 6th Gift is breaking down the barriers in human relationships, the 40th Gift is forging this new openness into our communities, the 47th Gift is allowing us to transform our old ways and the 64th Gift is opening us to a fresh set of possibilities for living in a new way. At a collective level therefore, every person manifesting the 6th Gift is a participant in this deep alchemical process of gradually bringing peace to earth.

These people realise how throttling defensiveness can be, both in individuals, cultures and whole races. As this Gift enters the world in a more widespread manner, as indeed it must, we shall begin to see the dismantling of the many types of borders and barriers that keep people apart from each other.

THE 6TH SIDDHI – PEACE

BUILDING THE BODY OF GLORY

The ultimate defence is emptiness. This is the essence of the wisdom taught by the great sages. Defence maintains the illusion that we are separate. As such it is truly futile because it protects against something that doesn't really exist in the first place. There is a famous story called the *Empty Boat* told by the Chinese sage Chuang Tzu. One day an old man was crossing a river in his boat when he mistakenly collided with another boat. The man in the second boat began shouting at the old man, cursing him. However, to the second man's dismay, the old man did not react in any way whatsoever but just stared at him impassively. The old man was an enlightened being who had transcended all sense of his own individuality. For him, there was no one in his boat to be shouted at, thus there was no one to defend and therefore no reaction within his being.

The 6th Siddhi of Peace is one of the truly great Siddhis. It is the underlying nature of all the Siddhis. In point of fact, it is the underlying nature of form itself once its true essence has been realised as consciousness. The 6th Siddhi is the final result of the process that occurs in the 6th Gift. The 6th Gift is marked by a constant balancing of opposites. In this sense, diplomacy requires a subtle effort since it is juggling within the realms of duality. However, at a certain point, this process spontaneously gives way to the 6th Siddhi and duality itself is transcended. You might say that whilst the 6th Gift is the activity that maintains peacefulness, the 6th Siddhi is peace itself. Peace is the reality that is experienced once all boundaries have dissolved. It is the true nature of humanity.

So many great teachers have spoken of the Peace of the 6th Siddhi. When Jesus stated that the Kingdom of Heaven has already arrived, but that people simply cannot see it, he was talking about this Siddhi. Peace is an auric emanation that surrounds anyone in a Siddhic state. In this siddhic state the awareness within your being separates from your emotional nature. It rises up and floats upon the vibrations of your emotional longing. This is a spontaneous process that cannot be caused by anything within your individuality, since it is that very individuality that has to die. The underlying emotional energy that was previously experienced as longing is now experienced as peace. Awareness is no longer trapped within the vicissitudes of the emotional human drama. In the deepest sense, something utterly wondrous happens inside you. The final boundary — the boundary of your body — dissolves. When this occurs awareness is no longer localised in the body, but is experienced as travelling through all forms. You realise that all of life is alive and that, even though the awareness within a form may dissolve as the form passes away, the consciousness behind this game of life and death always remains.

The ultimate defence is emptiness. This is the essence of the wisdom taught by the great sages.

The sense of peace that emanates from these revelations is indescribable. All your defences disappear and your boat becomes empty. Paradoxically, at the same time, this very emptiness is flooded by *everythingness*. Profound changes occur within your physical body before, during and after this realisation. Since your outer walls have dissolved, the process is mirrored within your body chemistry and all inner conflict comes to an end. It is as though all the cells within your body make peace with each other and experience their own inner Garden of Eden. In this sense, the true Garden of Eden is the body itself. The waves of peace that accompany this event travel deep into the collective consciousness of all humanity. To be in the physical presence of the 6th Siddhi is to bathe your body in its truth. It is to experience greater awareness than your own, travelling physically and emotionally through your body. Because the current human form is not designed for these heightened states of consciousness, some strange phenomena can occur within the body of someone experiencing this Siddhi.

The 6th Siddhi is the original and future state of humanity. From it stem our myths and memories of paradise and our intuitions and highest hopes for the future. When this state dawns inside you, your body begins to mutate. In a sense, it is trying to build a better vehicle to house this intense new frequency, but the raw materials are not fully present yet. Humankind is still evolving the new energetic circuitry that can carry these frequencies. This is why so many great sages suddenly become diseased. In particular they can experience problems and imbalances within the body's pH. These things occur because the body is trying to catch up to its future state. Regardless of what is experienced physically, the sense of utter peace does not even flicker. In certain rare cases, the elements of the future body are available through your genetics, and the process of building the future body can and does continue. This involves the 47th Siddhi of Transfiguration, part of the Ring of Alchemy mentioned earlier. The 6th Siddhi has a very special quality in this sense, in that it contains instructions on how to build a new genetic form. It is interesting to note that glycine, the amino acid for which the 6th Gene Key codes, is one of the only amino acids discovered in interstellar clouds in deep space. The suggestion is that this amino acid may play a key role in the formation of new life in the galaxy.

The new form of humanity is connected to this issue of the body's pH and also to the boundaries within the body itself. The ultimate boundary is our skin, and one effect of this Siddhi can be mutations in the skin. These mutations affect the way in which skin cells trap light, so that the skin of someone manifesting this Siddhi can appear to be translucent. If the 47th Siddhi is also involved in the process, an even more extraordinary phenomenon occurs — the Transfiguration of the human body into what the ancients called the *rainbow body* or the *body of glory*. This Transfiguration will eventually overtake our entire species. The builder of this process is the 6th Siddhi. As our skin cells learn to capture light, our digestive system will gradually become defunct, since light contains the ultimate nutrients for the subtle body. As far as the pH of the body is concerned, this spiritualisation of the physical form will gradually weaken the extremes of the pH scale, thereby reducing the acidity and alkalinity of the body. Eventually, as all the solutions within the body become neutralised, the hydrogen ions that form the very basis of the pH scale will evaporate and the body will disappear altogether.

On a collective level, the 6th Siddhi will be the very last Siddhi to fully mutate in the coming shift in global consciousness. Only when peace is recognised on earth as the natural state of our collective consciousness can the future vehicle of humanity be built. In other words, the myth of world peace is the pretext for building the future body that will take us beyond form itself.

☷ 7th Gene Key

Virtue is its own Reward

SIDDHI
VIRTUE

GIFT
GUIDANCE

SHADOW
DIVISION

Programming Partner: 13th Gene Key

Codon Ring: The Ring of Union
(4, 7, 29, 59)

Physiology: Diaphragm

Amino Acid: Valine

THE 7ᵀᴴ SHADOW – DIVISION

A WORLD DIVIDED

The 7th Shadow of Division is one of the main reasons human society operates in hierarchies. Not only is this Shadow responsible for the manifestation of hierarchy, but it also makes you *think* in terms of hierarchy. We have become so used to thinking this way that we no longer conceive that another way might exist. Hierarchy is based upon division — it divides human beings into social classes, economic classes, racial castes and political wings. The reason for this division is rooted far more deeply than our psychology — it is actually stamped upon our genes. It is through the 7th Shadow that we are programmed to follow leaders, and it is also through this Shadow that certain people are programmed to *behave* as leaders.

The entire issue of leadership and power are contained within this Shadow. The ancient Chinese named this 7th hexagram of the I Ching *The Army*, which is extremely fitting. The army represents the real power in a country on a political level and if you don't control the army you have no real power. This symbol of an army also represents the idea of power through force, rather than power through inspiration, which is the true nature of leadership expressed through the 7th Gift and Siddhi. This Shadow always rules by force, and today it runs the political systems across the whole expanse of our planet. Even modern democracy, which has at its heart a high ideal, does not entirely eradicate the concept of leadership through force. Now however, instead of military strength, democratic leaders have to lead through the force of numbers — they have to secure the majority of votes.

On a subtle level, even democracy encourages division. Modern political leaders can still force their way to the top — they can cheat, spin or bend the truth, manipulate and even buy their way into leadership. Until leadership demands its highest frequency, that of the 7th Siddhi of Virtue, we will never fully see the end of political division and hierarchy. The 7th Shadow cannot command true respect or loyalty because it has at its core the lust for power rather than the good of others. Those who are elected as our leaders are chosen because they have a genetic imprinting that marks them as the *alphas*.

*The real
leader
is the
ultimate
listener.*

However, this does not necessarily make them good leaders. Just as there are leaders, so there must also be followers, and these followers are equally influenced by the 7th Shadow. Since the mass consciousness of humanity operates at a low frequency, it does not recognise high frequency leaders, so it does not elect them to power.

On very rare occasions the mass consciousness does elect a high frequency leader to guide them into the future. This usually happens during extraordinary times. One example was the election of Vaclav Havel, a poet and playwright, to the presidency of the Czech and Slovak Federal Republic in 1989. At that time, the Fall of Communism created such an upsurge in planetary consciousness that it became possible for a man of true virtue to assume a position of leadership. However, throughout most of human history, our political leaders have been men or women of personal ambition rather than true unshakeable virtue, and the 7th Shadow of Division is not confined to the political world stage but operates at all levels of society. The other reason for this is to be found in the 13th Shadow of Discord, the programming partner of the 7th Shadow. The 13th Shadow concerns the inability to attune to or empathise with the heart of the people. It therefore undermines the principles of fellowship and trust between different groupings of human beings.

Wherever you recognise another as an authority or guide, the 7th Gene Key is at play. In the case of the Shadow frequency, like attracts like — in other words, someone at a victim level of consciousness will be drawn to someone who further strengthens that same energy frequency. If you are weak, you will magnetise those who reinforce your weakness and even play on it for their own benefit. It can be a huge shock to people when they realise for the first time that they have been playing out this victim consciousness all their lives. An even deeper shock is the discovery that most authority figures in the world today — our doctors, therapists, business advisors, even our spiritual teachers — are in the business of serving the Shadow of Division. Most people who are recognised as leaders do not want you to stop being a victim because they unconsciously fear that it will put them out of business. In this way, the leader is as much a victim of the 7th Shadow as the follower.

On an individual level, you must be ever watchful of your own tendency to give other people authority over you. The 7th Shadow usually does not see the hidden agendas of certain leaders until it is too late. It is all too easy to compromise your own authority to someone in a position of power or to someone with a great deal of charisma or personal charm. The mark of a true leader is one whose main interest is in empowering you to lead yourself rather than binding you to them. Ironically the false leader always tries to hold onto you whereas the real leader always tries to get rid of you! There is absolutely nothing wrong with looking to another with respect or reverence. It is a completely natural stage of the human journey. The trick is to find a person who can truly listen to you. The real leader is the ultimate listener — he or she will empathise so deeply with your suffering that you will finally give yourself permission to embrace it without fear, which will enable you to transcend it.

At the core of leadership at the Shadow frequency lies the fear of losing power, which keeps hierarchy intact. In the business world, the 7th Shadow of Division is the norm. Wherever money is concerned, hierarchy is found to be at its most rigid. Like the army, orders come from the top and they are to be obeyed. There is no real autonomy or two-way communication within a model such as this. There is very little room for trust or ordinary human intimacy in these kinds of businesses because the prime directive at their heart is to serve themselves and make money. The result can only be division. Division creates an attitude of *every man for himself*, or only slightly better *every business for itself*. This is still the essence of most modern business. As was stated at the beginning of this Gene Key, we humans really do not realise that there could be other ways to operate in the world. Our inability to think collectively is precisely what creates a world divided.

Only the leader who cannot be corrupted by power has true power at his or her disposal. As such leaders are recognised in the future we will see the gradual dissolution of the force that divides human beings from one another. Ultimately the driving force behind business, politics and all arenas of leadership will have to take a great quantum leap — it will have to move from fear to love, or in terms of business, it will have to switch from self-serving to *whole-serving*. This is the path indicated by the 7th Gift, which will finally blossom fully in the 7th Siddhi.

REPRESSIVE NATURE — HIDDEN

When the 7th Gene Key is repressed, it simply fails to manifest in the world. These people carry the genetic imprinting for recognition as leaders, but remain hidden. This creates immense pressure within them as well as frustration and resentment, all of which can manifest as physical symptoms in the body and emotions. The world today is actually full of hidden leaders living behind shadow veils that prevent others from seeing them and benefiting from them. The force that prevents a person's recognition comes from within that person and has nothing to do with what they do in the world. Thus for these people to step into leadership, they must first recognise the power within themselves. As they do, a great tide of optimism and intelligence is released back into the world, which instantly gains them recognition from society at large.

REACTIVE NATURE — DICTATORIAL

The reactive side of the 7th Shadow knows well that it is a leader, and takes full advantage of this by abusing its position. These people use those who follow them to their own advantage, thus reinforcing their status as followers. True leadership encourages people not to rely on others, whereas this style of leadership demands complete reliance, either through pure power of presence or through subtler means. These people are masters of patterns, and can manipulate others by locking them into certain patterns. These can be intellectual patterns of thinking or belief, powerful emotional games, or materialistic patterns involving money. The game is to trap followers into believing that they *need* leadership. Naturally such leaders only attract followers who want to remain at a victim level of consciousness.

THE 7ᵗʰ GIFT – GUIDANCE

THE POWER BEHIND THE THRONE

True leadership, like true education, does not impose itself on anyone. It is the gift of being able to help others find their own way forward in life, rather than taking away their individual power. This is precisely what is meant by the name of this 7ᵗʰ Gift — the Gift of Guidance. As was discussed in the 7ᵗʰ Shadow, it is the driving force behind leaders that determines what kind of followers or supporters they will attract. At the Gift level of frequency, we see the move away from fear towards service. Leaders operating at this frequency are able to think at a collective organisational level — they know that unless individuals are suitably self-empowered the organisation will not thrive. To this end, people with the 7ᵗʰ Gift are strong supporters and implementers of schemes that allow individuals more power, creativity and autonomy at all levels of society.

The 7ᵗʰ Gift in many ways is represented by the ideal of democracy. In the democratic ideal, every individual is free and leaders are chosen by the people to represent and guide them. Modern democratic government is designed to listen to the opinions of the mass consciousness and then use discernment to lead the nation forward. In this way, the 7ᵗʰ Gift and its programming partner the 13ᵗʰ Gift of Discernment knit the collective together in a single pattern where all the individuals within that collective structure work together. This at least, is the ideal of democracy. As we all know, it does not often work that way. Once in power, leaders tend to follow their own agendas, which may or may not reflect what the majority really wants. Politics tends to follow its own somewhat bent course, and that course greatly depends upon the core qualities and principles of those who lead. Having said all of this, modern democracy is a huge shift in the direction of higher consciousness when compared to more primitive modes of government that do not encourage freedom.

The 7ᵗʰ Gift of Guidance rests upon the ideal of service. To truly guide another person or group of people, you need to put aside your own opinions and judgments and listen intently to their needs. Great guides are great listeners. Sometimes simply by being listened to in the right way, a person finds the answer to their problem without direct input from the guide. People with the 7ᵗʰ Gift have a powerful magnetic presence, and just from being in their aura, you can come to deep clarity about your own direction. These people can in particular help others to see future patterns. This does not mean that they literally can see the future, but that their guidance is in alignment with future trends. It is this quality of being ahead of the crowd that marks these people as leaders. However, their recognition depends upon the times in which they live. History shows us again and again that many of the greatest leaders the world has known, in government, business, science or art, often pass unrecognised during their own times.

There is strong evidence for the current emergence of the 7ᵗʰ Gift at various levels within society. Since humanistic psychologist Abraham Maslow introduced his famous model of the Hierarchy of Human Needs in the 1940s, it has become the foundation of understanding organisational structure. Developed particularly in the business world, it enables us to understand large human frameworks in a wholly new way. Businesses can be seen as entire *cultures* with their own chemistry and life force. For the first time, people talk about the levels of consciousness of entire businesses. These are the early stages of a higher consciousness infiltrating the world of business. The greatest gestalt shift will come when the first truly holistic and service-based businesses prove that they can be more successful than the old greed-based empires.

Some new organisational models are also beginning to understand the levels of consciousness of different styles of leadership. In our exploration of the Gene Keys, we are only looking at the three levels encompassing the two great quantum leaps — the leap from the Shadow to the Gift and the leap from the Gift to the Siddhi. In reality there are subtler bands within the Spectrum of Consciousness, which give way to many other different levels. One new business model divides leaders into seven levels of consciousness according to the Hindu *chakra* system. Thus you have leadership styles ranging from Authoritarian (1st chakra) to Facilitator (4th chakra) to Visionary (7th chakra). It is this mid-range style of leadership (Facilitator) that corresponds to the 7th Gift. Facilitators, as the name suggests, make communication and implementation easier and smoother. People with the 7th Gift do not lead from the front, but guide the energy of the group itself. They create a space in which an organic team harmony can develop on its own, with minimal interference. They are often content to allow others with the requisite gifts to stand in the limelight while they lead quietly from behind the scenes, and in this sense the 7th Gene Key is an archetype of the power behind the throne. This is the true meaning of Guidance — trusting in the life process, rather than forcing matters by taking control. This ability to surrender to life itself is the foundation of true leadership.

THE 7TH SIDDHI – VIRTUE

REPAIRING THE WORLD

The 7th Siddhi is an energetic blueprint hidden inside each human being. Despite many interpretations of the word *virtue*, true virtue has nothing to do with our concepts of morals or behaviour. Anyone can behave in a pseudo virtuous fashion, yet have none of the power of this Siddhi. We saw throughout the 7th Gift how deeply this Gene Key is connected to the issue of leadership. In every pack of animals, there is an *alpha* — one animal that the others automatically follow. Among human beings this too is the case. Leadership is determined by genetics, but quality of leadership is determined by the frequency passing through those genetics. At the Gift level, we saw how the authoritarian style of leadership gave way to the more democratic style of the *facilitator*. Here in the 7th Siddhi, we find visionary leaders, but far more importantly, we find virtuous leaders.

Every truly virtuous deed done by human beings represents an upsurge in the consciousness of the totality.

True leadership is analogous with virtue. However, the world has seen very few true leaders. This 7th Siddhi waits for specific times in history when the frequency of the mass consciousness is elevated enough for it to be recognised. If the conditions are not right, then these people pass unrecognised at a collective level of society, even though they may have great influence on a local level. In the Tao Te Ching, the sage Lao Tzu speaks about true virtue as it impacts society through the *superior man*. Even though the language may be archaic, the message is pure — the secret of virtue lies in complete surrender to nature. Indeed, the word *Te* in the title of this profound work is a word generally translated as *virtue*. It was also through this book that a wonderful and oft-quoted saying was born: "Virtue is its own reward." Therein lies the other great secret of virtue; it lies beyond the need for recognition and beyond the need to be of service. Virtue is simply the unimpeded expression of nature through men or women living at their zenith.

Everywhere, every day, people commit small and often unseen acts of virtue, and the power of these acts is incalculable. They actually offset the powers of chaos.

Virtue has been a deeply misunderstood Siddhi over the centuries. Those beings in which this Siddhi has manifested have indeed led exemplary lives. The 7th Siddhi contains the seed of the future man and woman — an archetype of perfected behaviour that few *normal* people can live up to. The confusion has come about because people at lower levels of consciousness try to emulate the behaviour of those in whom the siddhic state is flowering. This creates deep tension because virtue is a final flowering, rather than a path in itself. The Siddhi of Virtue is a guiding purity that pulls humanity into the future. It constantly bubbles beneath the surface of humanity. Every truly virtuous deed done by human beings represents an upsurge in the consciousness of the totality. This Siddhi is in fact our insurance policy that we will never destroy ourselves. Everywhere, every day, people commit small and often unseen acts of virtue, and the power of these acts is incalculable. They actually offset the powers of chaos.

In the great Jewish book of mysticism, the Zohar, there is a powerful analogy to the journey through this 7th Gene Key. It is encapsulated in the Kabbalist concept known as *Tikkun Olam* — a phrase usually translated as *repairing the world*. The Kabbalists say that when the Creator made the world, he created a series of vessels to hold the Divine Light, but as the Light flowed down into these vessels, they shattered and fell towards the realm of matter. Thus the world in which we live is made up of countless shards of the original vessels in which the Divine Light is trapped. The Kabbalists go on to say that every virtuous act committed by a human being helps to repair one of these broken shards.

We can see from this beautiful metaphor how the 7th Shadow of Division represents the force that shatters the vessels and pulls the world apart, whereas the Gift of Guidance begins the process of reuniting the shards. The final result of reassembling all the shards is represented by the 7th Siddhi — when you see the completed picture deep within your being. The one in whom this Siddhi manifests becomes the *Adam Kadmon* — the perfected vehicle of divine realisation. At this level of consciousness, virtue becomes the intent of the universe expressed through a human being. At the root of all acts performed by such a being lies the pure force of virtue and this virtue has the remarkable effect of *repairing the world*. The 7th Siddhi allows a human being to see the certainty of the perfected future of mankind, and allows him or her to actually live *within* it. In this way, such people are truly ahead of their time even though for them time has ceased because they have remembered the future deep within their being.

There is an interesting fugue played between this 7th Siddhi and its programming partner, the 13th Siddhi of Empathy. This is best understood by the ancient symbol of the *ouroboros*, the serpent eating its own tail. Whereas the 7th Siddhi *pulls us* into the future, the 13th Siddhi *pushes* from the past. These two great Siddhis flower simultaneously in a human being, and they concern the collective destiny of mankind, with the 13th Siddhi representing the seed at the beginning of time, and the 7th representing the flowering at the end of time. These two Siddhis have a mass of esoteric mythology connected to them. People in whom these states flower act as energetic guides for the entire human race. They have been called many names by many cultures — the keepers of the world, the illuminati, the *shining* or *chosen* ones — but they have also been deeply misunderstood by many cultures.

As a vital link in the genetic chain known as the *Ring of Union*, this 7th Siddhi finds a chemical link with the 4th, 29th and 59th Gene Keys. Together, this chemical family contains the collective codes for purifying human relationships across our planet. The combined dynamic of virtue with forgiveness, devotion and transparency sets the stage for a completely new phenomenon to be seeded in humanity — collective leadership. Collective leadership is the stage beyond individual leadership in which leadership becomes an empathically shared energy field between individuals, thus bringing an end to hierarchy.

The 7th Siddhi waits for a specific moment in time to awaken in the collective. In the book of Revelation this 7th Siddhi is represented symbolically by the opening of the seventh seal, which precedes the return of Christ consciousness. There are even deeper secrets here concerning the mythology surrounding the number seven. However, at this deeper level we can perhaps see why the original Chinese hexagram was called *The Army*. The army refers to a collective group of beings — in the Book of Revelation, they are known as the 144,000, and you can learn more about them from reading the 44th Siddhi. Essentially, this group of beings represent the genetic equipment through which higher consciousness will first touch down at a collective level on our planet. The term *equipment* is used deliberately here in order to dispel any sense of glamour concerning these so-called *chosen ones*. These beings are a collective fractal of leaders in which the Siddhis will spontaneously flower over a cycle of many generations. They will be found at all levels of society. The core essence of each of these beings or leaders is Virtue, the 7th Siddhi, and the language that unites them is Empathy, the 13th Siddhi.

Ultimately, the 7th Siddhi concerns the future, and the future is about children. As the 7th Siddhi flowers in humanity, one of the first places it will manifest is through parents and teachers. Children who grow up among virtuous people need no other kind of education or guidance. Even now, this small piece of information is the greatest key for parents. Children who grow up in the aura of true virtue will take that energy into whatever sphere of life they move through, and in doing so they will slowly but surely transform the future of our planet.

☰ 8th Gene Key

Diamond of the Self

SIDDHI
EXQUISITENESS

GIFT
STYLE

SHADOW
MEDIOCRITY

Programming Partner: 14th Gene key

Codon Ring: The Ring of Water
(2, 8)

Physiology: Thyroid (Adam's Apple)

Amino Acid: Phenylalanine

THE 8ᵀᴴ SHADOW – MEDIOCRITY

BEYOND THE COMFORT ZONE

When you look at the world in which we live today, especially the western world, it is quite extraordinary how many people live such similar lives. The 8th Shadow, like all the Shadow frequencies, is founded upon a specific fear and in this case it is the fear of being different. The 8th Shadow prevents individuals from rising out of the mass consciousness and exploring the real adventure of life. The true nature of individuality is rebellion, but rebellion is unsafe, so the mass consciousness of humanity chooses the illusion of security instead. The 8th Shadow weaves a web across the world, and this web supports the planetary comfort zone. Only when life forces you to grow, through some kind of crisis or the death of a loved one, for example, do you come to experience your true nature outside the borders of this comfort zone.

In the western world in particular, your individuality is imprinted very early in life. One of the issues addressed repeatedly throughout this book is the influence of modern education, especially in the years leading up to the age of seven. Most educational systems encourage sameness rather than difference because difference threatens the system itself. In modern education, young children go through an accepted process of tests and examinations, the very nature of which is to regurgitate memorised information with little or no scope for spontaneous innovation. This indoctrination begins in your early childhood and lasts into your early twenties as you are cycled through the same system that shaped your parents and your parents' parents. By setting up a system where sameness is the rule, we have successfully turned individuals into outsiders and reactionaries.

One might rightly wonder what kind of education a child should have in order to keep his or her individuality intact? The radical question of the 8th Gene Key is this: is formal education really needed at all? In our modern world, the inflexibility inherent in our educational system is becoming more and more of a problem. Naturally there will always be certain children who show a disposition for formal learning — some in a variety of subjects and others in specific subjects.

*Mediocrity
prevents
people
from being
heroes or
heroines.*

However, many other children simply do not need a formal education and certainly do not respond well to one. Of course, the problem is also linked to many other aspects of our modern lifestyle. What is important to realise at this stage of our evolution is that this 8th Shadow is bred into you from an early age, and in most cases, it has to be *unlearned* later in life if you are to have the least chance of living out your gifts and finding your genius.

One of the deep fears emerging from the 8th Gene Key is the fear of success. This fear is reinforced through its programming partner, the 14th Shadow of Compromise. You compromise your dreams not because you fear you will fail, but because you know that to succeed you will have to rebel against the whole of society and its expectations of you. You fear what you might become because you do not know who you are. The highest frequency of the 14th Siddhi concerns Bounteousness, which is the reward for the individual who dares to break free from the trap of mediocrity. In other words, the path less travelled leads to treasure. The 8th Shadow gives you a recognisable stereotype in the world which not only makes you feel safe about who *you* think you are, but also makes others feel safe about who *they* think you are. Without this stereotypical facade, who might you be and how would others approach you? The answer is that the mainstream would look at you with a mixture of both fear and awe.

Mediocrity is defined by others rather than by yourself and mediocrity has two main functions. Firstly, it keeps you from thinking outside the box. Under the influence of this low frequency you think like everyone else, look like everyone else and more or less behave like everyone else, and you will do this based on what others may or may not think about you. The second function of Mediocrity is to serve the machinery of society rather than evolution. In other words, you become a cog in the wheel of the systems established by man. In so doing, you become a part of the background of life rather than a major player. Mediocrity prevents people from being heroes or heroines. Nowadays, most of us are simply content to dream about what we could have achieved in our lives. We may watch the films and movies of others and may even be deeply moved by them, but the 8th Shadow prevents us from believing in ourselves enough to embody such a life. And the reason is fear. Fear is endemic throughout the very infrastructure of our society. Those who move beyond this threshold of fear move in a world that seems incomprehensible to the mainstream culture.

The 8th Shadow makes you a follower of outside authorities. It even marries you to systems that then become your authorities. At every turn, this Shadow prevents you from breathing the cool fresh air of your true unbounded nature. It traps you in its twists and turns and prevents you from being a free thinker. What is a free thinker? It is someone who can see beyond the current structures imposed by life. It is a person who lives for the spontaneity of his or her own creativity. Free thinkers also live freely. They follow no one, even though they may have been influenced and inspired by others. The 8th Shadow represents the well-trodden paths through life — the compromised lifestyles, the unimaginative, conformist lives lived by the majority. It takes great energy and courage to push through the darkness and fear of this Shadow and find out who you really are and what you are truly capable of. If you wish to escape mediocrity, you will have to find your own path and invent your own identity in the world.

It will be like no other path and it will take you beyond your comfort zone into dangerous territory where there are no assurances that you will succeed except for your faith in your own deep and vibrant nature.

REPRESSIVE NATURE – WOODEN

Those who follow the mainstream path of mediocrity are essentially devoid of life force and lacking in a true sense of purpose. These are people whose words and actions may be of some service to the world but who, at the same time, lack fire and grit. Such lives are wooden lives and such people have allowed themselves to become hollow. They gave up on their dreams somewhere between being a child and growing into an adult. Without embracing their fears, they became swallowed up by their responsibilities and the many compromises they made throughout their lives. The result is that they live lives that do not belong to them or allow them space to breathe or create.

REACTIVE NATURE – ARTIFICIAL

The difference between the reactive and repressive nature concerns the human spirit. In the repressive nature, this spirit has collapsed inward at some point in life. In the reactive nature this spirit is channelled into the world where it creates an illusion of itself based upon its dreams. Such people live purely artificial lives. On the surface they may seem successful and even original, but beneath their patina they have compromised their spirit to some aspect of the system. You can spot the difference between the repressive and reactive nature through their relationships — the repressive side never leaves their relationships out of fear of change. The reactive side cannot stay in their relationships because their façade inevitably breaks down and rage rises to the surface, at which point they usually run away.

THE 8TH GIFT – STYLE

A REBEL WITH A CAUSE

To break out of the heavyweight frequency of the 8th Shadow you have to take an exciting leap of faith in yourself. This 8th Gift catches people's attention because it brings something fresh into the world. The Gift of Style has little to do with our popular interpretation of this word — it has more to do with following your own unique rebellious spirit out into the world. True style cannot be measured by materialistic trappings nor can it be faked. It is the natural flowering of your individuality. To find your own style is to be yourself without concern about what others may think. Since it bypasses the mind, style cannot be imitated or preconceived, always emerging spontaneously and naturally. It also involves travelling along paths that involve risk, but only as society defines risk.

Style is more than skin deep. It is the cutting edge of creation itself.

True Style takes great joy in shattering the grey world of mediocrity that is considered the *norm*. This Gift does not sit happily within society, although the pure joy and freedom of expressing your true nature far outweighs any repercussions that may come with the territory. Individual uniqueness is something that our modern societies uphold and idolise, but in reality we are afraid to have too many colourful individuals. Style is more than skin deep. It is the cutting edge of creation itself. Individuals manifesting this Gift have surrendered themselves to a creative process that controls them, rather than them controlling it. Creativity can be a thankless business. Oftentimes an individual is so ahead of his or her own time that their uniqueness is not appreciated until after their death. However, the sense of expansive freedom that comes from letting go into such a process is so fulfilling that success or failure is no longer a major driving force or consideration in their lives.

Style is dangerous to society and its logical structured system-based thinking and infrastructure. It is dangerous because it mimics nature, which so many systems try to control and explain. Like nature, style is given shape by a wild, organic and unpredictable energy. It is full of genius and quantum leaps. For the individual, the 8th Gift brings a deeply fulfilling sense of purpose, but so often these individuals find themselves unable to interface their Gift with the world. Style in and of itself does not make one an outsider — in fact it makes one an insider to the secret processes of life. But to the powers that try to maintain control of the world, style *is* considered dangerous or at best eccentric or quirky. This is why individual uniqueness is usually confined to those realms like art, fashion or music where it is acceptable and where there is some space for it to breathe. In most other spheres of society, individual style is generally suppressed because it is neither trusted nor understood.

Until a collective coalesces made up of individuals who are free thinkers, the 8th Gift of Style will remain on the fringes of society. In the meantime, anyone who breaks out of the lower frequencies of the Shadow states will have to confront the spectre of a collective that cannot allow individual freethinkers much leeway. Fortunately, freethinkers who find their own sense of style are not really concerned with *fitting in*. Their only concern is to free more freethinkers! Such freedom is truly contagious. Thus such people inherit a powerful mission in the world even though they may not see it that way.

Finally, this 8th Gift is about actually *manifesting* individual dreams rather than simply dreaming them. When the 8th Gift is released from inside your DNA, you suddenly begin to make things happen, and it may seem to you as though the rest of the world is simply stuck inside a dream world. This is the by-product of operating at a higher frequency. Now anything becomes possible for you because in yielding to your inner spirit, the sheer force of creativity that comes through you frees up channels and opportunities that were previously blocked. Such is the power of genius, for it carries within it more than just a new concept — it carries intent from a higher realm that is far beyond the individual through whom it manifests.

THE 8TH SIDDHI – EXQUISITENESS

THE ETERNAL LOVE KNOT

The 8th Siddhi of Exquisiteness is the natural revelation and manifestation of all the siddhic states. Each Siddhi is holographic in the sense of being reflected within all other Siddhis. Exquisiteness is experienced when the divine essence begins to shine through the individual. In this superlative state of bliss, you fall in love with your own pristine manifestation because it is through this uniqueness that the divine currents can be accessed. Exquisiteness hints at a beauty that is beyond all words. You shine like a diamond in the heart of creation and wherever you look, you see other diamonds in various states of clarity, each one unique and each one exquisite and incomparable.

At this rarefied level of consciousness, where all levels paradoxically cease, you come to realise the great joke of individuality and differentiation — despite its exquisiteness, individuality is an optical illusion created by the mind. You experience yourself riding in a genetic vehicle that is unique, but your *being* rests in the consciousness field behind all differentiated forms. This is the stage at which you achieve transcendence of your genetics. You remain differentiated at the level of form, but at the collective universal level your awareness permeates *all* forms, making you both a drop in the ocean as well as the ocean itself.

Your true nature is wildness. There is no taming where the divine is concerned, and thus this state of exquisiteness sees through mankind's many veils. All systems break down under this wild, ebullient energy that comes straight from the source of creation. The few who have manifested this Siddhi have been as fleeting and beautiful as light playing through clouds. Life never repeats itself and they are never the same from moment to moment. In every single second, they are new. This is the origin and meaning of the famous saying by Heraclitus: "You can never step in the same river twice." This saying points to the truth of the 8th Siddhi — the truth of human nature. Like water, it is always moving, changing and evolving but the river itself is unchanging, and the river is consciousness.

People in whom this truth is reflected shine as bright lights amidst the grey background of history. The great avatars and sages who walked among us expressed the true nature of this Siddhi. Our most common mistake lies in trying to emulate such people, which immediately takes us away from our own uniqueness and drops us back into the Shadow of Mediocrity.

The people of the 8th Siddhi are not leaders. They are examples. They do not want anyone to follow or mimic them. Wherever such people see the inauthentic or imitative, their very nature exposes its ugliness. Because of this trait, they have a very powerful liberating effect on individuals. At the same time they are often criticised and spurned by society. Like Socrates, they wish for people to find their own answers and their own questions. Their very presence acts as a light to free individuals from structured systems. Their language is the language of the rebel, and they use beauty as a means to reflect an individual's true nature. They are not limited to any single form of expression but will play with science or art, yoga or tantra, logic or poetry, since to them all expressions can be a means to display the exquisiteness of the Divine in form.

To the person immersed in the blissfulness of this 8th Siddhi, all life is absolutely clear. This clarity moreover is paradoxically found in the very uncertainty and wildness of life. These people leave no tracks for others to follow. They know life as a rush of contradictions and mysteries never designed to be solved.

As an aspect of the genetic family known as the Ring of Water, the Siddhi of Exquisiteness is chemically bonded to the 2nd Siddhi of Unity. These are the two great feminine Gene Keys in the human genome, drawing all human beings along their inevitable journey to self-realisation. One of the great mysteries of the 21 Codon Rings is found here. The Ring of Water forms a kind of eternal genetic knot with its polarity, the Ring of Fire. These two chemical families and their amino acids phenylalanine and lysine move us human beings along the trajectory of our destinies, ensuring that our genetic material finds its opposite match. Even deep within the body, these chemical families set the primal blueprint for the balance of all the opposing forces inside us. It is in the crossfire of these two Codon Rings that the mystical figure-of-eight is forged within each of us. This eternal love knot deep within our planetary gene pool sets each man and each woman squarely on the human journey. The true symbolism of the number eight represents the timeless quest for the buried treasure that lies inside each of us — that elusive diamond of our true nature.

People of the 8th Siddhi sparkle like exquisite rare jewels. Such people create their path as they walk it. Their legacy to the world is to dismantle all concepts of how one *should* live or what one needs to do in order to attain any particular state. These are the only people who are not strangers in the world, since they are inside existence itself. To them, the only outsiders in the world are those who look outside themselves for guidance or definition, and the only strangers in the world are those who spend their lives in imitation of someone else, because that is what makes them strangers unto themselves. These people give you no hope of following them at all. The only thing they give the world is their own exquisite lovingness — their absolute delight in being abandoned to the mystery itself, with no need for method or meaning. Theirs is the rush of existence, the aching pulse of the passing moment and the boundless joy that can only come from touching the core of your own existence.

People of the 8th Siddhi sparkle like exquisite rare jewels.

9th Gene Key

The Power of the Infinitesimal

SIDDHI
INVINCIBILITY

GIFT
DETERMINATION

SHADOW
INERTIA

Programming Partner: 16th Gene Key

Codon Ring: The Ring of Light
(5, 9, 11, 26)

Physiology: Sacral Plexus

Amino Acid: Threonine

THE 9ᵀᴴ SHADOW – INERTIA

THE DOMESTICATION OF DREAMS

In its original hexagram form in the Chinese I Ching, the 9th Gene Key has a rather unusual and cryptic name, which is commonly translated as *The Taming Power of the Small*. If you are familiar with the I Ching, you might recall another hexagram, the 26th hexagram, whose name is *The Taming Power of the Great*. Evidently there is a strong bond within these two archetypes and their Gene Keys. We can see that genetically they are indeed part of the same chemical Codon Ring and its amino acid threonine, which we will discuss later. As is often the case with these old Chinese names, they contain many layers of truth and possibility. In the case of the 9th Shadow, the *Taming Power of the Small* refers to the human tendency to become submerged in unnecessary and irrelevant details. Most human beings live lives where they simply get by, lives in which they become victims of all the details around them. At the higher frequencies, you *tame the small* by applying your energy only to that which serves your higher purpose. At the Shadow frequency however, the details *tame* you, sapping your life force, robbing you of your versatility (the 16th Gift and programming partner of the 9th Gene Key) and eventually pulling you into the common human state of inertia and indifference (the 16th Shadow).

The Chinese sage Lao Tzu uttered the famous statement: "The journey of a thousand miles begins with the first step", although a more accurate translation might be: *The journey of a thousand miles begins beneath one's feet*. This piece of timeless wisdom concerns focusing on what lies right in front of you rather than concerning yourself with where the future might or might not take you. The 9th Shadow is about where you place that focus — and primarily through your daily activity rather than your mind. There is something very magical about this 9th Gene Key, as we shall see. It holds one of the greatest of all secrets — how to stop your mind from undermining your natural destiny. An image representing both the 9th Shadow and the 9th Gift is of a pathway made of individual stepping-stones.

*Most human
beings live
lives where
they simply
get by, lives
in which
they become
victims of all
the details
around them.*

At the frequency of the 9th Shadow the stepping-stones go in a circle, so that as you look down at each step you fail to realise that you are simply following the same old footprints and your energy is going nowhere. This is the state of consciousness of the majority of human beings on our planet.

At the Gift level however, the stepping-stones go off into the distance and over the horizon. You do not know where they are going, but it doesn't matter — you know they are leading you forward. This makes every step you take all-important as well as an adventure. This 9th Gene Key is about finding the right activity in your daily life. Every step must lead you in the direction of your dream, whatever that may be. Included within this path are the many, many small acts we take on the mundane plane — eating, washing, shopping, cooking, etc. Because if all the steps, even the mundane daily chores, lead you in the direction of your dream, it is impossible for them to be unfulfilling. If your activity leaves you cold or bored, it does not necessarily mean it is the wrong activity. It probably means that you have lost contact with your greater dream — you have allowed the small to tame you. Every time you allow life to leave you bored or indifferent or you feel this lack of energy and inertia, it is up to you and you alone to reconnect with your dream.

Without a sense of higher purpose, human beings move in circles creating energy fields that prevent abundance. Even worse, the inertia of the 9th Shadow feeds off the victim mind, which also goes round and round in its patterns of either blaming or worrying. In our hearts however, all human beings are natural rebels. We are wild creatures. We are not here on earth to have our dreams *tamed*, clipped or domesticated. We are here to make magic happen and we cannot do that unless our every waking moment is directed and focused towards a single over-reaching vision or ideal. The 9th Shadow sucks all the hope and versatility out of you when you do not see immediate results and improvement in your situation. It takes you away from the focus and fulfilment of the moment and disturbs your concentration and patience. One of the modern expressions of this 9th Shadow is our addiction to trivia — details and trappings extraneous and unnecessary to our lives. Unless it is either beautiful or practical, it can safely be classified as trivia. The 9th Shadow draws your energy away from that which truly matters to you, which is beauty.

The energy field around this 9th Shadow and its programming partner the 16th Shadow of Indifference is a very intense cloud under which many humans are stuck. Together these two Gene Keys are a huge drain on your physical body. Lack of versatilirty leads to lack of energy and vice versa. You may even believe that you are taking steps to change when in fact all you are doing is going around in circles, still focused on details that are essentially irrelevant. The only way out of this field of inertia is to punch through it with one huge act of will. This first step out of the victim frequency resets your course onto a path that leads ahead rather than around. This 9th Shadow deeply affects the energy systems within the body, closing them down to higher voltages and cosmic high frequency energies. It also has an adverse jamming effect on your inner directional guidance system — your heart. If your heart is not behind your every act, not only do you choose an inappropriate course in life, but you also continually wear down your health.

In summary then, if your life force seems low or lacking in energy and you are finding it hard to get enthused about your life, the answer may well lie in the 9th Shadow. You are either too focused on the future instead of giving your full attention to whatever is right in front of you, or there is no sense of overall purpose moving beneath your daily acts and activities. Without this sense of inner focus, a great deal of your energy goes into complaining, whether vocally or mentally. All this energy needs to find a higher purpose — something it can serve that takes you beyond the mundane world and all its details. Most human beings are unaware of how much energy they are sitting on inside their own bodies. There truly is nothing in life you cannot accomplish if you put your heart squarely behind it.

REPRESSIVE NATURE – RELUCTANT

There is an inner reluctance in the repressed side of this 9th Shadow. It manifests as our seeming inability to do anything about our situation despite the fact that we understand it and even see our way out. The reluctance to move out of one's patterns is not a conscious choice but an inner dynamic where all one's life force remains frozen. This reluctance is essentially a paralysis of our will, brought about by following familiar, repetitive patterns that do not serve us. To break out of our inner reluctance is to leave our safety zone and move directly into our fears. It can be frustrating to onlookers that these people are unable to break their patterns, but it is just as frustrating for those in the grip of such deep-seated fear. Ultimately, it comes down to the power of the human will to either break through or fall into a continual and miserable decline.

REACTIVE NATURE – DIVERTED

The reactive nature of the 9th Shadow involves a totally different kind of inertia. These people can be highly restless and fidgety, as though nothing inside them can sit still. Their tactic is diversion, and they unconsciously seek any stimulus to draw some of their energy and fury out of their body. Naturally, such people cannot sustain this escapist pattern indefinitely and it takes an enormous toll on their health and often their finances. These people cannot find a fixed pattern in life at all. If they were to do so, all their rage would explode out of them. This makes it impossible for them to maintain serious commitments for any length of time. Although their lives are not inert in the strict sense of the word, they are inert in terms of fulfilment because they can simply never rest or relax.

THE 9TH GIFT – DETERMINATION

EVERY INTENTIONAL ACT IS A MAGICAL ACT

The 9th Gift fuels all determination. The Gift of Determination is built upon the rock of the very smallest acts. The controversial English magician Aleister Crowley once stated a deep truth of great relevance to this Gift: *"Every intentional act is a magical act."* Even the smallest action has a spin-off effect that travels out into the universe.

Actions made out of resentment or fear reinforce the Shadow frequency both in the world and in the individual. Indifferent acts reinforce indifference, whereas acts done in joy or service create more joy. Wherever we look into the matrix of this 9th Gene Key, it keeps pointing to the same truth — that a person without an ideal burning inside them is essentially destined to remain a follower of the crowd. However, it is important to distinguish that this is not a Gift of dreaming but of sustained activity and work towards a single powerful goal.

The power of the 9th Gift is the power of repetition. This gift creates a groove and once that groove is dug, all energy in your life will tend to follow the same pattern. This explains why the Gift of Determination is so powerful. It also explains why it is so hard to escape the inertia of the 9th Shadow at its lower frequency. When you break out of the lower frequency and connect with your vision or ideal as a feeling and a knowing deep inside, you have truly begun the journey of a thousand steps. Every single step you take from that moment on — which means all acts, no matter how inconsequential they may seem — lead in the direction of that core vision. As you continue in the direction of your heart, you begin to carve for yourself a powerful groove that becomes easier and easier to follow and is what the world knows as determination. This changes the whole shape of your life, as finally you will begin to feel the inner strength that comes from following your life purpose.

All those little acts done with heart begin to build up an inner momentum that eventually becomes unstoppable.

The strange thing about the Gift of Determination is that the more determined you become the less energy and willpower you have to use. This is contrary to the common view of determination, which is generally held to be a great battle or struggle. However, the secret of determination is about momentum. All those little acts done with heart begin to build up an inner momentum that eventually becomes unstoppable. The force of the entire universe begins to gather behind such a person. It is only at the beginning that you have to use great force because those first few steps out of the Shadow frequency take such enormous strength of will and courage. Thus the 9th Gift reveals one of the great secrets of the Gift frequency — the more you tread the path of heart the easier it gets. Instead of being tamed by life, you forge your own direction and destiny through taming the smallest and most inconsequential acts in your life.

The 9th Gift has an important connection to the power of magnetism. All of life is magnetic, and this Gift employs the use of magnetism through aligning itself with true north — the inner direction and rhythm sought by the universe as a whole. This is the groove we are speaking of — it is about moving down the force-lines of the universal energy grid instead of moving across or against them. Determination in this sense reveals another meaning. Your true course in life is already *predetermined* and thus all you have to do is find it and follow it. As was mentioned at the Shadow frequency, one of the most magical aspects of this 9th Gift is the effect it has on your mind. Once you have centered yourself in the groove and your course begins to feel more and more certain, your mind finally stops undermining you. The naturally flowing currents within your body begin to come into a universal harmony, and as they do so, your brainwave cycles slow down and you enter a higher consciousness field. It is one of the paradoxes of the Gene Keys that the higher your spiritual frequency rises, the lower your brainwave frequency falls.

These radical shifts in your mental functioning further serve to streamline the course of your life. With the mind operating at deeper levels of consciousness, you begin to let go of your mental constructs — your opinions, your fears, your beliefs and even eventually your hopes. Your mind begins to submerge itself in a wider and more collective awareness. Not only does it undermine you less and less, but it also confirms that your direction is logically valid. At the height of the Gift frequency as it begins to prepare for the leap into the siddhic consciousness, you become aware of the vast power that can move through the very smallest of things. As your vision of reality expands to contain universes, you realise how small you really are. At the same time you get to see how enormous your contribution to the whole is when you truly take the plunge and listen to your heart.

THE 9TH SIDDHI — INVINCIBILITY

INNER SPACE — THE FINAL FRONTIER

The world around us is filled with examples of the power that comes from taming the small. When humanity managed to tame the power of the atom — the basic magnetic unit of matter — we unlocked its tremendous energy and demonstrated a great universal law — the smaller something becomes, the more condensed is the universal energy within it. This law can be applied to your individual life as we learned through the 9th Gift. There is one final surprise lying in wait for us within this 9th Gene Key and it is found in its ultimate expression as the Siddhi of Invincibility. The 9th Siddhi is about the power of the infinitesimal — the infinitely small. The infinitesimal is also the paradoxical — if one goes on dividing a piece of string in half, one could theoretically go on forever. Thus the infinitesimal becomes the boundless and inner space leads to outer space.

At the highest level of consciousness, levels themselves disappear. Outer space becomes inner space, time becomes infinite yet absolutely present, and all boundaries are absorbed by consciousness itself. Now the moment something loses all boundaries, it becomes two contrary things — it becomes simultaneously defenceless and invincible. Invincibility can therefore be defined as merging your individual awareness back into the consciousness of the universe. To be invincible is to surrender to all possible foes by dissolving your entire reality *into* them. This is why these two closely named Siddhis, the 9th — the Taming Power of the Small, and the 26th — the Taming Power of the Great, are so intimately interconnected. The 9th represents *Invincibility* and the 26th represents *Invisibility*. To be invincible is also to be invisible, and vice versa. Invincibility means to dissolve into the will of the universe, and it is a Siddhi that has been deified in many different cultures. In Christianity for example, the archangel Michael holds the archetype of invincibility.

The only force in the universe that is truly invincible is love. Love knows only giving, so it creates a vacuum which in turn continually floods with more love. It is impossible to fight against such a force since it renders all other energies defunct by returning them to their source.

When a human being attains this Siddhi, their life becomes an expression of the invincible power of such love. These people have discovered that the entire universe exists in microcosmic form within the human body, and they may become teachers of this truth. In this respect they can be masters of techniques wherein the macrocosmic can be mapped onto the microcosmic. Such people become the focal point for an intense frequency of Divine light with a very specialised task to fulfil on earth. The Taming Power of the Small is like a laser that can pinpoint a very specific aspect of life, bringing enormous power to bear on that area. An ordinary person may find the aura of such a person almost unbearable, since it spotlights those aspects of your shadow nature that you most fear. If you are blessed to be karmically tied to such a being, it is highly likely that full realisation will occur within your lifetime.

It is fascinating to see how the higher programming matrix within our DNA is coded to awaken specific forces within us at specific times in our evolution. The 9th Siddhi belongs to a family of Siddhis that are interlinked through the human genetic structure by the amino acid known as threonine. Threonine brings together the 5th, 9th, 11th and 26th Siddhis, which collectively unlock two very powerful universal themes of humanity's higher nature — time and light. Known as *The Ring of Light*, these four themes — Invincibility, Timelessness, Light and Invisibility — form a kind of cross-coding message that is designed to operate across entire gene pools. There will come a time when humanity will awaken very quickly to its highest genetic frequency through the wavelengths of light carried by the aura. The end of our mental experience of time will therefore be triggered by human auras interacting with each other. In this respect, humanity will only become invincible when it realises its collective nature, at which point the individual becomes invisible in the sense of being completely porous to the indwelling group awareness.

You must capture the energy of your dream and hold onto it deep inside, because it is that dream that will act as the lens to focus the power of magic and manifestation in your life.

What the 9th Siddhi teaches us today is that every act you make is of vast importance to the whole of evolution. If your life takes on a cosmic focus, then life itself will intensify within you, moving you naturally into a far more co-operative pattern with your environment and with others. Every intentional act is a magical act, setting in motion either a force of creativity or a force of decay. Whenever you stand at the beginning of a journey, it is the first step that sets the tone for the entire journey and the next few steps that begin to create the groove. After even a relatively few steps then, it becomes very difficult to change your direction since that involves wrenching yourself out of your existing groove and creating a new one. Therefore whenever you reach a natural beginning — of a new cycle, a new relationship, a new home, even a new year — you would do well to remember this truth: those first few steps are critical to your evolution in the time ahead. You must capture the energy of your dream and hold onto it deep inside, because it is that dream that will act as the lens to focus the power of magic and manifestation in your life.

≣ 10th Gene Key

Being at Ease

SIDDHI
BEING

GIFT
NATURALNESS

SHADOW
SELF-OBSESSION

Programming Partner: 15th Gene Key

Codon Ring: The Ring of Humanity
(10, 17, 21, 25 38, 51)

Physiology: Chest (Heart)

Amino Acid: Arginine

THE 10ᵀᴴ SHADOW – SELF-OBSESSION

THE MAZY PATHWAYS OF THE SELF

As one of the keystones of human individuality, the 10th Gene Key and its frequency bands point towards one of the deepest of all human issues — the notion of self-love. This intangible force within human beings begins life here in the 10th Shadow, where it brings continual focus on your own immediate environment, which is your body. This is one of the most primitive of all the aspects and archetypes within the human genome. At its shadow frequency it tightens all your life force and forces it inward, which in the long run makes it one of the most mystical of the 64 Shadows. It is here that the individual journey towards awakening and transcendence truly begins. However, this genetic centripetal force excludes other beings from your immediate concern and attention. In early hominids, this 10th Shadow ensured individual survival since it put the safety of its own vehicle before anything else. In humanity, to see a person give his or her life for another or for a higher cause is to see this 10th Shadow transcended, as its prime purpose is to put oneself first.

In our modern world, the 10th Shadow still governs us on a collective level, even though it is showing signs of awakening today. The emphasis of the 10th Shadow is on the individual, which can be both a blessing and a curse. Individual differentiation is the cornerstone of evolution itself. If we humans do not discover our own identity and uniqueness, we cannot transcend it and move our society to a higher level. The blessing is that the more different we each allow ourselves to be, the more we operate as a unity. This is one of the most beautiful of all human paradoxes — that only through our very diversity can we arrive at our unity. But there are forces that tug against evolution, and these forces, coming from within, keep us from experiencing our true uniqueness.

The programming partner of this shadow is the 15th Shadow of Dullness, and since this 15th Shadow conceals a fear of being different, it shuts you down at a collective level. The 15th Shadow makes us into lemmings that follow the crowd, thus allowing our uniqueness to be over-ridden.

*Through the
lens of the
10th Shadow,
all you see
when you
look at others
is people you
would like to
change.*

Just as the 15th Shadow puts your attention on everything but your own uniqueness, the 10th Shadow does the opposite — it makes you obsessed with your own uniqueness and how to find it and follow it. Thus in the world today we can see two main types of people — those who follow the crowd and those who try to escape the crowd at all cost. The 10th Shadow does not and cannot consider anyone outside oneself. Through this shadow, you become so self-obsessed that you no longer see or hear the feelings of those around you. This makes it very difficult for other people to relate to you, even though you may feel that you can relate to them. Even though you may have many relationships, the truth is that you really don't have enough space within your psyche for the concept of others. Everything and everyone is viewed through your own subjective projection field and this loss of objectivity can lead to only one result — it creates havoc in all your relationships.

Through the lens of the 10th Shadow, all you see when you look at others is people you would like to change. Thus you find it extremely difficult to accept anyone else for *their* uniqueness. In psychology and psychiatry, such self-obsession is known as narcissism and in moderation it is considered to be an essential component of a healthy psyche. However, at the shadow frequency such narcissism, like the legend it is derived from, keeps human beings endlessly trapped by their own reflection. Ironically, the more of an expert you become in the subject of the lower self, the further away you travel from your higher self.

This self-obsession of the 10th Shadow is driven by fear, and it is a very specific unconscious fear — it is the fear of losing your identity. As one of the deepest of all human fears, this fear forces you into a pattern of trying to find out who you are in order that you might find some kind of permanent definition in life. This quest for your true identity is the greatest quest there is. It is the ageless journey of self-knowledge epitomised in the famous axiom cut into the stone above the door of the oracle in Delphi — "Know Thyself." However, at the low frequency of the shadow level, this quest for self-knowledge becomes an obsession that actually keeps you from defining who and what you are. In your thirst to escape your fear, you create a journey that has no end, filled with drama and adventure perhaps, but ultimately one in which you never have to face yourself. This is the craft of the 10th Shadow — it tricks you into chasing the shadow of your own reflection, and your very journey towards your true nature becomes your own net.

In the modern western world, self-obsession is everywhere. People are obsessed with how they feel and how they look, with what they wear, what they own and where they live. As long as you are looking at yourself you cannot see around you, and there is the rub. Until you come to recognise your own self-obsession you cannot transcend it, which is why it is a necessary evil. All inner journeys begin with this self-obsession and it can truly become an endless labyrinth. Even when self-obsession takes the form of a spiritual search, it can become a trap. In fact, in many ways the spiritual search for truth that is gripping so many individuals in our western world today is the greatest self-obsession of all. The path itself easily becomes an addiction that keeps you from truly being yourself. The more you look for your own true identity, the more ephemeral it becomes.

All circular paths do eventually arrive at breakthrough. Those who tread the path of self-knowledge (the ancient Chinese name for this 10th hexagram of the I Ching is *Treading*) will eventually escape the treadmill once they realise that what they are seeking cannot be found. This is a spontaneous revelation that can occur only when you have been well and truly lost in the mazy pathways of the self. No one can say how long it will take for such a revelation to dawn, since it differs from individual to individual, but it cannot be faked. The true revelation that comes through the 10th Shadow leads inevitably to the creative explosion that emerges through the 10th Gift of Naturalness.

REPRESSIVE NATURE – SELF-DENYING

The introverted nature of the Shadow of Self-Obsession is to deny oneself altogether. This kind of reversal of the energy of self-obsession becomes entirely centrifugal and focuses on everything and anything but oneself. These are people whose lives are lived for others and through others. However, such martyrdom does not have a positive spin and cannot serve evolution. These lives are made up entirely of compromise in which those who deny themselves become like zombies for the collective. Although this kind of language may sound shocking, a huge majority of the world population lives like this — with no real self-love and no true centre or heart within themselves. These people are the prime targets for the great world religions. Their self-denial does not allow them to recognise the Divine within themselves, finding it far easier to project such an authority outside them.

REACTIVE NATURE – NARCISSISTIC

When the self-obsessive force is externalised through the reactive nature, it becomes truly narcissistic in the sense that it excludes all others. Just as the repressive nature denies its own existence, the reactive denies the existence of others. These are people whose lives are based entirely around themselves. The fear within them lives as anger — they project their fear of losing their identity onto others and onto society in general. Such people cannot yield any part of themselves to others. They live in paranoia that the world or others will somehow rob them of their right to freedom. The reactive nature finds relationships the hardest of all, since they are essentially locked in a love affair with themselves, but not with their true self. The true self is everywhere and in everything, since its nature is love. However to realise such a concept, one has to drop the very thing that gives one the illusion of safety — the illusion of individual identity.

THE 10TH GIFT – NATURALNESS

LIVING YOUR OWN MYTH

The Gift of Naturalness is a Gift that waits within every single human being. It is the centre of your being and only through this centre can you express your own creative uniqueness. The life of every individual human being is a journey through the frequencies of this 10th Gene Key. To be natural means to be yourself. We are all trying to be who we are, but many of us have been conditioned to be other than who we are. It has been this way throughout our history, ever since we began to develop the complex neo-cortex.

The life of every individual human being is a journey through the frequencies of this 10th Gene Key.

The moment the mind was equipped to reflect on itself, the primal question *Who am I?* was born. Prior to this question, human beings did live in the state of naturalness, but only in the sense that they were more a part of the animal kingdom and had not yet fully developed into homo sapiens. Even so, it is still interesting how much we can learn about this 10th Gift through the animal kingdom, whose normal state is naturalness.

The human journey, of course, is unique and is obviously different from that of the animals. We have to solve the conundrum of having a mind that can look at itself. The primal question *Who am I?* has to be answered before the Gift of Naturalness can once again be realised. There are so many paradoxes in this 10th Gift — the main one being that you cannot be who you are as long as you think you are someone, but you still have to set off and search for this someone in order to realise that they do not exist! There is much tension inherent in this simple little question *Who am I?* and it is the releasing of this tension around your identity that is the essence of the 10th Gift. The 10th Gift of Naturalness can dawn only when self-obsession exhausts itself. The first great revelation that comes from the journey into your own unique nature is that you cannot be defined by any kind of label. Once you understand that you are neither your name nor your actions, feelings, thoughts or beliefs, you realise that human nature is something far greater and wider than you ever suspected.

At the Gift level of frequency, there is a huge releasing of energy through your being and out into your life as all attempts at defining who you are begin to subside. The usual manifestation of this energy is an intense creativity coupled with a new sense of playfulness. You lose your sense of identity and simultaneously begin to feel centred in a wider context. The Gift of Naturalness cannot be practiced, copied or systematised. It can only emerge through a rising sense of inner freedom and spaciousness. This sense of increasing ease and relaxedness in your life tends to follow an age-old archetypal sequence that appears to be universal. As we have seen, this process — which Jung called *individuation* — begins with the question *Who am I?* This question, which may also appear in different guises such as *Why am I here?* or *Why am I doing this?*, kick-starts a process of inner questioning about the purpose and meaning of your life.

The second phase of becoming natural tends to be a deep inner questioning and questing in which you isolate yourself in some sense from your previous responsibilities and give time and space to understanding your own nature more. For most people this phase lasts a long time and it is also the phase during which most people become stuck in the lower frequency of self-obsession, falling in love with the endless quest for self knowledge. At a certain point however this phase gives way to a natural collapse and a subsequent letting go of all searching as you realise the futility of trying to find something that is clearly indefinable. This realisation marks a huge transition point in your life and represents the release from your self-obsession. It can also be a very challenging time as you let go of all the constructs and techniques with which you may have identified and from which you derived security. Fortunately this third phase swiftly becomes consolidated and integrated into your being, and you enter a new dimension in which you become more and more deeply relaxed. This fourth phase is something like a rebirth as you externalise what is truly inside you for the first time in your life. It is a time of deep joyousness and purpose and it is only as you find this sense of relaxedness that you begin once again to feel truly natural.

The fifth and final phase of the process of becoming yourself manifests as the flowering of your individualised differentiated self. This is when you hit the peak of your own mythology and bring something entirely new into the world based on the high principle that has always lain latent within you. This final flowering of your inner being manifests as a challenge to the current norm, since your true nature is always found at the cutting edge of evolution. The idea, whatever it may be, represents the true beauty within the individual finding expression in the world.

Throughout this work there are references to the hidden chemical pathways within our DNA. These genetic networks are known as the 21 Codon Rings, and they contain many mysteries. The 10th Gene Key belongs to a chemical family known as *The Ring of Humanity* which includes the 10th, 17th, 21st, 25th, 38th and 51st Gene Keys. As one of the most complex codon groups, this chemical family holds the keys to the great mythic story lines from all cultures. These six Gene Keys encapsulate all the mythic elements of what it means to be a human being. Wounded from the outset (25), you must do battle with your shadows (38), overcome the limitations of your mind (17), surrender your need to control life (21) and find your true self (10) before you can awaken (51). You can see from this very profound grouping how deeply bonded we humans are by the same basic dramas. There is great beauty to be discovered here, and when you find it, you will finally realise that naturalness is the simplest thing of all. It is simply there inside you the moment you decide to stop arguing with life!

THE 10TH SIDDHI — BEING

DIVINE LAZINESS

When the differentiated self has manifested its highest expression in the world through the 10th Gift, one final surprise lies in store. There is a sixth phase to the process of becoming yourself that brings an end to the whole notion of self-knowledge. In a sense it is a return to the stage before you asked the fateful question *Who am I?* This sixth phase is the 10th Siddhi of Being. When the 10th Siddhi dawns in a human being, then for that person even evolution itself has ended. In the 10th Siddhi, the differentiated self spontaneously dissolves in a higher mirror of the third phase in which the constructs of your search for self-knowledge end. However, at this siddhic level, everything dissolves — self, not-self, mind, form, purpose and meaning. The only word that even gives a hint of what the 10th Siddhi holds is the word *Being*. In the expressions of the mystical traditions, this is sometimes known as the *I Am* consciousness, although the very use of the word *I* can be misleading. In the 10th Siddhi there is no sense of *I* — there is only pure consciousness expressed as being.

The 10th Siddhi is destined to be deeply misunderstood by all those who do not live in its embrace. Along with its programming partner, the 15th Siddhi of Florescence, it inspires one of the great metaphysical paradoxes of all time. This paradox is well expressed through a phenomenon particular to Buddhism — that of arhats and bodhisattvas, the two externalised mani-festations of the highest state of enlightenment or self-realisation. Without getting weighed down in hefty Buddhist dogma, these two expressions of human perfection can be understood as representing *being* and *becoming* respectively. The arhat is of the 10th Siddhi representing pure Being — a state in which evolution no longer exists or matters. For the arhat, once he or she is enlightened, the whole universe is also enlightened, so there is nothing more to be done. For the bodhisattva, who is of the 15th Siddhi, there is no end to life, which is a continually evolving state of flowering. Thus the bodhisattva takes a vow to deliberately hold back their enlighten-ment in favour of helping evolution complete itself in the form of guiding others towards the state of liberation.

These two expressions of human perfection — the arhat and the bodhisattva — have caused a great deal of confusion in mystical circles. The way of Being is represented strongly by the tradition of Advaita Vedanta, one of India's oldest spiritual streams. Through the 10th Siddhi, a great lightness is born, and a great humour concerning life itself. To experience life through the 10th Siddhi is to see everything that normal humans think of as important reduced to a game or illusion. To the arhat, life is meaningless, time is an illusion and therefore evolution itself is a game. Because this viewpoint is generally viewed by the unenlightened as selfish and threatening to their ongoing identification with evolution, the arhat has been mostly eschewed in favour of the bodhisattva. Thus flow the politics of enlightenment! To those outside these states, they appear utterly contrary to one another, but to the being manifesting them, both poles are experienced together. The only differentiation between them is the language that the person uses to describe his or her experience or revelation. The arhat has nothing left to do in the world whereas the bodhisattva has a deeply focused mission to help others.

The 10th Siddhi is truly a beautiful expression of consciousness coming through the human form. These are people whose awakening encompasses all existence. The intense focus on oneself that comes through the 10th Gene Key finally breaks its identification with form and experiences everything as Self, with love as the fluid in which the multifarious aspects of the Self are floating. You can see how easily misconstrued the mes-sage from this 10th Siddhi can become — especially in today's world, in which we humans are being called to participate more and more in our own evolution. The 10th Siddhi is a reminder that it is all but a game, a *leela* or play, in which even our loftiest ambitions are ultimately meaningless. Obviously if everyone held this view, evolution itself would stop in its tracks, since this view represents the ending of evolution. The 10th Siddhi sees the beauty of the game of evolu-tion but has no alternative but to undermine it. Being undermines everything — it does not allow identification with anything outside the wonder of the present moment.

The 10th Siddhi is a reminder that it is all but a game, a leela or play, in which even our loftiest ambitions are ultimately meaningless.

To the person who has entered the mystery of the 10th Siddhi, the two poles of being and becoming are one. It is their revelation to rest in the true nature of being and at the same time bear witness to the effervescent flow of form becoming more and more complex through evolu-tion. The outer destiny of each person is governed by the Primary Gene Keys that are genetically wired inside them. Thus it is the Primary Siddhi that determines the language and style of a particular enlightened manifestation. It is perhaps understandable then, and also a little sad, that the 10th Siddhi has been judged so unfavourably by metaphysicians for some time now. The powers that be do not want people to sit idly by and dream away their lives in the divine laziness of Being. Those quieter days of the arhats have been superseded. The modern-day view that holds favour is that of the evolutionist. Today we are obsessed by evolution — by where it is taking us and whether we can control our own direction as a species. At times of potential crisis, like the time we are moving through today, Being is seen as doing nothing. If we are to survive the next few centuries, there are things we *must* do.

To the one within the 10th Siddhi, there has never been anything to do because there is no one to do it, so why all the fuss about our future? Being is the nature of consciousness in form, and it has no agenda or direction. It simply is. In that simple statement lies a power beyond all comprehension, and it establishes the 10th Siddhi as the sleeping giant within the human genome. Nothing has more power than being itself. We would each do well to remember the nature of being that rests behind all the dramas of the world and our individual lives. To rest unattached in the supreme state of being, whilst at the same time participating in the adventure of our evolution is quite possibly the greatest task that the humans of the future will have to master.

11th Gene Key

The Light of Eden

SIDDHI
LIGHT

GIFT
IDEALISM

SHADOW
OBSCURITY

Programming Partner: 12th Gene Key
Codon Ring: The Ring of Light
    (5, 9, 11, 26)

Physiology: Pituitary Gland
Amino Acid: Threonine

THE 11ᵀᴴ SHADOW – OBSCURITY

THE FASCIST REGIME OF THE HUMAN EGO

The 11th Gene Key will open you to a whole new world — the world of light. Indeed, it is this Gene Key that gives its name to the important genetic and chemical group known as the Ring of Light. This Gene Key concerns human vision — both internal and external. As such it is deeply connected to the human eye and the way in which images are translated via the visual cortex into the brain as imagination. One of the most fascinating studies of light in all its potential can be seen through this genetic codon. The amino acid threonine programs your DNA through the 11th Gene Key. Threonine also codes for three other Gene Keys — the 5th, 26th and 9th. Each of these four Gene Keys concerns a different code through which human beings are connected with light. At the highest level of consciousness, the 5th Siddhi of Timelessness shows how time can be brought to an end through its connection with light via the medium of space. This is why transcendence of the speed of light also leads to the transcendence of time and therefore space. The 26th Siddhi of Invisibility concerns the supernatural ability to manipulate the human perception of light through magnetism, and the 9th Siddhi of Invincibility invokes the laser-like focusing of light in order to dissolve your physical reality thereby making you effectively omnipotent.

Each of these four Gene Keys can be viewed through the lens of their shadow frequencies, which further illustrates how deeply human suffering is linked to your ability or inability to tap into the powers of clarity through the medium of light. In the case of the 11th Shadow and Gift, we are looking at the interface between light and the human mind. The 11th Shadow places an interference frequency between light and the way in which the brain processes, translates and communicates that light. In other words, your whole experience of the world is pushed off kilter through the medium of the 11th Shadow. It therefore represents the field of illusion, delusion and obscurity.

The greater percentage of human beings on this planet lives within a very narrow band of light waves, which means that they do not see reality clearly. What most people think of as reality is a very dim and skewed view of the true reality. The 11th Shadow greatly limits a very specific functioning of the right hemisphere of the human brain — that aspect of your mind that does not see patterns and facts through language and number, but grasps reality through reams of interconnected and intuitively grasped fractal images emerging from the deep recesses of the brain. The right hemisphere of the brain has long been seen as the feminine side of the brain — it is the lateral thinking, intuitive and artistic side of your mind. If you could see how deeply limited your perception of reality is without the full functioning of this feminine side of your nature, you would be enormously shocked.

The 11th Shadow of Obscurity essentially places you inside a virtual reality — a construct created through a combination of programming via this 11th Shadow and its programming partner, the 12th Shadow of Vanity. This *reality* is a total obscuration in which you can only view life through a certain very limited set of parameters. Here is how it works: representing the feminine pole of the brain, the 11th Shadow creates a field of fear within human beings. The images that flood your mind from the right side of the brain can neither be controlled nor do they appear to make sense. In most cases, they are relegated to a backwater of your brain where they emerge as secret fantasies, repressed dreams, emotional issues and hidden agendas. The male oriented left side of the brain (encapsulated in the 17th Shadow) therefore becomes far more dominant, since it uses logic as a means of controlling your reality. Whereas the right brain seems chaotic, illogical and idealistic, the left brain is the voice of control and reason.

The greater percentage of human beings on this planet lives within a very narrow band of light waves.

The next part of the story involves the programming from the other side of our genome through the 12th Shadow. The 12th Shadow programs through sound rather than light. It translates the abstract oriented reality of the 11th Shadow into an inner *language* — a neuro-linguistic fabricated reality that you then project out onto the world. In the centre of this virtual world sits the separate self — a controlled illusion held together by a fascist inner regime operating through an inner media that continuously manipulates you though the medium of light and sound. In other words, the 11th and 12th Shadows only allow you to see and hear what they want you to see and hear. If this sounds familiar to you, it is because our outer world has a tendency to mirror the inner reality. Vanity is the name given to the false protagonist sitting in the middle of your screen, and it forms the basis of this planet's false reality. Many other systems and traditions have named this inner construct the *ego*.

The ego or separate self is thus a figment of our collective genetic conditioning, and just as our conditioning can be picked apart, so can the ego gradually be dissolved. This is an extremely delicate operation and is the basis of most mystical systems and certain types of psychoanalysis. The great fear for human beings remains the repressed right hemisphere of the brain. As you begin to unleash the floodgates of this part of your brain, whether through some form of shamanic training, mystical technique, drugs, therapy or art, you put your whole constructed reality in great danger. You may begin to feel overwhelmed by the flood of imagery that rains down on you from your repressed unconscious, and your inner language may not be able to handle and integrate the results of such dissolution. This is why such inner events are seen as a death and can often result in delusional states where your fear tries to identify with the archetypes emerging through the right brain.

The secret of the 11th Gene Key and the right brain is the secret of the archetypes. Every single element or image that floods you from your unconscious represents an archetype — a collectively held alchemical image that reflects the process of dissolution. There are archetypes that thrill you and archetypes that terrify you. In the modern world, the mass consciousness maintains its main outlet to this archetypal world through stories, television and cinema. You cannot escape the archetypes because they are the projections of your own psyche. But the real power of an archetype is its biophysical response within your body. These are not just images that you can view objectively, but are neurological links that stimulate your entire body via your glands. You try to avoid the particular archetypes that cause you the greatest fear, but you never can. The more you try to avoid them, the more they stalk you. Thus you keep recreating the very situations you hate, especially in your relationships, where the archetype often assumes the form of your partner.

The 11th Shadow is a veritable minefield of lost dreams, escapist behaviour, denial, guilt and repression. If you could just begin to trust in the images that come to you in your dreams and allow them to incubate in your imagination, you would begin to emerge from the false dream that you have placed yourself in. To awaken out of such a state is a vast experience. It is to step out of the illusion of the rest of the world and to tread a new kind of path that is at odds with the majority. It is the step that you must one day take to move from the 11th Shadow into the 11th Gift. When you finally make this courageous inner leap, the dream that has been hidden inside you for so long shakes you awake and introduces you to a new and boundless horizon.

REPRESSIVE NATURE – FANTASISING

When the archetypal imagery of the right brain is repressed, it turns inward and creates a world of fantasy. There is nothing wrong with fantasy so long as it finds a healthy and creative outlet, but in most cases it does not find such an outlet. The foundation of repression is fear, and when this fear is not embraced it creates an enormous drain on our whole body and being. Unlived lives gradually drain the physical energy from the body, leading to all manner of physical health problems. Even worse than this, inner fantasies that do not find a creative outlet block us from following our true destiny. Our destiny is always hidden away behind these fantasies. The longer we hold them within, the more twisted they become until what began as a simple archetype becomes perverted into a much darker form. Until we can own responsibility for such a form, we cannot release the creative power that lies within us.

REACTIVE NATURE – DELUDED

The reactive nature, rooted in denial and anger, turns the inner fantasy into a projection field and tries to manifest it in the world. If the anger were embraced and the deeper fear owned, then the archetypes within us would indeed manifest in the world, but the reactive nature never allows this. Reactive natures use the outer world to hide from what lies within them. These are the kind of people who have a great idea in their minds but they never manifest it. Because of the depth of their denial, they cannot let go of the inner image they have of what they would one day become. Such people are headed for eventual disappointment and even breakdown. Delusion is the attempted externalisation of a false dream that conceals a deeper and real dream. Until the true dream is uncovered, what emerges is simply a veneer with no power to manifest.

THE 11ᵀᴴ GIFT – IDEALISM

MAGICAL REALISM

The 11th Gift is one of the great keys of our present age. The more people who can be encouraged to play with the imagery and creative power coming from the right hemisphere of the brain, the more healthy the world will become. The historical repression of the feminine power and of women in general is a direct manifestation of the imbalance within our brain chemistry coming through the 11th Shadow. The imagery that lies trapped within us is the pressure of our ancestral past. In other words, these images are memories. Furthermore, these memories are not just personal memories, but collective memories that have lain repressed for millennia. Such memories exist within you as archetypes, and the moment you begin to understand what an archetype is capable of, you begin to work with the dynamic energy of the 11th Gift — the Gift of Idealism.

Idealism is given a bad rap in our modern world. It is seen as the opposite of realism, and realism is associated with the power of manifestation whereas idealism is generally seen as weaker. The 60th Gift of Realism however is actually founded upon a magical truth — *that the only thing needed for magic to occur is some form of structure and an open mind.* This is not in any way at odds with the true nature of idealism. What most people think of as idealism is really the Obscurity of the 11th Shadow that is unable to manifest its dreams. For idealism to manifest in the world, all it needs is a structure to manifest through. But, and it is a big but, first you have to discover what your true ideals and dreams really are.

The moment you begin to entertain the archetypes of the 11th Gift, you open a floodgate of seemingly chaotic imagery in your inner life. If you are not prepared for this and the chain reaction that will inevitably follow, all kinds of delusional problems can result. The power of working with an archetype is that you know beforehand that no matter what you feel or experience, it is a projection of your own inner psyche. For example, if you have an altered experience of becoming the Buddha or experience messianic powers, instead of identifying with this experience you can see it as an alchemical stage in your own psychic process. The danger comes with identification. The entire notion of past lives is based upon such identification, and although it seems perfectly harmless to identify yourself with a past historical character, it actually prevents the archetype from flowing deeper inside you. Archetypes move in fractal patterns, flowing from the past to the future and vice versa. Only in the present are you safe, because the present is the only thing you cannot identify with.

Idealism represents the steady flow of archetypal memory into the world of form. As long as it is allowed free movement, it will manifest your dreams in the world. One of the trickiest things about your dream is that you never know what it will become as it emerges into the world of form. You only know the feeling that it stirs deep inside your heart. Your mind conjures visual images around your dreams and ideals, and this is where potential blockages to the flow arise. You must believe in the power of your dream and at the same time you must give up what it looks like. Every image, archetype or mythical experience you have or see is a part of the river moving from the past towards the future and vice versa. Because of this, the real essence of the 11th Gift is the ability to *play* with the archetypes that pass before you and into your life. This very playfulness loosens your tendency to try to tightly grip your experiences.

The 11th Gift, as you may sense, is a world of magic and fairy tale, but make no mistake about its real power. With the right structure, this energy of idealism *will* manifest great things in your outer life. Through the 11th Gift, all the forms of the natural world swirl within us. This is the land of the tribal totems — creatures infused with particular and potent ancestral power. In every culture, such totems abound. Even in our modern world, we use symbols and animals to represent our businesses and lives. Each one of these symbols carries real power behind it when it resonates with your inner ideal. In the world of the 11th Gift, everything is a symbol of the endless pattern of the great archetypes moving from the world of the formless into the world of form — from the past into the future via the vital artery of the present.

As we humans begin to think once again with our right brains, so we will bring much-needed balance to the world of form. The result will manifest in the world around us as the dominance of the patriarchal forms subsides and the power of the feminine pole comes into balance with the male. This is the true meaning of the time we are living through. It is why so much ancient tribal knowledge is once again pouring into the collective consciousness of our planet. It is through this 11th Gift that the true art of magic is once again returning to our world.

THE 11TH SIDDHI — LIGHT

UPROOTING THE TREE OF KNOWLEDGE OF GOOD AND EVIL

As you follow the archetypal stream of the 11th Gift, your life follows a unique course of self-empowerment that encounters many twists and turns, swells and swoops and challenges to be confronted and ultimately embraced. There comes a time however when the archetypes you are encountering begin to fuse into fewer and fewer images until they become compounded into one single primary archetype. This image or being is the mark of your own nemesis. The primary archetype represents the distillation of all the aspects of your nature, and its appearance is so powerful that it literally brings an end to the process of evolution and growth. For some time, you must sit face to face with this inner daemon, feeling the feelings that it engenders — awe, terror, chaos and love. Whatever shape the mind tries to give to this creature is devoured by the extremely high frequency of the archetype itself. The primary archetype has been given many names by different cultures and creeds; it is the *doppelganger* of Freudian psychology,

It is through this 11th Gift that the true art of magic is once again returning to our world.

*That the light
is only to be
found within
the darkness
is the greatest
of mysteries.*

the *Gnostic Guardian of the Threshold*, or the *Christian devil in the wilderness*. It is a summation of the collective mythic projections of good and evil.

Confronted by this great inner daemon, the separate self begins to dissolve into the archetype itself. Every projection or urge to identify with anything is gradually purged from your psyche. Once the process has ended, true reality manifests within that human being for the first time. It is as though the world has become purified. Behind and through all forms a light shines — a clear pure light — not a physical light, but an intelligence beyond all identification. One enters the mind and body of the Divine. This rare occurrence is the expression of the 11th Siddhi and it is a hallmark of the one siddhic state. In linear terms, this 11th Siddhi represents the result of what happens when all the archetypal memory stored in the collective unconscious has been filtered and expressed through an individual's life. It is as though you have pierced the veil that separates man from his mythic fall. You once again experience the world through the eyes of a child — as though it were the mythic Garden of Eden.

Through the 11th Siddhi you will see the true nature of mind as emptiness or spaciousness, and that mind or no-mind is experienced through the brain as pure light. Light itself is seen as a metaphor within the world of form for the radiance of the Divine presence. The Light of the 11th Siddhi is nothing like the light that our eye perceives, although that is the closest words can come to conveying it to those outside of the siddhic state. With the light comes an ineffable breath of peace that continually breaks like a wave from within every single viewed form. What was viewed as magical from within the Gift level of frequency is now seen as the nature of reality. That the light is only to be found within the darkness is the greatest of mysteries. To the being within this Siddhi, nothing is obscured and no one can hide. Everything is now measured in terms of light and frequency and it is quite clear who is authentic and who is simply pretending.

For the being that attains the 11th Siddhi, light is everything. They appear enrobed in light because that is all they can see. If they become teachers, which is more than likely given their understanding of the mind, they continually reflect their own light but at the same time guide others into their own darkness. A great irony of those who understand the pure light of the 11th Siddhi is that they actually appear to lead people directly away from the light in order to draw them closer to that light. The one who brings this Siddhi into the world brings a vision of the true future of all humanity. In travelling back to the beginning of time in their consciousness, such people see the underlying fulfilment of evolution all around them. Even though time may not evidence this state of Eden, such people live within it all the time, which is their great paradox.

Throughout human history, certain generations are born in the world holding a key to specific evolutionary leaps in the planetary consciousness. Through the medium of certain cyclical astrological alignments, these generations have a high incidence of activation in the 11th Gift. They are the idealists, and when they come into the world, they change the world. The last time this occurred was at the time of the industrial revolution in Great Britain, which set the stage for the modern world we see around us today.

However, at such times, there are always one or two who come into the world and awaken through the 11th Siddhi, bringing a new spiritual vision of our future. Today, coinciding with the writing of this book, another such generation is being born in the world, but one with a different agenda. The future idealists will carry more of the siddhic frequency into the world, laying the foundation for another great revolution, but not a technological revolution. The future revolution will be a transformation in the human spirit itself.

The 11th Siddhi itself has confused many people over the years. In the modern new age, there is a huge rise in people who are actively seeking the light that this Siddhi promises. However, no one who attains this Siddhi does so except through travelling the darkest aspects of their psyche. The movement to embrace the inner light may be paved with good intentions, but so much of it simply emerges through the obscuration of the 11th Shadow, which wants the light without having to embrace the dark. As a result, there is a great Shadow dynamic at work in the world. This is especially true through the medium of the great religions, which seek the light *up there*. The light has never been outside us but resides deep within matter itself. It is a little known paradoxical law that the densest regions conceal the highest vibrations, and that true ascension moves downwards and inwards.

The great challenge in today's world is to bring clarity to the cosmic issue of good and evil. The mass consciousness, operating out of the left side of the brain, favours the light above all else. The feminine principle of darkness, coming through the right side of the brain, has been set against the light. This is why an upsurge in darkness always precedes the breaking through of the true light, as opposed to our projected image of the light. The pure light of the 11th Siddhi has nothing to do with good and evil. It represents the transcendence of duality. You can see how deeply this 11th Gene Key is wound up with the notion of spirituality and religion. You can also see how profound the Christian myth of Eden is with its tree of knowledge of good and evil. As long as we see things in terms of good and evil, we have embraced the fall. In the future, as the great mutation triggered by the 55th Gene Key comes to humanity, the 11th Siddhi will crack open our perception of the world, symbolically uprooting the tree of the knowledge of good and evil. At that point, we will feed directly from the Tree of Life, which streams forth light. This light will become our very food as it feeds our subtle body and slowly causes us to melt into the very heart of creation itself.

One final profound revelation coming from this 11th Gene Key and its Siddhi concerns the consciousness shift connected to the year 2012. When you read the 55th Gene Key, you will come to understand how this date relates to the Great Change that our human species is currently moving through. Known as the time of Melodic Resonance, 2012 brought our planet into geometric alignment with the centre of our galaxy. This event has been predicted and spoken of by many indigenous cultures for many thousands of years. In the great wheeling arc of the 64 Gene Keys, every Gene Key has an exact position in relationship to the cosmos and thus filters the cosmic stream as it enters our planetary system. It is the 11th Gene Key that relates geometrically to the galactic core itself. This is why it is from here through the 11th Siddhi that the core light emerges.

As we moved through and beyond 2012, the pure light from the galactic core has geometrically reach our planet and begun to rapidly transform our entire ecosystem. This will continuously trigger our DNA to release its inner light and eventually catalyse the siddhic states in many people. This in turn will exponentially increase the frequency field of the whole of humanity and Gaia herself.

12th Gene Key

A Pure Heart

SIDDHI
PURITY

GIFT
DISCRIMINATION

SHADOW
VANITY

Programming Partner: 11th Gene Key
Codon Ring: The Ring of Secrets
(The Ring of Trials - 12, 33, 56)

Physiology: Thyroid
Amino Acid: None (Terminator Codon)

THE 12ᵀᴴ SHADOW – VANITY

THE FINAL TRIAL

The 12th Gene Key, along with its Shadow and Siddhi, is one of the more extraordinary and far-reaching archetypes in the human genetic matrix. In the relationship between this knowledge and genetics, each of the 64 Gene Keys has a corresponding chemical family known as a *codon* in the genetic code. In order to decipher the genetic code, scientists have to find chemical marker points in the mass of coded information that lies in DNA and these marker points are known as *start codons* and *stop codons*. Such chemical punctuation marks have an unusual place of importance within the totality of the genetic code itself. This 12th Gene Key, along with the 56th and the 33rd, is related to what science terms the *stop* or *terminator* codons. On a purely archetypal level, the three stop codons — known collectively as the Ring of Trials — can be seen as three great mythic trials that test human beings on their road to self realisation. The 12th Shadow of Vanity marks the inner core of the Ring of Trials, which means to say that this shadow state represents the third and final aspect in this trilogy of human tests.

The 12th Gene Key is special. Within the mystery of the 21 Codon Rings, this 12th Gene Key forms a ring of its own *within* the Ring of Trials known as the Ring of Secrets. However, its secrets remain firmly locked away until you activate its highest frequency in the 12th Siddhi.

Vanity, like pride (the 26th Shadow), follows us to the mountain peaks of consciousness. It is an uncomfortable word for most of us, and in our vanity we do not usually like to associate ourselves with it at all. Unlike pride, which thrives before an audience, vanity is a far more internal Shadow. Vanity is like the lichens that cling to the rocks of the highest of mountain ranges. No matter how far your awareness advances, vanity will cling subtly to you even at the highest of vibrations. In one sense, vanity is the very first human vice, and it is also the last Shadow to let go of you.

The 12th Shadow is the love of your own uniqueness. It is about learning to love yourself, which is the true definition of vanity. However, vanity only stops being vanity when you realise that to love yourself is

*Vanity
can choose
the most
beautiful of
words, but
it can never
hide the
frequency of
its tone.*

actually to love everyone else, a revelation demanding a quantum leap out of your *self* altogether. This 12th Shadow is therefore deeply involved with issues of personal power as well as the human yearning to express the purest qualities latent within your soul. It allows you to progress to great intelligence and artistry, but at the same time prevents you from stepping into your wider heart. Vanity is afraid that if you come from your heart you will lose your power.

Because it involves the expression of the soul or heart, this Gene Key is deeply involved with your ability to connect with the power of your own breath and your emotions. It is connected to the thyroid and parathyroid glands and in particular with the human larynx — the prime organ of speech. It is humanity's vertical larynx that sets us apart from the animals. In esoteric lore, it is said that the animals with horizontal larynges operate under a group spirit, whereas the development of our upright larynx allowed the introduction of the ego. Indeed, it is this 12th Gene Key that allows your thoughts to be translated into language and vibration, giving you the illusion that your words have independent power. Out of this concept of independence were born two powerful human attributes — vanity and ego. In the ancient yoga systems there is a profound connection between the larynx, represented by the throat chakra, and the gonads, representing the sex centre. This is also reflected by the rapid growth of the larynx during puberty. The ancients say that long ago these two centres were in fact one but that over time they separated and the larynx slowly closed. Anatomically the larynx is enfolded by the thyroid and interestingly the Greek word for thyroid, *thyreos*, means *shield*. Similarly, the Dutch word for thyroid, *schildklier*, means *shield gland*, suggesting that the larynx is protected by the thyroid, concealing a great secret.

The great secret of the 12th Shadow is language. It is said that in the Garden of Eden, Adam swallowed the apple, which became stuck in his throat and remains there to this day as the Adam's apple in all men. The Adam's apple represents the masculine principle of the mind that becomes identified with your words, thoughts and actions and the larynx gives you the illusion of power through language. The 12th Shadow is about loving the sound of your own voice, and as such it is the root of language. However, as we will see at the highest levels of frequency, it is not what you say that matters, but the frequency that lies behind your voice. Vanity can choose the most beautiful of words, but it can never hide the frequency of its tone.

It is important to remember here that none of the 64 Shadows are inherently bad. If you declare something as bad or evil, you will miss the gift that lies hidden within it. Vanity is simply the lower frequency of the 12th Gene Key and, after all, vanity is the foundation of the Siddhi of Purity.

Vanity also goes beyond words and can hide in silence. Sometimes your vanity is hidden in what you do not say. It hides itself away in your thoughts and your feelings. Wherever there is self-identification, there is vanity. The reason that vanity is such a great trial for human beings is because it lies outside the scope of your ego. So you may well be thinking, what can I do about it? How do I transform this shadow state? Well, because this shadow is so elusive, it is better not to think about it at all. Even the very thought that you might be overcoming your own vanity leads to more vanity!

All you need to know is that as long as you experience yourself as separate from life — as long as you feel the strength and wonder of your individuality — vanity will still be there, quietly keeping pace with you. It is only late in your evolution, as you approach the highest frequencies of the Siddhis that vanity will finally and suddenly release its hold on you.

Vanity does however have an arch-enemy — which is love. Vanity keeps you from truly loving another because it keeps you isolated. You may be beautiful, intelligent and virtuous, but still with vanity you will remain defended against others. The more evolved your consciousness becomes, the subtler and therefore more powerful your vanity becomes. Vanity is a shield that, along with its programming partner the 11th Shadow of Obscurity, hides the truth from you. For the kind of people who are drawn to reading this material, vanity is one of your great challenges. As you refine your frequency, you naturally fall into the illusion that you are somehow different from others and are becoming purer than the rest of the world. You begin to identify more with your Higher Self, which your lower self takes great delight in! This is a most tricky time in your spiritual evolution because it is so easy for you to remain here, at a relatively high frequency. You feel powerful, individuated, wise and well meaning. However, you have yet to make the greatest leap there is — the leap into true purity — the leap into your own death.

REPRESSIVE NATURE – ELITIST

There are essentially two different types of vanity — coarse vanity and refined vanity. The repressive nature of the 12th Shadow is the refined version, which emerges in certain characters as elitism. Elitism is vanity working undercover. These characters may agree with you outwardly but inwardly feel very differently from what they say. Often they refrain from making any comment at all, preferring to remain detached. This is the domain of the *spiritually evolved* — those few who have *done a lot of work on themselves*. These people inwardly feel that they are clearer than most others around them. They can take great pride in being different, or in being beyond any creed or system. Such vanity ensures that such a person cannot make the leap that their inner being most longs for — the leap into permanent higher consciousness. This only begins when the awareness of our own vanity finally dawns.

REACTIVE NATURE – MALICIOUS

Malice stems from anger, which in turn stems from fear. The reactive nature of vanity can use the gift of this Gene Key, the Gift of Discrimination, as a means of hurting others. Whereas the elitist withdraws into silence out of fear of being seen as weak, these people openly use their substantial vocal gifts to inflict pain on others. Like all classic victim patterns, such people usually feel put upon in some way and react maliciously without thinking about the damage their words or actions might do. The 12th Gene Key has beneath it real emotional power — and it has a god-given talent for language and communication. These people really know how to push other people's buttons through the power of their voice. They can hurt people like no one else can. Their malice may not be premeditated, but it is usually brutal and ends in disaster for them.

THE 12TH GIFT – DISCRIMINATION

THE SECRETS OF HIGH ART

*Discrimination
gives humanity
a taste of a
higher order
operating
behind the scenes
of life.*

Discrimination may not sound like much of a gift, but when you truly understand the 12th Gene Key, you will see that it has a great power. To discriminate is to know inherently what and who is healthy for you in life. The energy of vanity is simply self-destructive unless it can be put to good use. Discrimination is exactly that. You take your vanity — the urge to be better or somehow purer than others — and you turn it into art. The 12th Gift is deeply associated with the arts — with music, with language, with dance, with drama and above all, with Love. The love within the 12th Gift is not universal love (as in the 25th Gene Key) — it is all about *falling* in love. This is human love with all its attendant drama, obsession, beauty and danger. Vanity is about loving only yourself, whereas discrimination is ultimately about loving the things and people outside yourself that make you feel good.

This 12th Gift is about feelings. If this Gift is a powerful aspect of your Hologenetic Profile, you will be strongly motivated and moved by feelings and emotions throughout your life. Your Gift is to communicate these feelings to others, and you may do this in a myriad of ways. If you are strongly influenced by this Gift you will recognise the beauty of true expression, which means that you will also recognise when something or someone is *not* expressing true soul. This can make you one of the best critics of others. However, this Gift is not about criticising flaws and details in others (which is a lower expression of the 18th Gene Key) — it is designed to spotlight that which is not authentic. Discrimination is about being attuned to a higher frequency, which means that it can metaphorically see through walls. Whenever a person is faking or has a concealed agenda, one with the Gift of Discrimination will immediately feel it as a deep discomfort. If they do not fall in love with something, they distrust it, and the same goes for the people in their lives. To such a person, authenticity is everything.

People with the 12th Gift cannot be taken in by the enchanters or idealists in the world. They have a deep respect for purity, and it is rooted in a natural prudence. The programming partner of this 12th Gift is the 11th Gift of Idealism, which means that they too are idealists, but they understand that idealism requires the balance of pragmatism and discrimination other-wise it is nothing more than a pipe dream. The Gift of Discrimination does set you apart from the crowd — it has no choice in doing so because it is naturally seeking a higher frequency. It represents an aspect of your DNA that is constantly striving for something higher and purer, which means that it directly challenges anything or anyone that is influenced by compromise. Discrimination gives humanity a taste of a higher order operating behind the scenes of life. This is why it so often manifests through true art. They are true art lovers. The 12th Gift does not shy away from anything that is authentic — no matter how messy it may be. These people are the great food discriminators, music discriminators, and language discriminators. They can become the great artists, virtuosos, poets, actors and educators of humanity. Their Gift is to enter unafraid into the drama of life, allowing it to flow through their veins and be expressed through their feelings.

Because of the depth of feeling that this Gift entails, it tells us something profound about the direction of the human species. We are here to learn to express the deepest yearnings and feelings within our souls. This is why we have to master language and the arts — because they are the transformational field through which we can transcend emotion and touch the higher planes. Out of this 12th Gift come humanity's great educators — those rare people who can allow art to touch their hearts and at the same time transmit that essence to others through their language and expression. Wherever you see true passion moving into the world, there you are seeing the influence of this 12th Gift. It is both sweeping and consuming yet at the same time highly refined. Ultimately these people are driven by the myth of true love — that is what they are yearning for deep within their souls, and it is why their acts and words, at their highest, are a reflection of the beauty and anguish of this human yearning.

Because of its connection with the thyroid system, the 12th Gift contains the great teachings of transformation and death. All high art contains the same codes that come through this 12th Gift — that life is about transformation, and death is the symbolic movement from one stage of consciousness to another. These truths have always been encoded in the great tragedies and comedies down the ages, and it is through your emotional nature that such truths are imbibed and passed on. It is the thyroid system that controls your metabolism and has such great influence on your general energy, mood and breathing patterns. When you laugh or cry you enter the sacred realm of transformation. It is through your laughter and your tears that the transcendent enters your body in order to alter your chemistry and the pattern of your breathing. Of all the Gene Keys, this 12th Gift represents the mythic passage from one state to a higher state in which the stop codon kills your identification with your past, allowing you to be transformed into something permanently and radically different.

THE 12TH SIDDHI – PURITY

SWALLOWED BY THE VOID

The human yearning for true love is in reality a lower frequency of a permanent state existing at a higher level of consciousness. This state, called many names by many cultures, is essentially your pure nature, untarnished by human desire and beyond the dualistic mechanism of the mind. Only the higher mind, which is another expression for our heart, can begin to understand true Purity, the 12th Siddhi.

Vanity and purity are mirrors at the two ends of the spectrum of human consciousness. With vanity your lower self falls in love with itself, and with purity the Higher Self falls in love with itself. You might say that purity is when the Divine falls in love with *You*. This can only happen when you come within the sphere of Divine love. Your behaviour, your thoughts, your feelings, the very air you breathe must resonate to one purpose — what the Sufis call *falling in love with the Beloved*. The Beloved is not something *out there* — it is the essence of your true nature, and you do *fall* into it because it resides deep within you.

Vanity, as we have seen, pursues you right to the end of your journey. Even when you are living from the higher frequency of your Gift, vanity is still there. Only when you have attained the highest peak of consciousness does the mystical event occur — you give up everything you have attained. You have to court annihilation. This is the final of the three great trials in The Ring of Trials. An opening from above and below then occurs simultaneously and you enter the hallowed sanctuary of the innermost of the Codon Rings — the Ring of Secrets. But for this miracle to occur, your vessel must be absolutely pure and flawless. Purity is a much misunderstood term. In human language it can be applied as an adjective to almost anything, but at its highest frequency it can only be applied to one word — the heart. When your heart finally remembers its own original purity, only then will you finally and willingly give up your existence.

Everything in the universe has at its core the same original purity — we are all shards of a Divine crystal, and as our forms go through their processes of polishing, so consciousness begins to remember itself through us. Even the most evil of beings has at its core a shining pure heart, which means that in truth, there is no evil. There are only progressive levels of density. This is the great secret you come to embody here in the Ring of Secrets. In the ancient alchemical systems, the throat centre was seen as the greatest of all initiations. In the Indian chakra system, the throat chakra known as *vishuddha* is the clearing house for the higher consciousness. All the lower chakras including the heart are synthesised and purified in the throat. In this respect it represents the boundary of the known and the unknown. Likewise in the Jewish Kabbalah, the throat is symbolised by an invisible sphere known as *daath*, the abyss. This abyss must be crossed in order for higher consciousness to dawn, and the crossing of death is a symbolic letting go of all your hard-earned knowledge. It is the ultimate purification in which you meet your own death and are reborn in a higher sphere.

The being who has passed through this abyss and entered the sphere of the 12th Siddhi becomes once again like a child. Through their heart, they can perceive the Divine — beyond longing, beyond concepts yet still deeply human and with a voice beyond words. Others feel such people truly do not belong in this world, even though they are the most natural expression of what it means to be human. In this state, nothing can contaminate their purity. Their bodies can be decrepit, even ugly, but their hearts can do nothing but sing out with the Truth of their true nature. Those who manifest this Siddhi often live the humblest of lives, unseen by the world at large. They often pass quietly through the world, living simply, yet reminding everyone who meets them that purity really can exist in a human form.

Those who manifest this Siddhi often live the humblest of lives, unseen by the world at large.

If you wish to play with the frequency of the 12th Siddhi, you need only keep reminding yourself of your own heart. Beneath the layers of karma, ancestral fear and inevitable childhood conditioning beats an aspect of the great universal Heart — and its purity can be remembered. Its colour is the white beyond white for it is the eternal child within you. It is the You that you cannot help but fall in love with. To look at the world from the 12th Siddhi is to see everyone through this crystalline vessel — it is all you can see in everyone you meet. However the very moment you view anything in a negative way, this presence within you will instantly vanish.

In language purity becomes poetry. In thought, purity becomes essence. When we combine language and thought, we have the master codes for ascending beyond the mental plane. The actual words for each of the 64 Siddhis are not really words in the normal sense — they are doorways to realms of higher frequency. The word *Purity* has an essence that is onomatopoeic at the level of vibration. In other words, if you sound the word *Purity* within your heart and mind over and over, you will actually feel its quality residing within your own heart. This is not about affirmation. You cannot simply do a technique and feel it. You have to be ready in your own heart — you have to love the word and all it stands for in order to feel this miracle. Words used poetically and vibrated within the chambers of the heart have the power to pierce through the layers of fear that envelop the hearts of others.

Purified thought has an even more powerful effect. Spoken language uses sound which restricts its impact to our solar system (sound waves eventually dissipate), but language cloaked in thought travels at an almost inconceivable speed and literally rebounds off the walls of our universe (only pure love can rupture the walls of a universe — see the 25th Gene Key). Therefore a pure thought will affect all levels of creation and all beings within creation almost instantaneously. A pure thought is like a sugar lump dropped into a teacup. Within a very short time, it permeates everything. At a certain level, purity of thinking will take you to the edge of hyperspace. The more you allow your thoughts to be imbued with the essence of divinity, the more your whole being reaches a kind of *escape velocity*. At a certain level of frequency, you simply dissolve into the great teacup! You transcend the mind by becoming one with mind. The paradox here is that in becoming mind, mind ceases to exist. This is the true symbolism of the genetic stop codon — it brings an end to your own individual existence, revealing vanity for what it truly is — the illusion that you are separate. You are so pure that it is not possible for you to exist as anything other than existence itself. In crossing the great cosmic throat centre, you are swallowed by life itself.

13th Gene Key

Listening through Love

SIDDHI
EMPATHY

GIFT
DISCERNMENT

SHADOW
DISCORD

Programming Partner: 7th Gene Key
Codon Ring: The Ring of Purification
   (13, 30)

Physiology: Amygdala
Amino Acid: Glutamine

THE 13TH SHADOW – DISCORD

THE CHEMISTRY OF PESSIMISM

The 13th Gene Key concerns a single theme — the theme of listening. Through this Gene Key we will see how many dimensions there are to the art of listening and how deeply tied it is with the expansion or contraction of human consciousness. At the Shadow frequency this is the Shadow of Discord, which is the inability to listen to and learn from your experiences in the world. Listening is entirely different from hearing. Hearing refers to the acoustic absorption of auditory information, whereas listening is something that can only be done with your whole being. Often listening requires withdrawal and time in order for it to function effectively. Listening is also highly linked to the way in which you process your life experiences on an emotional level. The link between listening and the emotions has profound implications for the future of this Gene Key and in particular its Shadow frequency. Because of the global mutation currently taking place in the solar plexus system of all human beings, our emotional chemistry is undergoing some extraordinary changes and the 13th Shadow will be affected by these changes.

Along with its programming partner, the 7th Shadow of Division, this genetic partnership exerts an enormous influence on the direction of humanity as a species. These two Gene Keys are prime programming agents for the way in which humans interact at a group level. They actually cut far deeper than the tribal programming archetypes within our genome, which also affect our interactive capacities. The 7th and 13th Gene Keys steer the one consciousness of humanity along the line of our destiny. Whilst the 7th Gene Key pulls you towards the future, the 13th urges you to listen to and learn from your past. This archetypal placement within your DNA makes these two Gene Keys different from all others, as though they were somehow outside the scope of human influence. It is the battle waged within the frequencies of these codes that decides your future. The 13th Gene Key in particular is one of the most important of all the 64 Gene Keys, since it involves the way in which you process your past.

Discord refers to the inability to escape your own past. Stacked within the human genome lies a library of collective human experience, and your inability to process all this memory is what keeps you locked into the same self-destructive patterns. The 13th Shadow shares a very important chemical connection to the 30th Shadow, through its link with the same genetic codon signature known as *The Ring of Purification*. The 30th Shadow — the Shadow of Desire — is where the connection between your ability to listen is weighed against the raw force of your human desire. This codon, which codes for the amino acid known as glutamine, is one of the great human genetic battlefields. Interestingly enough, there is now a large body of scientific evidence linking this amino acid to various functions and malfunctions in our gut. Symbolically, one might draw a connection between how effectively we human beings process our past and how effectively our body eliminates waste. The force of human desire found in the 30th Shadow usually outweighs our ability to listen to our past experience, and this leads us once again down roads that do not serve humanity as a whole.

The problem is rooted in the human emotional system and this 30th Shadow of Desire is the core of it. Because desire cannot be sated in its current form, it influences the direction of the whole human race. Despite what has happened in our past, we go on making the same poor decisions and judgments. This sets up a global frequency of discord in which we can clearly see where we have been going wrong, but are unable to remedy it at a collective level. An example of this is the current threat of global warming. We can see what damage our lifestyle is doing to the long-term future of our planet, but our desire outstrips the energy to do anything about it. Anyone who has a deep knowledge of human history will see the same cycles being repeated over and over but each time with different nuances. It is true that today there is more global awareness of the problem than ever before. However, this awareness does not change human behaviour. We hear the discord that we are creating, but we do not listen to it. In the end, our emotional urge to satiate the desire within always wins. That sums up the dilemma of the 13th Shadow.

We hear only discord in the world because we are unable to listen to the overall dilemma.

The ancient Chinese name for this 13th hexagram of the I Ching is *The Fellowship of Man*. It is a beautiful name, seemingly filled with hope, and it does in fact describe the higher frequencies of this 13th Gene Key. However at the lower frequencies, because of our inability to listen, we go on pursuing our own short-term agendas with disastrous consequences to the greater society. Ironically we find no solace from our own desires either. No matter how many of our desires we sate, more appear. We hear only discord in the world because we are unable to listen to the overall dilemma. We cannot yet see it at a collective level, so we are not ready to find out what the Fellowship of Man really is. This is why we keep programming our future direction (through the 7th Shadow) as a future of division. Not listening to the real problem, all our actions create more and more divisions within our societies.

Out of all this ancestral memory and our inability to create or find the *Fellowship of Man*, a huge pessimism emerges at an unconscious level. We no longer believe collectively that we can surmount our own nature and create a truly peaceful world. History provides compelling evidence of this. In fact our pessimism emerges directly from our genes because, shockingly, our pessimism is in actuality correctly founded — we *cannot* surmount our own nature. Only nature can overcome us, which is precisely what lies in the pipeline right now. Nature is preparing a new kind of human being. Nature as a whole has learned and listened to her past experience, and through us, she will create the giant leap that is needed to free us from a web that we have no hope of escaping. Nature will create a human in whom the art of listening will no longer have to compete with raw desire. In a single unprecedented and radical sweep, she will do this by killing off desire itself.

REPRESSIVE NATURE – PERMISSIVE

These people pretend to be empathic and sympathetic to others whilst at the same time doing nothing. The repressive side of this Shadow may make a show of listening to everyone but it soon becomes apparent that they have no real backbone. These people are permissive in the sense that they let others walk all over them without learning anything. They will agree with you no matter what you say. They mistake hearing for listening and in doing so they cut themselves off emotionally from others and from their environment. This is one of the deepest forms of emotional repression that does not want to take in the pain/pleasure cycle at any level, so it sacrifices both extremes and settles for a false sense of safety.

REACTIVE NATURE – NARROW-MINDED

When discord manifests through a reactive being, it becomes narrow-mindedness and/or bigotry. These people will *disagree* with you no matter what you say! Narrow-mindedness is about getting stuck in a reactionary emotional pattern and then making a lifestyle out of it. These people are literally unable to see beyond the limits of their own desires and they are filled with pessimism. Their philosophy is founded upon the fear patterns that have guided all human beings to where they are now, and they are not open to the possibility of real change. Such people harbour a deep bitterness about the nature of human beings and this comes out most frequently as anger, especially at those who do not see things their way. With such people it often becomes their mission in life to debunk the views of those who are optimistic about the future and about humanity.

THE 13TH GIFT – DISCERNMENT

THE FELLOWSHIP OF MAN

Discernment emerges as your emotional nature comes more and more into awareness. As you realise how deeply in the thrall of desire you are, you begin to understand the whole of humanity as well. Out of this incredible swelling of awareness arises the 13th Gift of Discernment. Discernment begins at an individual level as you come to see how deeply your view of others is connected to your feelings. Only when personal feelings are witnessed and examined can you begin to see things in a more objective manner. Over time, your personal agendas become more lucid and you reclaim the ability to listen to others and the world from a broader perspective.

At this level of frequency, you become fully aware of the desires coursing through your body and although you cannot do anything to stop them, you are no longer the victim of them. For the first time you see your own individuality clearly and you realise that something else exists behind that individuality — a kind of witnessing awareness that is greater than your sense of individuality. This is the birth of your ability to listen.

With the rising of your discernment comes many gifts. It may appear to you now as though a great veil has been drawn back from the world. As we have seen, the 13th Gene Key is a repository of the collective past experiences and memories of humanity. Instead of reacting out of this karmic library, you now begin to demystify your own past. What you may have thought of as a random series of experiences can be seen as following a discernible pattern. Without being caught in the drama of subjective emotional personification, you can begin to view life at a mythic level. This mythical thinking is the way in which the mind interprets your ability to listen behind the veil of your desire. To view life at a mythic level is to see the great archetypes playing out in our lives — both our personal lives and the greater collective history of man. Through understanding your own past in this way, you can see that the storylines of all individuals play out variations of the same themes. At this level, we can recognise these same archetypal themes in every culture's rituals, fairy tales, legends and myths.

To view life at a mythic level is to see the great archetypes playing out in our lives.

Once you stop viewing life on a subjective emotional level, a new landscape of feeling opens up inside you. This feeling is the early manifestation of your future ability to transcend your emotional system — and it is experienced as optimism. Optimism is the gradual swelling of awareness within your solar plexus system transferred and projected outward onto the world. One might also view it in reverse. The awareness connecting all life forms is received and drawn into your solar plexus system, thereby expanding your sense of Self to include a far wider reality — the true *Fellowship of Man*. Through our cultural heritage, whatever it happens to be, we are all embedded with the myths, stories and archetypes that we need to eventually transcend our own emotional dramas. All the world's stories, rituals and beliefs have grown out of the very infrastructure of our DNA, which is why the same patterns emerge over and over throughout history, regardless of culture, geography or even isolation.

These collective myths and fairy tales contain the alchemical codes of evolution, including its various mutations. The stories always pass through dark periods of transition ultimately arriving at transcendent states of consciousness. This is why discernment leads to optimism. With your Gift of Discernment you don't just see the symbols. You live them in your own life, which is why they bring you such optimism. Every character within the myths is an aspect of the world psyche and every circumstance they pass through is a part of our own genetic and spiritual evolution. All of this is particularly relevant today because we are passing through a mythic test as an entire species, which is perhaps the first time this has ever happened. Those with discernment know in their bellies that this phase of our history must and will lead to a transcendent leap in consciousness, regardless of the subjective interpretations of the fear-based culture in which we live.

THE 13ᵀᴴ SIDDHI – EMPATHY

THE GREAT COSMIC HUB

The 13ᵗʰ Siddhi, like its partner the 7ᵗʰ Siddhi, has much secret wisdom hidden within it. Our DNA is an incredible substance — it contains all the memories of our entire species and even those earlier life forms from which we descended. Like a fractal shard of God itself, DNA forms the living, vibrating link to the beginning of life on this planet. Even more profoundly, it connects us to the seed from which our very universe first sprouted. It is through this 13ᵗʰ Siddhi that information from our collective past can be recovered. Like a kind of cosmic librarian it has all the access codes to every volume every written on the pages of life. We are not just talking information here, but experience. The 13ᵗʰ Siddhi can unlock the essence of what it is to experience being a wild panther, or a strawberry, or a mollusc. This is the power of empathy.

These days most people use the word empathy without grasping its deeper connotations. As a fairly modern word, it is almost always confused with sympathy, a much older word coined by Aristotle. In the language of the 64 Gene Keys, empathy represents a siddhic vibration and as such cannot be grasped by the mind because true empathy has nothing whatsoever to do with the mind. The prefix *em* means *within* whereas the prefix *sym* means *with*. Herein lies the key to understanding the word. Whereas sympathy means to feel *with* another person, empathy means to be *inside* that person. Most people cannot conceive of being inside another person so the term has come to be seen as metaphorical and is generally understood as being a mental or emotional projection onto or into another person. Let us reclaim then the true power of this simple word. The Siddhi of Empathy requires complete dissolution of the individuated self. Once identification with one's form has been cut, empathy reveals itself as the background awareness of all living forms. In other words, we really are *inside* each other. Whereas sympathy requires two, empathy demands one.

We have seen that the foundation of this 13ᵗʰ Gene Key is based on listening. Indeed, at the level of the 13ᵗʰ Siddhi, listening becomes all there is. Like a black hole, this listening absorbs everything that moves into it — space, time, everything. Eventually listening becomes so total that it fuses together subject and object so that the concept of listening itself disappears. This gives a small hint of what the Siddhi of Empathy really means. It is interesting to note how vehemently mass consciousness has always feared the number thirteen. In traditional folklore, thirteen gained a bad name through the system of tarot, in which it stood for the Death Card. These ancient esoteric codes are always founded in an original truth. Empathy is indeed a death — the death of the separate self that reveals the underlying nature of humanity. Empathy is the truest sense organ of humanity, but it cannot function through the individual — only through the collective.

The 13ᵗʰ Siddhi and the 7ᵗʰ Siddhi program each other through time. The 13ᵗʰ, being the repository of the past, programs the future of humanity and the 7ᵗʰ programs the past. This last statement may appear to make no sense, but at the siddhic level it is a truism. At the siddhic level, the seed and the fruit exist within each other because of each other.

Whereas sympathy requires two, empathy demands one.

We think of the future being uncertain whereas the past is in fact just as uncertain because of how it is remembered. This is the power of myth — it can take an event or series of events in time and make them immortal. To be empathic is to have access to the egg out of which all the great myths have sprung. Empathy alone goes further and deeper than myth because it reunites the opposites and brings an end to the stories themselves. Empathy is a state of consciousness outside of time, even though it enfolds time. Again this is symbolised by the mystery of the number thirteen, which stands for the hub of the great wheel of twelve, the mystic number of the grail at the centre of the Round Table, or the Christ encircled by his twelve disciples. It is the number of Arachne, the spider, the mystical thirteenth sign of the zodiac. She sits at the heart of the web of creation and through her legs she remains in constant empathy with all that is.

There is huge fear to be absorbed before you attain a siddhic state. In the case of the 13th Siddhi, this involves complete absorption in the universal discord of the 13th Shadow. The human interpretation of this fear is the experience of feeling cut off from all that is alive and warm. It is an ice-cold zone in which aloneness is experienced as agony. However, through the purging of this experience the 13th Siddhi opens its eye on a new world — a world in which the past has been magically transformed into a radiant new future. We might recall here that this Gene Key is a part of *The Ring of Purification*. The human addiction to desire comes to an end in the siddhic state of consciousness, and it is said that one's spirit shall no longer return to the earth but take a birth in a higher sphere of existence. These are mythic rather than literal interpretations of what occurs to such a person. All the great esoteric interpretations — karma, rebirth and justice — must be mythic rather than literal, since the true siddhic state cannot be conveyed through words. In universal empathy, you enter the centre of the human family and become one with all that is.

Those who awaken to the truth of their consciousness through this Siddhi will become beacons that live on in our history. These people have learned the deepest art of all — the art of listening. To sit in the presence of such a person is to be healed through being listened to. Listening at this level is no longer a passive phenomenon. It is actually quite a forceful cleansing at an auric level. To be listened to by the great cosmic mother is to have all your fears brought to the surface and loved into oblivion. These people are the most powerful of all shamans — they can open the portal of the past within you so widely that you become aware of yourself as a great cosmic receptacle rather than a point of focused identity. They can show you that all creatures and beings dwell within you, and that if you open yourself widely enough to our collective past, you will also see our collective future. To such a being, the holy state of perfection known as the Fellowship of Man is a constant and manifest reality.

14th Gene Key

Radiating Prosperity

SIDDHI
BOUNTEOUSNESS

GIFT
COMPETENCE

SHADOW
COMPROMISE

Programming Partner: 8th Gene Key
Codon Ring: The Ring of Fire
  (1, 14)

Physiology: Small Intestine
Amino Acid: Lysine

THE 14TH SHADOW – COMPROMISE

THE LEAKING WAGON

In the Chinese I Ching, the word for the 14th hexagram is usually translated by the phrase *Possession in Great Measure* and it is traditionally symbolised by a large wagon laden down with goods. It is a symbol of wealth, health and prosperity, which are all themes linked to good fortune as well as hard work. This 14th Gene Key and its Shadow really concern the way in which humans work. It concerns the actual choice of work you do, the people you work with and above all, the *way* you work.

Every human being has a built-in genetic need to work. Interestingly, the very word *work* has become associated in the English language with the concept of effort, which shows just how deeply the collective Shadow of this 14th Gene Key has descended into our lives. Work can actually mean many different things. It all depends upon your attitude toward life. The source of the 14th Shadow is compromise, and compromise has become the norm for the majority of people on this planet. Compromise has become so embedded in our psyches that we don't even realise that we are doing it anymore.

Compromise is the by-product of living without a sense of personal freedom. It implies a lack of imagination and an inability to believe in the power of your own uniqueness and individuality. The 14th Shadow is paired with the 8th Shadow of Mediocrity, and thus we humans live, sandwiched between these two low frequency victim states, unable to envision a way out of our uninspiring life situations. Even those of us who can see a way out of our dissatisfaction rarely have the courage to actually follow through with our dreams because of a deep-seated fear that we lack the power and capacity to complete the long journey toward those dreams.

Compromise begins deep inside us. It begins when we are very young, as we inherit the world Shadows from our parents and our schooling. Many people dream and aspire to greatness when they are young, but most have given up on their dreams by the time they reach their forties. Many give up much earlier, and some of us never even get off the ground. Most young children growing up in today's world dream of being film stars or footballers or famous singers. And most adults see these dreams as passing phases in a young person's life. However, these young people are unconsciously projecting their own inherent urge to excel in an area of life. This kind of early dreamy aspiration actually burns at a very high frequency and if it can be maintained, harnessed and given direction, it will eventually lead these children in the direction of their excellence.

The sad truth is that the dreams of most children are stamped out by their schools in the endless monotony of the set educational curricula. It is in school that most children learn to associate work with boredom, effort and drudgery. The problem with such modern school systems is that they tend to homogenise children, treating them as one collective body that needs to be educated, instead of treating every individual differently. Many children simply do not suit schools at all. Added to this, if one's parents do not believe in themselves, then it will be very difficult for them to creatively inspire their children. It is small wonder then that we humans cannot sustain the enthusiasm of our dreams long into our adulthood, ironically at a time when we need that enthusiasm the most.

Whenever you compromise you are settling for second best, and so you can never fully enjoy what you do in life.

Compromise is bred into us through society's systems. Whenever you compromise you are settling for second best, and so you can never fully enjoy what you do in life. If you do not enjoy it you can never excel in it. Enjoyment and enthusiasm are the fuel in the turbine that drives you to excel. All children are born with a certain genius, and if allowed to develop in the right direction that genius will inevitably emerge and the work that they do will inspire others to achieve the same high standards. Genius itself is highly contagious, just as compromise is. Because genius is so contagious and empowering, individuals can truly change the collective frequency of the whole of humanity.

Compromise is often a very subtle thing at the beginning. It is not really about your action but about your spirit. If you do something that you don't really like but it is a stepping-stone towards your dreams then that will transform the way in which you do it. If however, you do something that you don't like because society has in some way pressured you to do it, then you are on a slippery slope. That kind of compromise can easily become a habit, and in the end it will sap your spirit and draw you further away from your true potential. You need to keep your enthusiasm alive by seeing what you do as a part of your overall direction in life. This is how you can bring meaning into even the dullest of jobs.

The 64 Gene Keys contain the genetic codes for all possible types of genius. This is what the 64 Gifts are — they form the collective matrix for true human genius. It is of note that the two words *genus* and *genius* are so closely related to each other, since genus encapsulates the entire human species and genius likewise is a hereditary trait that appears across all gene pools. However, contrary to widespread belief, genius is neither special nor rare. It is present as a seed in every single human being, just by dint of being born. The misunderstanding about genius concerns its definition, which has been narrowed down by mainstream use to refer mostly to the faculty of intellect. However, the etymology of the word genius points towards a far more mysterious faculty — a kind of guiding spirit or *daemon* within your life. Such a spirit might be seen to be your *higher self* — your Hologenetic Profile operating through you at a much higher frequency.

Genius denotes great creativity, originality and enthusiasm but it is in no way limited to people of high intellectual capacity. This 14th Gene Key forms a genetic codon group called *The Ring of Fire*, which suggests that genius is a spark that needs to be fanned into action. The other chemical component of this Ring of Fire is the 1st Gene Key with its Gift of Freshness. Genius is therefore intimately connected to this idea of freshness and newness. The greatest damage done to individuals in our world is the continual programming by the low frequency collective consciousness that forces people to compromise and give up on their dreams. The spark of genius is there at birth, and if recognised at an early age, the child can lead a childhood individually tailored to fan the flames of that particular genius.

Finally, compromise cannot bring you prosperity because it makes you a follower rather than a creator. True prosperity emerges from doing what we are here to do rather than being simply another cog in a collective machine. Prosperity is the by-product of individual creative endeavour and requires a sustained sense of empowered direction. The moment you make a compromise, your wagon begins to leak energy. The riches that you were born with begin to decay. The very frequency of compromise negates the possibility of good fortune and synchronicity — it is a no-man's land where nothing of beauty can ever occur, and no trace of inherent genius and purpose can be realised.

REPRESSIVE NATURE – IMPOTENT

For those who are unconsciously dominated by fear, the 14th Shadow represents an almost inescapable dilemma. The more one compromises in life, the more powerless one feels to escape. A great deal of our actual vital life energy is held within this 14th Gene Key, and when we do not release this energy through work that we find fulfilling, our power becomes choked up inside us. The word impotent has the double connotation of applying to our sexuality, which is also deeply connected to this 14th Gene Key. Sexual vibrancy and fertility in part rises and falls upon how much fulfilment we get out of our lives. Making compromises out of fear will actually deplete our fertility. Impotence does not necessarily mean that one appears weak. Repressed natures often do a good job of hiding their weakness by going along with whatever happens. True impotence in this respect is about not having the courage to stand alone and pursue one's unique path.

REACTIVE NATURE – ENSLAVED

The reactive nature of this Shadow is also based upon basic insecurity, but instead of collapsing into impotence, these people react by trying to prove themselves. This is a classic pattern in the world today — we see people working hard in jobs that are not really right for them in an attempt to prove to themselves and others that they are special. These people are enslaved to their own need for recognition. Ironically, even if they receive recognition, it doesn't fulfil them because they have never unleashed their true potential in life. Such people are merely pretending to be powerful in life. If you provoke them, they will soon show their insecurity through their rage. True power never needs to prove itself to others or the world. True power is only concerned with the job at hand.

THE 14TH GIFT – COMPETENCE

THE FIRE IN THE BELLY

As we have seen, within every human being lies a latent genius, and this genius emerges when you stop making compromises in life. Competence is the quality that accompanies a person who loves what they are doing. It is one of those gifts that cannot be taught because true competence hints at much more than simply doing things well. Competence carries within it efficiency, enthusiasm, flair and flexibility — the four keys to material success. It is efficient because it finds the quickest and simplest solutions to any obstacles along its path; it is enthusiastic because it is deeply fulfilled by what it does; it has flair because it is doing something in a manner that no one else can; and it is flexible because it can be applied to any sphere of human enterprise imaginable.

To be competent means to be able to think both logically and laterally whenever needed. The key to efficiency is the ability to switch spontaneously between creativity and receptivity, which means that it is self-adjusting. At the same time as acting, this Gift is simultaneously listening and responding to its environment. Inside your body, this awareness corresponds to the solar plexus area, which is the biophysical zone that dictates how centred you feel at any given moment. To feel centred is to move and breathe from the belly, and for centuries many cultures have understood this truth. The whole notion of the *solar plexus* and the *fire in the belly* comes from a profound understanding of the power of this region of our body.

Competence carries within it efficiency, enthusiasm, flair and flexibility — the four keys to material success.

The navel is seen symbolically as the font of human power and fertility, and the 14th Gift harnesses this vast internal power. The secret of the belly is about *drawing* rather than pushing. A good analogy of this can be found in the differing techniques used in carpentry around the world. In the Orient, the saw is pulled towards the belly rather than being forced away as is the western style. The resulting cut is finer than one made through pushing. It also uses far less energy to pull the saw towards the belly than it does to push it away with your muscles. The only advantage of the western style is that it is faster, but in the long run, quality is always a more economical investment than quantity. Competence in this sense is about performing a task with complete precision and elegance, in alignment with natural and harmonious rhythms.

The 14th Gift of Competence is one of those gifts that is extremely contagious. As noted, one of its manifestations is enthusiasm, which is such an essential quality within any group enterprise or business. Indeed, this 14th Gift is the main ingredient of any successful business venture or team. The enthusiasm that comes from this gift is the binding force that creates cohesion within groups. Those who have the 14th Gift strongly activated in their Hologenetic Profile are often the driving force that can implement a particular vision or idea. Theirs is the gift of rousing the team spirit that is necessary to get an idea off the paper and into the world. It is also interesting to note that a competent team cannot contain members who compromise. All team members must be joined together in mutual enthusiasm for the same ideal.

The programming partner of the 14th Gift is the 8th Gift of Style. Competence also has a unique approach to everything. To be competent in something is to innovate new and exciting ways of tackling things on the material plane. The 14th Gift also respects the unique and individual style and input of others. These people have real flair. They are not afraid of doing things in their own unique way, even though it may never have been seen or done before. At the same time, these people do not reject the advice of others if it helps in their task, making them powerful team players. If this 14th Gift is a part of your Hologenetic Profile or you simply feel drawn to it, you are probably someone who will work best within a small team. Whether you are the leader of the team does not matter — the only thing that matters is that your unique capacities are respected. These are people with an enormous capacity to work at something if they really love what they are involved in and those with whom they are working. In this respect, they are natural teachers of others, since they imbue people with a sense of their own independence and confidence.

The 14th Gift also has another latent power — the power of attraction. Competence creates a powerful magnetic field that not only draws in the right support but also has the ability to magnetise material wealth. Once you have attained the Gift frequency, the tendency is for consciousness to keep expanding. This is the origin of the many folk sayings about money and wealth — for example: "It takes wealth to become wealthy" or "Wealth begets wealth." Prosperity is an infectious energy field that grows exponentially as long as it is used in the service of empowerment. Competence is unafraid of success and power, and radiates confidence and assuredness wherever it goes. It is a force field that naturally creates prosperity. It cannot do otherwise because it has so much creative power within it. It has only to find the right outlet in the world.

Finally, the 14th Gift has great flexibility. You cannot be rigid and competent at the same time. Competence is the working aura of someone who loves the material work they are involved in — whether that be in the office or at home in the family. Mothers, in particular, who have the 14th Gift can create all the dynamics within a family for spiritual, emotional and material prosperity. As the backbone of the family, their powerful spirit will infuse and empower each child as they grow, as well as inspire strength and purpose in their partners or husbands. In fact, the moment you activate the 14th Gift, you release a certain aroma through your DNA that soon makes you the backbone and mainstay of any family or team. Moreover, your aura of confidence and capability can be easily transferred in any direction. This flexibility is not the same as having skill. It is a fearless openness to grapple with any task in a realistic and focused way. If you do not know how to do something, you learn, and once you have learned it, you can apply it to other areas of life. In this way, the 14th Gift continually extends its potential in many different directions simultaneously.

THE 14TH SIDDHI — BOUNTEOUSNESS

THE SINGLE-CELLED HEAVEN

At the siddhic level of consciousness, energy is transmitted through the individual via the medium of the morphogenetic field. Morphogenetic fields are energetic grids that connect all creatures and particles of matter at a sub-atomic level. In order for a human being to utilise the power of this energetic matrix, their body must emit a very high frequency wavelength. At the siddhic level, their whole being becomes like a vacuum. Something astounding occurs to the chemistry of a person at the siddhic level, and this transformation is so physiologically powerful that it literally drives the occupant, commonly known as *the ego*, right out of the vehicle.

It is interesting — and perhaps controversial to some — to note that the highest attainable state of consciousness in a human body is also akin to the lowest. It is the lowest in terms of the modus operandi of consciousness, for the human at the siddhic level most closely resembles the most basic life form on our planet — the eukaryote. Eukaryotes, such as amoeba, are incapable of awareness although they do have consciousness, and in this respect they are the building blocks of life on our planet. These primitive, single-celled organisms are an apt symbol of the siddhic level of consciousness — they have a nucleus containing genetic instructions and some form of protective membrane, and that's about it! Likewise for the one manifesting a siddhic state, there is no normal functioning of awareness. Naturally, human awareness continues, but it works mechanically and only when called upon by outside stimulus. Otherwise the siddhic state is very similar to that within the amoeba — there are genetic instructions to be followed from within the nucleus, but other than that, there is nothing!

When the 14th Siddhi flowers inside a person, they become a driving force for humanity itself. Their every word, thought and deed reaches deep into the morphogenetic field and pushes the whole of humanity in a new direction — the direction of abundance.

The 64 Gene Keys represent these genetic instructions within each human being — they form the archetypal alphabet that underlies the behaviour of all humans. How clearly the instructions are followed depends upon the frequency of the vehicle and its chemistry. At the Shadow level, there is a great deal of background noise to interfere with the instructions. This background noise comes from your past and the collective genetic past stored within your cells. As your genetic frequency moves toward the Gift level, the noise diminishes as you begin to live less from your own past and more within each moment. However, even at this level, the collective genetic memory still remains as an unconscious subtle conditioning element influencing your behaviour. At the true siddhic level all trace of personal and genetic history has to be burned from your body. This is no small matter.

One needs to try to understand the import of this last statement. Inside every human being's DNA is stored the collective memory of every human who has come before him or her. That means every feeling, fear, aspiration and desire dating back to the beginning of man's journey and, before this, every instinct within all the creatures from whom we have descended, going right back to the amoeba — all of the immeasurable genetic history of the evolutionary life on this planet has to be erased from our DNA.

Only when this has occurred can humanity's true power be experienced without limitation. This is what the 14th Siddhi is about — and it is encapsulated within the name the ancient Chinese gave this archetype — *Possession in Great Measure.*

The 14th Siddhi is the core archetype of your power as an individual. It is the Siddhi of Bounteousness — the ultimate legacy of what it means to be a human. This Siddhi is the true genetic inheritance of man — the ability to create in abundance. But the 14th Siddhi is not about accumulation — it is about fertility. Your true latent fertility lies mostly dormant in your genes. When the 14th Siddhi flowers inside a person, they become a driving force for humanity itself. Their every word, thought and deed reaches deep into the morphogenetic field and pushes the whole of humanity in a new direction — the direction of abundance. The overflowing fertile potential within humans is almost infinite. As this Siddhi testifies, if a single human can influence the entire direction of the species in a beneficial direction, one can only imagine what might happen if all humans were to manifest this Siddhi. The mass dawning of the 14th Siddhi would be what almost all cultures have at some time referred to — the uniting of heaven and earth. Not just a golden age, but a permanent state in which all humans take the helm together — a state in which individual power and collective power become one and the same.

As we saw with the 14th Gift, this is a contagious Gene Key. And we should remember that it is also contagious at its Shadow frequency. There are few things as contagious as compromise. However at its highest level the 14th Gene Key holds a great destiny for humanity. As people awaken through this Siddhi, it will begin to travel like wildfire through the world. Here is an example of how the morphogenetic field works. If someone truly is manifesting the 14th Siddhi, then merely thinking about them will trigger a flood of prosperity on many levels in your life. Even a single man or woman manifesting this Siddhi can create ripples of prosperity and spiritual well-being wherever they go. This is the origin of the ancient belief in the power of the guru. It is said that merely thinking of the guru or holding their photograph will bring about a beneficial transformation in your life.

To conclude this 14th Siddhi, we need to understand one thing more — a difficult truth for some to stomach. Individual power is nothing more than a myth. Despite the fact that this 14th Gene Key is all about personal power, the 14th Siddhi is about the end of all things personal. At this level, we need to view humanity as one large expanding genetic vehicle. Each individual human is simply another set of bio-energetic instructions in the greater body. In this sense, individuality is actually the same thing as interference. If a cell within the body behaves as though it has an identity of its own, it cannot function cleanly. It is as though the windows of that cell are dirty and information cannot pass efficiently and transparently from the greater body to the cell. Its illness is that it thinks it exists autonomously. Likewise we humans, complex though we like to think ourselves, are akin to single-celled organisms within a far greater body. We are merely building blocks for a higher life, and in that sense we are utterly expendable. It is not us that matter — there is no *us* — there is only the siddhic programming — to create more and more abundance, in every direction and at every level of consciousness. That is the true definition of bounteousness.

▦ 15th Gene Key

An Eternally Flowering Spring

SIDDHI
FLORESCENCE

GIFT
MAGNETISM

SHADOW
DULLNESS

Programming Partner: 10th Gene Key

Codon Ring: The Ring of Seeking
(15, 39, 52, 53, 54, 58)

Physiology: Liver

Amino Acid: Serine

THE 15ᵀᴴ SHADOW – DULLNESS

ANOTHER DAY IN HELL

The poet T.S. Elliot said that as far as literature is concerned: "The world is divided between Shakespeare and Dante — there is no third." Whilst most people are aware of Shakespeare and probably use many of his expressions in their everyday lives without realising it, Dante has for the most part remained a mystery to the wider world. And yet in his supreme work *The Divine Comedy*, Dante left us what is arguably the greatest map of human consciousness ever written. Whereas Shakespeare used drama, Dante used allegory as a means to communicate an immortal truth about human nature. The Divine Comedy essentially describes the geography of consciousness as it moves from the lower frequencies to the very highest. The 64 Gene Keys can be experienced at the three main frequency bands of the Shadow, the Gift and the Siddhi. Dante named these same levels of consciousness Hell, Purgatory and Paradise respectively.

This 15ᵗʰ Gene Key is the key aspect of a complex set of genetic blueprints in human beings known as *The Ring of Seeking*. It is this codon in your genome that initiates your evolutionary journey from being unaware of your true nature to your eventual awakening as an expression of Divine form. All your struggles, pains, agonies, triumphs and ecstasies are written here in this Codon Ring because, like the Divine Comedy, it lays down the geography and topography of your journey towards awakening. Within this genetic structure the 15ᵗʰ Gene Key plays perhaps the most vital role of all the Gene Keys. In a nutshell, it keeps us human. To be a human is to be a battlefield of opposing forces and frequencies, some of which pull you towards heaven and others which try to drag you downward into hell. Humanity is the bridge for consciousness to work out these many conflicting currents.

The 15ᵗʰ Shadow of Dullness describes a low frequency human attitude to being alive in a body. It is the fear of the ordinariness of life. One of the key elements in Dante's hell is repetition. Transgressors and evildoers are frequently depicted locked into endless cycles where the consequences of their evil deeds are played out over and over for eternity. One of the greatest fears of the 15ᵗʰ Shadow is just this — to be locked into

*Your attitude
affects life's
events,
which is the
first great
Law of
Magic.*

a repeating rhythm that never changes. And yet the huge irony here, which Dante captures perfectly in his great work, is that life does indeed consist of endless repeating patterns and rhythms. As you enter more and more deeply into the mysteries of the 64 Gene Keys, you will begin to realise the truth behind the latest revelations and suppositions of physics and quantum theory — that the universe appears to be a hologram in which the same patterns are repeated again and again in infinite fractal variations.

This 15th Gene Key is about the diversity of organic life. The 15th Shadow highlights the dullness of life. It represents an attitude of human awareness, and indeed of animal and plant awareness, although in the case of other life forms, one might not use the word *dullness*. A dog can sit on a doorstep for a whole month doing absolutely nothing and never be bored. Indeed when we humans observe animals, we often feel a sense of envy that they seem to have no worries — that their lives are so simple. With our self-reflective awareness, we humans are capable of something truly magical and at the same time quite terrifying — we are capable of attitude. Only the human neocortex makes dullness possible.

Two humans going through identical experiences can literally taste two entirely different worlds. One can be in heaven and the other in hell. Not only that, your attitude *affects* life's events, which is the first great Law of Magic. Experience mirrors attitude. Attitude is in fact one of the great mysteries of life, because its source is indefinable. You may think it is the way you think, but it is not your mind even though it operates through your thinking. You may think it is your unconscious, but it operates underneath even the deepest levels of your psychology and physiology. In short, your attitude denotes how consciousness is *using* your awareness at any given moment. It is the membrane that links the microcosm to the macrocosm and it is how DNA programs and is programmed by the environment. It is a natural response to life and its prime directive seems to consist of only one thing — growth. Finally, attitude is linked to your hormones, your brain chemistry and to something altogether beyond your conscious perception — your spirit.

The 15th Shadow manifests in human behaviour whenever a person's spirit feels low — for whatever reason. One of the great causes of shifts in your attitude is light or the lack of it — epitomised for example by changes in the weather. On a cloudy, grey day all human beings experience subtle changes in their physiology, as do animals and plants. It is how you interpret those changes that dictates your attitude. No human can escape the dullness of existence. It is all simply a matter of how you deal with it. It is in fact a deeply human experience and is at the core of our basic psychology. It can lead to extreme states — depression, violence, rage, and frustration. The moment your perspective shifts, whether because of inner or outer light, the feeling of dullness can actually become incredibly exciting and vibrant. We can see how this occurs at a higher frequency in the 15th Gift — the Gift of Magnetism. This word gives us a clue that helps to understand the true nature of what we are calling dullness — it is based upon the lack of magnetism or polarity. Dullness is neither the top nor the bottom of a spectrum — it is actually right in the middle where there is no polarity at all. Herein lies the rub — polarity is never dull. Rage is not dull. Violence is not dull. Only the *lack* of any charge — positive or negative — is dull.

This last truth — that the Spectrum of Consciousness is not a straight line with a negative pole at one end and a positive pole at the other — is a startling insight. What the 15th Shadow is telling us is that all the Shadow states have at their core this same dullness, and that the Shadows themselves only emerge according to how we respond to this dullness. It all depends upon how deeply you allow the dullness to sink in. The more deeply you embrace the dullness, the more mystical it becomes — like the emptiness described by the Buddha. The deepest truth of all is that life itself is meaningless. What you do with this truth determines what kind of universe you inhabit. If you get stuck at the level of your mind, you will probably try to find something — anything — to distract you from the dullness. If you get stuck at the deeper level of your feelings, you will probably fall into a kind of depression as your physical energy sinks. The programming partner of the 15th Shadow is the 10th Shadow, which is about self-obsession, and you cannot escape this polarity. You can only experience it from another frequency. If you can allow the dullness to permeate your entire being without being stuck in any reaction, it will cease to be dull. You will experience it as growth and, at even higher levels, as the flowering of consciousness known as Florescence.

This is truly the miracle of the Spectrum of Consciousness — that every Shadow state conceals a Divine Gift less than a breath away. It doesn't take a process to move from the Shadow to the Siddhi. It is simply a matter of pure acceptance. All three levels — the Shadow, the Gift and the Siddhi — are in reality one. From the Shadow, you cannot see the Gift or the Siddhi. From the Gift, you can see the Shadow (in fact, you are using its creative energy) but you cannot see the Siddhi. Only from the Siddhi can you see everything. You see that there is no difference whatsoever between the actual experience of the Shadow and the experience of the Siddhi. The only difference is that you do not resist the state in which you find yourself. Dullness is dullness. Life is meaningless. In order to accept this truth you have to live it totally. There is no trick or technique that will take away the feeling. Life is absolutely ordinary, and you too must be absolutely ordinary. The only way to true paradise is through this absolute ordinariness.

REPRESSIVE NATURE – EMPTY

When Dullness takes hold of someone with a naturally repressive nature it manifests as emptiness. This dullness is very different from boredom, which has a certain restless energetic quality hidden within it (see the 35th Shadow). This kind of emptiness is very close to depression and, in fact, will inevitably lead to depression. Such a person has at some level *given up*. It is a kind of negative acceptance or resignation. It is acceptance that has only gone to a certain depth within one's nature and has become stuck. Repressive natures are fear-based natures, and such people get stuck because at some level they become afraid of fear itself. This causes the locking mechanism within that leads to depression and other disorders of a similar nature. Such people can only escape these states when they discover the fear that is preventing the emptiness from moving deeper inside them. As we have seen, once someone allows this primordial emptiness to saturate them entirely, it is no longer experienced as something separate from them, neither is it experienced as emptiness. Rather the opposite — it unleashes a great surge of energy and vitality.

*One of
the great
challenges
for modern
humanity
is to learn
how to slow
down.*

REACTIVE NATURE – EXTREMIST

If you have a reactive nature, then you will rebel against the dullness. These people do not accept the dullness either, but their nature is to escape through denial. They often hide behind a kind of vagueness, moving from one experience to another with no real sense of rhythm or purpose. Being extremists they can find themselves in all kinds of diverse situations and places, but again, with no real directed sense of passion. These people are always on the move, never really able to commit to anything or anyone — running from their own shadows. Such people have a hidden rage within that prevents them from being with others for very long. Only when they see through their own patterns of reaction can they begin to accept what they are running from.

THE 15ᵀᴴ GIFT – MAGNETISM

HITTING THE SCHUMANN RESONANCE

The 15ᵗʰ Gene Key is such a powerful portal into life's mysteries. It is one of the Gene Keys that bridges humans with other life forms and in doing so it connects us profoundly to the living spirit of nature or Gaia. In 1977 a scientist by the name of Otto Schumann made a remarkable discovery. He mathematically predicted the exact frequency of the earth's heartbeat within the electromagnetic spectrum. Vibrating at 7.8 Hertz this wavelength, called the Schumann Resonance, is literally the vibrating pulse at the heart of all living organisms. This pulse binds us together as a single living organism. The 15ᵗʰ Gift represents the inner barometer within your DNA to the Schumann Resonance. All human disease comes about because of disturbances between your individual electromagnetic field and this greater field that emanates from the earth. Whenever your frequency ceases to correspond to the Schumann Resonance, you are out of kilter with your natural rhythms and the chemistry of your physiology experiences pressure and stress.

One of the great challenges for modern humanity is to learn how to slow down. The Schumann Resonance is a frequency oscillation that moves far more slowly than most human beings are used to, especially in our modern world. Time moves uniquely for all life on Gaia. She has never been in a hurry. The power of Gaia is the power of *greenness*. If we left our planet exactly as it is now for a thousand years, our cities and roads would once again become green forests. That is the speed, power and pace of Gaia. Since magnetism is the binding force of all creatures and forms, the more deeply you move into the Schumann Resonance, the more magnetic you become. This is all about trusting in the natural ebb and flow of life events. You can see how the agitation of the lower frequencies so easily experiences these slower frequencies as dullness. The fact is that when the frequency of your DNA hits the Schumann Resonance, your experience of time stops completely. These are truths that are still experienced and embodied by many of the indigenous cultures alive on our planet today. To live closely to the earth's natural rhythms is to experience the wisdom and clarity that comes of moving more slowly through the world.

As you ascend in consciousness to the Gift level of frequency, you experience for the first time the underlying beauty of life's diversity through the subtle mutations that take place in this 15th Gene Key. The very state that you feared was taking away your life force actually becomes an awesome font of vital magnetic energy. Dullness is no longer dull. It is as though you were peering into an empty vessel and suddenly realise that it isn't empty but full — full of potential. This simple shift in attitude allows you to unlock the higher magnetic power latent within your DNA. Such magnetic power is the foundation for the law of attraction — the universal law that draws towards you that which serves your higher purpose, be it money, people or resources. Until you hit the Schumann Resonance, the effortless power of this law cannot be realised in your life.

People displaying the 15th Gift as a powerful trait emanate a tangible physical presence. They stand out from others in this respect. The Gift of Magnetism makes a person physically glow with life force, literally making them appear larger than life. It is a Gift filled with enthusiasm and openness and above all it is a Gift of Love. As a universal unifying force, the magnetism unleashed by this 15th Gift allows you to harness all the power of the natural world. Such people often have a very strong connection with nature or with the animal, plant and organic kingdoms either personally or through their work. Being rooted in the natural world and its diversity of rhythms, the 15th Gift has a natural respect for all sentient life. Just as magnetism holds families of the same species together, people with the 15th Gift feel a deep kinship with humanity. This is a Gift that will readily accept and work with any extreme human behaviour or sphere. Magnetism does not exclude anyone, and the more diversity these people encounter in life, the more fulfilled they are.

The real beauty of the 15th Gift however lies in its acceptance of ordinariness. To experience life through this Gift is, in the words of William Blake: "To see heaven in a wildflower." It is to view the holographic vista of life as complete exactly as it appears before you. To the 15th Gene Key there can be no separation between the mystical and the mundane because life is experienced as an immortal journey through all the mythic strands and stories of human culture. Through the 15th Gene Key, you stand amidst the mundane as a warrior. The most boring everyday issue can seem to you no less than an opportunity for personal transformation. The great spiritual trials are not met in the extremes but in the everyday — in your relationships, whilst doing the dishes, cleaning the house or going to work. This is the origin of the mystical saying that in order to find heaven you never need even walk out of your own door.

You could summarise the Gift of Magnetism as being the power to influence and be influenced by means of the aura. Magnetism is all about the power of aura. As Goethe said: "Everything alive creates an atmosphere around itself." All life forms radiate bioenergetic energy fields, which interact with their environment through geometric laws. It is this 15th Gift that unlocks the power of the aura to conduct and relay information between different forms. The more resonant your aura becomes with the earth's natural frequencies, the wider your aura spreads, bringing you into contact with many of the hidden realities and kingdoms within nature.

Eventually, as the frequency of your aura comes into perfect resonance it interlocks with the earth grid itself and your consciousness expands exponentially, becoming one with Gaia and all creatures.

THE 15TH SIDDHI – FLORESCENCE

THE EMERGENCE OF SHAMBHALA

The 15th Siddhi is a quality known as Florescence, a word that essentially refers to the process of bursting into flower. Each of the Siddhis in fact represents a process of Florescence. That is what the Higher Self represents — your ultimate flowering. Florescence cannot be chased, hurried or enforced. It is a state that occurs to humans sporadically — meaning that no one can predict if or when it will occur to a particular person. One thing alone is certain — we humans recognise it when it occurs in another. A person in whom consciousness is spontaneously flowering is a person surrounded by magic, by light and by a tangible magnetic mystery. As the zenith of human magnetism, Florescence manifests like a supernova through the human aura, and once it has begun, it keeps expanding and expanding into deeper and deeper dimensions. Such a person becomes a magnet for the many others who hunger for such higher states.

Although Florescence cannot be predicted, there are certain signs that may appear before it occurs. If we look to the 10th Gene Key — the Gift of Naturalness, which is the programming partner of the 15th Gene Key — we find a clue. When a person becomes truly natural and comfortable with who they are, when they become truly accepting of theirs and others' lives, they are close to flowering. Florescence is experienced when awareness finally comes to rest because only when awareness stops seeking itself can it truly rest. When it does come to rest, consciousness — which was within it all along — finally shines through. The ancients and sages said that the mind must be still, or that all thought must stop, in order for reality to be perceived. This has actually been misinterpreted down the ages — it is not thought that must cease, but identification with thought. It is the thinker that must cease!

Enlightenment never occurs to just a single person, but successively jumps across generations during a certain time period.

Hearing the above truth, many have sought and continue to seek the source behind awareness. However, no amount of tinkering with awareness can reveal the truth. This is why there is no technique or system that can ensure enlightenment. Awareness has to come to rest of its own accord. It is a great and wondrous mystery. The fact that there is nothing that can be done to speed up our flowering can be seen as both a blessing and a curse. To the mind it seems a curse because the mind likes having something to do. To the inner being, it is a great blessing because the embodiment of this truth will lead to progressive levels of relaxedness. Ultimately, Florescence depends on nothing. It cannot be created — not by meditation, not by good karma, not by any form of effort, not even by effortlessness. Florescence is Grace. It occurs whenever and wherever it feels like occurring!

This 15th Gene Key forms the main stem of the Codon Ring of Seeking, a multidimensional genetic sub-program in all human beings. Each member of this chemical family ignites a different pressure that begs to be resolved in our lives. For example, the 54th Gene Key seeks to rise up, socially, spiritually and materially, whereas the 52nd Gene Key seeks a state of complete rest. It is through the mystery of the 15th Siddhi that all this pressure to seek is finally brought to an end. If you look at nature, which is the root of the word *florescence*, you will see that nature produces many buds on many different plants, but that all plants flower at unique times. Furthermore, the buds on each particular plant open collectively, in a great rush of dynamic energy. In human beings, enlightenment operates in the same way, through different time periods and epochs. Enlightenment never occurs to just a single person, but successively jumps across generations during a certain time period. It is a phenomenon that once ignited will play itself out through many different fractal flowerings in many different kinds of human lives.

From time immemorial, mankind has devised intuitive systems to understand the rhythms and timing of when people are born. There are obvious and fundamental truths that different cultures have discovered. Each of the 64 Gene Keys has a position within the arc of the year, creating a unique genetic astrology for each of us. People born with the 15th Gene Key prominent in their Hologenetic Profile are nearly always born at or close to either the Spring Equinox or the Summer Solstice. This is because the power of both the Equinox and the Solstice symbolises the power of Florescence — when all of nature is bursting forth and later reaching its zenith. A person manifesting the 15th Siddhi thus has a vibrancy that is simply irresistible.

Florescence as a word goes much further than the word *flowering*. Florescence refers to a process of exponential bursting into flower — as though the focus is not simply on a single flower, but an entire tree of flowers all bursting into bud at the same time. This is what such people do when the grace of Florescence occurs to them — they flower in many, many directions at once. Their potential seems endless; the diversity of their interests appears inexhaustible, as though they were flowering in every possible sphere of life at the same time. Such a being cannot stop anywhere and allow themselves to remain with any particular path or *way*. The awareness within them does what it must do — it reflects the energy of spring — always moving, growing, exploring, and delighting in whatever passes before it.

The 15th Siddhi can only occur when the mind has completely given up control. Florescence is utterly confusing to the mind — it is too spontaneous, too unpredictable and too wild for the mind to follow, let alone manage. It would be rather like a writer writing a hundred books simultaneously and jumping at random between them, whilst at the same time holding down a hundred other jobs and moving randomly between them. Florescence follows nature's own wild, organic rhythms, which demand an attitude of trust, surrender and childish delight. When you come into perfect harmony with the Schumann Resonance time itself will stop within your being. This is an impossible experience to relay in words. In the words of Christ, it is when we mystically *inherit the earth*. We come to experience the mystery of *philos anthropos* — the pure love of humanity embodied in the principle of philanthropy. All of life is now seen as philanthropic, constantly giving its own essence back into the whole.

This Codon Ring of Seeking is currently undergoing a deep mutative process that is chang-ing the essential rhythmic structure of all life on our planet. Above all it is bringing humans into greater awareness of the delicate balance of living biological systems, and in particular plants. As the pressure of human seeking burns away the greenness of our planet through pol-lution and industrial expansion, we are coming to realise that we are also responsible for these great changes in global rhythms. Our climate is simply the by-product of a deeper energetic change that is taking place. Deep within this genetic codon, the dynamic energy of liberation is moving via the electrical and explosive 39th Gene Key. This Gene Key in its higher frequency is actually a precursor of the great change that will transform all life on earth through the 55th Gene Key. But it all begins here in the Ring of Seeking with an amino acid known as serine. Human seeking will eventually burn itself out — that is the higher trajectory we as a species are following, whatever it looks like on the surface. Each of the six Gene Keys in this Codon Ring will awaken in a sequence, beginning with the 39th Gene Key and ending here with the 15th Gene Key. As this awakening moves through humanity and nature, we will experience power-ful, subtle Earth changes as our primary magnetic grids are recalibrated to a higher frequency.

What the 15th Siddhi really does for humanity is align us with a higher evolution. It is always difficult to see evolutions beyond your own. Nonetheless they exist. Our great teachers and mythologies have always been aware of them. The Earth itself contains subtle kingdoms of a far higher frequency than humanity — and these *devic* kingdoms are also preparing to take a leap in consciousness by merging with human consciousness. The greatest Florescence will occur when hidden higher forces latent within the Earth itself rise to the surface and absorb human consciousness. This awakening from within has long been a part of all human mythol-ogy, which tells of great cities within the Earth and a shining jewel at their centre known as *Shambhala* or *Agartha*. Such myths testify to the genetic change that is sweeping through all DNA on our planet. It is even impacting the evolution of the galaxy and cosmos itself. Flores-cence is a mirror of the supernovae we see in the heavens above, but it is a supernova that takes place deep within the structure of form itself. In a nutshell, Florescence is the explosive chain reaction of enlightenment of all forms to the true nature of consciousness. It is stamped into the destiny of every living creature, and it is coming to humanity very soon.

SIDDHI
MASTERY

GIFT
VERSATILITY

SHADOW
INDIFFERENCE

16th Gene Key

Magical Genius

Programming Partner: 9th Gene Key

Codon Ring: The Ring of Prosperity
(16, 45)

Physiology: Parathyroid

Amino Acid: Cysteine

THE 16ᵀᴴ SHADOW – INDIFFERENCE

THE DIFFUSION OF RESPONSIBILITY

At the Shadow frequency one of the most powerful and pervasive forces that keeps you from perceiving a higher reality is indifference. As long as you are *indifferent*, you can't ever be *different*. This is the key to the 16th Shadow, which concerns the very human fear of leaving your comfort zone and fully embracing change in your life. As an aspect of the Codon Ring of Prosperity, the 16th Gene Key is about excellence. To prosper truly in the world is to find the one thing in life at which you excel above all others. This is the true destiny of every human being, but to bring this dream to fruition you must first step out of the shadows and take the risk of being different. Indifference is an energy field that gets you to focus your precious time and life force on the inessential. The inessential in this context is anything that takes your attention away from the present moment and its limitless potential. As long as you are focused on the inessential, you will be indifferent to the things in life that really matter.

The holocaust survivor and writer Elie Wiesel had the following to say about this subject:

> *"The opposite of love is not hate, it's indifference.*
> *The opposite of art is not ugliness, it's indifference.*
> *The opposite of faith is not heresy, it's indifference.*
> *And the opposite of life is not death, it's indifference."*

The 16th Gene Key represents the collective expression of the state of health of humanity. In our modern world, this expression is clear. That the vast majority of the world lives in poverty whilst a select few thrive is a testament to the power of the 16th Shadow, which sucks all the life out of the world. Because human beings are not willing to stand up and be different, they become willing to sit by the side of life and observe. In social psychology this common human pattern is known as the *diffusion of responsibility* and refers to the human preponderance to look away when confronted by another in distress, often when there are a number

This 16th Shadow can be absolutely full of plans and good intentions about the world, but they rarely get off the ground.

of other people in the same vicinity. However, as a by-product of the 16th Shadow of Indifference, the diffusion of responsibility occurs across the width and breadth of our planet on a far subtler energetic level.

Indifference is an energy field created by all the human beings on this planet who are not doing what they would really love to be doing. The only reason for this state is fear. Whoever you are and whatever your life circumstances, if you transcend your fear you will suddenly become so much more capable than you were before. To break out of the field of indifference is to make the courageous move deep into your own fears and overcome the lethargy that prevents you from making something truly beautiful of your life. Indifference has many faces and excuses. One of the primary excuses human beings use for not doing what they truly want in life is the excuse that they don't have enough time. This habit of making yourself a victim of time is the core escape strategy of the 16th Shadow, but time actually has nothing to do with it. You have become a victim of your mind rather than of time. Time itself is as fluid as water, and as the 17th and 5th and 52nd Gene Keys testify, it can be bent, shortened, twisted, lengthened and even stopped. The moment you take a stand and begin to follow your dreams, time becomes your ally rather than your enemy, automatically adapting itself to fit your needs.

The programming partner to the 16th Shadow is the 9th Shadow of Inertia, and it is easy to see how these two genetic forces keep humans from actually getting anywhere. As with all the Shadow polarities, they create a biofeedback loop that traps energy at a low frequency. In this case, indifference cannot overcome inertia because it doesn't feel strongly enough to do anything about it. This 16th Shadow can be absolutely full of plans and good intentions about the world, but they rarely get off the ground. This is the Shadow of the *pipe dream* — the delusion that your dreams will one day become manifest without you ever actually having to initiate them. What is lacking in the lives of those in the grip of this Shadow is one vital factor — enthusiasm. Enthusiasm is the original Chinese name for this 16th Hexagram of the I Ching. Enthusiasm is the key that can pull you out of your inertia and snap the frozen pattern of your indifference. To be enthused, you must actually *have* an experience rather than simply dreaming *about* an experience.

One of the greatest challenges in surmounting the collective weight of indifference is your unique inability to identify with whatever your dreams really are. This is why experience is critical to this process of breaking free. It is not enough to daydream and plan how you might save the world — you have to actually begin the work. The other major excuse of those under the spell of this Shadow is: "I am not ready yet". Again, this excuse is based on a false notion of time. The truth is that you are always ready. You are ready right now. If you keep postponing your dreams, they will always *remain* dreams. If you actually have the courage to begin them, you will learn as you go along. When you have the courage to step onto the path of your destiny, two things immediately occur — the first is that you feel an enormous surge of new energy coming from your own enthusiasm and the second is that you stop compromising your path because of pressure coming from others.

Only your courage and your enthusiasm have the power to break through the collective energetic walls created by the diffusion of responsibility. Down at the Shadow frequencies you have no idea what is possible. You see only with your mind, which is incapable of encompassing full reality. Your enthusiasm has the power to overcome the mass propaganda telling you that your dreams are impossible, and your courage is the sword protecting you from individual pressure that threatens to choke you. With this newfound courage you can finally stand as a fully embodied human presence. There is a deep entrepreneurial spirit here in this genetic codon. It is, after all, called the Ring of Prosperity. It is the fusion of human talent coupled with the power of group synergy (the 45th Gene Key) that opens the field for true prosperity. The more you expand into your talent, the more the universal field of consciousness responds, coming to meet you halfway. Only by striking out alone will you find the right people to support, orient and bring your vision to fruition.

The 16th Shadow has much to do with the unique gifts concealed within every human being, but at the Shadow frequency this can be very confusing. The 16th Shadow puts all its emphasis on skills, techniques and systems rather than on the human spirit that enlivens them. These people can become addicted to information and techniques, but rarely transcend these techniques. The 16th Shadow creates experts, whereas the 16th Siddhi creates masters. The distance between the two is almost unfathomable. To be an expert is to remain indifferent and unenthused because once enthusiasm has been unleashed, skill gives way to something far more magical — it opens up the amazing Gift of Versatility. As you become more versatile, you for the first time become truly *different* — you become special, and the moment that happens, you have finally killed indifference.

REPRESSIVE NATURE – GULLIBLE

When indifference is repressed it manifests as gullibility. These people become victims of the mass propaganda of the collective. An example here is people who use the excuse that if governments cannot help the world situation, how can they? This gullibility essentially allows people to hide their indifference behind the indifference of others. At the deepest level this gullibility fosters individual weakness and a feeling of powerlessness. Every time you turn on the news on television for example, you meet this huge negative conditioning field. How you respond to this field determines your frequency. The response of the repressive nature is to bury your head in the sand in the belief that nothing can be done.

REACTIVE NATURE – SELF-DELUDED

The nature of indifference actually makes it impossible for a person to react, so the reactive nature in this instance appears to be something of a misnomer. These people hide their fear in their obsession with structure, systems and techniques. They become so deeply identified with the structures their minds are following that they forget about the reason for the structure in the first place. This gives rise to a deep self-delusion in which these people build powerful mental walls around themselves, keeping the world and others locked out. In a certain sense these people are also gullible like the repressive nature, but their gullibility concerns their own propaganda rather than anyone else's. They are convinced by their own minds. However, behind their mental walls seethes a huge anger that is far from indifferent, and which will eventually destroy them unless they can see through their own self-delusion.

THE 16ᵀᴴ GIFT – VERSATILITY

A TALENT FOR SUSTAINABILITY

In discussing the 16th Shadow, you need to realise something of great importance — indifference is an expression of the collective frequency of humanity operating through your genetics. Because of this, there can be no fault or blame placed on this trait. Indifference simply arises because the mass consciousness does not yet see its own true nature — that is, it doesn't yet realise itself as a holistic entity. However, our new awareness will enable us to identify as a single consciousness. This identification means, for example, that when someone turns on the television to watch the news and sees something unpleasant, they will no longer not identify with the people involved. There will be a deep-seated realisation that these people are an aspect of their own consciousness.

This new identification with humanity as your own greater body will lead you naturally into service. It will no longer be possible to be taken in by fear propaganda, as the victim consciousness on our planet will be splitting apart. It is at this stage that the powers inherent within the 16th Gift will become apparent. The strength of the 16th Gift is its remarkable versatility. The 16th Gift encompasses learning and acquiring techniques and skills, but true versatility occurs when you transcend technique. Whilst at the Shadow frequency your skills actually trap you, here at the Gift frequency they become a ladder into the true self. At the Shadow level the focus is on the skill or the technique itself, whereas here at the Gift level skill serves only one purpose — to act as a bridge to a higher state of consciousness.

Identification is an absolute requirement for the learning of any skill — you have to become utterly absorbed in the technique you are studying. All cycles of learning operate on a seven-year cellular pattern. This means that it takes seven years for a human being to really learn something, because that is how long it takes for the cells of the body to be physically imprinted by the skill or technique. So if someone has fully imbibed a skill, after seven years they will automatically transcend that skill. Naturally there are higher and higher levels to all skills, ultimately leading to mastery. However, it is at this key stage of transcendence that the Gift of Versatility engages. Versatility is not a skill — it is a state of frequency that can use a skill without being fully identified with it. This is the level at which skill becomes talent. Talent happens when a certain skill has been mastered. With talent you no longer have to think about the skill and a whole new world opens up to you.

In essence, the 16th Gift is not really about skill at all. It simply uses skill as its medium to reach this higher state. The Gift of being versatile is the ability to pick up any skill that is needed and use it for a single aim — for the betterment of humanity and the service of the whole. Versatility is driven by the dynamic energy of enthusiasm — the feeling that you are doing something thoroughly enjoyable, which also happens to improve people's lives and serve the whole. At the Gift level of frequency something truly remarkable happens — one Gift becomes interchangeable with another. No matter what your personal genetic makeup, if you can operate at this level of frequency, you can draw upon any of the 64 Gifts through the morphogenetic field that links all states of the same frequency together. This is the true meaning of the 16th Gift — it is the ability to tune into any Gift you need and make use of that Gift. That is true versatility.

The Gift level of frequency is an energy field that you *enter into*, rather than being defined by a particular set of skills. In the original Chinese I Ching, this 16th hexagram was related to the arts — to music, art and dance. Indeed, at the level of versatility, you will find that any branch of the arts or sciences is readily interchangeable with any other branch. The Gift level is the level that we refer to as genius and the nature of genius is versatility. One has the ability to understand an energy signature that pervades all creation. Leonardo Da Vinci is a good example of a man who displayed this versatile ability to move between the arts and sciences with great ease. A true genius rarely restricts him or herself to a single branch, because that would limit their Gift. The great joy of versatility is to spread your wings as widely as possible.

Versatility also happens to be a highly energy-efficient field of consciousness. If we look at the way in which humanity currently uses energy, we can see clear evidence of the power of the 16th Shadow in operation. Our current way of finding energy is to draw it from a single source and distribute that source to everyone. This is expressed through our obsession with fossil fuels — oil, coal and gas. You can see our huge indifference at play in the way we totally ignore the consequences of our actions. Thus huge tracts of the earth are raped in the name of progress. Modern business is utterly indifferent to the plight of the minority, not to mention the landscape, animals and creatures of the natural world. Indifference moves like a virus, draining one resource and then moving on. This is an unsustainable path. As the Gift of Versatility begins to take hold in the human genome during the coming century, we will learn to tackle our energy expenditure in a very different and far more prosperous way.

If we take the environment as a metaphor for how versatility works, we can see how it takes into account all aspects of any living system. It is interested in pure efficiency that does no harm to anyone or anything. The 16th Gift is therefore very connected to the notion of self-sustained living. Instead of draining one of the elements and then moving on to another, versatility uses them all at the same time, finding a fluid and natural energy exchange between them. In the case of energy use this means that every household, hamlet or village will eventually have to become self-sustaining. This might require a combination of solar, water, wind and even geo-thermal power to produce enough energy to maintain a home or small community. These days there are so many emerging *alternative* sources of power that in time the whole notion of government-run energy organisations will become unnecessary. The above cites a single example of the true potential locked up in this 16th Gift, and it can be applied to any

True prosperity is brought about by people following their genius whilst working together collectively.

field, from banking to bee keeping. Above all, versatility abhors waste — this is a Gift that can use or recycle anything and everything, which is why it can use any set of skills or techniques without being caught up or limited by them.

We have seen that the 16th Gene Key forms part of the Ring of Prosperity, and along with the 45th Gene Key — which is about group synergy — this chemical combination is the master-program in your body for activating true prosperity. True prosperity is brought about by people following their genius whilst working together collectively. This 16th Gene Key is hugely important for the further evolution and sustainability of humanity because it involves the correct education of our children. Through this Gene Key we can spot a child's inherent talents early in life, and we can place them in an environment that best supports nurturing those talents. In this Ring of Prosperity lies the mystical secrets of the Guilds — the genetic groupings of genius that when brought together can liberate prosperity exponentially through-out our civilisation.

THE 16ᵀᴴ SIDDHI — MASTERY

THE MIRACLE SIDDHIS

If we briefly review this 16th Gene Key, we recall that it is rooted in indifference caused by lack of enthusiasm and the inability to identify with what you are doing. No matter how skilled you are, if you don't have this enthusiasm you cannot attain the next level, which is talent. Furthermore, talent is something that must be earned over a period of time (no less than seven years), and its natural expression is versatility. You could say that skill is a vertical approach to learning, in that you have to focus great attention in one direction, and your whole impetus is to understand and learn as much in that field as you can. Versatility marks a shift in frequency in your whole being because it opens up the horizontal field at the same time as the vertical. You are no longer fixed to a single discipline but begin to view life at a holistic level, seeing how all systems and techniques interconnect in a truly integral way. You can thus apply your talent across the whole spectrum of human sciences and arts, without the dogmatic limitation of being tied to one single direction. At this level, you are tapping the natural laws behind all life processes, so your talents can be transferred easily between disciplines, even those that appear far apart and unrelated. For example, you might be an artist, but your talent also embraces business. At the Gift level, your talent is based upon the knack of *not identifying* with any single set of skills.

Just as the DNA in every cell of your body contains the blueprints for all life, so the 16ᵗʰ Siddhi activates the blueprints of mastery of all human fields of endeavour.

At the siddhic level a paradox opens up before us. We have gone from lack of identification at the Shadow level, to identification with a skill as you approach the Gift level and then a loosening of that identification in order that you become truly versatile. At the Siddhic level the issue of identifying with what you are doing is no longer relevant. At this heightened level of consciousness, your identity is erased and you become an empty vessel for life. Here all learning ends and there are no more levels.

With talent and even versatility, there are still levels. Even the greatest pianist can still improve and expand. However, with the 16th Siddhi of Mastery the whole game of expanding or evolving, either horizontally or vertically, is over.

The 16th Siddhi is a relatively rare expression in the world because it encompasses what some would call magical abilities. The blueprints behind every aspect of human genius come through the 16th Siddhi from the arts right through to the sciences. Just as the DNA in every cell of your body contains the blueprints for all life, so the 16th Siddhi activates the blueprints of mastery of all human fields of endeavour. The implications of this are really quite staggering. Consider an example from the field of music. Rachmaninov's 3rd Piano Concerto is considered one of the most difficult and awesome pieces of western classical music in existence. To learn to play this piece of music takes years of sustained practice and discipline, even if you have genuine talent. However, the kinaesthetic knowledge of how to play this piece is actually already contained in every single human being alive. The 16th Siddhi allows direct access to this and all other kinaesthetic knowledge, which means that, if the 16th Siddhi is activated, you could play this piece of music as well as any piano maestro who ever lived without ever having touched a piano in your life.

The 16th Siddhi has an interesting connection to the 35th Siddhi of Boundlessness, and on closer analysis one can see why. As one of the *Miracle Siddhis*, the 35th Siddhi's sole purpose is to expand the average human's perspective on life exponentially. It testifies that nothing is impossible if you can only move your mind out of the way. In traditional Eastern approaches to higher consciousness the *siddhis* are understood in a different context from this work on the 64 Gene Keys. There the Siddhis are traditionally seen as special powers that arise as *obstacles* on the path to true realisation. This kind of language actually derives from the cluster of Siddhis in the human genome designed to manifest miracles and special powers (obvious examples are the 14th, 16th, 35th and 60th Siddhis). When you attain a siddhic state, you attain a kind of cosmic versatility and your destiny continues to unfold according to your genetic predispositions. Much confusion exists in spiritual circles concerning the so-called *lure of the siddhis*. Those who have spoken out against them may also have been speaking through their own limitations. There is absolutely nothing wrong with the special powers of the siddhis. In fact they are an essential part of our evolution.

The enthusiasm inherent in the 16th Gene Key also has its manifestation in the Siddhi, where it shows its true nature. The word *enthusiasm* derives from the Latin and Greek meaning *to be possessed by the breath of God*. As the Divine currents flood human beings, shattering their identification with the world of form, so God has her play through us and one of the forms of this play is to display mastery over creation. It is impossible to be indifferent to a miracle, and this is the highest expression of the 16th Siddhi — after all, if God is not allowed to have fun, then who is? The special powers of the 16th Siddhi have only one purpose — to be of service to the whole, even if that manifests as a form of cosmic entertainment. True Mastery is to have absolutely no individual power whatsoever — it is to merge with the Divine Will and therefore become a master of nothingness. This is the ultimate theatre and the highest of all arts, and at the same time it has the incredible power of pulling humanity out of its indifference.

☰ 17th Gene Key

The Eye

SIDDHI
OMNISCIENCE

GIFT
FAR-SIGHTEDNESS

SHADOW
OPINION

Programming Partner: 18th Gene Key

Codon Ring: The Ring of Humanity
        (10, 17, 21, 25, 38, 51)

Physiology: Pituitary Gland

Amino Acid: Arginine

THE 17ᵀᴴ SHADOW – OPINION

THE CLOSING OF THE THIRD EYE

As we examine in depth the pairing of the 17th and 18th Shadows, we will see two of humanity's greatest mental gifts and dilemmas — opinion and judgment. These two qualities epitomise the logical left hemisphere of the human brain. Our ability to base an opinion on logical judgment is one of the great powers of human beings. It is this combination that sets us apart from all other creatures, and yet it is also what has pulled us away from the reality of our interconnectedness with all other creatures. So it may seem ironic that, at their very highest levels of frequency, these two attributes of the male polarity within the brain are actually destined to drive humanity towards the final stage of our evolution. Through the 17th Siddhi of Omniscience the human mind will finally see the perfection and beauty of our oneness with all existence, and through the 18th Siddhi of Perfection we will actually bring that vision into full manifestation.

At the shadow frequency however, these two attributes of opinion and judgment create nothing but division because they are based on seeing and challenging narrow aspects of the whole, rather than the whole itself. It is only when you have risen to the highest level of consciousness that these same attributes become capable of seeing every aspect of the whole simultaneously, which is the quality known as omniscience. Shadow frequencies always contain the seeds of greatness, despite the often-unpleasant ramifications of their lower manifestation. You can really see from the example of the 17th and 18th Shadows how deeply genetic binaries operate together to make our lives miserable. The 18th Shadow of Judgment is a drive rooted in dissatisfaction within every human being, and its single role is to try to find the cause of this deep uneasiness through the medium of the 17th Shadow.

As an intellectual capacity within the human brain, the 17th Shadow of Opinion corresponds to the digital left hemisphere of the brain that specialises in pattern recognition, the basis of logic. Fuelled by the restless nature of the 18th Shadow, this 17th Shadow seeks an answer to its unease through the mind, whose nature is to project itself out on the world. In other words, the Shadow of Opinion is programmed to look for flaws in the outer world, in society, in people and even in oneself. When a flaw is discovered it becomes the seed for an entire worldview and a whole story is built up inside the mind of that individual. The projected flaw, which is always based on some form of comparison, becomes the point of focus for all that dissatisfied energy coming through the 18th Shadow. This is how human opinions form — they crystallise around a single projection, which is born of early conditioning, and thus over time a molehill turns into a mountain.

This description is a simplified version of how early childhood conditioning impresses itself on the mental mechanism of every child. Your opinions sprout from seeds planted at some point within your first seven years, even though they do not begin to surface in the mind until your third seven-year cycle, some time during your teen years. If a child developed naturally, without any external synthetic pattern — physical, emotional or mental — imposed on them during these first seven years, it is highly unlikely that he or she would grow into an *opinionated* adult.

Your opinions sprout from seeds planted at some point within your first seven years.

In our modern society, opinions are actually considered to be a healthy thing, and it is true that they are not inherently unhealthy. The logical left brain thinks through taking sides and comparing. Problems occur when your dissatisfaction begins to identify with an opinion and turns it into a dogma. It is also at this point that another *disease* takes root inside you — the *dis-ease* of seriousness.

The healthy expression of opinion is rooted in a certain playfulness that comes of having an equally developed right hemisphere of the brain. Whereas the left brain sees the parts, the right hemisphere only sees the whole. If the inner structure of the brain is balanced, the male aspect will always serve the female, since seeing the part without seeing the whole is limiting, divisive and dangerous. Whenever you become over-serious about your opinions you immediately find yourself having to defend them. It is this dynamic that is at the root of violence. At the Gift level of frequency, as we shall see, opinion gives way to far-sightedness, which is based upon seeing both sides of a situation at the same time. If the right brain provides the backdrop to the whole picture, it becomes impossible for the left brain to become *fixed* on any single element of the whole. In this respect, if you do choose to take sides it is for the sole purpose of maintaining Integrity — the 18th Gift.

It is interesting to note that healthy brain development in early childhood has little to do with intellect and much more to do with non-interference. In their first seven years, children really only require an environment in which they can learn through play. Any imposition of an external rhythm that does not follow the biological and seasonal pulse will subtly disturb the delicate structure of the developing brain and nervous system. It is here in the 17th Gene Key that your inner timing is founded. This timing is anchored deep within the body and is connected to the frequency of the Earth itself. The seven-year cellular cycle is founded on a deep inner drumbeat that echoes throughout every system of the developing child from the point of conception forward. If these powerful natural rhythms are pushed out of kilter early in life, your inner timing will be distorted when you become an adult. This time distortion manifests most commonly as mental anxiety but can also lead to a wide range of physiological problems emerging later in life, all of which can be very difficult to diagnose or fix.

The 17th Shadow governs the way in which huge numbers of people view the world. On a collective level, this Shadow has created the world we see around us today. It has categorised humanity into boxes, countries, religions and hierarchies based on a left-brain view of the world. Almost every aspect of our society is based upon division, comparison and opinion. Most human beings make decisions on the strength of what the left brain tells them, without regard for the long term effects on the greater environment. It is because of this 17th Shadow that we have lost sight of the bigger picture. Because this Shadow only sees what it is programmed to see, it builds its life around that original germ of dissatisfaction. Our children, therefore, behave in the way the world programs them to behave, whether they conform or rebel. They are born into a world where the 17th Shadow is given precedent, and all their eyes see is their own point of view and how to maintain it.

However, as its programming partner the 18th Shadow demonstrates, the dissatisfaction inherent in human beings from birth is divine in nature. It is the very force that is here to challenge everything that goes against nature and the whole, which is why it will ultimately drive human beings to create a perfect world. It is only through the 17th Shadow that this basic drive of human dissatisfaction is twisted to serve the left brain, where instead of serving the whole it serves the parts, thus creating more and more division in the world. Because this Shadow is so adept at understanding structures, it is the prime organising agent within the human brain, allowing you to structure language, use numbers and see things in levels, bands and hierarchies. In its power it also creates the language of your individual reality and that language is built upon a dualistic linguistic structure that does not allow transcendence. The moment you open up the structure of your inner language to the third level that lies beyond the two contrary opinions — democratic or republican, conservative or labour, male or female — you ignite the power within your DNA to exit the Shadow frequency altogether.

In summary then, the 17th Shadow holds human thinking at a dualistic level, and it does this through over-emphasising the left brain approach. It makes the logical intellectual view the lynchpin of your neuro-linguistic reality. In this sense it compromises and represses the holistic right brain, a process described in detail through the 11th Shadow. It is only through combining and balancing these two viewpoints that a third transcendent view opens up to you. This is precisely what the 64 Gene Keys do inside your brain — they give you a new set of neuro-linguistic parameters that digitally reprogram your logic to transcend itself and find the third view. It is this third view that will ultimately allow you to perceive a wider and greater reality. It is this very same process of inner rewiring that generations of masters and mystics have called *the opening of the third eye*.

REPRESSIVE NATURE — SELF-CRITICAL

When the 17th Shadow turns inward it creates the tendency in the mind to consistently compare oneself to others, and furthermore to side with the others. This results in human beings who continually undermine themselves through self-criticism. As a pattern lodged deep within the unconscious, it also means they are rarely aware of what they are doing. The self-critical aspect of these personality types usually does not value itself enough to hold an opinion, much less express it. Such people lack the backbone necessary to stand up for themselves because they have no real sense of themselves. The power of the 17th Shadow is that once it has fixed a viewpoint, it can always find evidence to prove the validity of that view, since it is adept at pattern recognition. These people thus go on accumulating evidence throughout their lives to support the core belief that their life is of little value.

REACTIVE NATURE – OPINIONATED

Opinion can be understood as anything that requires defence to uphold. As the 17th Shadow turns outward, it creates the impressive systems of dogma seen all across the world. This Shadow has great appeal to the male population in particular since the male brain naturally favours the rational approach. These people use logic as a means to enforce their opinions and views in the world, thereby making others essentially *wrong*. The majority of the world is made up of those who create the opinion bases — sciences, systems and hierarchies, and those who then believe those opinions and unwittingly become their victims. Those who defend an opinion are ruled by their unconscious anger, which is the expression of their deeper fear. It is extremely rare to find a human being who is truly free from opinion. Such a person must have looked into their fears and embraced their rage in order to move into that magical place that sees both sides of life. Only in seeing both sides can balance be struck and control be yielded.

THE 17TH GIFT – FAR-SIGHTEDNESS

PRECOGNITION VIA THE HEART

It is ironic that what often passes for reason in the world is really someone's opinion dressed up in facts that happen to justify it. At the Shadow frequency the human mind distorts logic and facts for its own ends by building one argument and then concealing its counter-argument. The mass of humanity is easily influenced by one side or the other. It is out of the mind's duality that all human drama is born, and the mass consciousness of humanity loves drama. It is all a game played by spin doctors, and if you think you have escaped this game, then look deeply at your opinions and perhaps you will discover how easily conditioned your own mind really is.

    The fact is that human beings, like Skinner's rats, see only what they are programmed to see — which are flaws. All opinions are based upon seeing flaws in the geometry of life. It is only when you have transcended your fear of freedom that you can move beyond the *opinion game* and enter a higher sphere of existence. Moreover, to move beyond opinion you have to stop taking life seriously and personally. For most people the 17th Gift seems beyond their capacity, because to give up your viewpoint is to give up the deep-seated need of your mind to identify with something fixed. Your mind believes that in finding a fixed view or philosophy fear can be held in check and life will be under control. If you look within most human minds you will find layers and layers of conditioned dogma — from books, scientifically proven theories, religious beliefs and/or traditional views. You will find minds that are desperately trying to become certain about life.

And yet, the 17th Gift is not about having a great holistic vision of reality that takes no sides and is beyond opinion. The 17th Gift is a mentally dynamic Gift — it does not sit idly by with a philosophy of passive acceptance. This Gift has a purpose and that is to understand the minute mechanics of life *as well as* seeing the whole picture. Like its Shadow, the 17th Gift is driven by its programming partner the 18th Gift, which is Integrity. Thus the true purpose of the 17th Gift is to serve and uphold Integrity by challenging all misrepresentations of truth in the world. The very quality that leads to narrow-mindedness at the Shadow frequency has a mission to create open-mindedness at the Gift frequency. The wonderful truth of logic is to be found in the 63rd Gene Key, which demonstrates how self-defeating logic really is, albeit at the same time how mystical. This is the beauty of logic — that at a higher level it always defeats itself. Just as the 17th Shadow uses *mental spin* to condition people's minds in a certain direction, so the 17th Gift uses its own version of spin to undermine self-serving dogma.

The advantage of the 17th Gift over the mass-held opinions of the Shadow frequency is that it entails a higher form of cognition — far-sightedness. Because the 17th Gift doesn't dictate direction in a person's life, it is freed to do what it was always intended to do — differentiate between those patterns that lower your frequency and those that raise it. Not only does this Gift see beyond opinion, it also grasps the operating mechanism that gives rise to opinions and the human tendency to get caught up in them. It is through working with this Gift frequency that the process known as the *opening of the third eye* begins. The first stage in this process occurs when you see the futility of identifying with a specific opinion. This Gift brings access to a whole new inner language that circumvents the need for one-sided expression and highlights the deeper issues that underlie argument.

The 17th Gene Key is an aspect of one of the most complex codon groups in the human genetic matrix. Known as the Ring of Humanity, this chemical grouping and its associate amino acid arginine contain the very blueprints of human destiny. We can learn from this group that one of the primary causes of human suffering is rooted in the opinions of the 17th Shadow, which looks for and sees only differences. However, we can also find the future evolution of humankind coiled inside this Ring of Humanity. In particular, the Gift frequencies of these Gene Keys show us the six essential human attributes that are truly natural and inherent to our species and among these is the 17th Gift of Far-Sightedness. Many of the Gene Keys in this codon group concern love and the human capacity to trust in the authority of the heart. Thus far-sightedness can be said to arise directly out of the heart rather than from the mind. This 17th Gift is in fact the true instrument of human perception.

Far-sightedness can be said to arise directly out of the heart rather than from the mind.

We have already seen that the 17th Shadow is programmed to see only hierarchies. It simply cannot comprehend that other realities exist beyond its dualistic mental constructs. However, freed from the low frequency field of mass-created dogma, the 17th Gift begins to see higher realities and dimensions. It *sees* in a totally new way. Instead of serving the intellect, the 17th Gift *sees* through the awareness emerging from the solar plexus system. It allows universal knowing to settle within the logical construct of the left brain, which can then express the mechanics of oneness in totally new and fascinating ways.

In seeing the mechanics of life in this way, the 17th Gift also begins to demonstrate its other specific ability — prediction. The 17th Gift is peculiarly slanted towards looking ahead into the future. Wherever the focus of the 17th Gift is trained, it has the ability to see how one pattern gives birth to another related pattern, and since it can transform a transcendental view into a logical view it can see the logical progression of one phase of evolution to another.

If you have this 17th Gene Key as an integral part of your Hologenetic Profile then this specific Gift of prescience can be a powerful reality for you. However, you are not a visionary in the same way as someone with the 11th Gene Key. This 17th Gift does not see archetypes. Rather it sees patterns rooted in numbers, which can then be translated into letters and words. These people can grasp the divine mathematics of evolution and the progression of its plan. In seeing the whole plan as well as its parts, they are truly the real scientists. Unlimited by the traditional views and free to translate their deep knowing of the perfection of life however they wish, the people of the 17th Gift are the ones who will ultimately organise the next new human civilisation. Furthermore, because the Gift frequency is the frequency of the human heart, each person working with this 17th Gift holds a specific fractal view of the overall picture of our collective future, which means that not a single one of them will ever disagree with another.

THE 17TH SIDDHI – OMNISCIENCE

THE EYE

As you may have begun to grasp, the higher frequency dimensions of the 17th Gene Key offer a view of the collective future of humanity, and it is through this foresight that the various aspects of future reality will be organised and constructed. Every individual carrying the 17th Gift oversees a specific aspect of a great evolutionary pattern moving toward unity. However, at the Gift level, your view is still restricted to a specific aspect of the overall pattern. For example, one person with the 17th Gift might see how to restructure a new economic paradigm that will serve humanity instead of dividing it. Another might see how to create a community network based on a similar model. In this way, each 17th Gift holds a fractal part of the overall jigsaw puzzle of the evolving humanity.

It is only when we come to the 17th Siddhi that the entire picture is placed before a single human being. This Siddhi directly corresponds to the complete opening of the third eye. At this level of consciousness, the one who sees and the object of their seeing becomes one and the same. All that is left is a single eye through which consciousness endlessly pours. This is called the Siddhi of Omniscience. In traditional religious dogma, only God or the gods are seen as omniscient, which is a wonderful example of how effectively the lower frequencies keep the third eye firmly closed. This other eye inside human beings is located within the higher functioning of the pituitary gland and it can be opened in a number of ways. Various yogic disciplines are aimed specifically at the opening of the third eye and in the Hindu chakra system it corresponds to the 6th *ajna* chakra residing between the eyes.

Many drugs and psychotropic plants also affect the functioning of the third eye through the medium of the pituitary gland. For millennia, shamans have used natural psychotropic plants as a means of seeing through this incredible instrument of higher consciousness.

However, temporary activations of the third eye, whilst affording us a sneak preview of higher states, cannot compare to the Siddhi of Omniscience. In the alchemical sequence of awakening, the Siddhi of Omniscience can only flower after the full opening of the heart, and the heart-opening phase destroys all vestiges of self along the way. It is as though the eye can only open once the inner sun arises. This 17th Siddhi is something of an anomaly, since it is the only aspect of the mind to survive awakening. All other aspects of the mind are transmuted into the higher frequencies with the exception of this single eye — the mind's eye. Once the heart has bloomed, the inner eye provides a communicative link or interface between the brain and pure consciousness. The inner eye could be described as perfectly circular, seeing in all directions simultaneously. Whereas the 17th Gift can read the pattern of the future, the 17th Siddhi can see the future in precise detail. Just as time is distorted through the 17th Shadow, it is seen as illusory through the 17th Siddhi, which experiences past, present and future as a single eternal screen of consciousness.

It is through the Siddhi of Omniscience that the world gets its true seers and oracles. There is an ancient tradition that the true seer is always blind, or that the true oracle is deaf. This symbolises what it means for consciousness to see through you, rather than you through it. At such a high frequency, the human body becomes merely an instrument for consciousness. Through the 17th Siddhi consciousness plays the games of the maya even though it sits outside of the maya. If the 17th Siddhi offers an opinion or viewpoint it does so in order that a certain human drama can be played out. This is why certain masters sometimes say strange things to their disciples — in order that a specific opinion will pass through the awareness of that individual and bring about a particular transmutation. It is out of omniscience that such things are spoken, and anyone who speaks from within a siddhic state follows this same process. This is also a Siddhi that comes into the world to display itself. Sometimes and for certain people, it is a part of their karmic legacy to hear about their future. Sometimes hearing about your future actually creates that future and sometimes it creates the opposite effect of nullifying it.

Whenever you try to understand the 17th Siddhi with your logical mind, you will inevitably fail. The mind can only hold a single opinion — either that the future is predetermined or created as we live. The mind cannot grasp the paradox that both must be true, since both are interdependent. When the 17th Siddhi appears in the world, it comes to mess with the human mind! To omniscience, there can be no borders, levels or past lives. There is only the endless fractal pattern of life, repeating over and over — the same but always new — the serpent eating its own tail. This is a Siddhi that sees and knows only perfection (its programming partner the 18th Siddhi). As such it is beyond seeing flaws, because what were previously seen as flaws were really only devices to bring about the state of omniscience itself. Omniscience testifies that whatever is happening at this precise moment is absolutely perfect because it is following the direction of consciousness as it burrows into the world of form.

*Whatever
you decide
is absolutely
correct and in
full harmony
with all
that is.*

Perhaps this is the mystical meaning of the original name of this 17th archetype from the I Ching — *Following*. Everything that happens follows the plan and the plan is perfect, and the plan is both predetermined and created as we walk the path.

When the 17th Siddhi comes into the world it comes like a Divine afterthought. It is one of the seven powers of Grace, serving no purpose other than to offer the truth that nothing need be done in life. Whatever you decide is absolutely correct and in full harmony with all that is. To see perfection is to embody perfection, and such a state is an ending rather than a beginning. Wherever the grace of omniscience is seen, human denial is finished. If you are following a path of service, it is correct. If you are following a path of betrayal, murder or revenge, it is correct both when you are in denial and when you come out of denial. The truth of this 17th Siddhi creates upheaval in the world. We humans do not wish to hear a truth that so undermines our individual identities, but consciousness must express all its opinions through the world of form. The deepest irony of all life is caught in one famous mystic's words: "God can only come and visit you when you aren't there." To put it another way, if you want to know what omniscience looks like, you have to cease to exist.

18th Gene Key

The Healing Power of Mind

SIDDHI
PERFECTION

GIFT
INTEGRITY

SHADOW
JUDGEMENT

Programming Partner: 17th Gene key

Codon Ring: The Ring of Matter
     (18, 46, 48, 57)

Physiology: Lymphatic System

Amino Acid: Alanine

THE 18ᵀᴴ SHADOW – JUDGEMENT

THE VICTIM MIND

Built into the human genetic matrix is a deep sensitivity to imperfection, and it is this sensitivity that gives rise to the human qualities of criticism and judgement. As we shall see, the 18ᵗʰ Gene Key and its themes of Judgement and Integrity can have either an empowering or a disempowering effect on you and others. This theme of Judgement runs about as deep in human nature as any other trait you can imagine.

The 18ᵗʰ Shadow begins in your childhood. It has a built-in need to challenge authority, and the first real authority in your life is your parents. Challenging our parents is a fundamentally healthy thing to do, as it is a part of our innate urge to become differentiated. This process begins in earnest as we enter our third seven-year cycle and roughly spans the ages of 14 through 21. This stage of our development primarily concerns the growth and expansion of our mental faculties, and it is during this period that our future opinions are laid down and our capacity to judge is tested and forged. The key to this process lies more in the frequency of the parents than the child. If the parents make the mistake of taking this process personally, the child will never fully make the transition through this biological phase, instead becoming *stuck* at the same low-frequency as the parents. If, however, the parents do not get stuck in their own judgemental or self-judgemental patterns, then this phase will result in true adulthood. Sadly, most children never make it to true adulthood but stay solidly trapped in deep low frequency patterns of judgement for the rest of their lives.

The 18ᵗʰ Shadow gives rise to a collective phenomenon in the world known as the *victim mind*. The victim mind is a conglomeration of all the undermining, judgemental thought patterns throughout the world. If you sincerely examine your own thoughts during a typical day, you will probably discover that a great percentage of your thinking is affected by the victim mind. In other words you are allowing your mind to be influenced by the collective negative thought patterns of the whole of humanity. The true import of this last statement can come as a huge shock to many people. The world of the victim mind is an inner world of gossip, complaining and worrying.

*The more you
complain
the more
you damage
yourself and
the world.*

Most of us complain inwardly about all aspects of our lives, especially the people in our relationships, and we worry incessantly about mundane issues such as money and our health. Ironically, it is exactly this kind of thinking that keeps us from being abundant in terms of both our wealth and our health. You may think it is absolutely human to complain, but it creates a negative frequency in the human aura of both the complainer and the victim. In other words, the more you complain the more you damage yourself and the world.

Morally speaking, judgement gets a bad rap in the modern world. We talk about being *non-judgemental* as though it were one of the highest goals in life. In fact, it is impossible not to judge because judgement is the way in which the human mind thinks. What defines the low frequency of the victim mind is that you *identify* with what you think — in other words, your judgements define your identity and make you feel more secure. However, if you can make a judgement and at the same time be aware of judging then you are no longer trapped by your own mind and thus the frequency around that judgement changes. Because the 18th Shadow is twinned with the 17th Shadow of Opinion, these two genetic themes are inseparably linked through the structure of your DNA. All judgement is rooted in opinion and vice versa. The more you think you *are* your opinions the more you have to defend them, whereas the more lightly they are held the less attached you will be to being right. This 18th Shadow is about needing to be right. It takes a loosening and a lightening of your identity to let go of your need to be right. If this Gene Key is a part of your Hologenetic Profile then the most important question to ask yourself is: *Would I rather be right or happy?*

What is so fascinating about the 18th Shadow is that the details of the judgements themselves are utterly irrelevant. It doesn't matter what your political views are or how much you might detest or disapprove of someone. The only thing that matters is how seriously you take your own judgements. If your identity rests on your judgements then everywhere you go you will stir people up negatively — not through your opinions, but through the turbulent frequency you put out into your environment. Even if your judgement feels positive to you, the chances are it will still be misread by others just because you hold it so tightly. The themes of both the 17th and 18th Shadows can also become easily obsessed with details. The victim mind can fixate on the tiniest and most irrelevant details in life. At a low frequency these are the kinds of minds that blow things far out of proportion, espousing opinions that are actually far removed from the facts.

The 18th Shadow holds another of the key secrets of all human interaction — all judgement is self-judgement. Because we humans perceive ourselves as separate from each other, we miss this vital fact. To challenge another person is to create a division within yourself. This does not mean that you should never challenge others, but it does mean that you need to remember that the other represents something inside of you that you are being given an opportunity to resolve. The key to raising the frequency of the 18th Shadow lies in this perspective that everything external in your life is a mirror of an internal process seeking resolution. This is why the highest aspect of the 18th Gene Key is Perfection, because life is continually offering you a pathway to the realisation of your own perfection.

As an aspect of the Codon Ring of Matter, the 18th Gene Key is part of a genetic chain of low frequency fears buried in the human genome. These fears lie deep within the body's immune system and create the matrix for all human initiation into higher states. It is precisely your passage through these fears that allows you to gain access to the higher levels of frequency and realisation. It is through the cluster of Gene Keys within this codon that spirit first penetrates into the world of matter and releases its true radiance. In this inner world, the 18th Shadow governs the integrity of your mental body — your mind. The secret of this Shadow is the gradual realisation that your outer life *is* your greater body. The more deeply you accept responsibility for this, the less personally you take life and the easier it becomes. The 18th Shadow is thus a place inside your DNA that causes you to become mentally attached to the material world and all its distractions, which in the long run only serves to increase your suffering. However, as you escape the victim mind you begin to alter the coding structure of this aspect of your DNA, thereby attracting a higher energetic frequency stream into your life and changing everything for the better.

REPRESSIVE NATURE — INFERIORITY

The repressive side of the 18th Shadow is about feeling inferior. This pattern first develops through our relationship with our parents, particularly during our teen years. If our parents judge us too harshly during these formative years, a pattern of self-judgement sets in. If you are one of these people then your pattern is to turn judgement in on yourself. Out of your own self-judgement you become a conformist, going along with the majority instead of standing up for yourself and your own convictions. All judgement is based on comparison, and with self-judgement you compare yourself unfavourably to someone who you consider to be above you. This in turn gives rise to a deep-seated feeling of inferiority and a constant habit of undermining yourself.

REACTIVE NATURE — SUPERIORITY

The reactive side of the 18th Shadow manifests as the judgement of others based on feeling superior. This judgement emerges as a constant need to challenge any authority figure, beginning with the parents. If the process of challenging one's parents is not permitted a natural resolution, this behaviour pattern becomes a permanent part of one's character. It occurs when our parents do not maintain strong enough boundaries around us and instead fall victim to their own self-judgements. The urge to challenge then becomes cemented by an unconscious anger in the child that the parents were not strong enough to maintain their integrity. This translates in the teen's life to a deep disrespect for authority and a belief that one is somehow above it. These are volatile people whose very identity is then built around the undermining of others.

THE 18TH GIFT – INTEGRITY

TAKING ON THE WORLD

The 18th Gene Key thus determines whether a child enters adulthood as an integrated adult or as a wounded child posing as an adult.

The 18th Gene Key is psychologically extremely profound. It is actually the basis of psychology because it holds the keys to human conditioning. From the moment a child is born, an innate urge inside its DNA begins to explore the boundaries of its environment. This Gene Key is about the material, emotional and mental boundaries that you meet throughout your life. Because we are biologically imprinted in seven-year cycles, it is possible to say that we have not fully incarnated into our lives until the age of 21. The first seven years critically tests the integrity of the physical world through the development of our physical foundations — our skeletal structure, muscles, basic physiology and movement. The second seven-year cycle sees the development of our emotional life as we pass through puberty and learn to handle our sexuality and the way it governs our basic identity through attraction and repulsion. The third seven-year phase completes our incarnative cycle into adulthood by way of our teen years when our thinking develops extremely rapidly and we adopt a mental structure made up of many levels of judgement and opinion.

Each of the three phases of our early life is built upon each other in a logarithmic fashion. This means that a problem encountered in the first cycle will appear again at the same junction in the second and third cycles. Throughout each of these cycles, the Codon Ring of Matter and its associated Gene Keys play a crucial role through continued testing of the outer environment. Essentially, these Gene Keys are looking for constant biofeedback so that the child can orient itself in the world in a healthy manner. The three phases — physical, emotional and mental — are governed by the 46th, 48th and 18th Gene Keys respectively, whereas the 57th Gene Key lays down the even deeper seeds of these three phases through the three trimesters during gestation. Thus this codon ring is of huge importance at the developmental level as it governs the very infrastructure of our physical, emotional and mental health. As the Gene Keys are understood in more depth, these four in particular will bring a great revolution in the way in which we approach the whole notion of child education.

Since the 18th Gene Key governs your intellectual development through your teen years from 14 to 21, it answers a lot of questions that parents may have about this challenging phase of young adulthood. The patterns that emerge in this third cycle of your life are dependent on what happened in the earlier two cycles. If, for example, your parents split up during the middle of the second phase, then this same type of upheaval will appear in your life in the middle of your third cycle, but this time it will have a mental focus rather than an emotional one. At every stage in your early development you have many opportunities to heal aspects of earlier cycles, or depending on how you and those around you deal with these challenges, you will imprint these low frequency patterns even more deeply into your psyche.

The third seven-year cycle governed by the 18th Gene Key is therefore the most critical cycle because it represents the last chance parents have to help their child reach true adulthood. If the parents are not rooted deeply in their own integrity, they will fail to create the proper environment for the child to pass cleanly through this most challenging phase. The 18th Gene Key thus determines whether a child enters adulthood as an integrated adult or as a wounded child posing as an adult. Naturally parents have to set some strong boundaries during this teenage phase, but the 18th Gene Key is not really about this. The secret of the Gift of Integrity is to be able to hold your own space without *reacting* to your judgements or self-judgements. As a parent, you *will* be drawn in by the upheaval of this period in a child's life, but as a parent you also have to learn to stand back from it and trust in it. If parents truly understood how crucial their role is during this phase in their child's life, they might act with less judgement and feel much surer of themselves.

Integrity as a vibration means far more than just holding fast to your values. It is actually a word often used by architects and engineers to describe the strength of material structures. Integrity is a deeply physical attribute and is actually a function of your immune system in maintaining the tensile strength within your body. For many people who have to resolve issues from their teenage cycle, it will take a full seven years — whether through therapy, bodywork or some other sustained system that targets your conditioning. Integrity is born into the human vehicle, but it also has to be earned — whether during teen years or later on in life. When a person has attained the Gift of Integrity or grown into it through a healthy passage into adulthood, it will remain forever because it is the true nature of every human being. One of the greatest opportunities of the 64 Gene Keys is afforded to parents. If you can see the higher frequency of your child from the moment of birth and if you can continuously hold that frequency for them through thick and through thin, you will see a living miracle growing before your very eyes.

If the 18th Gene Key is prominent in your Hologenetic Profile, your life will continually bring you back to these recurring issues rooted in your childhood. You will have to understand what made you the way you are in order to release the aspects of your conditioning that do not belong to you. In time, this will make you a master of understanding the ways conditioning works and you can help others break free just as you have. When you are free from the trap of the victim mind, your judgement becomes Integrity — the archenemy of the victim mind. It is the same energy, the same archetype, but experienced from a higher level of consciousness. Judgement, criticism and correction are magnificent qualities when used in the right way. The Gift of Integrity is about demanding and maintaining a high standard in everything you do. As a fully healed adult, your purpose is to help others complete their childhoods so that they can finally enjoy their lives and pass on their integrity to their children.

To uphold Integrity, you have to be courageous — you have to challenge anything and anyone who does not meet your high standards. To live with Integrity is to take on the whole world — to challenge it to meet the high standards you are setting. Wherever you see someone living with Integrity, you are seeing someone using the power of judgement in an objective and impersonal way. This is the great Gift of the 18th Gene Key — not to use or take judgement personally, but to learn to judge from your heart. Judging from the heart can never be cruel because true Integrity has only one purpose — to serve the whole in the spirit of truth and compassion.

THE 18TH SIDDHI – PERFECTION

THE BODHISATTVA

Just as the 18th Gift completes our journey from childhood to adulthood, the 18th Siddhi shows us how to heal our mental anguish and assume our true place in the universe as whole adults. It does this through tireless compassionate service to the vision of perfection.

When you put your Integrity in service to the whole, an amazing thing happens — you become more and more dissatisfied! The more good you do, the more you realise how much more you could do. This is known as divine dissatisfaction. Everywhere you look, you see how the world could be improved. As you become more and more anchored in your integrity you aim higher and higher in your service. Perfection is the highest vision a human being can aim for, and you do begin to aim for this highest of ideals — not for yourself, but for the sake of the world. At this level you are using the archetype of judgement to challenge the fabric of reality itself. The higher levels are infinite and therefore seem impossible to attain, but like a Zen Koan that cannot be solved by the mind, if you keep aiming for perfection it will eventually blossom within you as a natural state. You will pass beyond the very definition of perfection.

The 18th Siddhi contains some profound paradoxes. Living in a state of perfection entails the death of the mind and, as such, perfection is an ending. When you realise perfection, evolution ends. But the 18th Siddhi is the source of the tradition of the bodhissatva vow. The bodhissatva is a being who forgoes his or her highest state of consciousness in order to stay in the world and help others attain that state. This tradition holds a deeply metaphorical genetic truth. Our genes are all bodhissatvas programmed to serve the whole until it has attained perfection.

Our genes are all bodhissatvas programmed to serve the whole until it has attained perfection.

We have seen that the 18th Gene Key is designed to ensure human health by constantly challenging our environment to come in line with higher principles. What then is perfect health? This is an interesting question because it involves the whole of creation. The true physical health of humanity cannot be attained until all human wounds have been healed. Even though a human may attain a state of perfection beyond all human understanding, the body is still a part of humanity and as long as there are still wounds in the world, almost nobody can experience perfect health. In other words, perfect health cannot occur unless the whole is healed. This is why the siddhic states of consciousness still have a purpose in the world — they reflect the state of inner perfection to all humanity so that in time we will recreate that state in the outer world as well. The 18th Siddhi will also one day bring a new science of mental healing into the world. This science will be built upon the understanding that the mind is an energy field existing on its own plane. As we begin to unlock this Siddhi, so our minds can be used to heal mental, emotional and even physical problems instantaneously. Even deep psychic disturbances like madness can be completely healed through contact with this 18th Siddhi. It will herald a new age of healing in which human perfection will come thicker and faster than we may believe.

The Codon Ring of Matter ensures that the whole of humanity must one day reach a state of perfection where evolution ends and we as a species finally enter the dimension of eternity.

Each of the Gene Keys within this Ring is responsible for an aspect of this spiritualisation of the material plane. The 18th Siddhi contains the knowledge of how to bring perfection to earth through the mental plane. To the one who lives within this Siddhi, perfection is already here as an imprint underlying all creation. These people live in the Eden that we all long for, but paradoxically they must help move that state from the inner reality to the outer.

The 18th Siddhi is not often seen in the world today. It involves the incarnation of an archetype of perfection. Such people are tireless in the service they give to humanity. Paradoxically, they know through personal spiritual experience that the world is perfect exactly as it is — there is nothing about the world that need be improved. Yet they are driven by this archetype of perfection to keep tirelessly improving both themselves and the world until the day they die. All beings within the siddhic state share this dilemma of having completed their inner evolution, yet find themselves still living in the world. The physical vehicle completely changes its programming after entering the highest siddhic state of enlightenment. If the 18th Gene Key is activated within your Hologenetic Profile then you will continue to lead humanity towards perfection in any and every way you can conceive of. Often such people are called *saints*.

Another aspect of the 18th Siddhi concerns the way in which we think, and especially the way in which we think about others. One of the great universal laws is that energy follows thought. This Gene Key has enormous power over the mental plane and can completely reformat a person's mental reality. Someone demonstrating the 18th Siddhi is above the victim mind — when they look at someone, even someone who appears genuinely evil, they see only the Higher Self hidden in that person. By holding this high thought frequency around that person, they actually influence hidden currents of energy that raise the frequency of the other. This is called *siddhic thinking*, and it involves thinking with the heart. This is much more than positive thinking. Siddhic thinking is really an energy field surrounding the aura of the one manifesting the 18th Siddhi, and in this sense it should not be considered thinking in the normal sense of the word.

Each Gene Key has its own higher programming function. Some are designed to just sit there awash in the bliss of their nirvana, whilst others have far more definitive roles to fulfil. The 18th Siddhi actually serves a very mystical purpose — it heals the split in the mental plane itself bringing humanity closer and closer to its realisation of oneness. It cannot and will not sit still until this great rift in the world mind has been healed. Higher reality perceives the mind not as a mere function of the human brain but as a *mental body* or *plane* woven into the living aura. This mental body is both an individual construct in the human auric field and a great plane that unites all human beings as one.

Perfection only exists within the timeless realm. It is beyond time and space because it is the end of evolution. The bodhissatva Gene Keys keep returning to our planet as individual human beings whose sole purpose is to build this collective vision of perfection on the material plane. All human beings are involved in this great dance, whether we are aware of it or not. We are all moving towards this single dream of a species that will one day attain perfection and reach the end of its story. As we will all one day see, the end of the story also happens to be the beginning of our universal life.

☰☱ 19th Gene Key

The Future Human Being

SIDDHI
SACRIFICE

GIFT
SENSITIVITY

SHADOW
CO-DEPENDENCE

Programming Partner: 33rd Gene Key

Codon Ring: The Ring of Gaia
(19, 60, 61)

Physiology: Body Hair

Amino Acid: Isoleucine

THE 19TH SHADOW – CO-DEPENDENCE

THE GREAT CHANGE

Along with the 49th and the 55th Shadows, the 19th Shadow is perhaps the most topical Shadow of our current age. We are living through a time of unprecedented global mutation. This mutation is on many different levels, from the most physical to the most spiritual. To truly capture the essence of what is occurring to humanity, one needs to look not so much at the spiritual dimension, but at the opposite end of the spectrum — the biogenetic. The mind of man is steeped in causality — as long as we recognise a cause we believe there is a purpose. If you are a geneticist, everything you look at appears to serve the purpose of the evolution of life, whereas if you are a mystic, everything serves the purpose of the evolution of consciousness.

For millennia humanity has tried to understand the purpose of life from the spiritual dimension, but until fairly recently we have not had the ability to understand the purpose of life from the material dimension. With the advent of genetics we are beginning to see the micro-processes that drive evolution itself. A scientist might say that man's spiritual evolution comes about as a result of his biological evolution. The mystic tends to see it the opposite way around. What is fascinating about the scientific view is that it focuses on the lower frequency realms — the realm of matter. The 19th Shadow represents an aspect of our genome that is currently undergoing an intense mutation. A mutation is a spontaneous and usually sudden quantum leap from one state to another. In genetics, mutations are often the *mistakes* that happen when genes copy one another. Such mistakes can lead to fascinating new chemical combinations which can in turn lead to the evolution of entirely new forms.

The 19th Shadow of Co-dependence has its roots deep within our tribal ancestral past. Co-dependence refers to the state of consciousness below independence. To be independent means to rely only upon yourself, whereas to be co-dependent means to rely on outside agencies. As primitives we primarily relied upon nature for our survival and because we relied upon an outside agency we anthropomorphised it. In other words, we created gods to represent that agency.

As long as man believes in a God outside himself, the frequency of our planet will remain at the level of the 19th Shadow.

Thus it is out of the 19th Shadow that all the world religions have been born. Our relationship with God or a set of gods is a purely co-dependent relationship because it is based on this need for outside authority. It is here in the 19th Gene Key that one of the great human stories is coded — the story of our relationship to God. As long as man believes in a God outside himself, the frequency of our planet will remain at the level of the 19th Shadow. The vibration of human suffering depends upon the existence of a separate authority of a higher frequency than us. This last sentence is the ultimate definition of what it means to be a victim, which is what characterises the shadow frequency. The programming partner to the 19th Shadow is the 33rd Shadow of Forgetting. In creating a God *out there*, we have forgotten the power that lies dormant within us.

The ultimate reliance we have outside ourselves is on food. God has always been about food and food has always been about territory. Food production was based upon tribal territory, which is why the different nations and cultures developed in the first place. But today, at least in the developed world, food no longer has to come from our own tribe. It can be flown in from anywhere in the world. For one thing, our dietary needs are changing as we learn to manipulate our environment with greater efficiency.

Through sciences such as nuclear physics and genetics, humanity is beginning to play god more and more and is thus moving from a state of co-dependence to a state of independence. Because we can now outwit the gods with modern technology we no longer need them as much. The more advanced the society, the deeper we question God as an outside agency.

However, the 19th Shadow is currently undergoing a huge genetic mutation, which means that man's reliance on religion is also undergoing a transformation. The old tribal fears of not having enough are dying, and with them the great religions. The breaking of such a deep-seated and ancient co-dependent relationship has powerful repercussions for our world. The old ways must die to the new — such is the purpose of mutation, and the process is one of destruction. Only as the dust finally settles will the new creation become fully realised. The reason all of this can seem so terrifying is that it represents a fork in our evolutionary development in which an entirely new path is opening, a path in which human beings will have to leave behind the old tribal co-dependent ways. The whole world is dividing into those who are becoming more independent, and those who cling to the comfort of the old ways. On a global level we are beginning to see this made manifest now in the battle between globalisation and factionalism — and between science and religion.

For us as individuals, the transformation of the 19th Shadow will see its deepest manifestation in our relationships. The old-style co-dependent relationships of the working husband and the homebound wife are giving way to a new level of independence. The liberation of women is changing the infrastructure of our civilisation and children are increasingly cared for collectively, so that both mother and father retain a higher level of independence. Whether we like this or not it is occurring all across the developed world. Our children are growing up as the children of society rather than the children of a single tribal family.

Because of the huge genetic shifts taking place across the board, male/female relationship dynamics are more challenging than ever. A great change is coming, and roles are changing to accommodate the birth of a new paradigm. Although it may be a difficult birth, in the not too distant future the 19th Shadow will have disappeared entirely from our world.

The contemporary mutation moving through this 19th Gene Key is having an unprecedented effect on all life on our planet. As a vital aspect of the Codon Ring of Gaia, along with the 60th and the 61st Gene Keys, it is breaking down the very patterns of the world psyche. The reactive Rigidity of the 60th Shadow and the Psychosis of the 61st Shadow have long held sway on our planet. A great reaction is occurring within the chemistry of our DNA as the old ways appear to tighten their grip on the only reality they have known. There is enormous fear being generated through the Shadows of this Codon Ring and enormous potential violence as our co-dependence is broken. However, the truth is that all life is and always has been interdependent because all life is one. Even independence is an illusion, and this realisation is bringing an end to the world psychosis that operates out of the 19th Shadow's low frequency survival-based reality. It is through this Ring of Gaia that we must and will eventually see and will live once again in union with all beings that share this planet earth.

REPRESSIVE NATURE – NEEDY

The repressive nature of the 19th Shadow emerges as neediness or clinging. These are people who cannot let go of the past out of a fear of being alone. This creates catastrophic relationship dynamics based on making other people victims of their need. Repressive natures can be very cunning with the way they transmit their Shadow patterns — these people will very likely use subtle tools such as guilt to get their own needs met. They need to feel needed, and they will act out all kinds of dramas, often totally unconsciously, to get the attention they crave. They are masters of *negative attention* — drawing other people's energy towards them without caring what expression it takes. Even violence is a form of attention. The only way to break out of such patterns is to move into independence.

REACTIVE NATURE – ISOLATED

The angry expression of this Gene Key is isolationist. These people refuse any attention, loudly proclaiming that they don't need anyone. Such a nature only pretends to be independent while beneath the surface they seethe with rage. Of course, people that isolate themselves like this always take great care to do it right before everyone else's eyes. They make a point of showing you how alone they are, craving the attention it brings them and becoming even more embittered when others leave them alone. Ironically, when others do try and support them or offer them friendship, they usually explode, projecting all their pent up anger onto the other person. It is easy to see how the repressive and the reactive nature together create the perfect dynamic of the typical dysfunctional co-dependent relationship.

THE 19TH GIFT – SENSITIVITY

THE WHISPERERS

The Gift of Sensitivity is about being highly attuned to the needs of others. In order to be able to sense others and their needs, you must first become independent from them, which is what this 19th Gift is about. The moment you reach the frequency of independence your natural energy becomes apparent. This 19th Gift is also a gift of touch. It does not have to be literal touch, although it can be — many of these people are gifted healers or therapists. It is more than just a physical sense of touch, but a *touch* with people as well as animals. As we learned from its Shadow, this Gene Key is rooted in material need, and when you elevate the frequency from your own need, you suddenly become aware of the needs of everything and everyone around you. This makes the 19th Gift a great environmental barometer.

There is a rare phenomenon in human beings known as *synaesthesia*. Synaesthesia is the genetic ability to link different senses internally — for example, to smell with your eyes, or sense colours with your hands. It is actually deeply connected to the 19th Gift and can often be activated when a higher frequency passes through this Gene Key. Synaesthesia is a by-product of increased sensitivity to your sensory environment. If this 19th Gift is an aspect of your Holo-genetic Profile then it is highly possible that you may discover latent abilities that allow you to feel deeply into your environment and in particular to sense the emotional patterns and needs of others through the living auric field. Many artists and healers have these kinds of gifts and can sense the hidden higher energy fields through their fingers or through their skin or hair. Sensing these fields in nature allows you to see a completely different world from most people — a world of energy fluctuations, intense colours, moods and internal pressure patterns. In the deepest sense, the higher frequencies of the 19th Gene Key have access to the realms of magic.

We may recall that the 19th Shadow finds its basis in the human dependence on external food. The 19th Gift simply shifts this perspective to a higher level. At a higher frequency *food* is really life energy, or what the ancients called *prana* or *chi*. This Gift allows you increased sensitivity to this living bio-energy that connects all creatures. When your heart and being opens to sense the abundance of this energy throughout nature, you become for the first time emotionally independent. Only the activation of the love inside your own DNA gives you this wider sense of being. Furthermore, because this 19th Gene Key is one of the first places in the human genome where the Great Change is having an immediate impact, the movement from co-dependence to independence is something we see increasingly throughout our human cultures. In its early stages, it is not an easy passage. Suddenly having increased sensitivity to your environment makes you more aware than ever of your old tendencies towards co-dependence. Even though *you* may have opened your heart to the Great Change, most of the world still has not done so, which puts you in a position of great responsibility, not to mention discomfort.

We are seeing such responses to the Great Change all across our culture today as human beings become more aware of how much damage our co-dependence is doing to our environment. The 19th Gene Key is a genetic portal into the unconscious and particularly into the collective unconscious. Interestingly, this Gift seems to have strong activation in cultures that live close to nature such as indigenous cultures. In such tribal groups there has always been an increased sensitivity to those regions that are beyond the five senses. What modern humans often interpret as naivety in such cultures is in fact a heightened genetic sensitivity to the quantum reality of the unconscious. As this Gene Key reawakens in humanity we see changes in our dream life and through this portal we can reconnect with our ancient sense of the magic that comes through dreams. People with the 19th Gift are often shamans because of their increased sensitivity to other worlds and other realms.

The 19th Gene Key is one of three primary portals (along with the 62nd and the 12th Gene Keys) that allow humanity access to other evolutionary kingdoms within nature. These kingdoms, which have often been referred to as the angelic or devic realms are planes of consciousness that follow a similar evolutionary pattern to humanity, but in parallel dimensions. The 19th Gene Key behaves like a genetic marker in human DNA, and only when you hit a certain genetic frequency does that marker activate the portal allowing information to cross clearly between these parallel worlds. Certain human beings have always been credited (or indeed discredited) with the ability to see faeries or hear voices of spirits or angels. This is a genetic ability that comes specifically through the 19th Gift. Of course, the 19th Shadow also has its low frequency counterparts, which can cause human beings to tune into the lower or subterranean kingdoms often known as the demon realm. As a matter of fact, most human beings are directly influenced by these parallel evolutions whether they realise it or not. Only a higher, virtuous frequency allows you to become independent of these Shadow forces, which otherwise pull you again and again into lower frequency patterns and emotional states.

The 19th Gift has a particularly powerful link with the mammal kingdom. Because it acts as a portal between the unconscious and conscious realms, those who know how to use this portal can access information from other realms apart from the human. Since this Gene Key has evolved from humanity's relationship to food, it also evolved out of our relationship to nature since traditionally we killed animals for our food, especially mammals. This ancient sacrificial relationship between humans and animals is actually based upon a timeless sacred pact between the two species. Those with sensitivity will know the future destiny of this cross-species contract. Most tribal cultures contain legends of a time when animals and humans were a single consciousness, and in the future this is our eventual destiny — to once again enter the collective quantum field where both humans and animals co-exist.

It is this ancient connection in our DNA between other mammals and humans that is the origin of the *whisperers* — specially gifted people who can communicate with animals, or who can act as a bridge between entirely different species. Such people can tune into the ancestral gene pools of certain species and often feel a profound connection to nature at a deeper level than *ordinary* people. In tribal societies, the shaman's specific skill was to bridge the ancestral spirit behind the tribe to the individuals within the tribe. This is a direct reflection of the function of the 19th Gift. People with these gifts have always been humanity's natural *interpreters* of other realms. Their increased awareness of the energetic pathways and portals between all realms, material, emotional, mental and divine, mark them as pioneers and initiators into these magical realms.

In today's world, people with the 19th Gift are likely to use their sensitivity in any field where people are working together in groups. Their ability to unconsciously sense the needs of others means that they are often seen as *psychic*. However, they can also be very grounded in the reality of the needs of the material plane — for example they can use their heightened sensitivity to bring balance to the spheres of money, work and relationships. In fact their very presence brings focus to these kinds of issues. This Gift truly spans all realms and its future function, as we shall see, is to collapse the barriers that separate them thus fusing together the ancient magical realms with the modern material plane.

THE 19TH SIDDHI — SACRIFICE

THE FIFTH INITIATION — THE ANNUNCIATION

Our awareness is like a series of Russian dolls — as we make each leap in awareness, we come to realise we are housed within a wider framework than we had previously understood.

The path of frequency through the 19th Gift goes from co-dependence to independence and finally to interdependence. Interdependence represents a quantum leap beyond the other two, and its realisation is the future destiny of our species. In many ways, it is the 19th Shadow of Co-dependence that contains the seed of the 19th Siddhi of Sacrifice. In a co-dependent relationship, both partners have sacrificed a part of their true selves to the relationship, and the resulting lack of synthesis they feel drives the negative patterns of the relationship. In a relationship that is truly interdependent, both partners also sacrifice their sense of individuality into a higher vision of their Divinity, holding nothing back for themselves at all. The true meaning of interdependence is about entering into a state of union with all beings in the cosmos, which involves the death of the separate self. This kind of sacrifice can only occur when you give your heart unconditionally to another. Instead of dying, you are actually reborn as a higher dimensional being. In surrendering your smaller self, you attain realisation of your greater Self.

Through the 19th Gene Key you can see how each level of frequency has to transcend itself entirely. Once humans overcome their co-dependence on outside agencies, they finally achieve independence. Likewise, once they attain independence, they have to take another great leap — to give up their hard-earned independence and trust in the totality itself. This surrender to the collective structure involves sacrifice at the highest level of the word's meaning. This is the sacrifice of your separate identity, and perhaps far more frightening for the separate self, it is the sacrifice of your body. The 19th Siddhi shares a profound connection with the 49th Siddhi of Rebirth. These two Siddhis represent the key mystical process that will eventually overtake the human species. Great secrets are held within the world's great myths, and here we can see the myth of sacrifice — the Norse Odin hanging upside down on the world tree, or Christ hanging on the cross. All sacrificial myths lead to a rebirth, and all such myths are anthropomorphises of deep genetic secrets hidden within human DNA.

As we have seen, this 19th Gene Key has a deep connection with the animal kingdoms. Humanity has evolved out of the animal realm. We are the result of a series of genetic mutations that took place in primates and finally led to the creation of a new species — Homo sapiens. Through this 19th Siddhi you can see how consciousness has travelled through form after form, each time creating a more complex form in order to house a higher frequency than the previous one. At every level within this chain, the higher form has lived off the form below it in order to keep evolving. Life is thus a living chain of sacrifice. Earth is really the breeding ground for a series of genetic leaps that parallel the awakening of our full spiritual realisation. Our awareness is like a series of Russian dolls — as we make each leap in awareness, we come to realise we are housed within a wider framework than we had previously understood. There are a total of nine dimensions that our earth has to move through, and as we pass through each of these initiations, we have to sacrifice our smaller, local self before finally being born as a true universal human being.

Consciousness is now beginning to outstrip man and is reaching out for a higher form. But the new form must emerge out of the old one, so deep within human DNA new mutations are being triggered. This is one of the main reasons for the vast population explosion on the planet — our genes require the maximum diversity in order to trigger a genetic mutation potent enough to reshape Homo sapiens into something quite different. It is also the reason why so many new diseases are emerging through our DNA. They are all early mutations — precursors of what is to follow. Through the 19th Siddhi it is not only the individual that must sacrifice him or herself into the whole, but also the whole human species that must be sacrificed. Everything we see happening in the world around us — from pollution to global warming to wars and social upheaval — is a result of the profound genetic process we are undergoing.

For those with the 19th Siddhi the focus is always on the future needs of humanity rather than the current needs. These people understand what is to come as well as what we must go through. Such people stand alone as heralds of a future consciousness and their lives are a magnificent example of sacrifice to that consciousness. They emerge during times of great mutation, since they themselves are in the grip of that mutation. With their hypersensitive mutated DNA, they see the new form emerging and do their utmost to prepare people for the coming consciousness shift. They are highly sensitised bridges to the new human and have the ability to siphon information about the new paradigm from behind the veil of the future into the present. Every Siddhi has to make its sacrifice in this way, because each of them represents a being from the future that works in the present.

The 19th Siddhi contains the secrets of mystical initiation. Every aspect of the earth's consciousness must move through the nine portals of initiation before our collective planetary evolution comes to an end. Each of these initiations is explored in more depth within the 22nd Gene Key.

THE NINE PORTALS OF PLANETARY INITIATION

1. Birth
2. Baptism
3. Confirmation
4. Matrimony
5. Annunciation
6. Communion
7. Ordination
8. Sanctification
9. Glorification

Each initiation brings us into a wider awareness of our interdependence with the whole. When the 19th Siddhi manifests in form, then a great being makes an individual sacrifice on behalf of the whole. This is the mystery and hidden meaning of the life of Christ. Through an individual's sacrifice, the 19th Siddhi allows the entire collective to pass through a group initiation. We can see how the Christian rites hold the codes of the great initiations, even though they have been effectively *frozen* into structures that have little or nothing to do with the initiations themselves, which occur organically and usually over the course of many incarnations. The Ring of Gaia then, of which this Gene Key is a vital aspect, connects all earth beings on this same initiatory journey. As the ultimate form manifestation of Gaia, humanity stands on the cusp of one of the greatest initiations — the fifth initiation of The Annunciation. Mystically speaking, this great initiation has to do with the conception of a holy child within the body of humanity. Thus the whole of humanity must sacrifice its independence for a higher vision.

This mystical Annunciation can only take place through the Synarchy, the communion of evolved souls who collectively pioneer this great sacrificial impulse. The 19th Siddhi will be one of the first Siddhis that awakens in man at a collective level. As soon as the great mutation has occurred within humanity and we pass through the Fifth Initiation, we will see what kind of shift has taken place. One of the traits of the new human will be an incredible sensitivity that is far beyond being psychic. The awareness in such a being will not recognise itself as being separate from other human beings, thus they will work for all of humanity without caring for themselves. Although we call that sacrifice it is not so for them, since they will know no other way to live. This 19th Siddhi is a herald of the future forms that will house higher frequencies of consciousness, and as such it displays the great inadequacy of our language. Just as our language has evolved out of our reliance on the five senses, future forms will operate through entirely different languages. Current human language is auditory, but future forms will communicate through their environment using a sense that is closer to what we call touch. This is the true language of Gaia — the interconnective auric tissue that brings all beings of the planetary sphere into full awareness of their inherent unity.

▤ 20th Gene Key

The Sacred Om

SIDDHI
PRESENCE

GIFT
SELF-ASSURANCE

SHADOW
SUPERFICIALITY

Programming Partner: 34th Gene Key

Codon Ring: The Ring of Life and Death
(3, 20, 23, 24, 27, 42)

Physiology: Brain Stem (Medulla)

Amino Acid: Leucine

THE 20ᵀᴴ SHADOW – SUPERFICIALITY

THE INSECT REVOLUTION

The language of the 20th Gene Key and its various frequencies is a purely existential language. It doesn't really involve thought or thinking in any way at all and as such it is quite challenging to grasp at an intellectual level. The 20th Shadow in particular concerns how deeply consciousness is able to incarnate into a human form. The more impeded consciousness is within the form, the less pure its expression will be. This Gene Key is about the *quantity* of consciousness that can express itself through a human being, and in this sense it is one of the most mystical of all the 64 Gene Keys. In the case of the 20th Shadow, very little consciousness can express itself at all, so that we see a very watered-down reflection of life's real potential coming through human beings operating at this frequency. The role of this Shadow is to keep human beings just on the fringe of life, without really plunging in. It is the Shadow of Superficiality.

It may sound as though the 20th Shadow is somehow the fault of human beings and that perhaps if we were to do things differently or better we might escape this plight. However, as we shall see, this aspect of our DNA is very ancient. In fact, it has led us to our current state of evolution. The 20th Gene Key and its Shadow represent a part of the human genome that developed through species that preceded human beings. Most notably, this Gene Key is the main aspect that humans share with the world of insects and opens up some striking parallels between these two worlds. When we observe insects, one of the things that we can see is how incredibly busy they always seem to be. Most insects only live for a very brief period of time during the summer months, and during those days, weeks or months they live pure existential lives whose only focus is to stay alive and reproduce. Early hominids lived in a similar existential reality, during a time when genetic programming was mostly expressed through the physical body — through breathing, eating, killing and having sex. The further we go back in our evolution, the more existential our lives appear to have been.

All of this changed with the development of the human neo-cortex, since this precipitated a shift in the geography of our awareness from our physical instincts to the rapidly burgeoning cognitive abilities in the brain. A strange thing happens when human beings begin to think — it appears that we leave the present moment. This illusion occurs because the mind can only think in a linear fashion, placing every object of thought into a temporal framework. This is the trade-off of the incredible abilities of the mind — for all its genius, it is limited to thinking within the domains of a false reality called time. To the extent that you are dominated by your thinking, consciousness is impeded from functioning fully. This is what we refer to in this Gene Key as superficiality. To live superficially means to live within the false illusions fabricated by the mind.

Many of us yearn for the so-called *old days*, when the human mind did not have such a grip on our reality and we lived a more primitive, simpler and somehow purer existence. The fact is that it wasn't purer in any way at all, and evolution cannot travel backwards. When we, like animals, functioned entirely from our instincts, we had very little sense of morality or conscience. If we were instantly projected there now, the *good old days* would probably seem both horrific and barbaric to us. The evolution of the human brain has changed the face of our planet, particularly over the past 100 years, and though we sometimes like to complain about the state of the world, this transformation has for the most part been hugely beneficial. The interesting thing about human beings is that we are a mixture of ancient instinctive animal awareness, our current rational mental awareness and a future holistic spiritual awareness that is about to come fully online. Our nostalgia about the *old days* actually stems from a genetic memory inside us of a very early epoch in human evolution in which our spiritual awareness was fully developed. However, all evidence of this *Eden* period was completely obliterated by a great cataclysm, which also mutated our DNA, eventually bringing about the current intellectual age. You can learn more about this from the 55th Gene Key.

What appears to be superficiality is really an adjustment in the way our awareness functions. The evolutionary phase that is now coming to an end is the superficial age — when mankind appears to have moved away from nature. But this is not really the case. Despite the fact that we are now dominated by an awareness that cannot exist in the present moment, we still live in the present moment whether we like it or not. Our current mental equipment simply prevents us from feeling one with life in the way that we once did. Evolution requires us to be superficial in order that we can complete this mental phase. However, it has led to a basic estrangement from the source of life itself, and this is reflected in our longing to return to that source. All our scientific and religious cravings have arisen due to the inability of consciousness to move beyond the human brain. Seen in this light, one can see how all the 64 Shadows have come about because of this single conundrum. The mind prevents us from feeling one with life.

When superficiality is translated into action, it becomes blind activity just for the sake of activity. The only time you are really one with life at the shadow frequency is when you are busy, even though by its very nature this *busyness* lacks real presence. Rather it is an absence of awareness. This makes human beings seem similar to insects; we are programmed to be incredibly busy on an individual level. However, unlike certain insect societies, we do not yet operate as a collective. The programming partner of the 20th Shadow is the 34th Shadow of Force, which is about self-absorption. When your activity lacks awareness it becomes a destructive force that plays havoc with your environment. These two Shadows manifest activity or lack of activity in response to the situations created by your mind.

Only when the mind has been overridden by our future awareness rooted in the solar plexus can consciousness begin to manifest activity that no longer meets resistance. This is what the Buddha referred to as *right action*. Right action occurs when the frequency of your DNA lifts you out of the Shadow states, at which point all your activity emerges naturally and harmoniously. Only in this final phase of our evolution will mankind begin to imitate the more complex insect societies such as ants or bees, as consciousness begins to breathe through every single individual, thus melding the collective into one.

REPRESSIVE NATURE – ABSENT

The introverted nature of the 20th Shadow gives the impression of being totally absent. This is primarily reflected in a person's eyes, which can appear staring or distant. This expression of the 20th Shadow can also be an intermittent theme rather than being permanent. The consciousness in such people is often frozen by an intense unconscious fear that temporarily displaces them. This withdrawing of consciousness from the body is actually a kind of mini death. It can also be brought about through an intense shock. These are people in whom life sporadically switches on and off according to the activity or inactivity in their minds.

REACTIVE NATURE – HECTIC

In the reactive nature, the mind will do the opposite to the repressed nature. Instead of numbing the fear, this kind of nature will immediately translate it into activity. Thus these people are ceaselessly on the move, unable to stop in their hectic self-absorption. Again, this is a state that is extremely widespread among human populations due to the deep connection between this Shadow and the way in which our mind translates it into activity. The world in which we live is increasingly hectic because our human programming makes it so. As new awareness begins to dawn in human beings, an enormous amount of human activity will simply cease, since it emerges as a by-product of the mind's restlessness.

THE 20TH GIFT – SELF-ASSURANCE

DIVINE RELAXATION

The Gift of Self-Assurance comes into being when a human being first learns to let go of their thinking as the authority in making decisions. With Self-Assurance must come a profound surrender to life in every moment. As you begin to accept that life has its own plans and flow, you also begin to stop interfering with the process at a mental level. No technique can lead to this Gift — only life itself can show you how to let go. This Gift is in fact the precursor to a higher state of spiritual awareness. You begin to discover that decisions in life simply get made, so there is really no point in agonising over them. Decisions at a higher level of frequency are made in the moment rather than being premeditated. Your mind may decide on a certain course of action, but when the decision is actually made at the Gift frequency, it emerges instantly and clearly through the entirety of your being.

The 20th Gift is the foundation of a happy life and is part of the genetic family known as the Ring of Life and Death. The Ring of Life and Death can actually teach us how to create the perfect environment for children to grow into happy, healthy and enlightened adults. For a child to grow up truly self-assured you need to model altruism (27) while remaining detached (42), and encourage invention (24), innovation (3) and above all, simplicity (23). These five keys can become the foundation of a complete life system for parents, and contemplation on them can lead to a profound sense of freedom for both children and parents. Self-Assurance is directly analogous to individual Strength through its programming partner the 34th Gift. Every child has their own natural strength that emerges organically as they develop, as long as they are given the right loving environment through their parents and peers. The ultimate loving environment for a child is to be surrounded by self-assured adults. The field of openness, integrity and relaxation created by a large family or community of adults manifesting right action is the perfect natural educational base for children.

Self-Assurance marks the end of the human tendency to worry about life. It also carries a certain sense of humour within it, although this is really an inner sense of humour or a sense of *lightness* about life. At this level of frequency, you begin to have the faintest inkling that you may exist at a higher level than you know. You also begin to have the feeling that some higher presence is actually looking after you. Human beings like to anthropomorphise this feeling in a whole host of ways — through spirit guides and angels, through contact with the souls of the departed or as the Godhead itself watching over us. In its purest sense, this state of Self-Assurance is actually your own greater nature emerging. It is a sign that more consciousness is being allowed to incarnate into the physical vehicle. Ironically, the more you anthropomorphise this feeling, the less it grows, because that is taking it back into the realm of the mind where it truly doesn't belong. What is occurring at this Gift level is that you are beginning to experiment with humanity's future instrument of awareness — the solar plexus system.

Within the human body are contained the genetic instructions for a higher mutation in the functioning of our solar plexus system. Once an individual mutation of this kind has been triggered (it is triggered by the 55th Gene Key), a whole new level of awareness functioning becomes available to you. The 20th Gift prepares the ground for this higher mutation, and we will see its full manifestation in the 20th Siddhi. Your Self-Assurance is directly equivalent to your ability to escape being a victim of your mind. As you begin to learn this knack, you experience the subtle expansion of a meditative awareness — an ability to witness your mind without being caught up in it. This sense of spaciousness continues to grow within, and as you let go more deeply into it, it seems as though you are in the grip of another force outside yourself. The by-product of this expansion is the quality of Self-Assurance — the continual feeling that everything is going to be all right. At this stage, a kind of relief dawns — you no longer have to try and control life. On the contrary, you now allow life to move you.

What actually occurs with this Gift is that the awareness within your solar plexus begins to take over from your mind. In China and Japan, there has always been a profound understanding based upon the solar plexus centre, which is known in the Orient as the *hara* or *dan tien*. The practitioners of the ancient yogic systems discovered this latent awareness within the hara and based their entire philosophy around it — from medicine to the martial arts. The awareness that emanates from the solar plexus centre is based upon surrender. In this sense it is about tapping a very powerful universal feminine power. It may come as a shock to some that the power of Self-Assurance is based upon surrender rather than assertion. Self-Assurance is far more than simple confidence, which can be built up through assertion or technique. The 20th Gift can only be cultivated through trust, patience and surrender, none of which are techniques. As such, Self-Assurance is based upon a philosophy of allowing everything to come to you rather than going out and chasing life down. It is because of this Divine *laziness* that the 20th Gift is the real foundation for inner relaxedness.

THE 20TH SIDDHI — PRESENCE

THE SACRED OCTAVE

The 20th Siddhi is so unique that there is very little that can be said about it. Mythically it is represented by the notion of the *breath of God* or the *word of God*. Presence is the underlying nature of being. In fact the word Presence does not do this Siddhi justice. It would actually be truer to name this Siddhi *The* Presence since the use of the definite article gives us the distinct impression that this is a state of consciousness that has nothing to do with any individual. It is the manifestation of The Presence of the Divine through a human instrument. Whenever a human being attains the siddhic state, he or she is suffused with The Presence. Pure consciousness floods their being, silencing the mental activity and drawing the person into the eternal present moment. When this occurs, the whole world created by the human mind is suddenly seen as utterly superficial. The smallest thing such as a leaf or stone is understood to have more life within it than all the greatest ideas of man. The Presence is experienced as everywhere and inherent in everything and is where we get the term *omnipresence*.

In the state of Presence, the individual is no more. You may find yourself sitting in the same spot for three days, and it seems as though not a single second has passed. Time dissolves into the background consciousness of all being. When you experience a brief moment of déjà vu, you are tasting the purity of a moment of true Presence in which the present moment becomes a funnel for both the past and the future, momentarily intensifying and transcending them. The Presence is also extraordinary in that it can be felt by anyone. When this Siddhi has flowered in a person, it actually creates a subtle atmosphere around itself — a kind of silent ease that pervades the aura of the person and radiates out to all creation. One of the great effects of being in The Presence is the deepening of your breath. The Presence links all human beings as one through the breath. Therefore people immersed in the atmosphere of Presence begin to breathe as one entity.

*Through the
mutation that
takes place
in the solar
plexus, the 20th
Siddhi will
finally allow
humanity to
be at ease with
our world.*

The true siddhic state is one of complete relaxation. As you release deeper and deeper levels of tension, the high frequencies of the Presence force you to sigh until your physical body comes into a state of great ease. The Presence can also be seen as an intense softness in a person's gaze. To a person immersed in The Presence, absolutely nothing matters — thoughts are irrelevant and suffering no longer exists because the mind has been cut at the root. There is nothing to say that is not superficial. Only silence can even approach the truth. We can see through the Ring of Life and Death — the codon ring associated with this Gene Key — that the Presence is also the same *Quintessence* sought by generations of seekers over millennia (the 23rd Gene Key). It also has a direct relationship to the experiences of Silence (24), Selflessness (27), Innocence (3) and Celebration (42).

The 20th Siddhi relates to the mystical nature of the octave — the transcendent note that begins, ends and links together all vibrations throughout the universe in which we live. It is also the eighth colour — the pure white containing the other seven colours, and that which also brings us back to the number zero, the blackness out of which all form arises. Above all, it is the sacred breath that unites the realms of both light and sound, allowing us to enter completely into consciousness as pure existence.

The 20th Siddhi also relates to the eighth plane of reality, the true ground of our Divinity known as the *Logoic plane*. As consciousness expands through each of the seven layers or sheaths that surround the human being (see the 22nd Gene Key for a full description of each) — the eternal cosmic ocean of consciousness in which we swim — the body of the logos or divine word is finally revealed in all its glory. This logoic sphere, the eighth transcendent plane, is beyond humanity, beyond words and beyond form.

THE SEVEN SACRED BODIES OF HUMANITY

1. Physical
2. Astral
3. Mental
4. Causal
5. Buddhic
6. Atmic
7. Monadic
8. Logoic

Once The Presence opens inside someone, it is difficult for them to maintain a role in the world because there is nothing of any importance to them in the world outside. Such people no longer fit into society. Neither can they explain their reality to anyone else. All they can do is go on existing, allowing life to bring them the experiences it brings, even though it is all simply uninteresting to them. The Presence sweeps everything before it — ending desire, sex, thought, even feeling. These people no longer feel anything but the Presence itself.

Even the deepest human genetic programs — the urge to eat and survive — are swallowed up in the majesty of the Presence. Some people who have entered this state actually have to be fed in order to stay alive. However, in most people the human habits remain and continue to function, even though they are really understood as non-essential to the continuation of consciousness.

In the world of the future, the 20th Siddhi will eventually absorb humanity. When it does we will enter an eternal world, a world in which mind will no longer have any role other than to help us communicate. Through the mutation that takes place in the solar plexus, the 20th Siddhi will finally allow humanity to be at ease with our world. In Eastern cultures, the primal sound at the heart of the world is known through the symbol of the Omkar, whose sound is Aum. When the 20th Siddhi dawns, this is the sound that is heard — a soundless sound, connecting all life, suffusing all creation and bringing everything back into its pure existential reality.

▤ 21st Gene Key

A Noble Life

SIDDHI
VALOUR

GIFT
AUTHORITY

SHADOW
CONTROL

Programming Partner: 48th Gene Key
Codon Ring: The Ring of Humanity
(10, 17, 21, 25, 38, 51)

Physiology: Lungs
Amino Acid: Arginine

THE 21ST SHADOW – CONTROL

THE DEMISE OF HIERARCHY

One of the major issues to plague human beings and the cause of enormous conflict and violation of basic human rights is the issue of control. All control is rooted entirely in a single theme — territory. As we shall see, there are various different ways in which we can understand territory. The first territory is yourself — the physical, emotional and intellectual confines of your own being and the boundary of your aura. If we extend that theme you might view your relationships and families as another form of territory. In turn, your home and land is obviously a further extension of your family territory. Then there is your entire community or race, the outer territory of which is made up of your nation. Finally there is the earth itself, which forms the current territory of all human beings. This constitutes a great deal of territory, which taken together creates the potential for a lot of human conflict rooted in control.

Another way of looking at territory is to view your life as territory and the events of your life as the aspects within that territory that you wish to control. The 21st Shadow certainly views life like this, and it does so at a deep genetic and unconscious level. Because humanity still operates collectively at the Shadow frequency, we are used to dividing life up into millions of individual territories and trying to control them. When you consider that we are really a single unified organism this seems rather a ridiculous way to behave, but it is nonetheless how the world works. At the Shadow frequency everyone is a victim — the controllers are victims of their need to remain in control, and the controlled are the victims of the controllers.

Wherever you find the 21st Shadow, you will find someone either too weak to control anything at all, or someone with a deep-seated need to control everything in their environment. In the past, control was about resources and food, and food depended on maintaining and defending your territory. In the modern world however, the battlefield has changed even though the genetic dynamic has not. Today the battleground is money, and this 21st Shadow has much to do with power and money. At the Shadow frequency, if you have money it appears that you also have power.

However, at the Gift level and beyond we shall see that true power has nothing to do with money. Control is based on tightness and fear. It creates tension and boundaries throughout all our environments. Even more crucially it creates the notion of hierarchy, because there are those who control and those who are controlled. In a contorted kind of way, this relationship between the controllers and the controlled can actually work quite well. It is the foundation of the idea of classes and castes, and in its ideal form it became the responsibility of the upper classes to feed and protect the lower classes.

This has been the way most societies on our planet have functioned for millennia and the greater part of the world still operates in this old way. It is the basis of our notion of royalty and ancestral lineages in families. It is only fairly recently that these archaic systems have really been questioned and are gradually losing their control and power. One of the manifestations of this decline of hierarchical control has been the rise of the middle class in the western world. However, the new rise of the middle classes does not offer a great deal more than the old system. It creates just as many problems. Families are now more cut off from each other than ever, and we have a world of *every family for itself.* The urge to control has simply shifted venues. Control now operates most powerfully through capitalism.

The issue of control is the issue of patriarchy. Patriarchal forms of government are the bedrock of our society, from politics to education to business. Most of those who are in control are only interested in power and money, and those who are not interested in power or money are generally too submissive to take any action. Apart from a few valiant individuals with true vision, the positions of true power on our planet are filled by those following personal agendas. The 21st Shadow makes it seem as though you simply cannot defeat the patriarchal system, so most people's true visions for a better world are choked before they are given a chance. However, the first ripples of a new frequency are emerging in the world. As those with the 21st Gift find positions of power the balance will begin to shift, because the higher frequencies of this Gene Key are not interested in power or money or control, despite having a Gift for all three. They are really interested in serving community. Even more than this, they have the courage to enact their visions, and that will make all the difference.

There are many misunderstandings about the issue of control and power. There will always be people who are naturally gifted leaders, but at the higher frequencies they see leadership as service, which means that those who serve with them or *under* them are never really under them — rather they are working alongside them. The problem with the old system is not the model but the frequency of the people in the model. The moment a system has everyone in exactly the right place, it no longer is patriarchal or matriarchal. It actually becomes synarchical. Synarchy is a model in which all people are equal but some still have more authority than others. This authority however is based upon frequency rather than fear. The reason that synarchy succeeds where hierarchy cannot is that every person in a synarchy is fulfilled by their role, regardless of how much or how little responsibility it carries. For a fuller description of such a model you can read the 44th Gene Key.

The ultimate root of territorial divisions across our planet is distrust in life itself. This is the real human disease. Territory and control through power and money are simply the manifestations of this disease. The programming partner of the 21st Shadow is the 48th Shadow of Inadequacy, which underpins all this fear of losing control of your territory. We simply do not yet know that we are a single entity. When the time comes that we see our true nature as a collective holistic human family, the need to control life will cease. Ultimately, the only ones who will be given positions of control will be those who have given up being in control. Those people are our future leaders — in business, in government and at all levels of human society. Those of us who continue to try and hold tight control over our territories and other's lives will eventually find that we are fighting only ourselves.

REPRESSIVE NATURE – SUBMISSIVE

All repressive natures are based upon the denial of personal power. Through the 21st Shadow this shows up as submission. These people let others remain in control without asserting their own authority. Added to this, the repressive nature has a tendency to defer control to life itself therefore not assuming responsibility for the direction in which it takes them. There is a fine line between surrendering to what life brings and influencing the path of one's own destiny. These people unconsciously blame life for whatever happens to them, closing down the centre of their willpower. The true nature of the 21st Gift is to control and manage situations, but the submissive side of this Shadow is afraid of being in control since it means they alone are accountable for their actions and potential success or failure. These people would rather not participate in life at all. They often masquerade as being *laid back*. However, the reality is that such people are hiding from their true responsibility.

Ultimately, the only ones who will be given positions of control will be those who have given up being in control.

REACTIVE NATURE – CONTROLLING

The other side of the 21st Shadow is the acute need for control. These are the people we sometimes label as *control freaks*. Their anger is so tightly coiled that they cannot allow any part of their environment to escape their control. As the repressed nature is loose, so the reactive nature is tight. These people cannot handle change unless they have instigated it. If others transgress the confines of their deeply controlled lives, all their pent up tension and anger is likely to be detonated. Such a nature strives to maintain control over others through hierarchical dominance or by taking the moral high ground. Unfortunately, such sustained insistence on control takes a great toll on your body and in particular on your heart. These are the kind of people who can only be humbled by a deep physical or emotional crisis, and because of their inability to let go, they often meet such crises in life.

THE 21ˢᵀ GIFT – AUTHORITY

THE AUTHORITY OF SUBMISSION

The Gift of Authority is a gift that is innate. If you have the 21ˢᵗ Gift in your Hologenetic Profile and you speak and act from your heart, you will inspire loyalty in others wherever you go. Authority is the true vibration of this 21ˢᵗ Gene Key when it has found that delicate balance between allowing things to go their own way and assuming control of the way things are going. Authority is a frequency that is determined by intent. If you assert authority through the hierarchical base of the 21ˢᵗ Shadow, then you rule through control and fear, which never inspires true loyalty. In such cases, the people around you may appear loyal, but given the right circumstances they will quickly and easily transfer their loyalties elsewhere. True loyalty can only be maintained when the frequency of love outweighs the frequency of fear. Such people are given authority by others, and this is the real meaning of authority. Authority cannot be asserted through will. It can only be afforded through trust.

Authority is a phenomenon within the auric field that takes place between one individual and another, or between an individual and a group. True authority unites rather than controls. This can be understood well by the archetype of the master and the servant, which is the bedrock of our societies. Some of the deepest relationships are formed between one person in a position of authority and another in a position that appears to be submissive. However, if both parties in such relationships are as deeply in service as each other then their relationship transcends its social stereotype of dominance and submission. In order for such relationships to work, both parties must be equally in service to the other. If this is the case, the relationship can be mutually beneficial and potentially very powerful.

This relationship between authority and submission is the relationship between yin and yang, between the male and female energies throughout our universe. The male polarity represents authority and the female represents the subject or servant of that authority. Despite the fact that this relationship has established itself culturally across most human societies, it has little to do with men and women. Women can equally be the authority and men the servants. It is just the difference between matriarchy and patriarchy, but neither works unless the intent at the root of the relationship is pure. If the female side is too submissive, then the male side will be too controlling and vice versa. Any imbalance in this kind of relationship is a manifestation of the Shadow aspect of this 21ˢᵗ Gene Key. There can be no need for power or resentment from either side. This balancing of the archetypes is a hallmark of the higher frequency of the 21ˢᵗ Gift. You will find that all relationships play out this same drama — it is the core relationship between parent and child, husband and wife, employer and employee.

The beauty and magic of the 21ˢᵗ Gift occurs when both sides of the relationship surrender to each other.

The beauty and magic of the 21ˢᵗ Gift occurs when both sides of the relationship surrender to each other. When the authority becomes the servant and the servant becomes the authority then the relationship really hums with power. On the outer plane, such a relationship may appear to be one-sided, but on the inner plane the balance of power is reversed.

Only when this happens does the real meaning of authority become clear. In large groups, communities, companies and even armies, if the authority figure represents and connects with all their subordinates, they will inspire the kind of loyalty that strongly binds the group. This kind of leadership is very different from that of the 7th or 31st Gene Keys, which is based upon the ability to completely let go of all control and trust the collective spirit of the group to make its own decisions. In the case of the 21st Gift, a pact is made that one person will become the decision maker for the group, and this person is thus responsible for the entire group.

There are many different styles of leadership within the human genetic matrix. Whereas the 31st Gift represents the voice of the group and the 7th Gift represents the heart of the group, the 21st Gift represents the *will* of the group. Thus, these people are designed to handle more responsibility than other human beings, since their will directly affects all those in submission to them. The key lies in the willingness of those in submission to be represented by the one who is chosen to be the authority. As has been shown, when these relationships work at a high enough frequency, the one in authority simply becomes the conduit of the will of his or her followers. When that happens, transcendence occurs. Such relationships are founded upon kinship rather than kingship. The authority figures, as we shall see at the very highest frequencies, must remain continually in communion with those below them in the hierarchy. Only when this merging and transcending of cultural stereotypes occurs can the highest spirit incarnate into a human group.

This 21st Gene Key can be further understood when seen in the context of its wider genetic family or Codon Ring. As part of the Ring of Humanity, it forms an integral aspect of all human wounding. The sacred wound at the heart of humanity and the reason for all of our suffering can be unlocked by the Gene Keys that make up the Ring of Humanity. Hierarchy is one of the oldest human wounds and like each aspect of our suffering it can only be healed by love. To activate the higher frequencies within this Gene Key you must have great courage. It takes a powerful human being to surrender completely to another, whether through authority or submission. The fact is that surrender makes authority submissive and submission authoritative, which is precisely what heals the wound and brings an end to hierarchy and control.

THE 21ST SIDDHI – VALOUR

THE NEW AGE OF CHIVALRY

In the 21st Gift we saw that true authority based on service inspires loyalty. At the highest level of consciousness, this coupling of love and power gives way to a great and peerless ideal — the ideal of Valour. We tend to associate the word valour with courage, particularly the courage shown by soldiers during battle. Although there is some truth to this image, the use of the word valour as a siddhic aspect of consciousness goes far beyond the idea of courage in the face of adversity. Valour is the highest frequency of the 21st Gene Key. It is a living energy field released into the world through a particular chemical signature within your genetics. Valour is the by-product of another potent word — nobility.

To act with Valour is to enter a higher world.

To understand nobility is to dive into the realm of human destiny. In the social history of humanity and in our collective unconscious, there has persisted an image of a royal human being — the King or Queen, the Emperor or Empress — the symbol of the highest potential within man and woman. Nobility is the quality usually associated with royalty or genetic pedigree, although down through the ages our human attempts to bestow such projections on certain personages has usually been shown severely wanting. Nobility, we have discovered, has little to do with breeding and much more to do with character. Indeed, most of our heroic myths centre on this notion of human nobility and valour. Valour then can be understood as nobility in action. It contains virtue, wisdom, love, courage and above all sacrifice. A truly valorous deed is an act of absolute self-surrender in which you lay your entire being on the line for a higher ideal. In our history books, this may have been recorded as dying for King and country, but in the language of the Siddhis, it is really about dying into a Divine ideal.

At the Shadow level the need to control fosters fear and reaction in others. At the Gift level, authority inspires loyalty. At the siddhic level, Valour invokes Communion. There is a deep genetic connection between the 45th Siddhi of Communion and the 21st Siddhi of Valour. Communion is about merging your individual being into a higher collective being, and this is precisely what happens through the frequency of Valour. Valour need not even act — it is a vibration of such intensity that it makes the hearts of others weep. It is the recognition of true nobility in another and the realisation that the other is your own mirror. As an aspect of the Ring of Humanity, we see that all human beings have this higher recognition as their final destiny. No matter who you are or what kind of life you lead, at certain points in your life you are given the opportunity to act with Valour. These moments are mythic moments and the script of your future life hinges upon them. To act with Valour is to enter a higher world.

Valour is the first great manifestation of the breakdown of the resistance between the opposite poles within relationships. It is the bowing of one being to the nobility within another. Mystically speaking, it is represented by the Fourth Initiation of Matrimony (see the 22nd Gene key), in which the quality of Valour actually courts annihilation through its self-sacrifice to another being. It is the absolute surrender of control into your hierarchical role. This means that socially you surrender to your position and mystically you surrender to your karma. Valour is the absolute courage and love of seeing the divine reflection within the face of another, no matter how unpleasant that other may appear to be. As an archetype Valour signifies the ending of all karma, even though the states preceding it often carry a great deal of karma. Valour must be forged on the anvil of life. You have to realise that no matter how muddy the waters you find yourself in, your pure nature can never be soiled. If the 21st Gene Key is a part of your personal Hologenetic Profile, the chances are you will need to get your hands dirty before you can understand this paradox.

To be a master — that is, a being who has attained realisation through this Siddhi — you must lead by example. You have to know the absolute horrors of living within the most tightly controlled hierarchical situations. You have to understand the human need for control and the depths of your own fear of losing control.

You have to be tested by the forces that will not relinquish control, and you have to continually be subjected to external controls. Once you see that no external form of control can rob you of your true nobility, this Siddhi of Valour will finally dawn deep inside your DNA. The being that has attained realisation through this Siddhi never feels like an authority figure even though they may be. These rare people will commune with you as a friend, despite the exquisite fragrance of their state of consciousness. It is this very humbling quality within the Siddhi of Valour that creates such deep levels of human communion.

Even though the 21st Siddhi may be incredibly humble and friendly, it also packs a serious punch when threatened by the forces of fear. This is the energy field of true knighthood, symbolised by the courageous and chivalrous acts of all great heroes and heroines. These forces will fight for the highest of ideals and hold that ideal for all those who follow them. Those with this Siddhi will gladly die for, and as an example of, the highest ideal — the ideal of communion between all beings. The lives of such people become myths, often because of the sacrificial nature of their deaths. Even so, awakening through this Siddhi does not obviously foreshadow such a death — it is simply symbolic of the highest expression of this 21st Gene Key. With the advent of the 55th Gene Key and its reawakening of the spirit of romance within humanity, this 21st Siddhi is actually ushering in a new myth and a new age of chivalry.

The essence of Valour is to be found in the symbol of the male pole surrendering to the female pole. The programming partner of the 21st Siddhi is the 48th Siddhi of Wisdom, which is one of the great archetypes of the Divine Feminine. Valour therefore represents the surrender of control (symbolised by man) to trust (symbolised by woman). This surrender results in the absolute annihilation of the male force by the female force, an ancient mythic enactment contained in many ancient creation myths. This surrender of the male into the female creates a reversal of roles and poles and thus the communion is fulfilled. Ironically, it is through this divine coupling that the male force is truly empowered through its opposite. In other words, the male force is knighted and empowered only through its surrender to the female. It is important to understand that this imagery is an internal truth rather than a literal representation of man and woman. The power of Valour could thus be summarised as the courage and love that must be found within the lower self to die into the unknown world of the higher Self.

☷ 22nd Gene Key

Grace under Pressure

SIDDHI
GRACE

GIFT
GRACIOUSNESS

SHADOW
DISHONOUR

Programming Partner: 47th Gene Key
Codon Ring: The Ring of Divinity
       (22, 36, 37, 63)

Physiology: Solar Plexus (Cranial Ganglia)
Amino Acid: Proline

INTRODUCTION TO THE 22ND GENE KEY

THE SWEETNESS OF SUFFERING

The 64 Gene Keys represent the seeds of a brand new synthesis coming into the world. It is important to clarify here that it is not the knowledge of the Gene Keys themselves that is new, but their revelation together as a complete matrix of the human evolutionary program. Each Gene Key is a portal to an encyclopedia of timeless knowledge and wisdom. Deep contemplation and meditation on the Gene Keys will open up a new world to you. There is no question that cannot be answered by them, since all the answers are inside you. Furthermore, as you enter the higher frequencies of the Gene Keys, the questions themselves begin to drop away and the higher states reveal themselves from within your very own DNA. At this stage in your evolution the knowledge itself ceases to hold any real interest and you see it as nothing more than a bridge that you can now discard. This is reflected in the Buddha's own timeless words:

> *"My teaching is a raft whereon men may reach the far shore.*
> *The sad fact is that so many mistake the raft for the shore."*

The 22nd Gene Key is special within the overall matrix of the 64 Gene Keys, containing a highly specific teaching and a powerful transmission. The transmission of this consciousness alone can alter the way in which your DNA operates. In many ways, the 22nd Gene Key is a sister transmission to the 55th Gene Key, and between the two of them a great mystery lies hidden. Just as the 55th Gene Key describes the process of awakening as a genetic evolutionary process rising up within your body, the 22nd Gene Key describes the process of awakening as the direct intervention of Divinity coming down into your body. Thus it is through these two Gene Keys that the forces of Evolution and Involution finally come together. As you enter the field of the 22nd Gene Key you are involving yourself in a magical process of invocation in which you directly invite a higher presence into your life. In this sense, the 22nd Gene Key needs to be approached in a prayerful and reverent manner, and in the spirit of nakedness. There is a great deal of information synthesised here.

Let it descend into your DNA and, rather than attempting to grasp it with your mind, simply appreciate what an awesome transmission it holds.

The subject of the 22nd Gene Key is the true meaning of suffering. As you begin to contemplate the suffering in your own life, you will perhaps come to see what an incredible blessing it holds. This simple and sweet realisation can and will transform your life. Welcome to the embrace of the Great Mother!

THE 22ND SHADOW – DISHONOUR

THE AKASHIC OCEAN

As we have already heard, the 22nd Gene Key is very special. There is no getting around this statement. Written into the evolutionary script are certain anomalies and Divine cosmic surprises. In this respect there is no other Gene Key to rival the 22nd. It is what makes the mythic drama of life so compelling. All great drama has but one pervasive universal theme — that of redemption. Whether or not a drama ends with redemption, it is always there as a longing inside our human hearts. Whenever we watch or hear a film or story, if there is no redemption at the end our hearts will feel cheated. Our minds may appreciate the art, but without a sense of atonement there is the sense that a great Truth has been misrepresented. The 22nd Gene Key is about the Truth of redemption. To those of a strong intellectual bias it will inevitably seem fantastic or romantic since it concerns the direct intervention of the Divine in the ordinary world.

The world that most of us see is not all that exists. We mostly live within very defined and closed-circuit parameters. Human beings generally have no notion of the great cosmic laws that exist behind the world of form. One of the greatest of these is the Law of Divine Memory. This law states that all thoughts, feelings and acts are recorded everywhere within the body of the universe. Science now shows that we live in a vast information field of subatomic particles, some of which are so tiny that they actually pass through matter. This ocean of consciousness exists in many dimensions and it responds to thoughts, acts, feelings, words and even intentions. It is a vast quantum field that acts like a great memory bank, holding and recording every impression ever made. In the language of the ancients, it is often referred to as the Akashic Record.

The 22nd Gene Key is deeply connected to this Law of Divine Memory. It is like a massive receiving dish that responds to the frequencies, sounds and vibrations that it hears, and it hears *everything*. Like a cosmic Aeolian wind harp, reception is determined by the way in which the strings are tuned. In the case of the 22nd Shadow, the strings of your DNA are twisted out of harmony and your behaviour and experience in the world is similarly misshapen. This is the Shadow of Dishonour. It only exists in the world because most human beings do not realise that all their acts are recorded. We do not realise that every act, thought or feeling creates a ripple in the Akashic Ocean, and that each ripple must and will one day return to its point of origin.

The 22nd Shadow is one of the most powerful emotional Shadow Gene Keys in the human genome. It is highly passionate and sexual with a huge emotional range encompassing extreme highs of sweetness and extreme lows of violence. Because of its situation in the genome it is directly or indirectly responsible for most of the relationship problems on our planet. However, before we journey any further into this profound Gene Key it should be borne in mind that negative emotions in and of themselves are a natural part of the world in its current state. If they can be usefully transformed or sublimated into art, creativity or service, their power is awesome. It is a matter of how much responsibility you can take for your own feelings. However, most people in the world today are utterly ruled by their emotions and whenever you project negative emotional power at another being, you dishonour both yourself and the other person.

Many so-called spiritual teachings suggest that you should subdue your negative emotional states in favour of sweeter more virtuous frequencies. In fact this is the basis of most of the great religions. But to subdue any state or feeling is to dishonour and distrust that feeling, which prevents acceptance. From the point of view of the 22nd Gene Key, every feeling, mood or thought you have is put there directly by God for you to trust in it. Trusting this process is obviously not the same as acting it out. Trust is a powerful internal process that requires great courage. One of the tricks of the 22nd Shadow is to con you into trying to change or *fix* your moods, rather than allowing them to simply pass through your system naturally. The fact is that you cannot reach higher states of consciousness without first passing through your own suffering.

This is the true purpose of the Akashic field and of the 22nd Gene Key — they invite you to receive your own slice of suffering. If you do not take responsibility for your own thoughts, words and deeds, the Akashic field simply sends the same forces back towards you again and again. This is the foundation of another of the great universal laws — the law of Karma, which we will explore in more depth below.

THE THREE PURE ONES

As we enter deeper into the transmission field of the 22nd Gene Key and its Shadow frequency, we encounter three streams of teachings concerning the nature of suffering left by three great world teachers or avatars. These three beings are actually one single being, which divided into three fractal aspects over the course of human evolution. Even though these were individual people or *Magi*, it is more helpful to see them as three fractal transmissions of the same Truth. The first is Hermes Trismegistus, whose legacy dates back to the Age of Atlantis and whose name (meaning *thrice great*) directly reflects the triple nature of this transmission. Hermes goes by many names — Thoth, Merlin and Fu Hsi to name a few. The teaching represented by this fractal is the teaching of Alchemy or High Magic. All true Alchemy concerns the transmutation of suffering through alignment with Divine Will. The second great teacher is Christ, whose fractal represents the transmutation of suffering through love and sacrifice. Finally, the third great teacher is the Buddha, whose fractal represents the transmutation of suffering through wisdom and compassion.

Because all the 64 Gene Keys are influenced either directly or indirectly by the 22nd Gene Key, these three great fractals and their teachings form the essence of the Gene Keys revelation. The Synthesis is made up of this great Trinity of: Divine Will, Love and Wisdom.

THE COSMIC MOTHER

Beyond, behind and between the great masculine Trinity lays a fourth transcendent field of consciousness that is born from their interaction. This is the field of the Divine Cosmic Mother whose embrace encircles, protects and contains all three of the great streams of the masculine Avatar consciousness. Within humanity, the only direct portal to this great being is the 22nd Gene Key. The great Cosmic Mother holds the master key to all suffering and stands behind the teachings themselves. She is the triple mystery of the transmission. Although she is beyond teachings, her way is that of Grace through Suffering. Those who enter deeply enough into the three great paths through suffering — the alchemical, the sacrificial, the mindful — will eventually meet the great Cosmic Mother, since she represents the very spirit of Grace that brings an end to human suffering.

The Mother is a field whose shakti or sexually liberated energy actually devours your separate self.

Contrary to the many depictions of the Holy Mother by religion, she is actually a highly ecstatic and sensuous field of energy. This is another way in which the 22nd Shadow dishonours the true nature of the feminine — through denying humanity the natural delights of sexual pleasure. As suffering comes to an end through her Grace, true pleasure will dawn. The Mother is a field whose *shakti* or sexually liberated energy actually devours your separate self and as she does so, you experience the highest of Divine ecstasies pouring through your body and aura. This is no staid Grandmother figure but a fully laden cosmic breast whose celestial milk will nourish the highest aspects of your being. As we explore the 22nd Siddhi of Grace in more depth we shall see how all pervading is this field throughout creation.

Once you come to realise that Hermes, Christ and Buddha are in fact three aspects of the same trinity, you will find that these three streams of teachings bring much clarity when fused together. Through Hermes and the Magi the teachings of Alchemy and Transmutation came into the world. Buddha brought the teachings of Karma and Rebirth, and the Christ brought the teachings of Forgiveness and Atonement. As these transmissions have travelled down the centuries they have become so distorted and confused that they barely resemble the simplicity of their original transmission. Over the following pages we will reunite these three great teachings and streams of wisdom, exploring the underlying fabric of the subtle worlds and processes that make up the human evolutionary journey.

THE CORPUS CHRISTI — THE SEVEN SACRED BODIES OF HUMANITY

There have been many systems devoted to the understanding of spiritual or occult science — those subtle layers of reality beyond the five senses. The great Oriental and Indian systems left us reams of intuited insight laid down over thousands of years of direct experience of the higher realms. Towards the end of the 19th century, much of this experience became accessible to the west and many new streams of thinking converged. Theosophy and Anthroposophy were born and a new era of spiritual science came into being, leading into the modern *New Age*, where so many ideas and lineages from East and West, both mystical and scientific, collided and merged. It is an exciting but also a confusing time as a grand new synthesis emerges from the cosmic soup.

One of the enduring insights of mysticism is the notion of the subtle bodies of the human aura. Depending on which system you follow there are between six and ten major subtle dimensions or planes upon which human beings function. These auric layers collectively make up what is known as the *Corpus Christi* or *Body of Christ*.

Listed below are the seven main layers of the human aura and their fundamental properties.

THE SEVEN SACRED BODIES OF HUMANITY

1. Physical
2. Astral
3. Mental
4. Causal
5. Buddhic
6. Atmic
7. Monadic

1. The Physical Body – The physical body forms the bedrock of incarnation. On the physical plane, the collective memory of humanity is stored in our DNA. The ultimate goal of human evolution is to merge the physical body completely with the monadic body, thus allowing the former to be assimilated back into its true essence. This corresponds to the 9th Initiation known as the *Glorification*, which is discussed at the end of this Gene Key. The physical body has a subtle twin counterpart known as the *etheric* body, around which the science of true health is built. Over time, the physical body more closely reflects the state of your astral body and its emotions.

2. The Astral Body – The astral body is the layer of the human aura that collects, stores and transmits all human emotion and desire, from the meanest to the loftiest. In the astral body, pleasure and pain are reflected as vibrational frequencies which effectively divide the astral plane into *hell* realms and *heaven* realms. The astral body is most active during sleep when it processes your daily urges through your dream life. As the next layer to the physical and etheric body, the astral body also has a huge effect on your health. After death, the astral body is directly confronted with the true nature of every single emotional impulse you had whilst alive in the physical body.

3. The Mental Body – The mental body exists at a higher frequency than your emotions and is constructed from your thinking life. The mental body is greatly influenced by the collective mental body of all of humanity. This tends to pull our thinking down into the unfulfilled desires of the astral body. As your thinking turns to higher impulses, the mental body gradually disentangles itself from the astral body and takes on greater power. The mental body can also be used by lower consciousness to repress the natural impulses of the astral body, which can also lead to health problems at all levels.

4. The Causal Body – Sometimes dubbed the *soul*, the causal body directly corresponds to the physical body but at a higher level. It stores the collected goodwill of the human soul as a memory signature written in light. This finely tuned vehicle forms the storage hub for all the high frequency thoughts, words and deeds that we have initiated during our many journeys in incarnation. After death, the lower three bodies disintegrate and only that which is refined and pure is drawn up and retained in the causal body. The causal body responds to higher visions and archetypes that lie beyond language but that can still be conveyed through direct transmission to the lower three planes. As your causal body develops more lucidity, the higher bodies can use it as a means of directing higher and higher frequencies to the lower three bodies. In this respect, the causal body is the bridge between the lower and higher planes.

5. The Buddhic Body – The buddhic body is the higher octave of the astral body. As such it reveals the pure Truth that humanity and all the earth planes are in fact one single organism. Once your awareness is fully anchored in the buddhic body, the causal body dissolves and reincarnation in the normal sense is no longer necessary. It is through the buddhic body that human beings have access to the field of universal love and the higher ecstasies associated with enlightenment. It represents the third feminine realm of the Holy Trinity — that of Divine Activity.

6. The Atmic Body – As the higher octave to the mental body, the atmic body allows human beings access to the higher evolutions outside the process of physical incarnation. Whilst the buddhic body retains its connection to humanity through its compassion, the atmic body brings awareness into the cosmic field of Christ consciousness, directly merging your awareness with Divine Mind and Heart, the second aspect of the Holy Trinity. It is through the atmic body that the great avatar streams enter the world. It is also the realm of the Siddhis — the many miraculous manifestations of the Divine.

7. The Monadic Body – Hardly a body in our normal sense of the word, the monad is the unbridled primal essence of Divine consciousness itself. It enters the world of the form through the causal body, which is the veil it takes on in order to enter the lower worlds, and corresponds to the first aspect of the Holy Trinity — Divine Will. The monadic body is present within every single atom on all planes right down to the physical plane. However, until awareness has risen to the atmic body, the monadic cannot be fully expressed. When it does express, it condenses the atmic body and all the others along with it, revealing true Divine essence as consciousness beyond understanding. At this stage, each of the lower three bodies, physical, astral and mental are absorbed into their high frequency counterparts the causal, buddhic and atmic, thus revealing the true mystical nature of the trinity as three in one.

KARMA AND REINCARNATION

The programming partner of the 22nd Gene Key is the 47th Gene Key, and there is much we can learn from this connection. The 47th Gene Key concerns the storage of world karma in human DNA. We have seen how the Akashic Ocean records all deeds through the seven subtle bodies, and how this physical storage takes place through DNA. It is here in the human genetic code that the world wound is to be found — the combined suffering and negative thoughts, deeds and words of every human being since the beginning of time are *wound* up in the non-coding or *junk* DNA inside your body. Depending on the unique genetic imprinting of your vehicle, certain aspects of the collective karma of humanity are highlighted in your DNA, and these determine your personal karma and the essential script of your life process. All this genetic storage takes place through the 47th Gene Key. The 22nd Gene Key, on the other hand, concerns those aspects of our subtle vehicles that survive death.

It is vital at this point to realise these higher subtle bodies that are said to *survive incarnation* are really aspects of the Akashic Ocean itself. They are like memory slates that overlap each other at higher and higher levels of frequency. At the highest levels, all the layers dissolve to reveal a single field of consciousness. This is why reincarnation is only a relative truth. It is relative to the body in which consciousness happens to be localised. With this basic understanding we can begin to understand one of the great keys to all human suffering — the inability to accept responsibility for our own thoughts, feelings and actions. Life gives us precisely the imprinting we can handle, and if we dishonour ourselves or others we actually increase our own suffering in the long run.

THE BARDO STATES

Many teachings exist from ancient cultures regarding the states of awareness that occur during and after the process of death. These states are often referred to as *bardo states*. It is only when we combine the teachings of the Christ and the Buddha that this process becomes simple and clear. At the point of death, your various bodies separate. Your physical body obviously returns to the earth, but your astral and mental bodies which contain all the feelings and thoughts from your current life begin an alchemical process of separation and refinement. The negative low frequency patterns are discarded whilst the higher frequency patterns are retained and drawn up into the causal body. Because you no longer have a physical body, emotions in the bardos after death are experienced more intensely than anything we can possibly imagine. In fact, emotions and thoughts actually assume a life of their own, appearing as entities — angelic or demonic — whose frequencies either cause intense agony and terror, or intense joy and ecstasy.

This process in the bardo states is one of pure redemption in which the subtle aspects of your being meet the consequences of your actions, thoughts and feelings while you were in form. Every aspect of your Shadow consciousness is purged and cleansed. Human intuition recalls enough of this process to integrate it into our various cultural and religious beliefs. However, most human thinking makes a fundamental error here between the concepts of retribution and redemption. At the Shadow consciousness, humans do not see the true operation of Grace

Emotions in the bardos after death are experienced more intensely than anything we can possibly imagine.

through Christ's teaching of forgiveness. We do indeed atone for our sins in the afterlife, but only in order that we can be given a clean slate before our causal body returns once again. Because there is no defined sense of linear time in the bardo states, it can indeed seem that our hell is eternal, just as it can seem that our heaven is eternal.

The 22nd Gene Key thus allows your causal body to become brighter and clearer from incarnation to incarnation as you learn from your own suffering, both in form and out of form. At the Shadow frequency, this process teaches you to accept your slice of group karma and offers you the opportunity to transmute it. This is what the Christ consciousness really is. It lies inside every single human being. We are forgiven over and over again, and the deeper we accept this Grace, the more powerful the impulse for goodness becomes when we are in form. Eventually our causal body becomes so resplendent that higher consciousness reaches down through it into the lower bodies — the mental, astral and physical bodies — and begins to impact them powerfully. Our thoughts turn more to God, our emotions and desires become sacrificed for a higher cause and eventually even our physical body becomes radiant as the various slates layered over it become transparent.

INSTANT KARMA

In light of the above, Karma can perhaps be understood in a new and beautiful way. Individual Karma does not travel beyond a single lifetime, even though it does pass into the human collective. Every negative act is recorded for future processing in the bardos, and it is stamped into the collective DNA of humanity where it must eventually be redressed. Our Karma at this level is shared, since humanity is really one entity. Contrary to certain popular beliefs, the conditions of your external physical life do not reflect your actions in past incarnations. The level of transparency of your causal body attracts the incarnative environment it needs to further its own evolution, whether those conditions are seen as good or bad. At the higher levels of transparency, the causal body will often take on greater suffering because there is a greater compassion manifesting through its vehicles. This evolutionary incarnative process follows a specific archetypal sequence known as the Nine Initiations, which we will explore at the end of this Gene Key.

Even though Karma is purified in the after-death state, it can manifest during a single lifetime. The law of cause and effect holds true on the material plane, too. However, the material plane is extremely dense, which means that we do not always see the results of our good or bad actions, thoughts or words quickly. Having said this, we are now living in a time towards the end of a great Epoch, and at such times natural laws are often seen to bend. The general collective consciousness of humanity has been gathering in the Akashic Ocean for millennia, and as such it has effectively programmed the way in which that Ocean functions. As our consciousness evolves more quickly, the turnaround for the Law of Divine Memory is altering. In other words, Karma is speeding up. Soon we will reach a point in evolution where redemption will manifest even on the physical plane. This is the epoch that is coming — the time of healing the sacred wound that causes human suffering.

This last point ought to give us all something to think about regarding the way in which we handle our own emotions and thoughts. Soon, none of us will be able to hide the truth of our acts or feelings. In the not too distant future, the 22nd Shadow will create almost instantaneous Karma whenever someone behaves dishonourably. This will utterly change the way we see ourselves and the world around us. Justice is a universal law. However, human beings at lower frequencies too often misinterpret this beautiful law as retribution or revenge. Because of Grace, evolution cannot travel backwards and it is not possible for form to devolve. It is all a matter of how finely tuned are the strings of your DNA. If you tune them to the lower frequencies, not only will you prevent yourself from experiencing joyousness, but you will also add to the weight of human karma stored in our collective ancestral DNA. In this sense, all human beings are given the feeling of free will in order to experience the consequences of their own deeds. But the key point here is that we do not learn through punishment and retribution, as so many religions would have us believe. We learn through joyousness and fulfilment, which comes through the 22nd Gift — the wonderful Gift of Graciousness.

REPRESSIVE NATURE – PROPER

The 22nd Shadow in its repressive phase gives rise to a deeply false sense of character. These people may appear outwardly very balanced, calm and proper. They can often be extremely socially adept. Inwardly however, their emotions are often seething. They can hide deep sexual lusts and often foster deep hatred and resentments. A good archetype of propriety is the Victorian Age in Great Britain. On the surface, the general culture was one of politeness and control, when in fact it concealed an underworld of repressed passion, sexuality and aggression. All the repressive Shadows are rooted in a deep-seated fear. The fear of the 22nd Shadow is the fear of losing control. We should remember that no Shadow is in itself bad. It matters how we deal with it. If you have a repressive nature, you can use it positively to transform inward negativity, rather than letting it stew and stew until it erupts. However, if there is no sense of virtue in these people, this Shadow can conceal the most violent and explosive of natures.

REACTIVE NATURE – INAPPROPRIATE

The reactive version of the 22nd Shadow manifests as inappropriate or antisocial behaviour. These people cannot control their emotional reactions. They often lead fairly disreputable but passionate lives, wearing their hearts on their sleeves. Their actions and behaviour are usually destructive, initially more towards others than themselves. Even in its Shadow, this archetype has such creative power that these people can produce wonderful art or music, but so often their inability to handle their own passions and treat others with respect leaves their private lives in tatters. Above all, these people have an inability to listen — either to others or to themselves. Thus even when their intentions are good, they are doomed to be mistimed and misunderstood.

THE 22ND GIFT – GRACIOUSNESS

THE MILK OF HUMAN KINDNESS

With the Gift of Graciousness, you begin to disperse your own Karma and that of your ancestral DNA.

The 22nd Gift is the Gift of Graciousness. It is a rare and beautiful quality that has a profound effect on everyone it touches. Graciousness means that whatever you do in life, you always consider the feelings of others. This is one of the great social gifts, and if it is a part of your Hologenetic Profile then your whole life is geared towards impacting people's emotions in a positive way. Even if this Gift is not a primary aspect of your profile, it still has a huge capacity to completely transform your life and the lives of everyone you meet. The 22nd Gift is not just about stirring people's feelings, but also about touching their hearts and even their souls. Graciousness means that you act with grace and consideration in everything you do.

Just like its Shadow, the 22nd Gift is extremely powerful in its effect upon others. As the Shadow can leave another feeling utterly dishonoured and disturbed, the Gift of Graciousness can greatly help others become free of their negatively charged emotions. There is a profound kindness at the core of this Gift that can elevate others beyond their normal consciousness to states of love, laughter or tears. For this reason, many people connected with this Gift assume artistic, musical or vocal roles in life, where they can influence others through their natural social grace. As we saw with the Shadow, people at the lower aspects of this Gene Key do not acknowledge their accountability for what they say or do. At the Gift level however, you begin to see that all is interconnected — all things and all people. You realise at a deep level that we are all being listened to, and you inherently know that if you do someone an injustice, it will come back to you. This in turn means that other people feel deeply understood and listened to around this Gift.

This profound awareness of Karma means that a great deal of your work in life will lie in the sphere of relationships and the emotions. At the Gift frequency of this Gene Key you learn to temper your own emotions, releasing them safely without disrespecting others or yourself. With the Gift of Graciousness, you begin to disperse your own Karma and that of your ancestral DNA. This is a huge task and it means that even though your relationships may be deeply challenging, you always maintain a frequency of respect around others. The 22nd Gift also ensures that you balance your respect of others with a healthy dose of self-respect by not allowing yourself to become a victim of someone else's emotions. This delicate balance between service and self-love marks you as someone who profoundly understands the power of emotional suffering. Because of this, others will look to you for guidance and authority.

The Gift of Graciousness might also be called the Gift of Soul. It is the ability to live life to the full, holding no feelings back, whilst at the same time having deep respect for the feelings of others. If you are fortunate enough to tap into the higher qualities of the 22nd Gift, your life can be filled with art, music, romance, deep relationships and enchantment. But above all else this is the Gift of living life from a place of deep love and soul.

THE VENUS SEQUENCE – A DIRECT TRANSMISSION OF GRACIOUSNESS

In the summer of 2004, during a rare transit of the planet Venus across the front of the sun, a profound knowledge came into the world. Called the Venus Sequence, it uses the 64 Gene Keys in combination with astrological data to pinpoint the exact patterns of karma that an individual takes on during each incarnation. The Venus Sequence reveals karma as a genetic sequence that unfolds throughout your life. Your ability to accept this karma with graciousness determines how quickly and easily you transcend your own suffering. Furthermore, as the sequence of your own karma unfolds and is transmuted, it reveals the higher frequencies that allow you to expand your consciousness and attain the higher states.

The Venus Sequence is the great science of human suffering, showing us precisely how every human being shares in world karma or the *sacred wound* through being genetically imprinted with it at the point of our conception. As we unravel our Venus Sequence, we discover an inner path of awakening behind our suffering. This culminates in our having to embrace one of the six essential core human wounds.

THE SIX CORE WOUNDS OF HUMANITY

1. Repression
2. Denial
3. Shame
4. Rejection
5. Guilt
6. Separation

These six patterns are laid down in a unique sequence in your DNA. Once you have access to your own sequence you will finally understand the basic script of your life process particularly as regards your relationships, where your karma is mostly played out. The wound itself is directly tied to the lower three bodies — the physical, emotional and mental. Through clear mental understanding and a gentle emotional process of self-forgiveness, the very patterns that cause human beings so much pain actually lead us directly to liberation through the higher three bodies. This is both a process of evolution (described in depth in the 55th Gene Key) and a process of involution whereby the higher frequencies within your DNA are being activated through the agency of Grace. However, the prime teaching of the Venus Sequence concerns showing another human being, through relationship, how to take responsibility for their own feelings instead of projecting them onto others. This ability is the very essence of the 22nd Gift of Graciousness and is preparing humanity to make the great leap into Freedom, the 55th Gift.

THE 22ND SIDDHI – GRACE

THE SEVEN SACRED SEALS AND THE BIOLOGICAL APOCALYPSE

In the sacred Book of Revelation, St. John The Divine wrote down his famous description of the Apocalypse — the so-called Judgement Day where all world karma is finally redeemed and all human suffering brought to an end. Despite centuries of misunderstanding by religion, the revelation of St. John contains some of the greatest initiatic secrets ever written down. One of these secret teachings is known as the Seven Sacred Seals. In allegorical form, St. John describes the sequential process of the opening of each of the seven seals by an angel, and the subsequent unfolding of the seven stages of the Apocalypse itself. It is not until the final seventh seal is opened that evil is conquered and humanity ascends to a higher plane.

When we decipher this allegory, we can see that each of the seven seals and its accompanying angel represents an agent of Grace — an involving spiritual force or *Siddhi* descending from the higher vehicles and directly affecting human DNA. The apocalypse is really a biological phenomena — a judgement day within our genes as a new species of human prepares to be born. Within the matrix of the 64 Gene Keys there are six Gene Keys that directly reflect this involving power of Grace, the seventh being the 22nd Siddhi of Grace itself:

THE SEVEN SACRED SEALS AND THEIR RESPECTIVE SIDDHIS

The First Seal – Divine Will (40th Siddhi)

The Second Seal – Omniscience (17th Siddhi)

The Third Seal – Universal Love (25th Siddhi)

The Fourth Seal – Epiphany (43rd Siddhi)

The Fifth Seal – Forgiveness (4th Siddhi)

The Sixth Seal – Truth (63rd Siddhi)

The Seventh Seal – Grace (22nd Siddhi)

The 22nd Siddhi of Grace always works through the field of these six Siddhis or Divine attributes. Thus we can see that each of these Sacred Seals is a Divine code sent down from higher consciousness to heal a specific aspect of the sacred wound. Just as there are six aspects to the sacred wound, so there are six Divine aspects responsible for healing that wound. This process takes place both individually and collectively and is described below.

THE OPENING OF THE FIRST SEAL – DIVINE WILL

The First Seal is opened by the Siddhi of Divine Will, which heals human repression. Repression is the primary wound of humanity, since it refers to the very storage of karma in the DNA of the physical body. It is because of the layers and layers of karma in your DNA that the higher subtle bodies are obscured from your awareness. Karma is a deep physical tension expressed as fear, and it inhabits every cell of your body. Only through the Grace of the 40th Siddhi can this karma be transmuted. The 40th Siddhi represents Divine Will, which is the only force powerful enough to transform all the layers of tension. Divine Will actually means complete relaxation, so as this seal opens throughout humanity, the physical body finally comes into complete relaxation. As it relaxes progressively into deeper and deeper states, the higher subtle bodies can express themselves fully. In the end, your body becomes nothing more than a completely relaxed instrument of Divine Will. Because this seal has to do with releasing core tension from the physical body, at a collective level it will eventually eradicate all disease from our planet.

THE OPENING OF THE SECOND SEAL — OMNISCIENCE

The Second Seal is opened by the 17th Siddhi of Omniscience and its target is the wound of denial. Denial is the external expression of fear as anger and aggression. If your core wound is denial then you are unable to see and take responsibility for your own negative behaviour. The more someone tries to show you your denial, the more powerful it becomes. Throughout humanity this wound expresses itself through fundamentalism, violence and sexuality. The only force that can break denial is Omniscience, which is what occurs when, even if only for a split second, your vision is opened and your higher bodies literally look down into your mental and astral bodies. This notion of *being seen through* comes as a deep shock to the recipient, who usually experiences a complete and permanent rebirth after the event. Once you can *see* your own denial, it is no longer denial. It is through the opening of this second seal that certain people experience sudden conversions or higher callings. At the collective level, this seal will bring great healing to human sexuality and will eventually extinguish violence.

THE OPENING OF THE THIRD SEAL — UNIVERSAL LOVE

The Third Seal is opened through the 25th Siddhi of Universal Love. This is one of the most pervasive forms of Divine Grace, and as it reaches down into human beings it triggers a huge wave of release that can spread from person to person like a positive virus. This seal heals the human wound of shame. Shame is brought about by a profound feeling of worthlessness. It is out of this feeling of deep shame that the whole world of hierarchy and competitiveness has come about. As the Siddhi of Universal Love descends into humanity, our urge to escape our own shame through selfishness and greed gives way to feelings of great joy and self love. It is this self love that leads to altruism and philanthropy rather than competition. Shame is obsessed with hiding, but Universal Love shows you that no matter where you hide, love is still there. It is through the opening of this third seal that you finally begin to enjoy life for what it is rather than always chasing some *ideal* future. As humanity experiences this opening, it will lead to a complete breakthrough in the way we use money and a final ending to human greed.

THE OPENING OF THE FOURTH SEAL — EPIPHANY

The Fourth Seal is opened by the 43rd Siddhi of Epiphany, symbolised by the descent of the dove of peace. Epiphany heals the wound of rejection, the wound that keeps human beings from opening their hearts fully to one another. Epiphany is actually a shattering experience in which the higher three bodies (symbolised by the Gifts of the three Magi of the Christian Epiphany) explode through into the lower bodies, opening the heart from the inside. As this seal is opened, the many barriers that humans have erected will begin to fall away — our countries, borders, armies and all aspects that attempt to protect and defend us from each other. On an individual level, the fourth seal opens the potential for you to lead a truly romantic life where you need hide nothing from anyone, and in which you wear your heart openly on your sleeve. Once human beings overcome the fear of external rejection, they become pure agents of Grace through their friendliness, openness and honesty. On a collective level, this seal opens the heart centre in humanity and manifests as kindness. This Siddhi will bring an end to world poverty.

THE OPENING OF THE FIFTH SEAL — FORGIVENESS

The Fifth Seal is opened through the 4th Siddhi of Forgiveness. As one of the great involving Siddhis of Grace, forgiveness has a special purpose — to work its way backwards into the collective DNA of humanity, releasing the many karmic blocks that plague the various gene pools. The Fifth Seal is specifically targeted to heal the patterns of unconscious guilt upon which karma is built. Guilt is a kind of karmic debt which exists between one person and another, or even between one race and another. As the power of forgiveness works its way into the human genome, so many ancestral curses will finally be lifted. This seal in particular has the capacity to create world peace, as individuals and whole nations forgive others the debts they are owed. The Grace that comes with forgiveness has unprecedented power and brings a sense of true justice back to humanity. Forgiveness is a direct manifestation of the fifth body, the Buddhic body, which has the capacity to literally burn karma out of our DNA. At a collective level, the releasing of all this world karma will bring an end to war.

THE OPENING OF THE SIXTH SEAL — TRUTH

As the final stage in the sequence (the seventh is the glorious afterglow), the Sixth Seal delivers the *coup de grace* as it heals the ultimate human wound — that of separation. Because to most of us the higher realities of our true nature are obscured; we have essentially been separated from the Divine for most of our lives. Because we feel this separation so keenly, we constantly seek fulfilment in the outer world. Ironically, it is this very seeking that keeps us from experiencing our true nature, which is to be found deep within our own suffering. The opening of the Sixth Seal is made possible through the 63rd Siddhi of Truth. Truth is something you become one with rather than something you find *out there*. Through the opening of this seal, every individual will come to know their true nature as an aspect of one vast consciousness. Absolute Truth is a collective phenomenon that will one day be fully embodied as the whole of humanity spontaneously recognises itself as a single Divine organism. Only then, at that marvellous moment, will all human seeking and striving come to an end. So it is that this 63rd Siddhi, through its direct realisation and expansive embodiment, will bring a crushing end to the greatest of human curses — indifference.

The Sixth Seal delivers the "coup de grace" as it heals the ultimate human wound — that of separation.

THE OPENING OF THE SEVENTH SEAL — GRACE

In the Book of Revelations, the opening of the seven seals is surrounded by layer upon layer of rich, apocalyptic imagery. Unless you are well versed in alchemical symbolism, it will be very difficult for you to penetrate to the true meaning of this wonderful prophetic transmission. There is, moreover, a demarcation between the opening of the first six seals and the opening of the Seventh Seal. The Seventh Seal involves seven angels and seven trumpets which sound the final judgement of humanity. The Seventh Seal is the spirit of Grace itself (represented by the 22nd Siddhi), and Grace descends only after great transmutation. It is like the rainbow that appears after the great storm, bringing complete transfiguration (the 47th Siddhi). On an individual level, the Seventh Seal represents the final absorption of all the previous six layers of the human aura into the monadic body, the primordial essence. At this level, even the flood of revelations and high frequencies of the atmic body must be surrendered into the void of what the mystics term the *seventh heaven*.

We are told in the Book of Genesis that God rested on the seventh day of creation, and this sevenfold pattern is reflected in many other cultural traditions. In the Hindu system, when the seventh chakra known as *sahashara* blooms, the Divine essence can finally reunite with the material plane. St. John describes this event as the coming of *a new heaven and a new earth*. As this seventh monadic plane absorbs the final vestiges of our separateness, each plane below and its frequencies disintegrate, to be reintegrated as the true monadic essence. This is the meaning of the sounding of the seven trumpets, which represent the seven layers of frequencies of the human aura. At the collective level, the opening of the Seventh Seal refers to the coming of the final human epoch and the return of the human race to its original *Edenic* state. It is the great trumpet fanfare announcing the redemption of all beings.

THE MEANING OF SPIRITUAL INITIATION

The final aspect of the transmission through the 22nd Gene Key follows the story of the individual human soul itself. All human stories enact the same mythic process or storyline, albeit in infinite and uniquely diverse forms. As our incarnative process recycles into new and different forms, each activating different inferences of human DNA during each lifetime, our great trans-generational story deepens. Like all great drama, we are weaving a multi-dimensional tapestry of different colours, tones and hues from the rich fabric of experience. Throughout every incarnative story, however, stands one immutable question that keeps us coming back — the question of our suffering. It is our changing relationship to this question that marks the various stages in our journey through time and space. There are nine major landmarks along the way, and they are known as the Nine Portals of Planetary Initiation.

Spiritual initiation is a natural, organic process that occurs to all human beings at a certain point in their lives.

Spiritual Initiation can mean many different things to people. The word *initiation* itself might conjure up all sorts of mystical rites you may have witnessed or heard of. Certainly in tribal societies, initiation is seen as a rite of passage, particularly for young men, who must perform some kind of test at a certain threshold age before entering full manhood.

Other ancient teachings and/or societies have systems of initiation that are performed as elaborate rituals at certain stages in an aspirant's life. In truth however, spiritual initiation is a natural, organic process that occurs to all human beings at a certain point in their lives. In essence, initiation refers to the naturally unfolding stages of all spiritual awakening. Initiation occurs to you no matter what you are doing in your life and once it has begun in earnest — when you have passed through the first Initiation — it is irreversible and inevitable.

THE NINE PORTALS OF PLANETARY INITIATION

The Nine Portals of Planetary Initiation are a synthesis of the initiation rites of many different cultures and lineages. Listed below are the nine stages, followed by brief introductions to each.

1. Birth
2. Baptism
3. Confirmation
4. Matrimony
5. Annunciation
6. Communion
7. Ordination
8. Sanctification
9. Glorification

THE FIRST INITIATION – BIRTH

The First Initiation marks the very beginning of your journey towards eventual transcendence and enlightenment. This Initiation may pass with little or no recognition that anything has occurred within your consciousness. There comes a time in every soul's life when it has to move beyond its basic mammalian survival programming. The first stage of passing this threshold really concerns the development of the mental body. In your early evolution, you are simply overwhelmed by the desire nature of the astral body, and this remains the central focus of your whole being. First it is interested in survival. Then, having mastered survival, it becomes interested only in pleasure. Lifetimes follow in which pleasure is pursued in any and every possible way. In one sense, the soul is trying to define what pleasure or happiness is and how it can be captured. Despite fleeting glimpses of happiness, the soul fails in finding true fulfilment through the outer senses and gradually turns its lens on the real prize — suffering itself.

The First Initiation is a birth out of selfishness and into service.

In looking into the nature of suffering, the mental body must first detach itself from the astral body, which means that for the first time the soul must consider its own nature. This turning inward marks a huge shift in focus in the life of the soul because in looking at its own nature it must also begin to consider the feelings and thoughts of others. In the deepest sense the First Initiation is a birth out of selfishness and into service in its broadest form. It is marked by the ability and willingness of an individual to accept responsibility for his or her actions.

It is the birth of a true morality — not in the sense of adhering to an outer code or set of laws, but of the natural human spirit of helpfulness and harmlessness. After the First Initiation, the soul becomes aware that greater pleasure is derived from giving than from taking, and this becomes the foundation of the higher life.

In the world around us today, many people have passed through the First Initiation. There is no set of beliefs or any common mission that unites these people save that of wanting to leave the world a better place for future generations. They can be spiritual, atheistic, opinionated, even dogmatic, but they cannot be indifferent to their own or another's suffering, and this is what makes them so powerful and so precious.

THE SECOND INITIATION – BAPTISM

The Second Initiation is very different from the First. Whereas the First Initiation is a general build up of basic human goodness over a long period, the Second Initiation of Baptism comes as a surprise. At the Second Initiation, the spirit of Grace descends through the layers or bodies of form and bestows a moment of higher contact upon the recipient. Baptism is a sudden immersion in the higher frequencies of your own higher vehicles and as such it always comes as something of a shock. As with all shock, it takes a long time to come to terms with what has happened to you. The duration of the experience varies greatly from person to person, and as the higher frequencies subside, the lower bodies are left with the task of recalibrating and readjusting to the influx of the new frequencies.

During this period of readjustment, many things may take place within an individual. The existing mental framework tries to place the experience within its old paradigm or within a paradigm recognised by society. Many people experience the Second Initiation as a higher calling into one of the great religions. Others continue to grapple with the experience and may even go through a mental or nervous breakdown. Another common response is prolonged depression and a longing to return to the high frequency state. Still others may even go into denial and try to forget the experience altogether. Baptism as such can be extremely challenging as it sets you apart from the rest of society in some way. It is in a sense a kind of purgatory, since you have tasted the higher life and can never entirely forget it.

If you are able to handle the frequencies of the higher bodies and can incorporate the experience cleanly in your life, Grace will periodically revisit you and baptise you in the higher frequencies of the causal body. The Second Initiation is an ongoing baptism in the higher reality, and the more readily the experience is digested, the more energy becomes available to you. We need to remember that the bodies above the causal body constitute your higher self. They know exactly how and when to allow the higher frequencies down into the lower vehicles. The period after the Second Initiation may also last many lifetimes as the causal body gains a stronger foothold over the astral and mental bodies — the sexual and intellectual faculties. Baptism is thus an Initiation into purification in which your lower essences are gradually refined in order to be able to hold a more sustained higher vibration.

Baptism is thus an Initiation into purification in which your lower essences are gradually refined.

Simply by heartfelt invocation, you can now call upon Grace.

THE THIRD INITIATION – CONFIRMATION

In some traditions, the Third Initiation is understood to be the first true Initiation because it is not until this Confirmation arrives that a person becomes stable in their purpose of seeking the Divine. Confirmation is another gift of Grace that is given as a kind of reward to the recipient. It is evident from their names that these Initiations are taken from the Christian tradition. Their mystery can be further understood through the original layout of the Christian church. The Birth represents the entry into the church itself, the body of the higher presence. The Baptism is always made at the font, which is usually located at the back of the church, between the entrance and the congregation and represents the introduction of the child as a member of the church. During Christian confirmation, a young person is officially initiated as a member of the congregation, whose arena is the main body of the church itself. Confirmation gives the young church member their first taste of a much higher Initiation — that of the Holy Communion.

In the list of the Nine Initiations above, you can see that they are laid out in threes. This is one of the great mysteries of Initiation, which is based upon final immersion into the Holy Trinity. At each of the three levels, the seeker enters into deeper communion with the triple nature of the trinity. Thus at the Third Initiation the seeker has a taste of the Sixth Initiation, and very faintly detects the echo of the Ninth Initiation. So it is with each of the other levels.

Confirmation is actually a fairly high vibratory Initiation. It denotes having reached a stable frequency in which your commitment has been tested and found strong and your lower nature has shown itself to be capable of some degree of sacrifice. At this level, the mystery of sacrifice is understood to be the deepest core of Initiation itself. Here at Confirmation you will grasp the certainty of your final goal in which your lower nature will be sacrificed to your higher. After the Third Initiation it is no longer possible for you to leave the path of Initiation, even though you will inevitably still stray from it at times. This is a level of frequency in which you become used to more regular periods of contact with the higher vehicles. The pathway between the causal body and the mental body is well trodden, which means that simply by heartfelt invocation, you can now call upon Grace.

THE FOURTH INITIATION – MATRIMONY

The Fourth Initiation represents the higher octave of the First Initiation of Birth. Matrimony is a birth into a higher dimension. It is a spontaneous inner commitment that greatly widens the pathways of awareness inside your being. The Initiation of Matrimony or Marriage is your first step into the collective way of life. In the Christian mystical traditions, matrimony represents the marriage of the seeker to Christ — a very deep level of commitment that actually led to the beginning of the monastic tradition. The modern cultural institution of marriage still contains many of the secret initiatic rites that mark this high stage of consciousness. The primary symbol of matrimony is the wedding ring — the emblem of sacred union and divine perfection.

The Fourth Initiation marks the beginning of the higher life in which your primary focus is to incarnate the causal body fully onto the physical plane. This means that your life's work is now offered in service of the whole, and there is no difference between your work and worship. The well known wedding vow, *until death us do part*, now becomes your living reality, and your only true beloved is the higher self of your Divine consciousness. Externally, this period of your life story is usually marked by deep service to humanity in which you are much concerned with improving the welfare of those less fortunate than yourself. You may recall that the causal body is the body of our enduring virtue, being the sum total of all that is good about us. After the Fourth Initiation, our lower nature becomes increasingly more organised by the causal body. Our sexual energies in particular are channelled into creative work of a higher nature. Just as physical marriage precedes the raising of a physical family, so this higher marriage can lead to a vast surge in creativity that can involve and awaken many others.

It is of interest to note that it is not just individuals who move through the great Initiations, but also entire species. Humanity has in fact passed through the Fourth Initiation during the past few hundred years. Our matrimony has vastly improved the state of the world in which we live. The steady growth of global awareness is outstripping individual greed. Politically the rise in democracy and social justice has changed the face of the planet. Gradually, human goodness is moving us forward, even though news headlines do not often reflect this truth. To understand Initiation, you must read between the lines and you must feel the truth with your heart as much as see it with your mind. As a species we are closer than we have ever been to the ideal of working together collectively as a single unified organism.

Matrimony has many hidden meanings. It can refer to the marriage of opposites — of East and West, science and religion, the masculine and the feminine. Inside your DNA it speaks of a fusion, an intense period in which the many opposites within your being mystically come together into greater harmony than ever before. In the analogy of the church, it is when you are singled out within the congregation and approach the high altar together with your betrothed. It is represented in the sacred geometry of the church by the *crossing* and the great side windows of the knave — the open arms of the church lying between the congregation and the choir. It is a place of expansion, where the body of the church opens out on both sides like two spreading wings. This is exactly what the Fourth Initiation denotes — a time in which you open your heart to the world and spread your wings to the winds of the Holy Spirit of Grace.

THE FIFTH INITIATION – ANNUNCIATION

The Fifth Initiation flows naturally from the Fourth, and at this level of expanded awareness the Initiations often follow each other in relatively quick succession, that is, within a single lifetime. Now that you are married to your higher consciousness, the next thing that happens is that you become pregnant. This is the Annunciation — the mystical announcement of the imminent birth of the Christ. Much of the symbolism surrounding this Fifth Initiation is feminine in origin, flowing as it does from the third aspect of the sacred Trinity — the Holy Mother. On the lower planes, the feminine is expressed through the astral plane — the desire nature and the emotions. At a higher level, this plane relates to the fifth or buddhic body.

You experience the sublimation of your sexuality into pure spiritual essence.

At the Fifth Initiation, the exquisite emanations and refined currents from the buddhic body begin to penetrate the lower astral nature. This is a deeply tantric phenomenon in which you experience the sublimation of your sexuality into pure spiritual essence.

The Annunciation is a chemical phenomenon that saturates your whole being. Just as a woman during pregnancy is flooded with hormones, so you become aware that your body is being purified and cleansed in preparation for a great inner event — the birth of Christ consciousness. In the analogy of the church, this Initiation relates to the situation of the choir who preside over the altar and who represent the voice of pure worship of the Godhead. The connection between the throat centre and the sexual centre also becomes clear during this Initiation. When a person spontaneously enters a higher ecstatic state that appears to be ongoing, it is a fair assumption that they are entering this fifth portal of the Annunciation. Many different mystical traditions speak of this magical time of higher pregnancy in which the immortal foetus is gestating inside our solar plexus centre. It is one of the great enduring mysteries of creation.

In terms of the Initiation of species, humanity stands at the threshold of the Fifth Initiation today. Even now, there are whispers in the air of a new form waiting to come into the world. Little do we realise that this form is gestating inside our very DNA. The timeline for a species to undergo any Initiation is obviously very different from an individual, and for humanity it may take many hundreds of years for the Annunciation to become realised. We are about to enter a time of great purification, in which the feminine spirit of Grace will be working actively in the world. The fact is that as a species we have already been fertilised and are in the very first stages of pregnancy. Just as the body of a woman takes several weeks and months to register the changes externally, humanity will not become generally aware of the great change that is upon it for some considerable time. Only those who are sensitive to the subtler currents moving beneath the world of matter will detect the first signs of the new human that has now begun its period of gestation.

THE SIXTH INITIATION — COMMUNION

The Sixth Initiation is the greatest experience a human being can have. It represents the zenith of our human development and the end of our evolution on the earth. The mystery of Holy Communion is the mystery of sacrifice. This is the full awakening of the Christ consciousness in a human vehicle, which requires the death of that which identifies with the principle of form itself. This stage is often referred to as enlightenment. The light referred to is the pure light of the sixth body — the atmic body, which is born from within the buddhic body and enlightens the lower three bodies. As it does so, it causes the causal body to dissolve thus severing the link or bridge between the lower and higher worlds. In mystical terms this involves the dissolution of the soul — that aspect of human consciousness that is drawn again and again into incarnation. Thus it is said that at the point of enlightenment the indwelling awareness can no longer take incarnation and forever escapes the wheel of *samsara* or illusion.

The Sixth Initiation is the greatest experience a human being can have.

The Initiation of Communion also shares its name with the 45th Siddhi, which describes the great mystery of the taking of the sacred sacrament. The Communion involves the direct imbibing of Divine consciousness at the altar. In entering this field of frequency, you are transcending all sense of being separate from others. This is symbolised by the blood of Christ and marks the final breaking up of the karmic residues held in your DNA. For the Grace of Christ to enter you, you must be willing to make the ultimate sacrifice — to give up your lower bodies and their desires, feelings, memories, dreams and knowledge and to be taken over by the greater being who has been waiting all along within you. To enter this great Initiation is to die into the second aspect of the sacred Trinity — the Christ.

THE HIGHER EVOLUTION AND THE SEVENTH, EIGHTH AND NINTH INITIATIONS

The remaining three Initiations are part of what is called *The Higher Evolution* which lies beyond humanity and our human story. As such they are difficult to describe in words and certain binding oaths surround them in order to protect humans from their dangerous frequencies. These frequencies are no longer conveyable other than through sound, or through direct transmission in silence. Even so, The Higher Evolution has always been known about and fragments of truth concerning it can be found throughout human culture. In the Christian church the higher evolution is represented through the priesthood and its hierarchies.

The Seventh Initiation of Ordination has much to do with the notion of *co-ordination* and these incarnations point humanity in specific new directions. It is through these beings that the higher secrets of Initiation may be released. The presence of such beings in the world always precipitates a great shift in the consciousness of the whole planet. Interestingly, there are always five avatars present on the earth at any given moment. They form a unified force-field that steers the balance of humanity towards evolution, rather than destruction.

The Eighth Initiation of Sanctification is an extremely rare event on our planet. Initiations at this level are really beyond human understanding as they have to do with the flow of essence from the seventh Monadic body into the Atmic and Buddhic bodies. Although it does occasionally occur on an individual level (with spectacular results), the Eighth Initiation is a collective Initiation that can only be taken towards the end of humanity's evolution.

The Ninth and final Initiation brings an end to the story of consciousness. On an individual level, the Ninth Initiation can only take place through silence. In certain esoteric traditions it is also referred to as The Refusal. After this Initiation, the indwelling awareness *refuses* to materialise and dissolves back into the primordial essence from which it was born. Tradition tells us that only a very few great initiates have taken the Ninth Initiation to date. It is during this final Initiation that the first body and the seventh body fuse together. The physical form then ascends and completes its final great destiny.

Initiations at this level are really beyond human understanding.

THE NEW HEAVEN AND THE NEW EARTH

One can see from the depth of its transmission that the 22nd Siddhi is difficult to describe in words. The reason for this is that you need to experience it in order to know what it is. Although Grace is a word that is often used in spiritual circles these days, it should not be used lightly. Rather it needs to be treated with the utmost respect.

We have learned that Grace has to be earned through Graciousness. This is the great message of the 22nd Gene Key — to find graciousness in the face of suffering, and perhaps even to find something more — holiness itself wearing a disguise. If the 22nd Gene Key happens to be an aspect of your Hologenetic Profile, then the theme of Grace will be very strong in your life, and you must not turn your face away from the pain life offers you. We are all here to be tested, over and over, until we show that our faith in nature herself can never again be lost.

Grace is a presence that descends on humanity, and like all the Siddhis, it requires that we meet it halfway, which for us humans may seem a very long way. This is after all a perfected state in which everything about you and your life will be changed permanently. When true Grace descends, it wipes out all your past Karma in a flash. It also wipes out the Karma of all your ancestors and all their ancestors. Grace softens your rough edges, puts a permanent end to your fear and leaves you in no doubt whatsoever about your divinity. It also ensures that you never again forget. It is impossible to measure in words the sheer number of blessings that Grace bestows when it alights on us.

One who has been touched by Grace is always touched by Grace. If it happened to you one millennia ago in another universe or in another incarnation, it will never leave you. It will go on bathing you again and again. To be in the presence of someone manifesting this Siddhi is to be entranced by the aura of love that surrounds them. It is something you can never forget, and it will stir your own soul to seek it until you find it. Grace is the very breath of the Divine. It is always there, waiting for us high above, if we only persist in our sacrifices. Wherever there is oppression, there is the possibility of Grace. If you can face oppression with a gracious spirit and a forgiving heart, Grace will come to you sooner or later. Grace is a feminine spirit, and she cannot resist giving herself to those who smile in the face of adversity.

This is the great message of the 22nd Gene Key — to find graciousness in the face of suffering, and perhaps even to find something more — holiness itself wearing a disguise.

As we saw with the 22nd Shadow, there is nowhere you can hide in this universe. Everything is heard and recorded. Neither can you hide from Grace. Grace is your true nature. It is your inheritance. It is the soul of the world. It is also a state that is beyond the laws of our world. If Grace touches you, you no longer create Karma. If Grace touches you, you no longer have your own destiny but become a musical instrument tuned and played by the gods. With Grace, all human emotion is instantly transformed into love. It is not a state with which most human beings are very familiar. As a species however, we are moving into a new epoch that will be marked by Grace. As each of the seven seals is opened, the world we have become used to will begin to drop away. In its place will rise, shining and resplendent as the summer sun, the new heaven and the new earth that St. John talked about in his great revelation.

Now that you have imbibed this profound transmission that is the 22nd Gene Key, it is recommended that you give it some time to digest within the many layers of your being. As a feminine spirit, Grace calls upon each of us to listen and receive her message and blessings. Above all, remember this: through Grace, the universe has but a single wish — for you to remember that you are love, and there is nothing in you but love.

▦ 23rd Gene Key

The Alchemy of Simplicity

SIDDHI
QUINTESSENCE

GIFT
SIMPLICITY

SHADOW
COMPLEXITY

Programming Partner: 43rd Gene Key
Codon Ring: The Ring of Life and Death
(3, 20, 23, 24, 27, 42)

Physiology: Throat (Thyroid)
Amino Acid: Leucine

THE 23RD SHADOW – COMPLEXITY

SPLITTING THE WORLD APART

The philosopher Ludwig Wittgenstein once said: "You cannot think something unless you can talk about it." What he was suggesting is that thinking and language appear to be inseparable. The 23rd Shadow directly concerns the connection between thinking, knowing and the expression of that knowing through language. As a functioning, or indeed malfunctioning, of our genetics, the 64 Shadows govern humanity in different cycles. This is to say that at certain points in history a specific Shadow will appear to suddenly dominate the human species, whilst at other times the same Shadow will appear to lie relatively dormant. As each aspect of the Shadow consciousness rises into prominence, history is written accordingly. The overall direction of the evolution of humanity is governed by these regular inner cycles and impulses that emerge from within our DNA.

The 23rd Shadow is one of the most potent contemporary Shadows driving humanity in our present post-modern age. It is the Shadow of Complexity. Complexity is the result of the human mind trying to control its environment. The more humans try to use their minds to create a feeling of security in the world, the more complicated and unsafe the world becomes. At an individual level this Shadow creates misunderstanding and division in two different ways — through individuals saying the wrong things, and through saying them at the wrong time. This human trait has been around for as long as language has been around, and it is responsible for some of the most horrendous events in human history. Some of the bloodiest wars on our planet have begun because of a few simple misconstrued words. When the Buddha introduced his great teaching known as the Eightfold Path, he spoke of *Right Understanding*, which leads to *Right Speech*. He hit upon this great and simple truth — that language flows directly out of your state of consciousness. The more profoundly you understand and embrace the Shadow, the freer of it you become.

The challenge within the 23rd Shadow is found in its programming partner — the 43rd Shadow of Deafness. The 23rd Shadow represents an overwhelming human urge to express yourself. Coupled with the inability to hear either yourself or others, this creates a lethal cocktail.

*Whenever
this 23rd
Shadow
speaks, it
complicates
the situation.*

One of the great problems for individuals is to communicate clearly with others. Hearing yourself means being aware of what is going on inside you. If you have no such self-awareness, then you are also unable to really hear what is going on outside you, which means that you will not know how to relate to others. Language is immensely powerful as a medium for stirring up the volatile human emotional spirit. Whenever this 23rd Shadow speaks, it complicates the situation, which in turn catalyses an exponential process of misunderstanding that can soon escalate from a mental process to an emotional one. The fear beneath this 23rd Shadow is of meeting intolerance and being excluded by others. Ironically, this very fear drives the behaviour that manifests the fear.

The difficulty for the 23rd Shadow is that it tends to make human beings *know* that they are right, which firmly closes the door on their being open to the views of others. This is where the deafness comes into play. In conversation with someone strongly influenced by this Shadow they can seem like a stuck record. They simply cannot give you a straight answer to a question. You can even ask them specifically to answer in a single word, and they cannot do it! There is an unconscious assumption on their part that you are somehow inside their head with them and understand everything they are thinking with crystal clarity. Because of this pattern you often have the feeling that such people are simply talking *at* you, rather than communicating *with* you. This generally causes the listener to feel physically uncomfortable and back away or try to interrupt the stream. Either way, the situation can become agitated and complex when a misunderstanding could have been easily avoided.

Within this 23rd Gene Key lies a secret of timing. Because the 23rd Shadow concerns the way in which your brain translates thoughts into linguistic patterns it is through this Gene Key that all manner of speech difficulties can occur — from verbal diarrhoea to dumbness. Such problems are rooted in the subtle timing mechanism hidden within speech and language. Human speech is made up of many layers of intonations with spaces in between. Even though your brain may *hear* the language inside your head, translating that to the vocal chords is another matter. How successful this process is depends upon your overall level of frequency. If you are caught in the victim patterns of the Shadows, then the process of translation slightly misfires, as though a code has been copied incorrectly. This leads to a language pattern that is not harmonised to the situation or listener.

This process of timing and mistiming happens at a level well below your conscious aware-ness. It is not the words themselves that cause the static between the speaker and the listener, but the subtle cadences of intonation itself. There are people who can give speeches in the most engaging language, but which actually make no sense to the audience whatsoever. The gift of communication does not lie in the speaker's linguistic skills, but somewhere much subtler — in the speaker's heart. If even the subtlest trace of fear drives your speech, then that speech can never be fully imbibed by the listener. However, when someone speaks or writes from their heart, you will understand the gist of what they are saying, regardless of *how* they say it. Con-versely, when you read or hear something that leaves you feeling cold, you are probably hearing someone's deep-seated fear.

People driven by the 23rd Shadow speak in order to gain approval or recognition — and this always, without exception, sets them up to be misunderstood.

At a collective level, the misfiring of human language patterns has led to intolerance and schisms. It has, in fact, led to organised religion. All the major religions have evolved from simple language patterns spoken by individuals in the now, which were then mis-translated by others who were not physically present when the words were spoken. As an example, the words of Christ are beautiful in their pure simplicity, but the hundreds of schisms caused by their many different interpretations have transformed the original simplicity into something incredibly complicated and ultimately ugly. The moment someone speaks out of a frequency of fear or anger, even the most beautiful sounding words can become dangerous and divisive.

It is easy to see from this Shadow why the ancient Chinese referred to the 23rd hexagram of the I Ching using the phrase *Splitting Apart*. It is absolutely true that the frequency of this Shadow does cause human beings to distrust each other, driving us further and further apart and creating more and more complexity and fragmentation. As was suggested earlier, the 23rd Shadow of Complexity is responsible for our planet becoming more and more unsafe. Ultimately however, this splitting apart has a hidden goal — to make human beings aware that they are the cause of their own suffering and in time to bring about the gift of *Right Understanding*.

REPRESSIVE NATURE — DUMB

In a repressive nature or society, the expression of the 23rd Shadow will rarely be tolerated. Wherever you see societies that keep others silent, you are seeing fear at work through this Shadow. Depending on how acute the level of fear is the individual or group will probably never express what they really think. Language will be held back, either through internal repression or external oppression. It is interesting that the modern meaning of the word *dumb* has come to refer to a lack of intelligence, because that is often how such people are perceived. If you are choked by fear, you cannot speak clearly, if at all, and so these people slowly learn not to say what they really think, instead lapsing into silence or superficiality. This is particularly true of children who have very oppressive parents.

REACTIVE NATURE — FRAGMENTED

The other side of this Shadow is the expressive nature, which often cannot stop talking. However, because there is a general *mis-timing* within their neurology, they create enormous interference and fragmentation wherever they go. These are people who always say the wrong thing, or they say the right thing but at the wrong time. The language pattern of such people tends to over-complicate and miss the essential meaning of things. They spend an enormous amount of energy trying to be heard only to find that they are constantly pushed away. These people tend to talk in circles, over-explaining everything in an unconscious attempt to conceal the anger that lies within them at not being understood on a much deeper level.

THE 23ᴿᴰ GIFT – SIMPLICITY

THE NOBLEST TRUTH

One of the hardest things for most modern human beings is to live simply. Sometimes simplicity is even considered synonymous with stupidity. Actually, the lower the frequency of your genetics, the more complicated you will tend to make things. The reason for this is the human mind. The mind does not trust simplicity because it thrives on complexity. The more complicated something is, the more the mind can think about it. So, when we come to this 23ʳᵈ Gift, we are learning one of the great secrets of a happy life — keep it simple.

The 23ʳᵈ Gift abhors clutter and jargon. It communicates precisely, clearly and with great economy. The power of simplicity is to create efficiency wherever it goes. People exhibiting the Gift of Simplicity waste nothing in life. Their living areas usually reflect their thinking, with plenty of open space and room to breathe. They are able to cut through all the dross and get straight to the essential point. Simplicity is a state of being — an attitude towards life — and as such it cannot really be taught. It takes great sensitivity to be this simple. It is more of a knack, an aura — even a love. If you love simplicity, you will manifest it around you.

The Gift level of frequency is a clearing-house for the process of preparing you to realise siddhic awareness. As such, the 23ʳᵈ Gift is an ongoing process in which more and more clutter is gradually removed from your inner and outer life. As your mind clears, spaciousness opens inside you, allowing you to see things with great clarity. Another manifestation of this frequency is a slowing down of your internal system and a gradual lessening of your need to resolve everything in your life. Emotions are allowed to follow their natural course, thoughts begin to have more gaps between them and physical impulses are either observed objectively and dispassionately or are indulged without guilt. Everything inside your nature begins to become clear. The many problems of your life are seen as phantasms created by your mind acting upon your desires. You naturally begin to turn within and contemplate your own essence.

A good metaphor for Simplicity is flying an airplane through and then above the clouds. In the clouds the mind finds only complexity and it goes around in circles trying to find the way out. In these lower frequency clouds you pick up static all the time. At higher levels of frequency your mind sees further and more clearly. At the level beyond the clouds, the world mind is left below and you begin to move into more silence. This 23ʳᵈ Gift and its programming partner the 43ʳᵈ Gift of Insight are deeply connected to the acoustic field of sound. These two Gifts are about clear hearing and clear translation of that hearing. On an inner level clear hearing really refers to knowing. We noted in the 43ʳᵈ Shadow that there is a theme of deafness underlying these two Gene Keys. As your frequency rises however, this very deafness becomes an ally because it acts as a truth filter. Through the 23ʳᵈ Gift you hear only the essence, and all extraneous low frequency noise is edited out.

As part of the marvellous Codon Ring known as the Ring of Life and Death, the 23ʳᵈ Gene Key captures the very same insights that were left for humanity by Gautama Buddha.

*If you love
simplicity,
you will
manifest it
around you.*

The beautiful simplicity of the Four Noble Truths and the Eightfold Path — the wisdom left by the Buddha — reflect the secrets of this Codon Ring. Through the 3rd Gene Key, we see that all life is change and is thus subject to cessation. Through the 20th Gene Key the Buddha discovered that all truth lies in the awareness of the present moment. The 24th Gene Key describes the continual process of rebirth and the wheel of samsara. The 27th Gene Key speaks of the essential moral code of goodness lying in the heart of humanity and the 42nd Gene Key teaches the power of detachment. Once you have discovered each of these secrets in your life you will arrive at a point of beautiful simplicity in which the higher level of the 23rd Gene Key — the Quintessence — is seen as the centre of centres lying deep within your belly and your being. Enlightenment comes only through this deep immersion in the core acceptance of life and death.

To be around someone exhibiting the 23rd Gift is a wonderful and powerful experience. Things that you thought were problems somehow dissolve in their presence, and through their insight and clear language, difficult things become easy. Above all, you begin to relax physically in their presence, and as this occurs the mind lets go of its need to constantly resolve complexities. Simplicity is deeply practical in the material world and these people are often very good at handling money. It is not that they know how to make money, but they know where not to waste it. They are not tight in any way but they find the simplest solutions to things. They are masters of efficiency in that they can come up with highly original yet utterly practical ideas that lead to quantum leaps in productivity.

The 23rd Gift does not look at things in a logical way, nor would you describe it as artistic or abstract. It simply *knows* things spontaneously, without knowing where the knowing came from. It is the essence of genius. The quality that most often accompanies this 23rd Gift is playfulness. Although this can be a very intense Gift, it is not a serious one. These people think in very lateral ways, not in the normal way in which we understand thinking. They observe silently until the solution jumps into their awareness, at which point they can communicate it effectively and beautifully. The combination of all these traits makes them wonderful teachers.

A final aspect to mention about this 23rd Gift is its incredible sense of humour. The way in which this Gift operates can give rise to hilarious spontaneous expressions that in some way capture the essence of a person or a particular moment. These people are so on the ball. Because they do not plan what they are going to say, they often take everyone by surprise. At lower frequencies, this trait can cause hurt and rejection. At a high frequency however, it usually causes laughter or awe. It is all a matter of frequency.

Above all, people operating out of the 23rd Gift are wonderfully clear communicators and advisers. Their true genius lies in the economy of their language and the originality of their expression. Theirs is the noblest truth of all — the Gift of Simplicity.

THE 23ᴿᴰ SIDDHI – QUINTESSENCE

THE BUDDHA FEVER

At it highest level the 23ʳᵈ Gift gives way to the Siddhi of Quintessence. Quintessence is a word that derives mainly from ancient and medieval philosophy. It refers to the so-called magical fifth element or *aether* that was believed to be inherent in all things. At the level of the 64 Siddhis, words are vibrations rather than words with single specific meanings. This word *Quintessence* carries secrets within it that give us many clues about the highest aspect of this 23ʳᵈ Gene Key.

The 23ʳᵈ Siddhi has an alchemical flavour to it. At a metaphoric level this refers to the ability to extract gold from base metal — the goal of all alchemy. Therefore people of this Siddhi are able to touch the gold hidden within other human beings. They can transmit the power of the awakened state through a word or a look or a gesture. This is the Midas Touch. Every person has a unique key that opens them to higher states of consciousness, and the people of this Siddhi hold all the keys. They do not necessarily know that they hold the keys — they simply respond spontaneously to each person, and in so doing, they touch the core of that person.

These are highly unpredictable people. As the ancient Chinese name for this archetype suggests, this Siddhi is about splitting people apart. At the frequency of the Siddhis, thought is revealed for what it is — atomic energy. Every thought that is not absolutely necessary instantaneously undergoes a process of nuclear fission — it splits apart and releases its pure organic energy into the physical aura of the being. At the highest frequency, this *splitting apart* refers to the separation of your inner essence from the illusion of your separateness.

Anyone in a siddhic state undergoes a process of alchemical transformation. In the earlier stages of the siddhic state, your physiology undergoes a deep genetic transmutation. This change is caused primarily by the *fallout* from the splitting apart of your thoughts. The entire body often undergoes a period of profound physical sensation, sometimes experienced as pleasure and sometimes as pain, whilst the continuity of your thinking is destroyed. Once the dust has settled, what remains is the quintessence — consciousness itself speaking or acting through the shell of the personality. This is the true explanation of the secrets hinted at by alchemy — that the physical body itself contains the seed of the siddhic consciousness. It is concealed in your DNA and controlled by a hidden timing mechanism that is utterly spontaneous and out of your individual reach. There is no way for you to trigger this event because it lies beyond the reach of your mind, and anything you try to do to encourage the event actually gets in the way of your natural process. In the same way that the 23ʳᵈ Shadow unconsciously *blurts* out its inappropriate expression, so the higher consciousness suddenly and unexpectedly detonates within your body.

The sheer beauty of the ultimate realisation is that it is always discovered by mistake.

Because your awakening is acausal — in other words beyond human intervention — there can never exist a technology of enlightenment. Enlightenment is beyond the grasp of technique, lying thankfully within the realm of eternal mystery.

The 23rd Siddhi represents the transcendence of the yin path and the yang path. This is why the Buddha has referred to it as *The Middle Way*. The yang path is that of science, which can only go as far as paradox, the absolute limit of scientific or logical thinking. The yin path is that of the artist or poet, and although this path comes closer to the centre than the scientist, it too always falls short. The poet moves beyond the mind and strives to approach the mystery through his or her heart. But heart and soul are bound by the limits of their own longing, and even though they may taste the centre for short periods, they fall short of the ultimate.

The third way is that of the mystic, who neither takes the path of seeking nor the path of longing. The mystic enters bodily into the mystery itself, not seeking resolution or acting out of a thirst for truth. The mystic holds such a deep respect for the mystery that he or she has no agenda or need to *understand* it but simply revels in the mystery of the ultimate, drinking it in through the pores of his or her being. Only the mystic can extract the quintessence from the ultimate. He or she simply enters the narrow gate in a state of wonder. The sheer beauty of the ultimate realisation is that it is always discovered by mistake. After realisation has occurred to someone, they can see that there are simple conditions for awakening and, like the Buddha, they can point out the exact nature of those conditions. However, the awakened one also knows that these conditions must come about in their own way, rather than becoming synthetic goals or targets for future seekers. Herein lies the dilemma of the awakened one. How many people will be led astray by the teachings, and would it be better to say nothing? But if just one person intuitively grasps the dharma (teachings), then it will be worth all those who have gone astray.

The real power of the 23rd Siddhi is therefore in its direct transmission in the flesh. The words that emerge from this Siddhi are phenomenally potent at the moment of their utterance. Once the 23rd Siddhi has fully dawned inside a person, they become an alchemical agent. Like quicksilver, they bind themselves to people and find their way into the cracks within their mental structure. Around such a person, you will undergo a series of spontaneous simplifications as the continued presence of their aura slowly begins to extract the quintessence from the dross placed within you by your culture and conditioning. As a recipient of this vibration, you will inevitably go through a complete deconstruction, which may be very difficult for you. Like all such alchemical processes, it can be extremely dangerous to your mental health unless you complete the entire process.

This 23rd Siddhi is custodian of a sacred truth: trust your own inner path before any external teaching or teacher. In the case of the Buddha, many millions of Buddhists have followed his teachings, and very few have been able to read between the words and extract the living quintessence. However, the truths left by the great masters are far easier to embody than your mind believes. To follow the mystical path is to surrender utterly to that which is within you, no matter where it leads you. The Middle Way is not, as it sounds, a delicate path threading between the opposites. It is a path of complete abandon that is formed as each foot falls before you. It is a path sculpted from the void, untrodden by anyone before you and therefore without law, rhyme nor reason. To walk this path, you will have to dig deep within yourself and revel in that true quintessence that only you can recognise. This is the Buddha Fever.

≣ 24th Gene Key

Silence – The Ultimate Addiction

SIDDHI
SILENCE
GIFT
INVENTION
SHADOW
ADDICTION

Programming Partner: 44th Gene Key
Codon Ring: The Ring of Life and Death
(3, 20, 23, 24, 27, 42)

Physiology: Neocortex
Amino Acid: Leucine

THE 24TH SHADOW – ADDICTION

THE GREAT GENETIC GLITCH

The 24th Shadow, when correctly understood, explains much about the Shadow state itself as well as why human beings find it so difficult to resolve the deeper repetitive problems in their lives. This is the Shadow that keeps the psychology profession in business, and it is the Shadow that big advertising companies take full advantage of. We humans come pre-programmed for addiction, and the main culprit responsible for this is our minds. A well-known urban myth states that we only use a very small percentage of our brains, but any neuroscientist will tell you that this is untrue. In an average day we use almost all of our brains. It is not how much of our brain we use; it is how efficiently we use it. As it stands today, the human brain is still such unexplored territory for consciousness.

The way in which you use your brain is determined by your genetics, so that some people think more logically and others more laterally. Let's compare the circuitry of the brain to the 88 keys on a piano. We could say that the piano keys offer almost infinite potential for creating music and melody. However, in the case of our brains, we tend to find the tunes that we like and play them over and over. Many people don't really think for themselves at all, but have learned conditioned patterns from their parents and their environment and keep following those patterns their whole lives. Of course everyone uses their brain all the time, but it is relatively rare for someone to find an entirely new way of thinking. Synaptic routes within the ganglia of the brain become as trodden and familiar as public pathways and the 24th Shadow works hard to keep it that way.

The mass consciousness of humanity is still dominated by the archaic fear and survival-based aspects of our brain. This fear is a powerfully pervasive force within the chemistry of the human body. The 24th Shadow creates a low frequency rhythm throughout your whole being, and this frequency prevents you from thinking outside your comfort zone, which exists on a physical, emotional and mental level. We generally follow the same predictable paths in our external lives that our brains follow neurologically. This gives rise to the phenomenon known as addictive behaviour. When we talk here of addiction we are not talking about specific psychological disorders, but of an entire human behavioural code that is so self-limiting it does not recognise its own addiction. The fact is that all modes of thinking are addictive. You can be a left-brain addict or a right-brain addict, and you will be equally addicted. The only thing that breaks addictive thinking is silence — real silence. As we shall see, it is silence that gives birth to the 24th Gift of Invention — the art of thinking and acting in totally original ways.

Every cycle of addiction has natural *gaps* within it and human beings experience these gaps in awareness. They can occur at any time and they confront you directly with your own suffering. At such times you will generally feel a profound sense of emptiness that is deeply uncomfortable. Our general response to these gaps in awareness is to try to avoid them, either through numbing ourselves or distracting ourselves. The Shadow of Addiction ensures that people don't really change. Even though we may seek new ways of being, we usually end up following our addiction on the outside and never changing on the inside. You only have to look at a person's actions in the world to see whether they are living in a state of addiction or the higher state of invention. If they are influenced by the shadow state, there is always an anxious part of them that never seems to rest. Even though they may make cosmetic changes in their outer life, they never create anything substantially original. This pattern also tends to play out in their relationships, where they often find themselves repeatedly reliving the same scenarios without realising why. Even if they change the outer relationship for one that promises to be completely different, the same tactics show up again, manifesting an inner neurological patterning that they seem unable to escape.

It is important to understand the mechanics of the Shadow of Addiction in order to find the root of this deepest of human suffering. One way to enter into this mystery is using the model of the Corpus Christi mentioned in the 22nd Gene Key. The Corpus Christi refers to the seven bodies, layers or sheaths of a human being. To really understand the mechanics of your suffering you have to consider the three densest bodies and their relationship to each other — the physical, the astral and the mental. The root of human suffering is wired into your DNA in the physical body, the very densest body of the seven. It is the primal sacred wound running throughout humanity, and its single purpose is to eventually trigger your awakening.

The next body to the physical, in terms of vibration, is the astral body out of which all your desires emerge — your sexuality, emotions, longings, cravings and feelings. Whereas the physical body simply has essential needs such as food and warmth, the astral has cravings, which are inessential to survival. The primal wound is our perceived separation from the totality, and the astral body responds to this wound through desire. All desires are really rooted in a single desire — the desire to escape the suffering caused by the primal wound and return to the pure state of unity. If you could look deeply enough into this desire without acting on it, it would actually burn itself out, which is the underlying purpose of meditation.

The next aspect of the story of your suffering is found in the mental body — the mind. Your mind reacts to the desires of the astral body and tries to think its way out of suffering.

Addiction begins in the interaction of these three lower bodies. Your mind builds images, stories and projections around the desires of the astral body, which set you on an addictive course of behaviour aimed at relieving the suffering. Because the mind functions across time, its main tendency is to base hope of happiness on the future rather than accepting the real conditions of the present moment. The external civilisation designed by humanity feeds the mind's strategy of trying to escape suffering and find happiness in the future. It is designed *by* the Shadow *for* the Shadow, which is why it is so challenging to transcend the Shadow consciousness in everyday life.

Nothing external can bring an end to your suffering, since it is rooted deep within your DNA. Only when you turn inward and look for the source of your suffering will you finally face the addictive quality of your own mind. The programming partner of the 24th Shadow is the 44th Shadow of Interference, which is responsible for the dysfunctional relationships that occur across our planet and which are the current norm. This dysfunction is the by-product of a universal glitch in the genetic operating system of humanity and is reinforced by the 24th Shadow. Addiction is the constant replaying of the same perceptive frame with no pause between frames and this is what happens at a synaptic level within the brain chemistry of an average human. Like mice in spinning wheels, we simply re-enact the same scenarios over and over without realising what we are doing.

The question then is: how do you escape the wheel of addiction and reset the program to run without the glitch? The full answer to this question lies within the 24th Gift, but in a nutshell the very realisation that you are in an addictive pattern begins to shift the programming of the 24th Shadow. Your willingness to confront your mind creates the necessary pause between the frames, allowing the mental body to disentangle itself from the astral body. When this happens, then for the first time the astral body and its cravings are no longer fuelled by the grasping of your mind, and you go much deeper into the source of your suffering. You purify the astral body through starving it of external stimulation and thus you enter into the holy ground of pure desire — the desire to return to your true source.

REPRESSIVE NATURE – FROZEN

When repressive natures come across a *gap* in their awareness, the fear inside them causes them to freeze up. This freezing can manifest in many ways — physically as a complete lack of energy, emotionally as depression and mentally as a narrow-minded and guarded perception of reality. The secret to all addiction lies in how we respond to these gaps in the functioning of our awareness. The danger of addiction is that we seal our fate in these precious moments without realising we are doing it. At the Shadow frequency, we simply do not allow ourselves to experience the void that precedes a shift in awareness. The repressive nature shrinks from feeling that empty state of silence. If we were to face such times in our lives without either shrinking or reacting, something truly amazing would seed itself in us.

REACTIVE NATURE – ANXIOUS

The reactive side of the 24th Shadow, like the repressive side, is unwilling to fully experience their feelings of emptiness. This feeling of emptiness occurs to all human beings at different times in their lives. The *gaps* that open up before us can throw us into panic. If we react actively instead of passively, we do so out of anxiety. In order to escape the feeling of falling into a bottomless pit we translate our fear into activity, and that activity obscures the potential magic that could have occurred inside us. Essentially there are two kinds of addicts — there are those who numb themselves (the repressed nature) and those who stimulate themselves (the reactive nature). The reactive nature is more typically a workaholic or gambling type as opposed to the repressive nature, which might exhibit an alcoholic type of addiction. These people who cannot sit still, but are flooded with the anxiety that comes from avoiding the powerful and absolutely natural chemical process that is taking place inside them.

THE 24TH GIFT – INVENTION

RESTING IN THE GAP

In discussing the 24th Shadow, we have examined the idea of being caught in a set of addictive behavioural patterns whose very nature conceals those patterns. Added to this, we have seen that such addictions can mutate into different patterns if the individual recognises a *gap* between the repeating neurological pathways. Lasting from as little as a second to a week or more, these *gaps* actually occur spontaneously to everyone. Such gaps can allow an entire gestalt shift to take place in your life, and they can feel very destabilising if you do not surrender fully to them. Your response to them determines whether they become bridges to another level of awareness or you simply fall back into your familiar ways of thinking and acting. In the case of the 24th Gift, the gap in awareness is fully embraced and therefore reveals its magic. People with this Gift either catch themselves avoiding the gap or fully recognise such moments when they arrive. Either way, they do not cower before the fear that precedes these moments and thus they discover that these gaps in awareness are not frightening at all.

So what happens in the gap when you do embrace it? The answer is: absolutely anything. The 24th Gift is truly magical and contains the secret to genius. Genius is far more than lateral thinking — it is the ability to make quantum leaps. For example, the difference between a very good tennis player and a truly great tennis player is to be found in the 24th Gift. The great player follows no set pattern that their opponent can break down, counteract and defeat. They will throw in shots that are totally unexpected and come at the most intense moments. This is the Gift of Invention, of bringing new things into the world. The 24th Gift is a birth canal for originality and surprises itself as much as others.

The brain can be loosely divided into gray matter and white matter, the former being largely involved in the processing of information and the latter in communicating that information. The 24th Gift involves the mysterious processes that take place within the deep grey areas of the brain, and this Gift seems to act as some kind of neural trigger that allows original thinking to emerge from these areas. The human act of pondering might be an apt description for what occurs through this Gift. As you ponder a subject, you first travel familiar neurological pathways around that subject until you arrive at one of those magical *gaps*. The moment you hit a gap, it is as though you change neurological gear and suddenly a new, more efficient network of synapses opens up in your brain. This allows you to suddenly see things in a totally new light.

Ironically, as your brain becomes more efficient you may well use less of it rather than more. If you can simplify the amount of neural firing in your brain, then ideas and insights may actually become sharper and clearer. The great joke may well be that the smaller the percentage of the brain you use, the more intelligent you become! One of the best ways of exploring the gaps is through the art of contemplation. Contemplation on the mysteries of life and death, on the nature of change and suffering can lead you into sudden heightened states of awareness. To contemplate is to surrender yourself into the great mystery until it suddenly and unexpectedly reveals itself to you through a process of deep insight.

Whether you can communicate your insights successfully is a matter for the 23rd Gift, which, like the 24th Gift, is a part of the genetic family known as the Ring of Life and Death. The Ring of Life and Death is a complex genetic codon group governing the many processes that bring human beings in and out of form. One of its major symbols is the wheel. It speaks of the mechanical processes that connect the wheeling of the constellations and galaxies to the rotation of the atomic structures deep within our bodies. It is all about those gaps between the cycles — the intervals between lives, the spaces between atoms, the silence between notes. Magic and mutation occur within those gaps. In human beings this codon allows you to make quantum leaps in terms of the evolution of your awareness.

The 24th Gift is the heart of the human creative process. The secret of the 24th Gift is really one of the secrets of creativity itself. It is in fact an acoustic field involving the raising of vibrational frequency through your genetics. Each time you hit one of these magical gaps, you have the opportunity to either shift up an octave in frequency or remain in the same loop. One of the truths about addiction is that it can be used creatively to elevate your frequency. Invention is really creative addiction. You simply experience a spontaneous rising of the frequency of the mind's tendency to think in circles. Addiction thinks in circles whereas invention thinks in spirals. All the great insights that have driven human evolution, from the arts to the sciences, have come across the bridge of the 24th Gift. Only one who is willing to face their own ignorance can cross this bridge. It begins with your willingness to admit that you do not know and that you may never know. This inner honesty creates the environment for invention to occur even though *when* it may occur cannot be predicted. It usually occurs when you are at rest, manifesting the gap in some external way yourself — sitting silently, sleeping, dreaming or simply doing nothing at all.

Addiction thinks in circles whereas invention thinks in spirals.

THE 24ᵀᴴ SIDDHI – SILENCE

EXITING THE WHEEL

The 24th Siddhi is a challenging Siddhi to write about because how do you describe silence? Obviously you cannot. What you can do is consider the manifestations of this Siddhi and build a context around it. Thus far we have looked at the nature of addiction and how it can be elevated or transformed into creative invention. We have discussed the way that the brain follows patterns and how these patterns can mutate into newer, more efficient and more original patterns. When we come to the siddhic level, we really have to put away the mind altogether. Silence is the natural background state of all human awareness, and silence can only occur when thinking ceases entirely.

Over the centuries, human beings have tried all manner of techniques to stop the mind from thinking. Thinking can in fact be masked by certain techniques, but that temporary quietness is not the same as the pure silence of the 24th Siddhi. The silence of the 24th Siddhi is a silence that descends on you, even though it already lives inside you. It occurs as a result of the shift in the inner geography of human awareness. True silence reigns when the mechanism controlling your awareness moves from your head down into your solar plexus. You cannot say that silence is experienced because silence invokes the paradox of there no longer being an *experiencer* to have an experience. Silence negates everything, fusing together both subject and object. For awareness to shift in this way, some kind of physical mutation has to occur within the human body. Certain chemicals must be created by the endocrine system that prevent the process of normal acoustic thinking. You no longer think but are *thought by life*. This is not something that can be described. At the siddhic level there is no longer thinking but there is both knowing and not knowing simultaneously. Not knowing is there when awareness is at rest and knowing is there when awareness engages in some form of communication with the outside world. Both expressions really amount to the same thing.

When the mind ceases to think, then all addictive behaviour also ceases. The ultimate addiction of believing yourself to be separate from life is eradicated. In this sense, silence does not necessarily denote physical silence. The silence of this Siddhi is an internal silence that is permanently engaged. Just as addiction can be transformed through natural *gaps* in the functioning of awareness, so the siddhic state exponentially expands this gap. Awareness itself is shown to be a veil that obscures the true nature of these gaps, which are really windows into the nature of pure consciousness. Pure consciousness is silence or emptiness — it is represented temporally as the time before the beginning of the world. Only true inner silence can bring an end to the *maya* or illusion of your separateness. It is this silence that lies between and behind the activity of your thoughts that systems of meditation attempt to induce.

The 24th Siddhi has other surprises too. The ancient Chinese sages named this 24th hexagram of the I Ching *Returning*, a name with many interpretations and implications. We have already seen how the constantly revolving nature of this Gene Key is played out. It is rather like watching a film reel going round and round over the same series of frames until, at the Gift level, a new set of frames is spliced in before the revolution continues. The circular movement of life returns to the gap of silence — that magical place where evolution and mutation occur. This is the silence between the words of life, or the gaps between the atoms that we have already mentioned — the so-called elusive *dark matter* of modern physics. In terms of human destiny, our sages have ever sought to explain this returning movement at a metaphysical level and it has given rise to many of the great mystical theories, two of the most enduring being that of karma and reincarnation.

Of particular relevance to this Siddhi is the entire doctrine of reincarnation. The sages have long spoken of the greater destiny of the human soul as it incarnates into the world of form, lives out its karmic cycle and then reincarnates again and again until such time as it transcends form altogether. It is as though human beings have a kind of cosmic addiction to form itself, and until this addiction is broken once and for all, humanity can never be truly free. At this final stage in which the connection with material form is ruptured, the soul is said to become enlightened and returns to the ocean of being, or Godhead. It is easy to see how such doctrines have emerged from the way in which the human mind interprets life through this 24th Gene Key. In the post-modern age, the west has picked up this primarily oriental idea of reincarnation and adopted it as a mainstream new age dogma. Most of the better known contemporary teachers and gurus speak openly about the *fact* of reincarnation. Indeed, it is discussed in some considerable depth through the 22nd Gene Key. However, as the 24th Siddhi bears witness, reincarnation is simply one interpretation of a much simpler truth and from the point of view of this Siddhi at least, it is nothing but an illusion.

Despite what the teachers may say, most people like the idea of reincarnation because it gives them the sense that there is something about themselves they can hang onto after death. It gives us a sense of continuity and justice underlying all creation. From the point of view of the 24th Siddhi however, reincarnation is simply another human concept born out of the language of the maya itself. This theme is explored in some depth in the 44th Siddhi, which is the programming partner of this 24th Siddhi. Both these Siddhis strip the human story down to its purest and simplest form in which human beings are viewed quite coolly as simply genetic equipment for consciousness to play in. When one piece of equipment dies, it is gone, but it played a part in writing a great transcendent script. Even though the equipment has died the play must go on, as the saying goes.

The only thing that really reincarnates is silence itself.

Therefore, what appears to keep returning to the world of form actually never leaves. Bodies are born and die and their specific awareness functioning dies with them, but the overall consciousness beneath continues. It is silent, undying, intangible and beyond form. Thus when you identify with or remember a past life, or even a future life, you are simply reading the information within your *fractal* line. It is contained within your blood, but no aspect of *me* survives death except the silence of consciousness itself. How can there be anything other than silence, when true silence negates awareness? There is indeed relative truth within the concept of reincarnation and its culmination in enlightenment. However, the whole attempt to identify with continuity of awareness is a watered-down truth. The only thing that really reincarnates is silence itself.

In the 22nd Gene Key you can read all about reincarnation — the gradual enlightenment of the various subtle bodies as they incarnate into form, move through the bardo states in between lives and then return again. You can also learn about the causal body, that subtle layer of your being that survives death to carry the essence of your evolution on into the next life. But the causal body is seen as an illusion at a certain point in the process, and as it dissolves or shatters, so your enlightenment dawns. At this point reincarnation ends as you find that gap in the great wheel and exit the whole drama permanently. All these wonderful descriptions are in fact part of the script of our celestial drama and their real purpose is to help settle your mind and give it a sense of logical continuity. Nevertheless, stay cognisant of the nature of the vast paradox that hits you at the siddhic level. All such descriptions are essentially devices within the maya that may or may not help you to relax within the maya. The key is always to relax, because only as you relax can you find the magic gaps and experience the truth directly — not through your mind but through your innermost being.

Within the flowing story of the Gene Keys, the 24th Siddhi is the great trigger for the enlightenment experience. At a certain point in the evolution of every genetic and karmic fractal line, a certain piece of genetic equipment comes into the world that represents the culmination of that line, and as it explodes into its full blossoming it brings an end to a whole mythic aspect of the human story. As more and more of these people are born, the overall story of humanity will slowly come to an end. When all the stories have been told, all that will remain will be the silence that was there all along, in those magical gaps between the atoms, the notes, the words and the lives of all human beings.

25th Gene Key

The Myth of the Sacred Wound

SIDDHI
UNIVERSAL LOVE

GIFT
ACCEPTANCE

SHADOW
CONSTRICTION

Programming Partner: 46th Gene Key

Codon Ring: The Ring of Humanity
(10, 17, 21, 25, 38, 51)

Physiology: Heart

Amino Acid: Arginine

THE 25ᵀᴴ SHADOW — CONSTRICTION

CONSTRICTION TRAINING

If there is a single Gene Key among the 64 Gene Keys that captures the essence of the entire work, it is this 25ᵗʰ Gene Key. Here lies the secret that men and women have always sought — the secret of love. Here too lies the foil of love, the 25ᵗʰ Shadow of Constriction. Constriction exists wherever love is absent, and it is the underlying source of all human suffering. It is self-perpetuating because to constrict life in yourself or in another, is to welcome more suffering into your life.

The Shadow of Constriction operates at all levels of society. In the individual, constriction occurs first and foremost in your breathing. It creates tightness around the chest and compresses tension into the abdomen. Most of us enter constriction training at a very young age. We learn it through the breathing patterns of our parents, even while we are still in the womb. It is even there at the point of conception, passed subtly from the sperm to the egg as a code containing the genetic blueprints of the constriction felt by your ancestors as they struggled from their births to their deaths. Here in the 25ᵗʰ Shadow we find the basis for the myth of the *sacred wound* — the source of all human suffering, and a trans-genetic anomaly literally *wound* around the helix of all human DNA.

The 25ᵗʰ Shadow is the great turbine of human seeking. Everywhere you go you carry this constriction with you. If you tune into your body you will sense it deep inside you, and your response or reaction to the discomfort it brings defines the shape of your life. If you recoil from the pain, you will live a life of denial and distraction — a half-life, lost in the pale shadowlands of mediocrity. The harder you choke down the pain inside, the tighter it engulfs you. If, however, you have the courage to honour and listen to the wound within, everything will change for you. You will discover that if something is wound around you, constricting you, it must have a purpose, and that purpose is to *unwind*. Thus as you face your pain through the Gift of Acceptance, the wound begins to unwind and another higher destiny opens before you.

Constriction is also found in our communities, through our need for laws, territory, barbed wire fences, passports and money. Above all perhaps, are the massive constrictions we unconsciously place upon ourselves through our measuring of time. Our total dependence on time creates an enormous energy field of tension and pressure on a global level. The other great agent of the 25th Shadow is the human mind. Almost all systems of thought create more constriction inside you, with the exception of those that lead you into deeper acceptance of your true nature.

Just as the 25th Gift opens avenues at all levels of creation, the 25th Shadow closes them down. The programming partner of the 25th Shadow is the 46th Shadow of Seriousness, which teaches you that the tighter you hold onto anything, whether an opinion, a person or an object, the more constricting it becomes for you. The 25th Shadow distorts the universal love of the Siddhi into the lust for matter. This manifests as obsession with materialism — the most obvious direction for the flow of constricted fear to take. The urge to allay the fear becomes the need to reduce the universe to objects and to hold onto them as tightly as possible. Anyone who clings to the objects around them is deeply under the influence of the 25th Shadow. This can also apply to relationships. In relationships, this human tendency to try to hold onto others as physical anchors distorts and restricts the natural flow of love between people. Love grows through freedom and dies through constriction.

It is very important that you understand the 25th Shadow because it represents the beginning of your journey into form. To be alive in a physical body can be experienced as the ultimate constriction, especially if your reality is rooted in fear. Fear is the by-product of all low frequency energy fields. Constriction is the physical manifestation of fear at an individual and universal level. Moreover, fear creates a very effective biofeedback loop that ensures its own survival. It's really very clever when you think about it — fear is afraid of itself, which ensures that it never accepts itself, which ensures that it always survives. There are different kinds of fear as well. There is the fear that is hotwired into all physical creatures — the genetic fear that ensures our survival as individuals and groups or tribes. Then there are the more pervasive manifestations of fear — the fear of war or chaos or cataclysm, which exists at a collective level within the human unconscious. The deepest fear of all is the fear of not existing — the fear of death on the individual level, or the fear of extinction on the collective level. Such fears form the general backdrop awareness field of our planet. Then we come to pure fear, which does not even have a target. Pure fear is simply a collective thought form that hangs like a grey fog across our world.

This is the 25th Shadow — a wound beyond understanding, an endless chasm leading downward, a vast constricting pressure choking the life out of you and yet, ultimately, nothing more than pure illusion. This Shadow has a divine purpose — it holds a great design within it and ever draws you on toward your inevitable destiny in the 25th Siddhi. To really grasp what the 25th Shadow is you have to see how radical the 25th Siddhi is at the other end of the spectrum. It advocates a life without boundaries or constrictions of any kind. The process of entering into the sacred wound is therefore a process of unravelling and unwinding the karmic knots and cords that come with our lives and our relationships. As we learn to face our wounds, so our journey gradually becomes clearer and easier. This is the epitome of the journey through the 25th Gene Key — the path from fear to love.

To be alive in a physical body can be experienced as the ultimate constriction, especially if your reality is rooted in fear.

REPRESSIVE NATURE – IGNORANT

The 25th Shadow teaches us something profound about the nature of ignorance. In reality it is a form of repression. All repressive natures essentially use their energy to maintain a state of ignorance. Ignorance in this context refers to the inability to look at your own pain. The deeper your personal wound is held captive inside you, the more your higher faculties shut down. Ignorance is not bliss. Ignorance is misery, but it doesn't recognise this unless something momentous happens. Because of the pervasiveness of the Shadow frequency, ignorance is still one of the greatest diseases in our world. It takes a massive collective effort to constrict the life force wanting to burst from our bodies. The moment something triggers the letting go of your pain, there is a flood of release and relief around your heart as the Shadow of Constriction is weakened. Whether you repress this pain again after the event is another matter entirely.

REACTIVE NATURE – COLD

Just as the repressive nature is *unable* to own the depth of the sacred wound within, the reactive nature is *unwilling* to own it. These people express their pain through projection. Thus they are cut off from their hearts in a different way. The repressive nature does not know how it feels, whereas the reactive nature *hates* how it feels. It expresses that hatred by being cold-hearted. As we repeatedly see through the 64 Shadows, all reactive natures express fear through anger, so these people take out their pain on the world, especially on those closest to them. In effect it is impossible for anyone to come very close to these people. Their volatile nature quickly pushes away any form of genuine warmth because it reminds them of their own pain. As with all reactive natures this often leads to abusive or very short-lived relationships.

THE 25TH GIFT – ACCEPTANCE

ACCEPTING LOVE

Here in the 25th Gift we arrive at one of the greatest and most powerful of all human Gifts, the Gift of Acceptance. Because this 25th Gift represents the portal to the true nature of the universe as love, it has incredible relevance for all human beings. The flowering of love follows the burgeoning of acceptance. Acceptance is based on taking the *soft* approach to life, as the 25th Gift's programming partner, the 46th Gift of Delight, testifies. The path of love is the path of acceptance, which is not a technique but more of a *seeing*. In order to accept something about yourself, and especially something uncomfortable, it first must be recognised. This kind of acceptance takes place when you build up the courage to look into your own Shadow.

In the 25th Shadow we saw how fear creates its own biofeedback loop, which maintains this constriction of the life force inside you. The only way of breaking this loop is to have the courage to simply feel the fear. As soon as you reduce fear down to its essentials — that is a physical sensation deeply entrenched within the chemistry of your body — you discover the great secret of acceptance. You no longer need to be afraid of fear.

Fear is the very vibration of the sacred wound, which at a deeper level makes it holy ground. In allowing yourself to experience the true constriction of this primal wound, you begin to feel a subtle softening around your chest, and slowly, almost imperceptibly, your breathing deepens. This softening process of accepting whatever you feel at any given moment builds its own momentum in your life, and after a period of time you will really feel different. You are moving out of the lower frequency fields of the victim consciousness and experiencing more of life itself.

As the Constriction of the 25th Shadow loosens its hold, you also access far more energy and optimism. Optimism is not the opposite of pessimism but its true underlying nature when freed from constriction. The state of acceptance is akin to spiritual springtime — everything seems possible again as everything in your life begins to flow more freely and easily. This will manifest outside you as universal synchronicity is activated through the 25th Gift. However, even at the Gift level of frequency suffering still exists, and it keeps returning to test your level of acceptance. There are many layers of frequency bands between the Shadow, Gift and Siddhi levels. Over time your acceptance deepens even further until you finally let go of your need to escape the wound altogether. At this heightened stage your acceptance becomes complete and you spontaneously make the leap to the siddhic level. It is important to understand that this is all a process, but there is no real technique to it, even though it may *begin* with technique. Any attempt at *trying* to accept your nature only reveals a further level of subtle unacceptance. What you are really accepting therefore is your own utter helplessness.

If you have the Gift of Acceptance in your Hologenetic Profile or feel a strong connection to it, you will probably be the kind of person who feels like you belong everywhere. You will likely not judge others in the same way that most people do. Acceptance is not something you can easily learn. It is often something you are born with, and the more accepting you are the more life will tend to test you. You will quite possibly undergo challenges to test and deepen your sense of innocence and trust. The 25th Gift makes it difficult to hold a grudge or worry too much about life. You will walk through life with an air of another world about you, while at the same time being deeply grounded and open to others. In a nutshell, you carry the seed of love.

The 25th Gene Key is the master key within the genetic codon group known as the Ring of Humanity. The 25th Gene Key is the core of your humanity. It is the irritation inside the oyster that eventually causes the formation of the pearl, and the pearl is acceptance. Acceptance is the grail that you are seeking. When you can finally accept everything in your life just as it is right now, you will have embraced the human wound. Acceptance comes in layers — layers upon layers — just like the tightly wound genetic double helix itself. You have to relax these layers deep inside you so that you can feel the flow of love once more moving within your being. The more you can accept yourself and others, the more love will bloom in your life. It is as beautiful and as simple as that.

You have to relax these layers deep inside you so that you can feel the flow of love once more moving within your being.

THE 25TH SIDDHI – UNIVERSAL LOVE

THE ROSE AND THE CHALICE

The 25th Siddhi is a special Siddhi, whether it happens to be within your Hologenetic Profile or not. Every Gene Key exists inside each of us, and the 25th Gene Key is the primary archetype of love. Behind the mystery of the 25th Gift lies the mystery of suffering, and as we have seen through the 25th Shadow, it is a theme integral to the evolution of human consciousness. This connection to human suffering brings the 25th Gene Key into close relationship with the 22nd Gene Key, by whose Grace deep acceptance and love may be found. The sacred wound only reveals its true purpose through the 25th Siddhi of Universal Love.

Within the 64 Siddhis, there are many other variations of love, and in fact every Siddhi is a fractal aspect of Universal Love. Examples are the ecstatic sensual love of the 46th Siddhi, the devotional heart love of the 29th Siddhi, the Intoxication of the 56th Siddhi and the Compassion of the 36th Siddhi. The 25th Siddhi is Love itself as the source of all, and in this sense it can be called Universal Love. In all cultures, this love is represented through the great myths and in many of these myths it is symbolised by the sacred symbol of blood. The symbol of blood has many meanings and levels. It represents the conduit for the sacred wound itself, the blood that passes from human to human down the ages and that contains the codes for our ultimate healing. On a more universal level, it symbolises consciousness, that which moves through and behind all forms, knitting them together into a great cosmic pattern.

Perhaps the best known of all myths concerning blood is the myth of Christ, whose blood is said to have been shed for all humanity. There is a deep secret concealed in this myth. The sacred wound that hides within each human being can be understood at the three main levels of consciousness — at the Shadow frequency the wound maintains human suffering, at the Gift frequency the wound provokes humanity to evolve, and at the Siddhic frequency the wound reveals humanity's true nature as an expression of Universal Love. Only at the siddhic level of consciousness can you understand the true meaning of the blood of Christ. When your frequency is lifted up to the siddhic level everything takes on a cosmic dimension and you have no choice but to take into yourself the suffering of all beings.

In terms of our genetics, the blood of Christ represents the absolute acceptance of the suffering of all men and women from the beginning of time, all of which is encoded in the human genome. This is why it is said that a being at the highest frequency takes on the sins, the suffering, of humanity, whether they are termed a Christ or a Bodhisattva. This is the only way to attain the highest state of Universal Love. At the Gift level of frequency, you begin to take responsibility for your own suffering. As your experience of suffering deepens it is experienced as unending and you begin to transform the ancestral wounds of those who have come before you. The deeper you move into acceptance the more you have to open your heart to human pain, and the more pain you transform the more love you feel. At a certain level, the process loses its personal flavour and takes on a universal dimension.

*The bodies of
these people
have undergone
a radical
transformation
in order to be
the receiver or
chalice for pure
consciousness.*

At the level of the 25th Siddhi, a permanent leap is made whereby everything becomes acceptance and the rose of Universal Love blooms. This is the true beauty and purpose of all suffering.

There is another mystery surrounding the symbol of blood, symbolised by the vessel that holds the blood. This vessel has been given a variety of names in many cultures. Sometimes it is seen as a cauldron or vial containing an elixir, and other times simply a cup or chalice. In the grail legends, it is said that whomever drinks from the cup of Christ will live forever, and if the King drinks from it the land will be restored. The 25th Siddhi is the ultimate grail of all seekers. That which we seek is our original nature, and it lies within us, hiding within our shadows and wounds. Everyone in your life is part of your own wound and the healing of that wound. Through gracious acceptance you will begin to feel the extent of the pain that lies within and all about you. You need never be afraid of this pain as it is the direct and only route leading to the heart of your heart. We can all take heart from this greatest of truths.

The other truth that emerges from this 25th Gene Key is the resolution of an old mystical chestnut about seeking. It is said that the greatest obstacle to realisation is the seeking of it. All spiritual paths begin with the urge to bring an end to personal suffering. It is your unacceptance of suffering that makes you seek, and as you seek you eventually realise that your seeking is based on the urge to avoid your wound. This revelation gives rise to the quantum leap to the siddhic level of consciousness. Thus it is said that only when all hope of finding the grail is lost can the grail finally be found.

Those beings who have attained realisation through this 25th Siddhi become, in some form, legends. Their lives follow a familiar mythical pattern. They are people whose lives reflect the grappling with their wound, whether that is the wound of duality transcended by the Buddha, or the wound of Christ hanging on the cross. They take paths that others have never travelled, and in doing so they take into themselves the suffering of the world. The aura of such a person shines like a light down the centuries and is felt within the bloodlines of all future generations. Every time someone attains the 25th Siddhi, a great genetic constriction is removed throughout humanity. The love emanating within such people has an otherworldly quality — this is not human love as we know it, but Universal Love. The bodies of these people have undergone a radical transformation in order to be the receiver or chalice for pure consciousness — the uncontaminated blood of the universe.

SIDDHI
INVISIBILITY

GIFT
ARTFULNESS

SHADOW
PRIDE

≣ 26th Gene Key

Sacred Tricksters

Programming Partner: 45th Gene Key

Codon Ring: The Ring of Light
(5, 9, 11, 26)

Physiology: Thymus Gland

Amino Acid: Threonine

THE 26TH SHADOW – PRIDE

WIELDING THE WILL

Deep within the substructure of your physical body lies a set of four chemical codes whose ultimate role is to determine how your body captures, stores and transmutes light waves into energy. This chemical family (which includes the 5th, 9th, 11th and 26th Gene Keys) is collectively known as *The Ring of Light* and in genetics it codes for the amino acid threonine. Depending on the frequency of light you allow to enter your DNA, different biochemical processes are initiated, affected and/or prevented inside your body by means of these codes. It is already well known scientifically how the ultraviolet frequencies within sunlight catalyse your body to produce vitamin D, a vital component of your physical health. Light contains many such catalytic codes within its spectrum that determine not just physical health but also emotional, mental and ultimately spiritual well being. The central message of the Gene Keys revelation concerns the power that you as a human being wield to consciously and unconsciously raise or lower the frequency of light entering your body, thereby altering your reality through the medium of your DNA.

At the lower frequencies, those frequencies governed by fear, your DNA relays instructions throughout your body that are based upon individual survival. This very narrow set of parameters is the primary paradigm ruling our planet today. The 26th Gene Key is unique within the threonine group because it has to do with utilising light waves via the medium of the individual will. In other words, through your willpower you can bend light and turn it to your advantage. This Gene Key therefore has a great deal to do with the correct and harmonious use of will. There is a huge conditioning field on this planet telling you that nothing will ever come to you in life unless you reach out and grab it. At the Shadow frequency, you allow your unconscious fear to make you distrust the natural and easy flow of life. Out of this fear, you try to control life by pitting your individual will against the whole, and so the 26th Shadow of Pride begins to rule your life.

The power of will is actually a magical power — it is quite literally your ability to harness the power of light and project it through your body as action, thought and words. It is the key to manifesting dreams on the material plane. If you have enough willpower, you can achieve almost anything. This is the language of the 26th Shadow, and it is important to remember that there is nothing wrong with this belief. It is a vital stepping-stone to a far higher frequency. However, there are two types of will — there is *your will*, which as we shall see is an illusion, and then there is *thy will*, which we will look at when we examine the higher frequencies of this 26th Gene Key. *Your will* is the foundation of human pride — it is the belief that you as an individual can control the forces of nature and come out on top. Nowhere in our society is this Shadow of wilful pride more dominant than in the sphere of business. All modern business is built on this rationale of individual willpower. If you are driven by the 26th Shadow then you, like many others, use your willpower for the sake of personal gain and recognition. In business this means that to climb to the top you must consciously or unconsciously push others down.

You can identify the 26th Shadow in a person because they simply cannot relax, even for a single moment.

The word most commonly used to describe the energy and effect of the 26th Shadow is the word *ego*. In spiritual circles, the ego is widely regarded as the archenemy of the higher self, but when we come to explore it through the 26th Gift, we will learn that it has a higher purpose. When you allow yourself to be governed by fear, you are giving instructions to your DNA to try to establish and maintain control over others or your environment by means of your ego. In doing so, you effectively cut yourself off from those others as well as from your environment. At the Shadow frequency, this is the only way you can see to create success and security for yourself in life, so the 26th Gene Key tells you to get on the ladder of competitiveness just like everyone else around you.

In the modern world, hierarchy, competitiveness and ego are considered normal and even healthy. Unfortunately, the current definition of health on this planet is based on a very narrow set of criteria. Real health is only found when your inner being is completely at ease with life's inherent uncertainty. The 26th Gene Key has a strong link to the function of the thymus gland, which is vital to your immune system. Societal conditioning tells you that if you really want something badly enough, you make it happen. If you base your life on this premise, the lower frequencies of the 26th Shadow will gradually grind down your immune response, leading to increased stress and premature ageing, not to mention a host of potential problems or diseases associated with the weakening of your immune system. If you try to force life to follow your will, you may well succeed in achieving your goals, but at what cost? By moving against nature, you deny yourself true happiness and replace it with an addiction to stress. You can identify the 26th Shadow in a person because they simply cannot relax, even for a single moment.

If however you look deeply enough into the possibilities lying dormant within your DNA, you will find that other models for success and security exist — models that most people on this planet can scarcely entertain as realistic, so distant they seem from our modern reality. These higher models are waiting to be activated inside your DNA by the higher frequencies within the light spectrum. As we shall see at these higher frequencies, the 26th Gene Key reveals many hidden pathways within the quantum web that connects all beings though time and space.

Within the Ring of Light the 5th Gene Key allows you to synchronise your individual biorhythm into the wider cosmic pattern or grid, but it is the 26th Gene Key that finds all the shortcuts through the matrix. Such shortcuts or wormholes allow human beings operating at higher frequencies to break the normal laws of the material plane. However, at the shadow frequency, none of these shortcuts are available to you, and your only resource is to use brute force of will to get what you want. The philosophy of the 26th Shadow is based on *every man for himself* and in such a system, someone always loses. We can also see this reflected in the 45th Gene Key of Dominance, which is the programming partner of the 26th Gene Key.

As we have seen, the 26th Shadow is about ego and pride. One of the underlying fears of this Shadow is being perceived as powerless. It has much to do with projecting an image of success and confidence. The activation of this Gene Key at its lower frequencies leaves you craving security in your identity and your natural reaction to this discomfort will be to seek that security by setting yourself above others through an act of will. Even deeper lies the core human fear of non-existence. In a great cosmic irony, the more solidly you become attached to a powerful sense of ego and identity in the world, the more profoundly this fear affects and undermines you. You look back on your life and consider how many people you must have subdued to get where you are now. The deepest sadness of all is that this 26th Gene Key contains such effortless magic, but it is effectively negated by the deep-seated and false belief that you cannot get anywhere in life except by sheer force of will.

REPRESSIVE NATURE – MANIPULATIVE

The 26th Shadow can be really sneaky. When pride manifests through a repressive human nature it becomes manipulation, but not in an obvious way. This is covert manipulation. These people manipulate others by putting them down, or use guilt and shame as subtle weapons. Often these patterns are acted out unconsciously, which means that such people rarely feel accountable for their actions or the effects they have on others. The 26th Shadow uses its natural cunning to make others feel bad or inferior — behaviour patterns rooted in fear. Pride that is rooted in fear always leads to manipulation. It is a simple and lethal equation.

REACTIVE NATURE – BOASTFUL

The more up-front version of pride is boastfulness. This is the pride that we all know so well as ego. All lower frequencies are inherently isolating. To boast is to try and draw attention to oneself — to render oneself more visible. When someone is boastful, they do not realise that their acts or words have the reverse effect of their intention. There are many ways of boasting, both conscious and unconscious. One of the most obvious is seen in displays of personal wealth, power and property. While the higher frequencies of the 26th Gene Key naturally draw positive recognition, the 26th Shadow and all its displays draw jealousy, resentment or at worst, inspire greed. These people suppress great anger, and when it inevitably explodes, it manifests as the ugliest form of pride and creates a strong repulsion in others.

THE 26TH GIFT – THE GIFT OF ARTFULNESS

HEART MARKETING

*The 26th Gift
celebrates
your ego,
without self-
judgement
and in full
awareness.*

When you get to the Gift level of the 26th Gene Key, you learn a very great secret in life. You learn the difference between will and intention. At the lower frequency, the 26th Shadow leads you to believe that you must exert huge willpower to attain your dreams in life. The problem here is that at the Shadow frequency you cannot possibly know what your dreams really are. Inside the coils of your DNA your higher purpose lies concealed. You cannot force that purpose out into the world by sheer force of will. Your higher purpose can only manifest when you surrender your individual will to nature as a whole. This process of surrender begins with the understanding that your higher purpose is something you must tune into, rather than impose upon the world.

When you first come into contact with your higher purpose in life you will most likely experience it inside as a subtle intention. The more you listen inwardly, the more deeply you will realise that this intention lies behind everything you do and say in life. Your attitude about this intention determines the success of its manifestation in the world. At the Shadow frequency, you respond to this intention through your pride and thus distort its manifestation by trying to pre-empt and force it. At the Gift frequency, your attitude invites a higher waveband of light into your DNA. This more refined frequency catalyses new micro-processes inside your body. For one thing, your immune system becomes stronger and your general level of health improves considerably. When the 26th Gift is activated, your thymus gland engages at a higher level and some curious things occur throughout your being. The first thing you may notice is that you begin to feel physically and emotionally warmer on the inside. The higher functioning of the thymus gland releases a soft vibration throughout your chest area that brings with it a wonderful feeling of open-heartedness and warmth.

You may well wonder what it is that you are doing to activate these higher frequencies inside your body, and the answer is simple — you are listening to your intention. You are listening to the higher purpose inside you. Simply by listening in this way without needing to do anything, you begin to absorb a higher frequency of light. Over time, prolonged attunement to these frequencies results in a *frameshift mutation* within your DNA. This spontaneous change completely reorganises the way in which your genetic code is translated, and another hidden code reveals itself for the first time in your life. It is this hidden code that corresponds to your higher purpose. Once it has been unlocked your life will change irreversibly.

The 26th Gift celebrates your ego, without self-judgement and in full awareness. When this Gift awakens, you realise that there is absolutely nothing wrong with pride. Pride is simply a low frequency word for the same energy that can be called *Artfulness*. When you learn how to use pride creatively, it becomes something powerful and even beautiful. The 26th Gift loves attention. It is designed to draw attention. This Gift is about the love of selling something to someone — whether a product, yourself, or a truth. The 26th Gift represents the love of marketing — of dressing something up so that others will buy into it.

In order to sell a product or a truth, you have to put yourself in the limelight. You have to embrace the energy of pride and ego that lies within all humans and use it in the service of your higher purpose.

The 26th Gift includes natural-born shrewdness. Through this Gift, you can use the power of your ego to deliver your message. To do so, you must fully embrace it. We have seen that ego has a negative connotation in many spiritual circles, where it is often seen as something to be conquered and transcended. Actually, nothing can be transcended through conquest. Only through absorption, acceptance and even enjoyment can the ego be transcended. This is the gift of enjoying your ego. Through the Gift of Artfulness, your ego actually becomes an art form. Out of this Gift emerges the ability to manipulate racial memory — in other words, you instinctively know how to speak the exact language of the person before you. This ability to manipulate your audience can be devastating at the lower frequencies, since it is rooted in fear and sells itself through fear. But at a higher frequency, freed from fear, the 26th Gift sells itself through love. This is heart marketing.

This word *manipulation* is another *bête noir* in the spiritual domain. However, as long as manipulation is open and honest, it can be a beautiful thing. Art is a subtle form of manipulation, as is music. Human beings can be moved from lower frequencies to higher frequencies through manipulation, and this is where the 26th Gift excels. It lets you know you are being manipulated so that you can give in and allow yourself to be swept away, or reject what is being offered. It is a game for the 26th Gift. Through this Gift you manipulate your own ego in order to manipulate someone else's ego. The difference between the frequency of the Shadow and the Gift is that when you are operating from the 26th Shadow you are effectively consumed by your own ego and its voracious need to feed off of success, recognition and dominance. When you are manifesting the 26th Gift you are not identified with your own ego — it is simply something you use, like a costume you put on before a show.

THE 26TH SIDDHI – INVISIBILITY

THE GREAT CINNABAR FIELD

There is a saying on the mystical path, which more or less translates as: *In order to transcend the ego, one must first have an ego worth giving up.* This delightful statement beautifully encapsulates the teaching of this entire 26th Gene Key. In the original Chinese I Ching, the name of this hexagram was *The Taming Power of the Great*, which at the Gift level could well refer to the process of taming one's ego. When the ego runs wild in our lives, as it does at the lower frequencies, it causes havoc. As you become aware of your own ego, it releases its hold on you and you can learn to play with it. However, there exists an even higher mutation waiting to occur within your DNA, one in which this *taming of the power of the great* refers to a magical process that only occurs through the 26th Siddhi. As you open your heart to life more and more, your thymus gland triggers a vibration of such refined intensity that it spontaneously ignites the higher functioning of the pineal gland.

As most esoteric systems testify, the pineal gland opens a chemical pathway within your brain that allows you to access cosmic consciousness. In this way, we may understand the *taming power of the great* to refer to the meeting of the macrocosm and the microcosm deep within your physical body.

The 26th Siddhi of Invisibility is a very unusual manifestation of the higher states of consciousness. Invisibility at this level can mean many different things. The ancient Chinese Taoists had another special name for this 26th Gene Key, referring to it as *The Great Cinnabar Field*. For the Taoists, cinnabar was a substance that represented the mercurial aspect in alchemy. The qualities of mercury or quicksilver are the qualities of the 26th Siddhi. In this respect, mercury represents the ability to become one with your environment, to camouflage yourself so as to appear invisible. This is the difference between the embodiment of *my will* and *thy will*, which was mentioned earlier. This 26th Siddhi dissolves all sense of individual will. Everything in the cosmos is driven by a single turbine and surrender to this impulse renders the individual ego invisible — not just your own ego, but everyone else's egos as well. The whole game just falls away.

Those who manifest this Siddhi teach humanity to take life less seriously. They trick us into the great Truths.

One manifesting this Siddhi stands nowhere. These beings cannot be pinned down to a single concept — wherever you look, they are there and whenever you try and pin them down, they are gone. This is the meaning of invisibility in this context — it refers to one who has become one with existence. The Great Cinnabar Field is the energetic grid that links all aspects of existence. It is the quantum ocean of fluctuating, interchangeable energy and matter, and it is the play of the waves of existence upon this ocean. One who becomes master of this ocean is one who can move within the grid without separate existence — one in whom God is at play.

Play is a large part of the 26th Siddhi. We human beings have always sensed the secrets of the 64 Siddhis, but we have rarely considered such divine manifestations to be concealed within our own genetic code. Throughout human history, the Siddhis have been anthropomorphised and projected outside our bodies into the figures of our gods, archetypes and mythologies. The gods of the 26th Gene Key are the trickster gods — the Nordic Loki, the American Indian Coyote or the Hindu monkey god Hanuman. These are the great archetypes of shape shifting. The Celtic figure of Merlin is another of the archetypes of invisibility, and the 26th Siddhi certainly shares his wizard-like, playful qualities. Through these archetypes the Divine is seen as being playful and sometimes mischievous. Those who manifest this Siddhi teach humanity to take life less seriously. They trick us into the great Truths.

The 26th Gene Key, as we have seen, carries the quality of ego within it and is adept at marketing. At the siddhic level, these gifts are played with for the sole purpose of Divine fun. These people will use whatever faculties they possess to transmit their sense of love for creation. From their vast understanding of the language of the universal quantum field, they may create extremely complex teachings just to lure the mind through its addiction to finding answers. Whatever tricks, shortcuts and wormholes they use to draw you in, they do so to draw your attention to a higher truth — that nothing really matters. Nothing can alter consciousness.

To truly integrate this simple truth is to realise that all human life is essentially meaningless. But this meaninglessness does not detract from the wonder of existence; rather it enhances the beauty around us. Above all, it leaves us alive to play.

To dance with the 26th Siddhi is to let go of all agendas. Such people are invisible in a way that most cannot understand. They are invisible because they do not care what others project on them. They do not seek to enlighten anyone; they do not really want to influence anyone at all. They truly have no agenda. They are simply here as loose cogs within the machinery of existence. They love to defy the laws that we humans cling to. They are the tricksters who love to twist and turn in the currents of existence, for no other reason than that they can.

Ironically, such people with no agenda have often left the greatest mark on the history of our consciousness. Because we cannot pin them down, because we cannot understand them with our minds, we either have to reject them or laugh with them. Laughter is the true legacy of the 26th Siddhi. Their laughter peals like an endless string of bells throughout the Great Cinnabar Field of existence.

27th Gene Key

Food of the Gods

SIDDHI
SELFLESSNESS

GIFT
ALTRUISM

SHADOW
SELFISHNESS

Programming Partner: 28th Gene Key

Codon Ring: The Ring of Life and Death
(3, 20, 23, 24, 27, 42)

Physiology: Sacral Plexus

Amino Acid: Leucine

THE 27TH SHADOW – SELFISHNESS

THE MATHEMATICS OF LOVE AND SELFISHNESS

The 27th Gene Key is truly vast in its implications at a planetary level. It governs the structure of the food chain, the preservation of gene pools both human and animal, and is key to understanding the precise mathematical laws that maintain an overall balance between the different species on our planet. It even controls the subtle shifts and changes underlying global climate and the weather. The ancient Chinese called this 27th hexagram of the I Ching *Nourishment* with good reason. It represents a built-in planetary law that rules all sentient life — to give is to receive.

Seen from a higher level of frequency, the 27th Shadow of Selfishness is a distortion of this fundamental law. When we look at nature through a macrocosmic lens, we see that all the different systems on this planet are interconnected. All life forms and matter, both organic and inorganic, are essentially porous at the subatomic level. There is an entire mathematics of giving and receiving that unites all forms and is primarily based upon food. We are using the word *food* here in the broadest possible sense — if you are a bacteria for example, your definition of *food* might be anything from gasoline to wood. The point is that life is a living chain of birth and decay — of creatures living off each other and transforming one thing's death into another thing's birth. At the profoundest level, nothing exists unless it can be eaten by something else.

We might refer to this principle within the 27th Gene Key as hologenetic. It is present in all creatures at a genetic level, but can also be replicated as a set of laws governing any and every system of life. In humans, for example, this law forms the basic thread of our morality — of what we consider good and bad. This 27th Shadow of Selfishness in particular is labelled as morally *bad* or undesirable. Through the 64 Gene Keys, however, all morality can be understood as simply the movement of frequency through a certain archetype. Seen in this objective way, there is no moral agenda. The 64 Shadows are not *bad*, even though their external manifestations are usually labelled so. All forms on our planet are constantly evolving in frequency, so among humans we see higher frequencies dominating in some places and lower frequencies in others.

Selfishness is where this 27th Gene Key begins its evolutionary journey in humans. The so-called *selfish gene* has been a requirement for us to survive, particularly for blood ties and close genetic groupings. However, selfishness must be transcended in order for the next form to mutate out of this existing form of Homo sapiens. This is how the mathematics of nature works. As the frequency within a species reaches its zenith, it pushes and stretches that form to make a quantum leap. For some time, the new form needs the care of the old form, even though ironically, the new form will ultimately eradicate the old form. Selfishness then is a stage in our evolution that is in the process of being transcended at a collective level. If it weren't, humans would die out. This is not a question of morality. It is a question of evolution.

When we look at our world today, particularly through the world media, we have a tendency to focus on the negative aspects of life. That is due to the general low frequency of mass consciousness. But as a collective body we humans have already ascended far beyond individual or tribal selfishness. The structures that we have built into our society create opportunities for so many more people to be fed and nourished than ever before. It is true that an enormous amount of the world population still lives in poverty and suffers from malnutrition, just as it is true that selfishness is primarily responsible for this. However, collectively we have moved a long way from being monkeys. Even so, the current human vehicle is still designed for selfishness and is not easily suited to the higher frequencies. It is still relatively rare for humans to raise the frequency of their genetics even to the Gift level, let alone the Siddhic level. Thus our species must prepare itself to make the genetic quantum leap, because that is the only way that it can truly transcend its built-in selfishness.

Selfishness then is a stage in our evolution that is in the process of being transcended at a collective level. If it weren't, humans would die out.

By viewing selfishness in this way, perhaps we humans can begin to see through to the higher frequencies of the 27th Gene Key. Selfish acts cause devolution whereas selfless acts cause evolution. If we add to this a further equation from nature's mathematics, we will see that the 27th Shadow of Selfishness is equal to purposelessness. This equation is created out of the twin binary of the 27th Shadow and its programming partner the 28th Shadow of Purposelessness. This binary coding leads to a dead-end. Selfishness does not pay well since it makes us imporous rather than porous. In the long run it closes down the possibility of our being nourished, whether by food or love. Selfishness cuts us off from the collective. Even though it may ensure individual survival, in order for our species to take the next evolutionary leap survival must become absolutely communal.

As an aspect of the chemical family known as the Ring of Life and Death, the 27th Shadow reminds us of the cosmic forces of creation and destruction. Every Shadow is destructive and leads to death, whereas every Gift leads to life. It is only at the highest siddhic level that life and death are finally transcended. Each of the codon groups function collectively across the entire gene pool to establish a field of frequency that pervades and influences the whole planet. Because the 27th Shadow is bound to the other Gene Keys in this Codon Ring, it is easy to see the true nature of selfishness.

Through the 24th Shadow we can see how addictive it is, through the 3rd Shadow how it sets up chaos and through the 20th Shadow how it requires an absence of basic human awareness. Through the 23rd Shadow we can see how complicated selfishness makes life and finally through the 42nd Shadow how it is rooted in the false expectation that it will bring an end to suffering.

REPRESSIVE NATURE – SELF-SACRIFICING

The repressed nature of this Shadow manifests as self-sacrifice, in the sense of giving away your personal power rather than giving from your heart. You give to others but without any natural sense of boundary, which either leads to your being taken advantage of or to a feeling of resentment from the receiver. The laws of life state that there must be a mutually beneficial exchange in order for a relationship to remain healthy. The repressive nature is afraid of its own dark side and tries to gloss it over by expending all its energy on others. Such self sacrifice also contains a subtle trace of guilt. The frequency of the giving does not come from one's true heart and can only be received in the manner in which it is given, without real gratitude. Giving in this way causes more harm than good, because you inevitably deplete your own resources and gradually wear down your own health.

REACTIVE NATURE – SELF-CENTRED

The reactive side of this Shadow is about giving with an agenda and not about being purely selfish in the sense of holding back your energy. These people give to others in order to get something back for themselves. This kind of political giving creates its own aura of manipulation and encourages distrust. When such people give to others and do not get back what they expected, the latent anger of their reactive nature suddenly explodes to the surface. All reactive natures have this capacity to lash out at others, and the 27th Shadow can often seem the most shocking because in the beginning it appears to be so generous and giving. This kind of giving comes entirely from the mind rather than from the heart.

THE 27TH GIFT – ALTRUISM

THE POD MIND

The 27th Gift can be understood most clearly by observing it in the animal kingdom. Among the 64 Gene Keys there are certain Gifts that have a close connection with other species. In the case of the 27th Gift, the connection is with other mammals. This Gift represents the communal bond that exists between members of a group or family of mammals. For example, in a pod of dolphins, the 27th Gift is seen reflected in the *pod mind* — the invisible psychic force that unites these creatures and keeps them together. The pod mind monitors the safety of the pod as a whole. It functions through each individual dolphin but is instantly communicated to all members of the pod.

*To give to
others simply
for the sake
of giving
activates
healthy
currents deep
within your
body.*

The nature of this pod mind is Altruism, in the sense that if an individual member the pod is in danger, all dolphins will turn to assist it. Sometimes in mammals, an older member will even sacrifice itself to save a younger member, thus ensuring the continuation of the lineage.

At a higher frequency, this communal altruism that exists between family groups extends to the species as a whole. In humans, altruism ensures our survival as a species. It ensures a happier, healthier life, even though it may not always be the life you were expecting. Giving from the heart can stir up unexpected cosmic forces that work to your favour. To give to others simply for the sake of giving activates healthy currents deep within your body. Through selfishness you may acquire much for yourself, but you will not acquire a true sense of higher purpose. Purpose flows from altruism like water from a spring — it bubbles up inside you making you feel warm as well as spreading this feeling to others.

Another aspect of altruism is its ability to be detached. We have already seen through the Ring of Life and Death how closely linked the 27th Gift is to the 42nd Gift of Detachment. The primary difference between the 27th Shadow and the 27th Gift is that altruism gives without expectation as well as giving to causes that it knows will be fertile. The repressive side of the 27th Shadow — self-sacrifice — is about giving to the wrong people, exemplified in Christ's parable of the farmer whose seed falls on infertile ground. Altruism is actually a form of intelligence that knows, through its vital connection to the group mind, what is worth giving to whom. It does not feed victim consciousness. Essentially it supports the empowerment of individuals through a process of deep communal bonding.

The 27th Gift's ability to give in a detached way also means that these people are willing to bend or even break laws or moral codes in order to give others the support that they need. When caring comes from the heart, it has no trace of morality. This kind of caring is akin to a parent looking after its child, and indeed this 27th Gift has much to do with the education and nurturing of children. Those who have children of their own understand this urge to protect their young at a genetic level. It is one of the strongest forces on our planet. Because it has such a close resonance with all life forms, this Gift also has much to do with natural seven year cycles. The seven year cycle is the foundation of the educational and nurturing process. In human beings, this Gift creates enormous genetic pressure to stay with your offspring for a minimum of seven years. If you have this Gift as a part of your Hologenetic Profile, as a parent you will actually cause yourself physical, emotional and mental damage if you are not an integral part of your child's first seven years of life. This genetic pressure is inherently healthy for the family. Even if the mother and father are no longer intimate, the child must become the primary focus. Every child needs the balance of both the male and female aura consistently during that crucial first seven years of life.

Everything in the child's psyche is forged in the first seven year cycle. Beyond these seven years, other cycles of imprinting do exist, but the stage is already set. Any child who receives true nurturing from both the male and female in their first seven years will have a strong physical, emotional and mental constitution. In effect, they will always have the capacity to find their own inner strength. Such a child is likely to grow up with an altruistic nature rather than a selfish one. Of course, destiny arranges all manner of events that may split a mother and father apart, but nothing is ever lost. The opportunity for healing a split in these first seven years will always recur in your life at a later date. You may indeed have to heal your own childhood wounds by becoming a parent yourself. All relationships give us the opportunity to heal old wounds through sustained nurturing. This is in fact the secret of a happy relationship. If you are not content in your relationships it may be because you have not nurtured yourself fully. Every relationship provides this mirror.

The true nature of the 27th Gift is generosity, which is primarily about caring for others and for nature in general. These people can also make wonderful gardeners since they have a natural connection to the cycles of life and the ebb and flow of nature's rhythms. This Gift also has a particular soft spot for those who are weak or afflicted. It is quite natural for those with this Gift in their Hologenetic Profile to find themselves involved in the service professions where they can offer nurturing to others. At a very high frequency the 27th Gift gives off a powerful aura of trust, which is immediately felt by others. Such an aura of trust often allows others to let down their guard and open themselves up to nurturing, sometimes for the first time in their lives. At a deep genetic level, the presence of this 27th Gift engages a feeling of communal security through its strong resonance to the group mind. As such it is one of the most powerful healing Gifts in the whole genetic matrix.

THE 27TH SIDDHI – SELFLESSNESS

LOVE – THE NEW SUPER-FOOD

The 27th Siddhi is about as mystical a Siddhi as you can get. It is very challenging to explain this archetype in our common language. The knowledge that comprises the 64 Gene Keys is reflected in the substructure of life at a genetic level. Until humans fully understand the science of fractals, we will not really grasp the nature of the universe. Everything we look at is a hologram containing the blueprints to everything else. The current evolution of humanity as a whole is set to make the transition from the frequency of the 64 Shadows to the 64 Gifts. The highest frequency of the 64 Siddhis does not really concern the current stage of our species and has more to do with our future as a consciousness. This is why in certain rare humans a siddhic state will spontaneously blossom but cannot yet blossom among the totality. Those people in whom the Siddhi is made manifest represent a state that is not really designed for the current version of Homo sapiens. In this sense, all siddhic states appear as anomalies rather than the norm.

It is extraordinary to reflect that incredible information about our future hides within certain places in our genes. The 27th Siddhi is one of these secret places. In its present cycle this part of our genome is dormant, waiting for its time. It is operating at a mere fraction of its full capacity. In our current cycle it manifests within us as a kind of deep unconscious yearning for harmony. The outer symbol of this yearning is food, which is why the original Chinese name for this archetype was *nourishment*.

The fact that our current planetary evolution is built upon one creature having to physically consume another to survive shows our deep limitation. We are not a species that can survive without food, but in the future we will be. At that far off stage in our evolution the true nature of 27ᵗʰ Siddhi will become fully apparent.

However, there are stages in our evolution where our genetics mutate and each one of the 64 Siddhis subtly shifts its capacity. We stand at the cusp of one of the greatest of these changes now. In the next few hundred years, the Siddhi of Selflessness will become widespread across our planet. Eventually this Siddhi will take hold of all life on earth and bind it together as one creature, although our current carbon based life forms are not capable of sustaining such a high band frequency. This will presage the time of the ending of food and the dawn of what we now understand as immortality — the 28ᵗʰ Siddhi — the programming partner of the 27ᵗʰ Siddhi. From where we stand now on the evolutionary ladder, this future stage seems like science fiction. When humanity stands in its Truth, devoid of any sense of self, true planetary awareness will dawn. In ancient cultures such as the aboriginals of Australia, ancient tales exist of a time when the awareness in animals and humans was indistinguishable and we operated as a single unified field of consciousness. This is the direction in which our human species is headed. At a higher level of reality we will in essence become extinct. We will no longer experience ourselves as a separate species but will function more like the neurological network for Gaia — the earth.

By taking a walk through each of the 64 Siddhis you may see the pattern of humanity's eternal quest to meet its divine nature. There have been many different paths towards God. One of the greatest of these paths is the path of service, known in India as Bhakti Yoga — the path of compassion. Also known as the *doctrine of harmlessness,* this 27ᵗʰ Siddhi of Selflessness has manifested in the lives of many saints throughout history. This is not really about being a do-gooder. There are those who try to help others for the wrong reasons, out of the subtlest desire for recognition, or to cover up their own deep pain. Selflessness is an absolutely pure yearning to provide succour to others. There is no trace of self-consciousness in it. Such people have made a quantum leap, and the siddhic energy available to them is enormous. They can work in situations that normal people would find impossible. Despite endlessly giving themselves to others, they don't become depleted but are continually nourished by the refined currents of the aura of love surrounding them.

The Mathematics of Love are built into the structure of all life, but only when we discover these laws will human beings realise their highest potential. Inherent in the 27ᵗʰ Siddhi is the realisation of the divine law of philanthropy — the love of humanity. Philanthropy unlocks the secret of free energy because it gives selflessly of itself, and it gives intelligently. To give without awareness is charity, but to give with awareness is philanthropy. The 27ᵗʰ Siddhi knows how to distinguish between that which is alive and that which is decaying, and it gives unremittingly to that which is alive. At the siddhic level the 27ᵗʰ Gene Key becomes capable of miraculous healing. As long as there is an ounce of life force within something, the vast love coming through the 27ᵗʰ Siddhi can revive it.

*As long as
there is an
ounce of life
force within
something,
the vast
love coming
through the
27ᵗʰ Siddhi
can revive it.*

Because this Siddhi draws upon the entire bio-energetic field of Gaia, it has the capacity to heal any disease or sickness, so long as the indwelling awareness is aligned more strongly with life than with death.

One other way of approaching this Siddhi is through the analogy of music. There are certain Siddhis that have a deep relationship to music as the binding force behind all life. This 27th Siddhi concerns the music of the elements as they combine and recombine, endlessly feeding off each other. There are indeed precise fractal mathematical laws that govern the cycles of the elements on our planet. The way in which water and air combine, both through the global weather patterns and within the digestive tracts and respiratory systems of all creatures, creates an exquisite harmonic if we could but hear it. To reverse an old metaphor, on our planet love is the food of music, and it can be found everywhere. To those listening through the ears of the 27th Siddhi, this love is all they hear. One day you too will hear this music.

SIDDHI
IMMORTALITY
GIFT
TOTALITY
SHADOW
PURPOSELESSNESS

⚏ **28th** Gene Key

Embracing the Dark Side

Programming Partner: 27th Gene Key

Codon Ring: The Ring of Illusion
      (28, 32)

Physiology: Kidneys

Amino Acid: Asparaginic Acid

THE 28ᵀᴴ SHADOW – PURPOSELESSNESS

THE EGREGOR OF FEAR

The theme of purpose is a notion that lies central to this entire work on the 64 Gene Keys. In following your true purpose you unlock the manifesting power of your specific Gifts. However, there are forces in the world that directly challenge your ability to find your purpose, much less follow it. The 28ᵗʰ Shadow of Purposelessness represents your potential nemesis in this regard, because it can ensure that you either never find your purpose or you never follow it through. This Shadow cuts to the core of the deepest of all human fears — the fear of death. All fears can be reduced to this one fear. It is the fear of being extinguished — the prime emanation of human fear. Humans will do all manner of things to escape it. We tend to have one of two main patterns that emerge from this 28ᵗʰ Shadow. We either deny death altogether until it finally catches up with us, or we allow it to consume us and in so doing we live in constant reaction to our fear.

Our unconscious symbols of the dark side of human nature have arisen out of this Gene Key. As such this Gene Key contains some of the darkest coding in the whole genetic matrix. All the demonic archetypes from around the world emerge as a direct personification of the unconscious fear of death that lies within every human being. Using the word *darkest* need not necessarily be construed as negative. The 28ᵗʰ Shadow simply sets us humans up in a falsely constructed reality peopled by good forces and by evil forces. The 28ᵗʰ Shadow is a major aspect of the genetic survival equipment that has led to the success of the human species. This Shadow with its fear of death has sharpened our individual instincts for millennia, and it has led us out of the darker ages of our prehistory into the current age in which individual survival is more ensured than ever before.

The fact of death leads directly to the question of the purpose of an individual's life. The prime purpose at a physical level is to maintain your health for as long as possible, but there is another core purpose built into us — to evolve. For a human being evolution means creative uniqueness.

We are each born with a creative purpose that no other human being carries. If you are to release your true creativity into the world, you must meet your own dark side. In other words, you must at some point face your deepest fear of death.

The fact of your death actually gives an edge to your life — putting you under pressure to find your life purpose and to take the risk of following your individual dreams. The amount of life you feel is directly proportionate to your willingness to face the fears that threaten your dreams. Actually, fear never threatens your dreams, only your mind sees it that way. If you are under the spell of your mind, as most people are, then you will fail to see the true nature of your fear because of your attempts to avoid it. As all our great myths testify, we must pass into the underworld in order to be reborn into the light. We must face our unconscious fears in the outside world.

The most common way of avoiding your fear of death is to adopt a fixed mental philosophy and then live within that philosophy rather than living with your fear. These philosophies are our religions, beliefs, sciences and systems — anything that becomes dogma for you and numbs your fear. The human mind does not like surprises! To live in the continual acceptance of death means to live with the continual threat of the unexpected. The human mind would have you believe that your life's purpose lies in the future rather than here and now, so you keep postponing your life until you meet all your fears face to face at the end. You must look deeply into your fears now in order to find your purpose, because your purpose actually lies *within* your fears. This is why in the great mystical traditions, it is said that you must die before you can live.

The theme of purposelessness is really a contemporary theme that has become more and more pronounced as humanity has learned to master the material world. Survival gives you a powerful purpose. In the West we no longer fear our survival because we have created a society that supports everyone at a collective level. Almost no one in a wealthy advanced country will die of starvation. Because of this, our fear has shifted to the fear of purposelessness. Now instead of being afraid to die, people are more afraid to live. The fear of not finding your purpose is still really a dressed up fear of death. The majority of people do not even want to think about whether they are fulfilling their true purpose or not because to do so is to look right into their deepest fears. The majority compromise and fall in line with the collective belief that they are trapped by the system — by money, by responsibilities, by taxes. In this regard it is interesting to ponder the 27th Shadow of Selfishness, the programming partner to this 28th Shadow. People are afraid to appear selfish and follow their dreams even though those dreams, if they are true dreams, will be of far greater service to the planet than anything else they do.

The 28th Shadow represents a deep attunement to the acoustic field and is rooted in frequency and sound. In this Shadow every fear can be perceived as a vibration. These vibrations have been personified by many cultures as demons or entities with a separate existence from their host. This very interesting phenomenon forms the basis of many systems that explore the darker unconscious side of human nature, from shamanism to psychoanalysis. The shaman operates in the world of vibration, and he or she identifies fear patterns as entities, which must be either dislodged from your inner being or transmuted. This is the foundation of true shamanic practice. Psychoanalysis on the other hand examines your mental and emotional world and names these same fear patterns *neuroses*. Other systems give other names to these fear frequencies. The true shaman or therapist also knows that he or she can never take away another person's fears. He or she can only help the person identify those fears so that they can be accepted. Wholeness comes about as all your deepest fears are embraced, thus it is said that the only way to kill a demon is to absorb it into the light inside yourself.

All your inner demons emanate from a single source — an egregor or collectivity of all your fears rolled into one. This is the arch-demon, the antichrist or the *doppelganger* within each of us. The 28th Shadow truly represents everything within your psyche that you do not wish to accept, right down to your core fear of non-existence. Only as you reclaim each of these dark aspects of your inner being can you begin to assemble and manifest the true purpose of your life. This is the magic and true purpose of the 28th Shadow.

REPRESSIVE NATURE – HOLLOW

When you repress the darker sides of your nature, your life appears hollow and devoid of real juice. To turn away from your fears is to live a lacklustre existence with no deeply felt sense of purpose. Your life may be glamorous and successful or dull and mundane, but it has no centre. Such people often try hard to appear happy and easy-going, even to themselves, but to anyone who knows their own dark side, they hide nothing. The deeper you go into your fears, the more you feel the authenticity of others. Those who will not face their own demons live a half-life, unaware of how transparent they are. These people often pretend to be happy or evolved, but they lack the depth and deep understanding of those who have dared to look into the deep, dark mirrors of their souls.

REACTIVE NATURE – GAMBLING

The reactive side of this 28th Shadow is about risk-taking. These people react to the fear that they feel by transforming their fear into activity. This results in rash and impetuous acts that temporarily cloak the fear. Such people quickly become addicted to this risk-taking pattern and live their lives at the other extreme — at such a fast and unpredictable pace that they cannot stop and examine the fear that is driving them. These people will try anything and everything to give them a sense of purpose, but they cannot stop moving. Their most terrifying space is their own inner silence and stillness.

THE 28TH GIFT – TOTALITY

ALL OF LIFE'S A STAGE

The Gift of Totality is a wonderful Gift to have, and it belongs to anyone who really knows how to trust in life. Totality means to live alongside fear — to live with the unexpected and to stay open to continual change whilst remaining committed. Totality is the balance between the extremes of the 28th Shadow — one extreme that cannot change and the other that cannot

*There is a
deep sense
of thrill that
comes as you
progressively
face your
inner demons.*

commit. The Gift of Totality means to embrace the whole of your nature and the whole of life — the pleasure and the pain. To be total in the sense of this Gift also means to live without allowing your mind to dictate your life. This is life lived for the moment, in the full knowledge that life purpose can only be found in the present moment rather than in the distant future.

To have the Gift of Totality is to follow a mythic path. As you steadily embrace the various challenges that life brings, you gather and compound the various aspects of your psyche until you reach what Jung called the state of individuation. The shaman might call this same process the retrieval or incarnation of your full soul.

The state of Totality appears to be a continual state of taking risks — not the unfounded risks of the 28th Shadow but the risk of building something that you cannot really see until it is complete. What you are building of course is the path of your true destiny. This is a path of deep trust in which the individual strikes out on their path, surrendering their whole being into the mystery of life and its hidden rhythms. To be total is to be alive in every sense — it is to be acoustically alert to the vitality of every moment as it comes. In the resonant vessel of each moment fear cannot survive and thus you experience a deepening calm and quietness growing naturally within.

When you meet life through the Gift of Totality you meet life as a game to be played or a stage on which to act. This is life lived as a romance — a romance that includes both comedy and tragedy. There is a deep sense of thrill that comes as you progressively face your inner demons. Your demons, as we have learned, are really your angels in disguise. Every situation in your life is devised as a form of initiation that allows you either to remain as you are or to evolve. At the individual level this Gift gives you a profound sense of freedom even when external forces appear to obstruct, challenge or entrap you. On the inner planes, Totality gives itself to every situation, allowing the game to play itself out with no expectation but with absolute conviction. When lived in this intuitive way, life shows you that it has an underlying purpose in everything that it brings you. You simply have to align yourself to the dramatic plotline as it unwinds before you. People aligned with the 28th Gift have the wonderful knack of handling adversity lightly. The more deeply they accept each feeling of fear that comes the lighter they grow and the more love they feel towards life.

As you begin to feel life's purpose moving like a wave beneath you, you will feel the programming partner of this Gene Key, the 27th Gift of Altruism, become more influential. As your own issues begin to subside into the background, your life energy begins to direct itself towards others. One of the great mysteries of existence is that the only true sense of purpose in life comes from the impulse to serve something greater than one's self. These are people whose acts and deeds immortalise them because their lives burn so brightly with the fire of their deep sense of purpose. In overcoming the fear of death, you begin to realise that one of the only things in life that lasts forever is the human spirit itself. This realisation about the eternal nature of the human spirit paves the way for the ultimate flowering of human consciousness — the Siddhi of Immortality.

The 28th Gene Key and the 32nd Gene Key make up the binary genetic codon group known as the Ring of Illusion. These two Gene Keys both share a common theme of fear connected to death. Whilst the 28th Shadow fears death itself, the 32nd Shadow fears dying unfulfilled. The Ring of Illusion causes human beings to postpone their lives, seeking a sense of purpose in the future. Your thinking is based upon fulfilment then, rather than now. The Gift within this codon however is that the illusion can easily be broken through a simple understanding: true purpose is found in giving yourself one hundred percent to every moment, rather than having to do with any kind of achievement. When your fulfilment comes from simply being alive in the thrill of the role life is bringing you in the moment, only then are you total. Whatever role you are playing in the game of life — lover, villain, master, disciple or seeker — as long as your commitment is total, you will discover a mysterious detachment running beneath the role itself, and this detachment is the reward for your courageous totality.

THE 28TH SIDDHI — IMMORTALITY

THE TRUE NATURE OF THE BEAST

Since the beginning of time humanity has wondered about the possibilities of immortality. Alchemists have long sought the precious elixir vitae — the spiritual essence that when drunk will restore eternal youth. Modern medicine has in its turn extended the human life span and will probably continue to do so. With the promise of the new genetic science, many scientists are already talking about being able to expand human life indefinitely. When we think of immortality we also think in terms of the soul. Certainly the dream of the great world religions is that our soul will survive death and live on in an eternal dimension or heaven realm. The fear built into the lower frequencies of the 28th Gene Key also gives rise to the opposite side of this duality, the notion of hell and eternal damnation in the underworld.

The 28th Siddhi has very little to do with such things, which are largely projections of the lower frequencies of this Gene Key and its Shadow. It is true however that the human genetic matrix does contain the seed of physical immortality, although our current biophysical vehicle will not properly support this transformation. It might even be possible to alter our genetics to make this current vehicle immortal, but the consequences would not be pleasant. To create a new body out of the fear of the 28th Shadow means that the indwelling awareness will not have evolved naturally to suit such a body. Such a being would be a genetic freak and even though the body might never die, the awareness within could not cope with such a concept, remaining rooted in the fear of death. Just because a body can genetically go on living doesn't eliminate the possibility of death from other causes. Instead of taking away the fear of death, such a circumstance would more than likely increase the fear of death to obsessive proportions. Without the grounding of the acceptance of your dark side, the results of such an occurrence would very likely be catastrophic.

The mind cannot really grasp what immortality means. True immortality is actually the cessation of time altogether.

When the human mind imagines the concept of immortality it does so from within its own limitations. The mind can only conceptualise something within time, so it sees immortality as time that simply stretches into the future forever. This is why the mind cannot really grasp what immortality means. True immortality is actually the cessation of time altogether. This is the only way to escape death — to live so entirely within the present moment that death cannot exist. This is why the Gift of Totality must eventually lead to such a state. Whereas Totality means to live life to the utmost, Immortality means to die into the everlasting moment. To do this, your sense of identity and separateness must first die, leaving only life in its place. Once there is no localised centre of awareness, there is no death because there is nothing to die. Only consciousness remains, moving from one form to another endlessly.

In Christian mythology, the embodiment of fear is expressed through the Antichrist, Lucifer — the embodiment of evil. There are some curious secrets hiding within this myth. The destiny of Lucifer at a mythic level is actually to become one with the Godhead. Lucifer was originally the favourite and strongest of God's angels. In mythology, the strongest is always chosen to fall and forget his true nature. This is the higher mythic meaning behind betrayal. This wonderful anthropomorphism contains the great secret of the meaning of evil and the dark side itself. Evil is simply everything about life that has not yet been accepted and embraced. The only misinterpretation of the old legends is seen in the battles between good and evil in which good triumphs. The ancient symbols and images of dragons being killed is a projection from the 28th Shadow. In the end, the archangel Michael, who represents good, must embrace Lucifer in his arms rather than killing him. Only in this way can he fulfil the myth by transmuting Lucifer's true essence into a force even higher than himself, thus revealing Lucifer as God himself. That is how the Christian myth should really read! Many other more ancient myths from other cultures also describe the very same dynamic.

The Siddhi of Immortality requires that the individual surrender him or herself into his own deepest fears and in dying, he or she is reborn as pure consciousness. Such a being realises their true purpose as the purpose of life itself — to live in the immortal truth of its own nature, beyond time and form. When a being attains this state through the 28th Siddhi, they take on a particular mythology. Their specific gift is to highlight the fears of others wherever they go. This is simply an aspect of their awakening operating through their genetics — thus they are said to have the gift of casting out demons, because this is precisely what their aura does. Through its grace, it highlights the darker, unaccepted aspects of one's nature and absorbs them into its deathless state of consciousness. As all siddhic states are really one and the same state, this is an aspect common to them all, but it carries specific mythological power in the 28th Siddhi's destiny.

Finally, a few words about the future role of the 28th Siddhi. As was stated earlier, this Siddhi contains the seed of the manifestation of immortality in form. After all the Siddhis have dawned in humanity and our collective body has begun to transmute into its future form, the 60th Siddhi will flower and the laws that hold our world together will dissolve.

At that time, elements of our future vehicle will coalesce and begin to form another kind of vehicle to house the higher consciousness of humanity. In this vehicle, the 28th Siddhi will finally come to fruition and in doing so it will synthesise the animal soul of Gaia with the human soul, thus making an immortal body. Herein lies the secret behind all the codes within our mythologies where man and beast are combined as one. The animal kingdoms of our planet constitute an awareness that already operates in the immortal field and whose sacrifice suggests an evolution even higher than our own. At a physical level, man must absorb his entire animal nature into himself in order for its true purpose to be shown. Only then will we see for the first time the true nature of the beast.

🀫 29th Gene Key

Leaping into the Void

SIDDHI
DEVOTION

GIFT
COMMITMENT

SHADOW
HALF-HEARTEDNESS

Programming Partner: 30th Gene Key

Codon Ring: The Ring of Union
      (4, 7, 29, 59)

Physiology: Sacral Plexus

Amino Acid: Valine

THE 29ᵀᴴ SHADOW – HALF-HEARTEDNESS

THE LIFE HALF-LIVED

The 29th Shadow of Half-Heartedness in combination with its programming partner, the 30th Shadow of Desire, creates quite possibly the greatest emotional chaos of any two of the Shadow pairings. This is a deep and ancient genetic programming within human beings. More than anything it concerns a basic lack of trust in human desire. Desire, as we can learn from the 30th Shadow and its Gift of Lightness, serves a greater purpose than simply creating emotional confusion. Desire that is fully embraced always leads to a beneficial outcome. It is after all a pure life energy within us. The problems with desire arise through the 29th Shadow and its dynamics. In essence this Shadow is about two things that ultimately stem from the same source — over-commitment and lack of commitment.

The secret to all commitment lies in the way in which you begin. It is the energy behind your actions that creates your future rather than the actions themselves. There is nothing that is worth doing in life unless it is done with absolute commitment. It doesn't matter whether you love it or hate it, if you attempt anything half-heartedly you might as well not attempt it at all. Without commitment, action lacks power or direction and above all, it lacks luck. This last comment may sound rather odd, but there exists a universal law that anything that is done with full commitment carries within it the seed of good fortune. Likewise, anything done without full commitment carries the seed of misfortune. All life is a continuum and every action you take leads you down a certain path. It must also be said that there is no morality behind this universal law of commitment. It simply represents an invitation for you to trust in life.

Half-heartedness robs you of your opportunity to participate in life's mystery. It stalls life's natural tendency towards the magical and the profound. This Shadow keeps you a victim of fate rather than being a player in the great play. It keeps you in the sidelines, ensuring that you play roles in life that are dull and monotonous or else fraught with emotional agony. In a nutshell, when you enter into something half-heartedly you invite misery into your life.

This 29th Gene Key is about human feelings. It's about sex and relating, failure and success, desire and expectation. No matter who you are, your life depends upon your honouring the laws of this Gene Key. When you do something half-heartedly you are actually behaving dishonestly. You may not be literally dishonest with others, but you are being dishonest with yourself and with life and this always has unpleasant consequences.

The 29th Shadow is a wake-up call to all human beings. How clearly you hear the message depends on how deeply asleep you are. Commitment operates within a cycle and at the end of its cycle it either automatically renews itself or it lets go and commits to something else. These cycles can be of many different lengths. The cellular cycle lasts for seven full years, since that is how long it takes the body to replace all of its renewable cells. True commitment within a cycle of time therefore lasts seven years or more. Cycles of desire can last much less time, but each has its own built-in timing mechanism. Human beings must work through their desire cycles until they reach a natural completion. Unfortunately there is no simple way to know when something will end. You need to remain committed until the play has ended of its own accord. If you break out of a cycle prematurely your life will reconstruct the same patterns of experience until you actually finish the cycle and learn the lessons it holds for you.

Half-heartedness robs you of your opportunity to participate in life's mystery.

The 29th Shadow keeps most human beings in repeating patterns because they do not follow things through to their natural endings. True commitment includes the energy to overcome obstacles and adversity. Half-heartedness is about giving up at the first sign of trouble or discomfort and, ultimately, all half-heartedness is rooted in deep un-embraced fear. The lesson that comes from the 29th Shadow is really simple. If you quit something too soon you will stay in the same old loop, but if you follow experiences through to the end you will make quantum leaps, both in terms of your good fortune and your fulfilment. You need to see that this Shadow, like all the Shadows, has a beneficial purpose in the long run. It teaches you the value of your experiences when you look back upon them with hindsight. If you look back and keep seeing the same old emotional traumas repeating in your life, you will eventually learn what you are doing or not doing that is causing these patterns.

The ancient Chinese had a wonderful and somewhat disturbing name for this 29th Gene Key or hexagram. They called it *The Abysmal* and it was considered one of the major symbols for predicting danger on life's path. This is exactly what the 29th Shadow does — it keeps throwing you into very challenging situations in which your commitment is tested. Once you have embarked on a certain path in life, you are really flying blind. You have only the power of your commitment to carry you through the *abyss*. With half-heartedness, you are constantly concerned about where this path will take you and whether you have made the right decision in the first place. Your fears heckle you from within and threaten to undermine your commitment. If you yield to them, you actually create the conditions for misfortune. But if you can weather your doubts, especially at those critical moments, then you create the conditions for transcendence.

In this 29th Shadow lies the secret of what is generally seen as success. Success in life hinges upon two things — commitment and luck — and commitment actually engenders luck.

Failure means that you have remained stuck in the same old cycle, and nowhere does this Shadow have more relevance than in the field of human relationships. Since it is paired with the 30th Shadow of Desire, it leads to the beginning of many human bonds and alliances. It is through this coupling that so much confusion enters into relationships. All desire operates within clear cycles and these cycles must be honoured, even if they are not fulfilled. If a desire is honestly embraced, then its cycle will soon reveal itself. It may last a day or a year, but the cycle can never be wrong. This is not about social morality but life energy. In marriages (whether formalised or internal), commitment is a built-in requirement. If sexual desire for another rears its head, it signals one of two things — either the marriage is nearing the end of its cycle of commitment, or it is about to be strengthened through an honest cooperative grappling with the desire cycle and whatever that entails. The 29th Shadow responds to most sexual desire out of fear, which usually manifests as behaviour rooted in either guilt or shame. In this sense, half-heartedness means to hide your true feelings or follow them through in secret. Thus this 29th Shadow can lead to all manner of unfortunate emotional situations and relationship disasters in life.

As the old folk saying goes: "A life lived in fear is a life half-lived." It is an apt saying for this 29th Shadow, which can cause all kinds of emotional trauma for human beings, especially in the domain of relationships and material success. To live half-heartedly means that you never fully embrace and trust your decisions. This Shadow makes you constantly worry about your decisions and where they might or might not lead you. The great illusion of failure and success is that they are simply inner attitudes linked to your beliefs about yourself. To move beyond the domain of this 29th Shadow you will have to let go of all such ideas and allow life to catapult you into and through the abyss of the unknown. You must hold back nothing and be totally honest, both with yourself and others. There are so many rewards and fruits waiting for you if you can stand by your decisions and follow them through to their natural, organic conclusion.

REPRESSIVE NATURE – OVER-COMMITTING

These people are conditioned not only to make commitments but also to keep them no matter what happens. In other words, these people are unable or unwilling to recognise when natural cycles are over. Such a nature takes on far more than it is capable of handling and then becomes gradually exhausted by the magnitude of its commitments. These people often become victims of others, or they become slaves of large organisations. Because of the fear in their nature they do not have the courage to admit when something needs to end and they continue to allow others to abuse them, either consciously or unconsciously.

REACTIVE NATURE – UNRELIABLE

When the 29th Shadow is played out through a reactive nature, it hides a deep fear of commitment that is seen by others as a lack of reliability. When one does something without true commitment, then one can rarely follow through an action or a cycle with confidence and competence. The usual result is the breaking off of the cycle and a consequent disappointment and sense of failure or shame. These are natures that may say yes to all kinds of things but then react to the pressure by pulling out of their commitments. The anger inherent within their nature is usually triggered by someone else's expectation of them, and thus they tend to talk things up whilst lacking the ability to deliver on their promises.

THE 29ᵀᴴ GIFT – COMMITMENT

THE BUSINESS OF LUCK

As your frequency becomes clearer and more refined, your decision-making process naturally becomes cleaner and quicker as well. The 29th Gift is not subject to the pressures of condition-ing and expectation from others, but opens into a deep connection to the direction of the life force within. This Gift inherently attunes to the cyclic flow of life. People with the 29th Gift have the knack of getting out of the way of life, and through this Gift they can watch their lives unwinding in powerful and mythical ways. Without the 29th Gift and its ability to make crystal clear commitments, life becomes choked and confused, and nowhere is this truer than at the emotional and sexual level.

Commitment is akin to trust, which can neither be forced nor willed. It flows like a great river from deep within your being and out into your actions. With commitment you have no need to think about the future or the goal because the commitment contains the seed of the goal within it. Only time will show where the river of each cycle of experience will lead. Thus for the 29th Gift the goal is not what is important. What really matters is the commitment to keep following the journey until its end. Life is latticed with cycles within cycles — some journeys last five minutes and some last a lifetime. The ultimate journey is your entire life, and the shape of your life is formed by the millions of tiny decisions that are made over the course of your life. To live your entire life with this profound level of commitment means to make every decision with the same commitment — from the way you have sex to the way you wash the dishes!

As an integral part of a chemical family known as the Ring of Union, the 29th Gift shares a common theme with the 4th, 7th and 59th Gene Keys. This Codon Ring is currently undergoing a great deal of spontaneous mutation in our DNA and is directly responsible for a huge shift in the way in which we humans relate, particularly through our sexuality and gender. Much of the impetus for these genetic changes is coming through the 59th Gene Key and its programming partner the 55th Gene Key. The very role of human sexuality is about to change, which means that at present the world is deeply confused about the value of its morality and its age-old insti-tutions such as marriage. Through the 29th Gift we can experience a new definition of the word commitment, which has less to do with social expectations and everything to do with saying yes to life. The only real commitment is commitment to your own inner guidance in the now (the 7th Gift). Finding this guidance is dependent on your surrendering to the life force within you, which involves utter trust in life's natural cycles of commitment. It is this trust that is moving into the world today, and as it comes it is smashing apart all our false moralities.

True commitment is an energetic dynamic felt within your whole being rather than a social requirement. Many people view commitment through morality. You see this particularly in human relationships, where commitment is generally enforced by social pressure. If, for example, a relationship breaks up or a marriage ends in divorce, it is still often thought of as a failure. True commitment is not moralistic. It lasts as long as it lasts. When the cycle is over, it is over, and both parties will feel this at the same time. Anyone who begins a relationship alliance out of true commitment knows this truth. Relationships that begin with this kind of clean commitment generally end in a clean way, without all the usual emotional turbulence that goes with breaking up. Some relationship commitment cycles really do last for one night, and others forever. The length of the cycle has nothing to do with success or failure. At the frequency level of the 29th Gift, all relationships form a part of the evolving storyline of your life, and thus they are appreciated for adding richness and depth to life rather than being seen in terms of failure or success.

People with the 29th Gift in their Hologenetic Profile can be exceedingly lucky people. Their clear, committed decisions create the conditions for their own good fortune. These people cannot afford to be led by others — they cannot listen to teachers, gurus, oracles or systems. Neither can they succumb to pressure or expectation from others. Their decisions flow from deep within their bellies and brook no argument. With the 29th Gift, a clear decision is felt as a quiet and powerful warmth that courses through your whole being. These are not emotional decisions, neither are they excited or nervous or explosive. Commitment is a wholesome energy, as though nature itself were taking control of your destiny and showing you the way ahead. It is at this stage that you begin to understand that to commit is also to surrender. Rather than expending a huge effort to maintain your commitment, you simply surrender yourself to it. Sometimes if you feel a lack of commitment, it is because you need to surrender even more deeply into your process.

A clear decision is felt as a quiet and powerful warmth that courses through your whole being.

Whether you have the 29th Gene Key as an aspect of your Hologenetic Profile or not, luck is made or unmade here, every time you make a clear decision in life. This can be especially true in business. Like a microcosm of life, a business is a journey with ups and downs. Prosperity is directly linked to clear commitment — both in your relationships and in your daily work. Within business there are many cycles that begin, end and begin again. Financial success cannot be measured by a single cycle, but by continued commitment and certainty in your decision-making. For example, sometimes when you stay committed to a direction that seems to be unsuccessful, it opens into opportunities that later *become* successful. You cannot think your way through life. You can only align your inner direction truly, trust in it come what may, and allow nature to do the rest. That's the pure magic of the 29th Gift.

THE 29ᵀᴴ SIDDHI – DEVOTION

TANTRIC CONTAGION

When the Gift of Commitment takes on a universal level of functioning, it becomes transmuted into the Siddhi of Devotion. Out of this consciousness all the great paths known in the East as Bhakti Yoga have arisen. Bhakti Yoga refers to the path of devotion, or the path of the heart. Devotional paths are all paths of self-surrender in which you completely lose your own sense of self in another. The other can be a mission, as in the case of Mother Teresa's devotion to the poor, or the other can be an ideal or symbol such as a god or a guru. The path of devotion is far removed from society. Devotion is commitment gone mad. It is mad in the sense that you have to leave the order of your mind to enter into the wildness of your heart.

At the Gift level, commitment can still have traces of selfishness even though it is surrendered and contains great power. However, the more the frequency rises through this 29ᵗʰ Gene Key, the more you find your commitments directed towards the service of others. As the frequency gets higher and higher, your commitment takes on a devotional quality and begins to activate the heart centre. At a certain stage in this process, you cannot help but become a devotee of some higher cause or being, and yet still the process goes on. As the energy of love pours outward into service, it requires you to surrender your very identity to what appears to be an outside being or symbol. To those who do not understand the true matters of the heart, such devotion appears misguided at best. People approaching this Siddhi can appear to worship gurus or idols without a care for themselves. To the devoted one however, the only thing that exists is the object of their devotion. If the devotional energy is towards a guru, then the guru is seen everywhere and in every thing. If it is towards a mission, then that mission is all that matters in your life and everything else must bend in that single-pointed direction.

When the quantum leap from the 29ᵗʰ Gift into the 29ᵗʰ Siddhi occurs, an extraordinary thing happens. All the love that has been poured into the object of devotion suddenly begins pouring back into the devoted one from everything in their universe. At this point, the one manifesting the Siddhi often refers to everything, including themselves, as the Beloved. Thus it is said by such people that the very rocks and the trees bleed their love towards the beloved. Wherever such people go, their heart is in constant meltdown from everything and everyone they meet. These people often become poets or divine drunks or servers of others. The programming partner to this Siddhi is the 30ᵗʰ Siddhi of Rapture, and these two words — devotion and rapture — are inseparable. These people are literally seized by love. The aura of such a being is so soft that it can draw in almost anyone as a devotee. When meeting this frequency in the flesh, it is near-impossible to say no to such a being.

Another aspect concerning this Siddhi, and in fact one partially born out of this Siddhi is the path of tantra. Tantra loosely refers to the transmutation of sexual or dense frequency energy into Divine energy. Even at the Gift level of the 29ᵗʰ Gene Key you begin to enter the stream of tantra. As you allow the energy of commitment to act itself out in the world, you begin to realise that an energy separate from your body is moving through you.

This high frequency energy is actually moving into your body from your higher bodies and in particular from the 5th body known as the Buddhic body. The more refined your frequency becomes, the more you can sense this energy or vitality moving through you. At these higher levels, people often become interested in practices such as yoga that can help you become even more sensitised to these life currents within the body. When the life currents begin to move through the chambers of your chest and heart, the devotional energy is activated. This is the essence of tantra — the spontaneous surrender of your being to its higher evolution.

This 29th Siddhi remains deeply immersed in human relationships. In many tantric practises, the devotee visualises him or herself having sex with a Divine consort, or he or she experiences alchemical shifts within their body through sustained loving intercourse with another person. As a path opposite to that of yoga, which is based upon discipline, tantra is about letting life take you wherever it wants to. Such paths draw a great deal of moral judgement from society because they are essentially amoral and given the nature of the sexual energy inherent in this Gene Key, taboos are often broken. However, if such paths are followed through with absolute commitment, they do eventually raise the consciousness into its devotional aspect. As you may recall, the ancient Chinese referred to this archetype as *The Abysmal*. There is no greater symbol for diving into the abyss than that of living your life entirely from your heart.

The path of devotion is one of the simplest ways of touching the Divine and is why you find it in so many of the great religions. Here it is based on the notion of sacred relationship and its method is worship and prayer. There is a certain safety in worship because there always remains a worshipper and an object of worship. However, between the realm of the Gifts and the realm of the Siddhis lies the great abyss. It is this abyss that brings an end to worship. The great challenge for the devotee is to allow him or herself to be annihilated and take the great leap into full embodiment. When the devotee crosses the abyss, he or she will never emerge on the other side. Only the Divine will remain. This is the dilemma of the siddhic realm — once such a person speaks, they will always sound as though they are God talking, rather than a humble worshipper of God. This is where most religions end. Once the 29th Siddhi comes into being, all life is witnessed as prayerful so there is no sense in praying anymore. Since you are the embodiment of the Divine, to whom will you now pray?

The 29th Siddhi of Devotion is an absolutely contagious Siddhi. Wherever it goes it inspires devotion in others. This electrical, almost sexual aura creates considerable waves when it appears in the world. These are masters who care nothing for morality or taboos. They are only interested in one thing — surrender into the heart. When it first pours into you, this 29th Siddhi often appears to be a chaotic energy without structure or rhythm. As your devotion deepens you adjust your inner being to the organic nature of the heart, which knows no rules and takes its own strange twists and turns. What appears to be chaos to the outsider is really a deep state of wild, harmonic transcendence in which the boundaries of *normal* reality have utterly melted. This is the Siddhi of *saying yes* to life — it gives itself to absolutely anything or anyone that comes its way. The message of the 29th Siddhi is exactly this — trust your heart above all else, and never worry about the consequences. To be devoted means to lie forever in the lap of the Divine.

To be devoted means to lie forever in the lap of the Divine.

30th Gene Key

Celestial Fire

SIDDHI
RAPTURE

GIFT
LIGHTNESS

SHADOW
DESIRE

Programming Partner: 29th Gene Key

Codon Ring: The Ring of Purification
      (13, 30)

Physiology: Solar Plexus/Digestion

Amino Acid: Glutamine

THE 30ᵀᴴ SHADOW – DESIRE

NATURE'S GREATEST CON

Deep within the matrix of the DNA molecule lies a vital code that is greatly responsible for the building of our human civilisation. This is the 30th Shadow of human Desire. Often when we think of desire we think first of sexual desire, which is in fact only one direction that desire can take. To understand this 30th Shadow we will have to strip desire down to its essence, and this means detaching the force of desire from its projection into the world. As a pure phenomenon, desire is simply genetic hunger. It does have a deep connection to our physical need to eat, but desire alone is not responsible for our individual survival. This aspect of our genetics does not influence individuality at all. If anything, desire is far more likely to get us killed than protect us. However, desire does serve a far wider purpose when viewed at a collective genetic level.

The real purpose of desire is to get human beings to make mistakes so that we can evolve. Let us clarify this statement — desire does not serve the individual but it does teach us something valuable at a collective level. The real hunger coming from this 30th Shadow is the hunger for experience itself. In order for human beings to master their environment, they have to taste all aspects of it which means that they also have to explore the darker side of experience as well as the lighter. The fact that individuals or even large populations are killed in the process is of no consequence to the awareness operating through the whole human gene pool. Humans are expendable to the collective — even whole races are expendable — but humanity itself is not expendable. We are programmed through the 30th Gene Key to learn and evolve through experience, so there is nothing we can avoid experiencing. If there is anything that is still untried by humanity, then somewhere, inside someone, the hunger of the 30th Shadow will push them into trying it — no matter how outlandish or depraved it may be.

For consciousness to enter into form, it must immerse itself deep *within* the form. In human beings, consciousness has at its disposal a very powerful and sensitive system of awareness — an awareness only overshadowed by the spirit of Gaia herself, of which humanity is essentially a sense organ.

There is something very profound to understand here — the genetic instructions that guide humanity do not come from humanity as a separate entity, even though it appears that way to us. We are only a part of a wider organism with its own genetic imperative, and desire has formed an integral part of humanity's evolution as the driving force behind our sensory apparatus. It is as though humanity operates like nature's brain, with desire providing the force that opens up all manner of neural networks within that brain. Some neural connections lead to short circuits while others open us up to great leaps in awareness. All possible connections must be tried and tested in order to discover which are of the greatest service to the totality.

The greater part of human experience is actually non-essential to our future, but it still has to be explored. As part of a chemical family known as The Ring of Purification, this 30th Gene Key naturally bonds with the 13th Gene Key whose shadow frequency creates a powerful energetic field of discord and pessimism. Humanity is thus designed to move through a vast evolutionary cycle in which it is progressively purified. Such purification can only take place through the shadow frequency, since this provides the raw material for a higher state of consciousness. Most of the fallout from this incredible genetic process is found on the human emotional plane, sometimes also known as the *astral plane*.

As every human being learns, the cycle of desire is eternal.

The astral plane is a subtle electromagnetic frequency field arising out of the sum total of all human desires and feelings. At the lower frequency bands of the astral plane the overall tone is one of Discord, the 13th Shadow. Discord is a collective frequency that is sounded as a result of human beings inexorably expressing what they think to be their individual desires. At higher levels of the astral plane, human desire and emotion begin to turn inward and upward to seek its own source, moving through progressive levels of purity before it is experienced in its purest form — ecstatic rapture.

When seen in this macrocosmic way, it almost seems as if nature has conned humanity — as though we have been set up as guinea pigs for evolution itself. This is exactly the case — that even though desire in itself is a pure impulse of nature, it tends to drive human beings mad! In Eastern Buddhist thought, desire is said to be the source of all human suffering. In fact it is not desire itself but your reaction to it that creates your suffering. In every human being the force of desire is translated differently and projected out upon the world in a unique way. Thus some people experience it through their sexuality, others through their yearning to be rich, famous, in love or spiritually enlightened. The point is that the desire itself is pure. It isn't *for* anything. Human beings are simply creatures who are by design supposed to feel a yearning. What the 30th Shadow of Desire does is emotionally fuel your mind to try and find a way of escaping the yearning, or at the very least venting it in some way. However, as every human being learns, the cycle of desire is eternal. The moment you have sated your hunger, you once again find yourself empty as the yearning begins a new cycle.

The interpretation of the original Chinese name for the 30th hexagram of the I Ching literally means *Clinging Fire*. This is a wonderfully evocative name for this 30th Gene Key. It causes you to burn with longing or desire but despite what you do to try to fulfil that desire, it persists in clinging to you. It continually drives you out into the world of experience, which of course is exactly what is intended. Nature's con is that we humans cannot escape the fire of our desire, no matter what we do. It must be accepted and embraced as a part of our mortality. Moreover, the more you think of desire as *unspiritual*, the more powerful you make it. Many of the great religious and spiritual traditions have been taken in by the great con — having us believe that desire can be somehow transcended or fought. Fighting an evolutionary force as powerful as desire is what causes human beings so much suffering in the first place. The great heretical irony about desire is that the desire to be one with God and the desire to kill your enemy is the same desire. Both lead to metaphorical hell.

Alongside its programming partner, the 29th Shadow of Half-Heartedness, the Shadow of Desire programmes our world into a deep unconscious pessimism. Somewhere deep within every human being a very uncomfortable truth resides — that you can never transcend desire. Our denial of this truth is what causes us to live half-heartedly. We do not fully indulge our desire because it is too terrifying. Those who do so externally usually end in destroying themselves, while those who hold it back also destroy themselves on the inside. It seems then that desire has human beings outflanked and outgunned. Whatever we do, we are caught like rats in a maze. Even the most sophisticated esoteric traditions promise us transcendence from desire, but then warn us that our very desire to transcend in the first place will prevent such a state. What is one supposed to do with this information? What is the future of a race that is caught in such a paradox, if desire itself prevents us from ever evolving beyond desire? As ever, the answers lie within the higher frequencies of this 30th Gene Key.

REPRESSIVE NATURE – OVER-SERIOUS

When desire is repressed, life force is also repressed and this leads to a stiffening of one's whole being — physically, emotionally and psychologically. We begin to take life very seriously. As we have seen, desire equates to fire and passion. When it isn't allowed to burn within us, our inner fire fizzles out. Many people deal with their desires in this way, particularly in repressive societies and religions. Over-seriousness manifests through religion itself, which almost always imposes moral laws over the top of our natural desires. One might even go so far as to say that the majority of human civilisation has repressed desire and become over-serious. It is the hallmark of our whole modern world. The great fear within the 30th Shadow is the fear of being burned by our feelings, a collective fear that has been repressed on a collective level. True feelings unleashed on a collective level might lead to anarchy — that is the fear that is held in check by this aspect of the 30th Shadow.

REACTIVE NATURE – FLIPPANT

Those people in the world who indulge their desires without care always run the very real risk of being ostracised by society. They cannot subscribe to any moral framework and abhor all forms of religion or imposed control. They then become flippant about life as a reaction to that society. The result is that they often become embittered by the projections of the world and lash out by abandoning themselves even further into their desires. Just as the repressed nature wants to end desire through controlling it, the reactive nature wants to end desire though exhausting it. The result of this is that such people burn themselves out, and often early on in life. By totally externalising all your desires, you are in fact becoming their victim. This kind of flippancy is akin to paganism, the extreme pole to organised religion. Just as one is a repression, the other is a reaction.

THE 30TH GIFT – LIGHTNESS

THE FINAL DESIRE

There are two possible human responses to a paradox — either you become tense or you surrender to it. The mind has great difficulty with paradoxes because it is not built to handle them. It can only resolve things through logical reason. The higher mind, which is really akin to an awareness operating outside your brain and body, adores paradoxes because it knows that they represent Truth. When you surrender to the helplessness of being a normal human being, then something remarkable occurs — you experience a shift in frequency of your whole being. You begin to lighten up. The human mind is incredibly serious about life because it so wants to understand and control existence. The 30th Gift however is a new operating wavelength for humanity, and it really involves an inner surrender that cannot be forced or faked.

It is not that you become helpless in a victim sense, but that you realise you are beyond needing help.

When we talk about *lightness* in respect to this 30th Gift, we are not talking about escaping life by making light of life. On the contrary, we are talking about entering more deeply into life's suffering than ever before. We are talking about a kind of suicidal tendency that simply holds up its hands and says to creation: "OK, I give in, do your worst!" What is actually committing suicide is your distrust in life, or what some traditions label the ego. You have to come to a deep realisation of your own mortality and weakness in order to realise that true strength lies in simply trusting. Life is having its game through you and you are simply a genetic pawn in that game. You have to become utterly helpless. It is not that you become helpless in a victim sense, but that you realise you are beyond needing help. And then the magic happens — you discover you can enter into that expanded state of consciousness that is guiding the overall game. This gives you access to a higher level of functioning and you realise it was simply your distrust all along that was making your life seem so difficult.

The Gift of Lightness does not change your destiny. It simply allows you to view it from a different level of awareness. However, the very shift to this higher awareness functioning always signifies a change in the script of your life. It is a breakthrough that occurs, or one might describe it better as a meltdown. All life follows a mythical script or storyline, and you can only see your life from this expanded awareness when you have escaped feeling that you are a victim of it. The moment you see the joke that life must be the way it is, you also see your place in the overall script and you immediately feel at home. More than this, your whole being lightens even if your body continues to suffer. This lightening then finds its way into all your actions.

Whatever you do from the Gift of Lightness there is always a glint in your eye, because at some deep level you know it is all really just a game, and the worst thing anyone can do is take it too seriously. There is a great difference between one who pretends to live life lightly and one who has real lightness of being. This difference is always found in the emotional nature — one who pretends is afraid of his or her true feelings, whereas one with real lightness is never afraid of being overwhelmed.

The Gift of Lightness does not make you immune to desire, but neither does it cause you to react to desire. It allows you to *become* your desires in all their mystery. This is a Gift that knows desires do not necessarily have to be followed, but they do have to be felt. Sometimes they *do* have to be followed in order that something is learned, but generally what this Gift knows is that the fulfilment of desire is a sham. When human awareness penetrates down into the emotional system in this depth, then a huge sense of freedom emerges. This is the freedom of having a wide-angle view of desire. You know that whether or not you follow through with the desire, it will not lead to any lasting sense of peace. This means that desires no longer have such an addictive hold over you. In fact, desires become like guests you invite in for tea; they either leave in their own time or they stay and insist on being followed. In this sense true lightness can be seen as the letting go of the need to escape desire itself.

The other vital hallmark of the Gift of Lightness is a sense of humour. Everything is viewed in a detached way even though it may still be deeply and sensually felt, so everything is seen lightly. The humour that comes from this Gift is not a clever or sarcastic kind of humour and neither is it ever directed personally at another. It always manifests as the ability to laugh at yourself above all else. Your own life becomes a great tragi-comedy, since it incorporates both sides of the experiential spectrum. You learn to see through all human behaviour. You see both the depth of suffering that lies in the false belief that your desires can ever be fulfilled, and the great pleasure that comes from the build up and release of your desires. The humour that comes through this 30th Gift is a very compassionate humour — it is not about laughing *at* anything — it is simply the true response of a human being who has surrendered to the higher self.

As you move into the higher frequencies of the 30th Gene Key, you finally begin to understand the mystery of the cycle of desire. Beneath the thousands of desires that pass through your emotional system each day, you begin to determine a single underlying desire that gradually becomes stronger and stronger — this is the desire to end your own suffering. The desire to end your own suffering is the same desire that leads all human beings on the path of spirituality and inner inquiry. This desire to escape or to ascend or to be free is the final great desire within human beings. It is the urge of evolution itself longing to transcend the form. As you step fully into the pure field of this longing, you enter the purifying fires of consciousness. This codon group, the Ring of Purification, takes you on a journey in which your whole being begins to relinquish the hold of desire. You will have to trust implicitly in your desire to transcend, even as you understand that the very desire itself prevents you from transcending. This final desire must be followed, tracked, embraced and allowed, and, as it is, the fire of your longing becomes intensely bright. This is the other meaning of lightness in the context of the 30th Gift — your body actually begins to fill with light as your lower vehicles and subtle bodies are purified by the strength of your longing.

THE 30TH SIDDHI — RAPTURE

FROM BHAKTI TO SHAKTI

The 30th Siddhi is an unusual Siddhi in that it manifests as one of the great Divine ecstatic states. Along with its programming partner, the 29th Siddhi, these are aspects of our genetics that really terrify the majority of human beings. In our western culture in particular, these kinds of ecstatic states are deeply distrusted, as we no longer have any clear cultural vent for them. In times gone by it was the ancient shaman who could enter these states of consciousness. Nowadays, the closest we come is through the burgeoning drug and dance culture. We have grown so far away from being a devotional culture that we can no longer understand it at all. Certain religions such as Islam are rooted in these devotional codes, and they therefore recognise them more readily, although they can also misinterpret them just as easily — the contemporary cult of the suicide bomber is clearly rooted in the lower frequencies of the 30th Shadow.

The 30th and 29th Siddhis represent an archetypal stream within your DNA that causes powerful mutations in your endocrine system. These result in the ongoing production of certain rarefied hormones within the brain chemistry of the pineal gland that induce states of great devotion and divine rapture. The 30th Siddhi of Rapture only occurs when you step willingly into the fires of annihilation. We may recall that the ancient name for this 30th hexagram of the I Ching is the *Clinging Fire*. At the Siddhic level, you dissolve entirely into the fire. Everything about this 30th Siddhi looks insane to ordinary waking consciousness. It involves mystical suicide — a complete immersion in the fires of Divine longing. You finally give up everything, even the desire to transcend. All desires merge into the one primal desire — the desire with no target — the essence of all creation and the pure longing at the heart of the life force itself. There is a mystery here that is carved into our English language and captured in the word *belonging*. We can only truly belong in the world when we can utterly *be our longing*.

The state of Divine Rapture is really akin to being burned by the fires of bliss over and over again. These are people who are so flammable that the smallest thing will set them off, and whoever comes near them will very likely catch their devotional energy. The nature of the mutated solar plexus system, which is where this 30th Siddhi lies, is to carry awareness outside of the physical body through the aura. Thus the 30th Siddhi is what creates the devotees of its programming partner — the 29th Siddhi of Devotion. These two siddhis radiate their power and shakti through the morphogenetic field of the enraptured one. This is why certain teachers and masters can literally transform the hearts of their devotees forever. The aura of such a being is as palpable as it is dangerous. It is dangerous to the human mind that cannot comprehend such a phenomenon and will not let go of trying to be in control. The very lifeblood of this 30th Siddhi is about dissolution into the primal chaos of the Divine frequencies.

The 30th Siddhi of Rapture only occurs when you step willingly into the fires of annihilation.

These Siddhis are not such a common phenomenon in the world. Wherever they erupt, they are often misunderstood. If this were to happen to someone in the West, they would most likely be heavily sedated and locked away.

In India, both the Divine ecstatics and madmen were traditionally given the same treatment and were both revered, since there is a fine line between the two states. Early manifestations of this Siddhi and the 29th Siddhi can also cause all kinds of difficulties in the human body since we have not yet evolved to fully carry these extremely high emotional frequencies. In this respect the 30th Siddhi has a particular role to play within the coming evolution of the human solar plexus system as described in the 55th Gene Key. The role of the 30th Siddhi is literally to burn out all human desire from our DNA. This means that humans who attain this Siddhi are really performing an important genetic task on behalf of the collective. They are allowing their vehicle to deliberately short circuit and burn out the collective desire of our past. The perk that goes with this role is that they experience the Divine Rapture!

In the 22nd Gene Key there is a detailed description of the seven subtle bodies of the human aura. The process of transcendence through this 30th Gene Key directly mirrors the purification of the second astral body and its assimilation into the higher buddhic body (the fifth body). It is the human longing of the 30th Gene Key that creates enough *bhakti* to catalyse the higher currents of grace from within the higher self. The *bhakti* is the subtle fluidic emanation that rises up from this purification of human longing. It is an emanation that reaches up into the buddhic body and activates its counterpart known as *shakti*. Shakti is the Divine Essence that descends or rains down upon the initiate and causes the state of Rapture. It is this interchange of bhakti and shakti that characterises the 30th Siddhi. The Divine Longing expressed as bhakti is an evolutionary force, whereas the Divine Grace expressed as shakti is the involutionary force.

In the future genetic vehicle of humanity, this 30th Siddhi will cease to exist, having deliberately burned itself out. In this sense it will become redundant. The experience of Divine Rapture is a genetic anomaly within the evolutionary programme for humanity. It has one purpose alone — to kill desire in order that a new awareness can dawn. Interestingly, the 29th Siddhi of Devotion will not share the same fate as the 30th Siddhi but will go on to form the basis of human relationships and therefore whole communities. In the meantime, all those who have a deep affinity with this 30th Gene Key will experience certain degrees of this burning within their lives. We can easily see it working its way out of humanity at its lower frequencies through all those in whom devotion and rapture are translated into external destruction and devotional fanaticism. The higher your frequency rises however, the more you must surrender to the Divine currents that are waiting to burn you into a higher state of consciousness.

▤ 31st Gene Key

Sounding your Truth

SIDDHI
HUMILITY

GIFT
LEADERSHIP

SHADOW
ARROGANCE

Programming Partner: 41st Gene Key
Codon Ring: The Ring of No Return
    (31, 62)

Physiology: Throat/Thyroid
Amino Acid: Tyrosine

THE 31ST SHADOW – ARROGANCE

THE WORLD WIDE WEB OF WORDS

This 31st Shadow and its various frequency bands really turn some fundamental human concepts upside down. You may note that these bands track all human evolution as it moves from a state of arrogance to a state of humility. The chances are that you, like the rest of humanity, have been conditioned to think of arrogance and humility in a certain way. The general conditioning says that arrogance is a negative trait and that humility is a positive trait. Since this 31st Gift directly concerns leadership and influence, we are going to have to explore the real definitions of these two words, because the 31st Gift of Leadership sits like the fulcrum of a seesaw between the two of them.

As one side of a binary genetic codon known as the Ring of No Return, the 31st Gene Key is chemically bonded to the 62nd Gene Key — they both code for an amino acid in the body known as Tyrosine. This genetic coupling is very interesting when we explore it a little further. The 62nd Shadow is the Shadow of Intellect and concerns the mental skill of manipulating language in order to try to understand our environment. In effect, this Shadow encases you inside a world of language and words from an early age, projecting a mental neuro-linguistic map onto reality as you perceive it. It is because of the 62nd Shadow that when you look at a tree, the word *tree* unconsciously forms in your mind. The 31st Shadow extends this cognitive ability beyond the simple creation of a neurological map of reality — it uses this skill in order to control and manipulate others. The nice word we have for this is leadership. Because of the chemical bridge between these two Shadows, humanity comes pre-programmed to follow those who can best manipulate language, facts and words.

In terms of leadership, we human beings operate very differently from animals. Animals choose their leaders instinctively based upon one animal having an *alpha gene* which marks it as the leader. In humans, the *alpha gene* is present in the one who can best manipulate others through language.

Such people may or may not have a strong moralistic character — that is immaterial. What is important here is that language is the medium through which human leadership is manifested. Furthermore, because language is the means through which frequency can be conveyed, at higher levels of frequency it can never lie. However, at the Shadow frequency, language is the ultimate programming tool and can be used for example to *blind* an entire race of people to the truth, thereby locking them at a certain level of frequency.

If we extend this concept of language as a programming tool even wider, we might say it is language that controls us rather than we who control it. There is a big difference between leadership and authority. Authority, the Gift of the 21st Gene Key, is based upon the control of power through strength of will. These are people who do not lead through words, but whose sheer presence controls the will of others. In the case of the 31st Shadow, words can have a far deeper effect than presence, since words endure forever. Concepts spoken thousands of years ago can still influence the whole inner reality of entire nations and peoples. We see this most clearly through the great religions. It is incredible to contemplate how even a single misplaced or mistranslated word spoken by a particular person can lead to the deaths of millions of people over thousands of years. That is the power of the 31st Shadow at work.

Why then is the 31st Shadow called the Shadow of Arrogance if it is language that really controls humans rather than the other way around? The reason is that the collective Shadow frequency itself is arrogant. Human arrogance is rooted in the belief that we have any independent control over reality at all! We shall see when we come to the Gift frequency that the real leaders among human beings are those who understand how deeply programmed and limited we are by our own intellect. At the Shadow frequency human beings are completely convinced by the mass programming of the society in which they live and its collective history, beliefs and culture. It is always the frequency that creates the language patterns rather than the other way around. Our arrogance is based upon our belief that we can think, talk or act our way out of the Shadow frequencies, when in fact those very frequencies are created through our intellect.

The real arrogance comes from being cut off from our Divine source. Words that are not rooted in a profound sense of wonder are always at some level arrogant. Only the human heart can truly offer an answer to the mystery of existence, and unless your words carry the perfume of love somewhere behind them, they are still wading around in the Shadow frequency. At the Shadow frequency humans think, speak and act as if they were separate from nature, rather than being an organ of it. Nature controls human beings, nature has evolved our intellect, and nature has ensnared us in our false mental reality because that is where nature wants us to be at our current stage in evolution. Nature has even made us arrogant in believing that we have some control over nature!

These truths may well seem difficult to us human beings with our deep mental investment in individual freedom. As we shall see, true humility has no basis whatsoever in behaviour, but in understanding. You can only ever be free when you have escaped language — when you are no longer a victim of your own mental constructs and words. When the heart finally begins to speak, it organises the words without you having to think about their meaning.

The high frequency of the heart is what conveys the true meaning. The Ring of No Return describes a state of consciousness that lies far beyond words, even though it may use words as a means of vibrational communication. Arrogance then is a kind of addiction to words and language rather than the frequency of the intention that hides between and behind the words.

The modern world we see around us is nothing more than a mental construct created out of billions and billions of words. At the Shadow frequency, we are trapped in this world wide web of words. Matter itself may well feel solid to our touch, but we now know it to be ephemeral. The news headlines may seem to be deeply important, but even they are simply the working out of an evolutionary scheme of which humanity is but one small component. The 31st Shadow therefore has drawn human beings so deeply into their own illusion or *maya* that we cannot see how little influence we really have. Our new obsession with the environment and with trying to slow down our inexorable progress may well turn out to be unfounded. Our arrogance persists in assuming that we are affecting our environment instead of seeing that there is no distinction between the environment and us. Perhaps the environment needs us to mutate it so that it can mutate us? Perhaps nature is forcing us in the direction we are going because she has an idea that we have not yet glimpsed?

The modern world we see around us is nothing more than a mental construct created out of billions and billions of words.

Whenever a human being speaks without understanding the false mental fabrication in which they live, they speak from the 31st Shadow. The maya is only interested in strengthening itself through our thoughts and words. Rare indeed are words which come from a person who understands how false words themselves are. The programming partner of the 31st Shadow is the 41st Shadow of Fantasy, and perhaps that is the word that really says it all. All words, opinions and thoughts with which you identify are agents of the great illusion of your separate existence.

REPRESSIVE NATURE – DEFERRING

There are essentially two forms of arrogance and the repressive form manifests as false humility. These are people who defer their power to others, thus deliberately setting themselves below them. These are really people who care deeply what others think of them and more than anything fear to be thought of as arrogant. Ironically, such behaviour is designed to draw attention even though it claims the opposite to be the case. In many ways, these people are even more arrogant than people with the reactive nature. Humility is seen by society as a highly prized and noble accolade, but this kind of false humility hides nothing more than fear.

REACTIVE NATURE – SCORNFUL

The reactive side to arrogance is based on anger rather than fear and emerges as a kind of haughty scorn. This more *traditional* brand of arrogance assumes itself to be above others because it sees how easily they are conditioned and can therefore be manipulated. What such people miss is how deeply they themselves are caught up in the same conditioning, since they have a deep-seated need for recognition. Unfortunately, no amount of recognition can ever feel good when it comes from those whom you take to be inferior, so these people remain scornful of the very people they influence. This in turn makes them even more angry and disrespectful of human beings.

THE 31ST GIFT – LEADERSHIP

HEART BRANDING

The Gift of Leadership is really a gift of influence rather than an inherent predisposition to be a leader per se. The real leaders in the 64 Gift matrix are the people with a strong activation of the 7th Gift of Guidance at a genetic level. The 31st Gift is known as the Gift of Leadership only because that is the projection of the collective upon this Gene Key. Deep inside such a person, there is little or no inclination to be a leader and even less identification with feeling like a leader. It is however this very reluctance that sets such people up as leaders. The mass consciousness, as we have seen, is programmed to be led but has no idea of how to pick a leader. Most of the time, the mass consciousness will choose leaders not for their policies or beliefs, but for their style. The 31st Gift knows all about trends and patterns because it has an inherent understanding and mastery of language. These people therefore understand the openness of the collective and its need to be influenced and led.

The difference between the 31st Gift and the 31st Shadow is that the 31st Gift does not believe its own propaganda. This Gift is no longer caught up in the fear of what others think, and if you want to influence people in a certain direction, this is a distinct advantage. The modern day term for this is *the spin doctor*. The spin-doctor spins the truth whichever way happens to suit one's audience. At the Shadow frequency, this is all rooted in getting more personal recognition and/or more control or wealth. At the Gift level, the higher frequency has already moved the person beyond the tight co-dependent dynamic of the leader and the led or the shepherd and the sheep. At this level, you still play the game but for a different purpose — now your primary influence is to help others break out of the very same matrix in which you yourself were once caught.

The 31st Gift, like all the Gifts, represents a huge step beyond the intellect and into the heart. Only the heart really perceives the collective suffering that comes from being marooned in your intellect at the Shadow frequency. Thus your natural urge is to help others escape that suffering in some way. Furthermore, since the 31st Gift is the gift of understanding the masses and the language of the Shadow frequency, it naturally uses this gift to help draw others out of the co-dependent victim patterns and relationships of the lower frequencies. The 31st Gift still carries an agenda, but that agenda is to help people out of the narrow, neurolinguistic confines of their own conditioning. No one understands the specifics of conditioning quite like the 31st Gift. The 62nd Gift of Precision, which we have seen is chemically linked to this Gift, allows a person operating at this frequency to use very specific programming words and languages to *de-programme* other people's conditioning.

Because the 31st Gift has its finger on the pulse of the collective patterns of humanity, it can have enormous influence over large numbers of people. An artist or writer expressing him or herself through this Gift can create a work of art whose influence far outweighs its content. In our modern world, we can see how the specific release of a particular book or film at a particular time can have a vast impact on the whole planetary consciousness. In our modern communication-rich global culture, the 31st Gift is really opening up more and more to its true potential. Certain human beings operating out of the 31st Gift can literally become the voice of the collective. At the Shadow frequency they become the voice of our fear, whereas at the Gift level they become the voice of our creativity as well as the messengers of our future evolution. Along with its programming partner the 41st Gift of Anticipation, the 31st Gift naturally has an affinity with the future and is always expressing the cutting edge of consciousness in form.

The secret to the 31st Gift is the secret of branding. These are people who inherently know that the medium they use — whether it be music, art, science, literature or simply the naked human voice — is really only the means of expressing a transmission that lies beyond words. The art itself carries the frequency encoded within it, but the frequency lies well beyond the words, colours or tones themselves. In other words, all human expression is really nothing but branding. If you release the right brand at the right time in history, you will have a huge influence. This is why the 31st Gift is founded upon the rock of anticipation — it always has to sense what is coming into the world next, and to do this it has to be free from the old conditioning. It truly is the heart that separates the Shadow frequencies from the Gift frequencies and, in our current stage of evolution, the heart is the hottest frequency there is. If you want to really see the nature of success on the material plane, then the heart represents the cutting edge today. From the level of the heart, success is now reaching for a new definition. Life is no longer about individual success but success as a collective organism. This is the direction in which evolution is currently taking us.

One can therefore see from the 31st Gift that a new era is opening up before humanity, and it is an era in which individual leadership is declining while something we have never seen before is also occurring — the collective is beginning to lead itself. Those frequencies that really catch the heart of the collective are coming through as collective voices and expressions, even though they are transmitted by individuals. The point is that those individuals have little or no self-interest. It is truly an extraordinary age that we are moving towards, and it is only those ready to make the leap into the heart who will inherit the huge benefits coming to us all.

THE 31ST SIDDHI – HUMILITY

THE RING OF NO RETURN

If you happen to be excited at the words coming out of the 31st Gift, then you may be in for a shock when you take the plunge into the 31st Siddhi! Humility, as we shall see, is not for the faint-hearted. It is therefore highly ironic that humility is a word that has come to be associated with people of a saintly or virtuous countenance.

We can see how the specific release of a particular book or film at a particular time can have a vast impact on the whole planetary consciousness.

The irony here lies in the misunderstanding that humility has anything to do with behaviour, because it does not. That kind of humility is usually arrogance in disguise. True humility can only occur at the siddhic frequency because it requires the complete obliteration of your individuality. The words humility and arrogance reflect a behavioural duality found in human beings, but they are also words with a highly moralistic charge. At the siddhic level, all such morality is finished because your separation from existence is finished.

If we look even more deeply into these two words, we can see that one — arrogance — represents an archetypal male characteristic, and the other — humility — represents an archetypal female characteristic. Such terms describe the universal polarity of being rather than descriptions specific to men and women. One could even say that all energy behaves in an arrogant, probing way, whereas all matter is humble and submissive. Looking at life forms in this way without any morality allows both expressions their own integrity and beauty. It is in this sense that the 31st Siddhi is called humble because, like the feminine pole, it represents non-existence. Out of non-existence, existence is born. Because the human intellect cannot transcend language at this level, we are using the language of the duality to express something that lies beyond the duality. The feminine pole comes the closest to describing this transcendence.

To the 31st Siddhi, there is no longer the possibility of achieving anything independently from life; thus many human words and expressions cease to have real meaning. If a human being behaves in what others describe as an arrogant manner, the 31st Siddhi simply cannot comprehend this. To the 31st Siddhi, all behaviour is seen as an impersonal manifestation of the totality, and there is no such concept as an independent doer. This is what arrogance really means — you mis-identify with your individuality and assume you have power over life. Arrogance is really concerned with what others think, whereas humility truly doesn't care. This is why arrogance always tries to be humble. However, to be truly humble, you wouldn't mind others thinking you were arrogant!

If this is giving you a headache, then that is what the siddhic frequency always does to the intellect. Someone outside its sphere cannot possibly understand the 31st Siddhi. It is just too simple for the poor human mind and it is deeply disturbing to human concepts of morality. The 31st Siddhi has exceeded all concepts involving levels and movement and even the need for being in the heart. There is no need to be anywhere or do anything other than what you are doing. True humility arises from the truth that you cannot ever do anything wrong.

Consciousness is enjoying its game through your life. No matter what you believe, achieve or do, it is not you attaining it because you are the great illusion. The person manifesting the 31st Siddhi does not even see human beings as divided into those who are awake and those who are asleep. There is no such thing as being enlightened or unenlightened to such a person. If there were, that would assume the existence of separate beings rather than the reality of one unbroken chain of consciousness.

Because the 31st Siddhi is truly humble, it has no agenda. It is not necessarily interested in freeing people from their illusions, even though it may unwittingly do so. Such a being realises that it is not possible to influence anyone in this world.

To be truly humble, you wouldn't mind others thinking you were arrogant!

It doesn't matter what such a person does in life. They are content to leave the world exactly as it is. Nonetheless, such a person still carries the same genetic coding that requires they speak from the cutting edge of consciousness. The only difference between the Gift and the Siddhi is that the Gift level still identifies with evolution. They are still caught up in the game, even though it is at a level beyond fear. The 31st Siddhi says whatever it must say, knowing that the words come from the collective and go back into the collective. It is not concerned with the effect of its words because it has no concept of words conveying any real meaning.

The 31st Siddhi is rooted in the ancient tradition of the Oracle. The oracle is simply a voice box for the Divine. It speaks without being attached or even interested in what it says. Whatever the oracle says is perfect and its translation is also perfect. Words and language can be interpreted in so many different ways. This mysteriously named codon, the Ring of No Return, refers to the great initiatic leap from the 4th causal body into the 5th buddhic body (see the 22nd Gene Key for further details). Once this leap in consciousness has been made, there is no possibility of return because your identity or ego will be erased. With your identity go all language and all words, concepts, facts and names. Ultimately this is the great step in which your name is dissolved. Once you have allowed the siddhic frequency to command your voice, you have entered the first stage of Divine embodiment. From this moment onward, whenever you use the word *I*, it is the voice of Divinity itself speaking. Such speech is pure transmission. It uses words to indicate the path beyond words. That is the beauty of its paradox.

▥ **32nd** Gene Key

Ancestral Reverence

SIDDHI
VENERATION
GIFT
PRESERVATION
SHADOW
FAILURE

Programming Partner: 42nd Gene Key
Codon Ring: The Ring of Illusion
(28, 32)

Physiology: Spleen
Amino Acid: Asparaginic Acid

THE 32ND SHADOW – FAILURE

THE MYTH OF FAILURE

One of the greatest fears that haunts humanity is found in the 32nd Shadow — the fear of failure. This fear has a deep biological root since it is built into your very DNA. Our early hominid ancestors had exactly the same fears as we do today, even though they may have manifested differently. Your individual fear of failure is rooted in the collective fear of failure to survive as a species. One of the first realisations that our ancestors had was that we had a better chance of survival if we stayed together and operated in packs. The pack or tribe consisted of a network of individuals and families all with different skills and responsibilities which, when pooled together, enormously improved the chances of survival. In prehistoric times if an individual became isolated or ostracised from the group, it almost certainly spelled death.

At the core of this genetic reflex to stay together in tribal groupings lies another related fear — the fear of not passing on your genetic material — in other words, the fear that the tribe or family itself might die out. For a woman, this is the fear of not being able to give birth or find a mate, and for a man, it is the fear of infertility. Let's translate this ancient fear for a moment right into the present. Obviously, for most people in the world, this genetic fear is responsible for keeping alive many tribal lines and traditions. However, in the West and in the developing world, we can see something different occurring — we no longer have such strong tribal family structures. Now most young men and women leave their families seeking opportunities outside the old family structures, which leads to the dismembering of those older structures and the support they offered. The reason for this trend is that the modern world revolves around money.

Our collective fear of survival is almost entirely projected onto the amount of money we have. The fear of failure as a feeling inside your physical body is intimately linked with money. The 32nd Shadow drives our modern society and keeps humans operating at a low level of frequency by means of this fear. We have created a world in which individual groups operate independently of each other in their need to compete and maintain their genetic lines. It may not seem that way, but the fear is that same old fear — your desire to have

*True success
means no
longer being
ruled by the
concept of
success and
failure.*

a bigger house or a faster car is actually born out of a huge collective energy field of fear and competition, which is in turn rooted in a much older fear. The larger your bank balance, the less chance there is of failure — that is how your mind thinks at the Shadow frequency. This is how money has become the great symbol of success in the modern world.

But, and it's a big but, the whole money game is an illusion and a sham — it is nothing more than a symbolic mirage created by the 32nd Shadow. The very existence and notion of money feeds the same ancient genetic fear. And that old fear is always somewhere in the background of our lives, no matter how many millions we have earned or inherited. Money remains one huge issue for most people on our planet. Why is this? The answer is that fear always needs something on which to feed. Even if we were to eradicate money from our civilisation, fear would simply find another place to hide. It is fear itself we have to conquer — not money! True success means no longer being ruled by the concept of success and failure. Thus the 32nd Shadow ensures that human beings will always remain selfish, restricting themselves to small elitist groups, families, businesses and fiefdoms. Until we can raise the consciousness of humanity as a whole, we humans will always remain essentially stingy in spirit, confining ourselves to our own gene pools and our own little tribal circles and cliques.

As our ancient ancestors discovered, failure really means one thing — to be isolated. The moment you cut yourself off from your tribal support network, you lose touch with the living chain that supports and nourishes you. Nowadays, we have become so adept at survival that if you have enough money you can flourish in isolation, without ever seeing anyone! But the 32nd Shadow is not only about people; it's also about life as a whole. We humans now live isolated from the earth herself. We still think in terms of the survival of our own families and at best, our cultures, but we have not raised our group consciousness enough to think in terms of our species. Yes, there are individuals who think ahead, and these days there are many more of them, but we still have not transformed the 32nd Shadow and its myth of failure.

Although the fear of this Shadow has its roots deep within the human immune system, it is the mind that reacts to it and feeds off it. If you are not in control of your mind or aware of its power then your mind will run your life, which means that fear will run your life. To raise the consciousness of the mind is to escape the grip of all fear. This doesn't mean that we will never feel these ancient fears — at the Gift frequencies we probably will, because they are still a part of our planetary consciousness — but we no longer have to *react* to them. That is the key. These fears have actually served their purpose — they have kept the human species alive and allowed us to flourish. We can see from the 42nd Shadow — the programming partner to the 32nd Shadow — how strongly conditioned we are to be stingy and competitive, not only financially but also in our thinking. The 42nd Shadow represents the inability to let go and it is connected to the theme of death itself. This very codon group — the Ring of Illusion — is based on the illusion of death through its ally, the 28th Gene Key. There is a direct genetic linking between the themes of death and money (or death and taxes, as the old adage goes!). Therefore, until we begin to think in far wider terms, beyond our own little lives and out into the whole, we will remain isolated in our own little boxes with our own little bank accounts.

Failure is only an outcome when you cut yourself off from the whole. When you raise your frequency beyond the reach of concepts such as success and failure, you remember that all of life moves in a great cosmic pattern. As you let go into this pattern you always find your natural support within it. The more evolved among humanity have found this truth reflected by the way in which individual finances operate. When you surrender to the greater pattern and raise your consciousness above the fear threshold, money always arrives just when it is needed. Money actually provides a wonderful lesson in letting go of fear, and in many respects it has become one of the new spiritual teachers on our planet. Whilst it is here (which is not forever), we should make the fullest use of it as an outer symbol of our ability to surrender to higher forms of consciousness. Every time you feel yourself worrying about money, smile, take a breath, thank your ancestors, and relax. When you truly need it, it always comes.

REPRESSIVE NATURE – FUNDAMENTALIST

The repressive side of the 32nd Shadow is an extreme form of conservatism. The 32nd Shadow is a deeply contracting energy in itself, so that when it manifests through a repressive and fearful nature it becomes extremely tight and fundamentalist. Such people are literally choking themselves physically, emotionally and financially. They are starving themselves of breath and of support from others. The tendency of such people is to isolate themselves in tight little communities that do not interface with the wider world. Such communities, groupings or cults can easily become paranoid about the rest of the world, and it is usually only a matter of time before they die out altogether.

REACTIVE NATURE – DISJOINTED

The reactive nature of the 32nd Shadow is about losing all sense of the continuity in life. This is a state based on anger — the anger that there is nothing to support you but yourself. This means that your anger will drive you into a self-destructive life pattern that will probably escalate at an alarming rate. If you have lost touch with the flow of life, then nothing really goes smoothly for you — you have cut yourself off from the source. People who live such disjointed lives, with no real rhythm or purpose, put themselves in great physical danger. The decisions they make cannot follow the natural flow that leads to health and wealth. Every decision we take in life either connects us to something greater than ourselves or cuts us off from our true and vital inheritance, leaving us feeling isolated and alone.

THE 32ND GIFT – PRESERVATION

THE ART OF GRAFTING

The 32nd Gift is called the Gift of Preservation. It is a truly noble Gift because it is about seeing beyond your own little world, which means going beyond selfishness. The 32nd Gift is about keeping things alive. However, it is not about keeping just anything alive, it is about knowing *what* to keep alive. As we saw with the repressive side of the 32nd Shadow, this Gene Key can just as easily lead to the preservation of things that really do not serve the human race, such as fundamentalism. However, the person who has raised the frequency of this Gene Key can see beyond the confines of their fear-based thinking, and they find that they have an instinctive gift for investment.

Investment can be understood on many levels. If the 32nd Gene Key is a major aspect of your Hologenetic Profile, then you have the potential of a powerful instinct for sensing the long-term picture in all situations. People with this gift also often have a deep capacity for restraint. They have the strength to withhold their energy (or money) from situations that appear tempting but that in the long term would not serve them. By the same token, this Gift enables people to trust in an instinct (especially about other people) that often does not appear to be at all logical, but that in the end will be extremely beneficial to them and to many others.

The secret to this Gift is the instinctive ability to balance restraint (what to keep alive) and risk (what to change). These are people who inherently know that if you want to maintain success in life, you have to have an unflinching set of principles coupled with a constant need to update, revolutionise and expand your original investment. The parable of the Talents from the New Testament is an excellent metaphor for the 32nd Gift. Here is a loose translation: A landowner gave three of his tenants ten, five and one talent each (a talent being an ancient unit of currency) and bade them make something of their investment. The first man (who had been given ten talents) returned with twenty talents, the second man (who had been given five talents) returned with ten talents and the third man (who had been given one talent) returned with only his one talent having buried it in the ground for fear of losing it. The landowner rewarded the first two men but took away the third man's single talent.

The lesson from the above parable concerns overcoming the fear of failure. The 32nd Gift is not about self-preservation, but is about the Preservation of Life. Only that which will adapt itself can survive and flourish. The 32nd Gift has the ability to assess the past, weeding out weaknesses and building upon strengths. These people have a natural understanding of the ebb and flow of the seasons and the rhythms of nature. It allows them to instinctively recognise when something is dying and to decide whether it needs to be pruned right back or disposed of altogether. Since the 42nd Gift (its programming partner) is that of Detachment, we can also see another of the strengths of this Gift — the ability to let go of whatever no longer serves the expanding vision of the 32nd Gift.

This 32nd Gift is the Gift of grafting, which is the essence of true Preservation. You have to retain that which is strong — the rootstock — and then you have to graft the new onto that which is strong. In this way, you will always maximise your energies. This Gift of grafting can be applied across any and every field of human endeavour. Detachment is also an essential aspect of this process since you will have to let go of your own notion of failure. It is usually the fear of failure that prevents human beings from adapting to a new system. The Gift of Preservation is mirrored everywhere you look in nature.

The more human beings align themselves with nature, the more successful we will become as a species. Success at the Gift level is about economy, and economy comes from being in harmony rather than in competition.

From a profound understanding of the Ring of Illusion, you can see the dilemma that humanity currently faces. We have created the modern world in the image of the 32nd Shadow. Our two great fears are death and failure. As we move into the 32nd Gift as a species, we shall once again return to nature. Nature represents the old rootstock. Its very wildness is its strength, and we, humanity, are the vibrant young shoot with its dreams of transcendence. As we learn once again to honour where we have come from, the earth will teach us how to move in harmony with its natural rhythms and cycles. As we listen to the great wisdom of our ancestors and of the indigenous tribal cultures, we will once again find our correct inner spirit. Once we have this as our anchor, we can graft our modern technologies onto the old wisdom and the result will be truly transcendent. This is the great secret of preservation.

As we listen to the great wisdom of our ancestors and of the indigenous tribal cultures, we will once again find our correct inner spirit.

The other great domain of the 32nd Gift is relationships. People with this Gift in their Hologenetic Profile have an instinct for who will make a good ally and who will not. They don't just see individuals — they see the inter-relationships between many different people as well. Thus they have a natural understanding of hierarchies and the continuity of those relationships within the hierarchy. Because of this they can be invaluable within any business or community. Given their gift of restraint, they can also sometimes appear at first glance to be quite conservative. However, each of the 64 Gifts is essentially a balancing act between the two extremes of the Shadow — hence people with this gift are neither too conservative nor too haphazard — they simply have the gift of knowing when to be one or the other. In this respect they really hold the future of the planet in their hands. If these people cannot overcome their fear of personal failure and selfishness we, humanity, are all at risk. However, if they rise above that fear and move beyond the selfish tendencies of the Shadow frequency, they can be the most vehement defenders and preservers of our earth.

THE 32ND SIDDHI – VENERATION

THE PERFUME OF CONSCIOUSNESS

To attain a siddhic state presupposes that fear has been entirely transcended. It has left the building altogether! Through the 32nd Gift we can see how a negative pattern such as fear or anger can be transformed into a beneficial force simply by raising the frequency of that Gene Key and pressing it into service of the whole. It is this very transformation that will eventually lead to the higher state of the 32nd Siddhi of Veneration. Each of the 64 Gene Keys is essentially a process in which fear or anger is progressively transformed by being put to good use in the service of the totality.

At a certain point, however, the ancient genetic fears can be transcended altogether. When you dedicate your life to the service of others this creates a progressive build up of frequency within your subtle bodies. The first obstacle that has to be overcome is the karma of your past. If your commitment is consistent enough (usually over many lives), eventually you will burn up all the karma accumulated in your ancestral DNA.

The power of service should never be underestimated. Service is the expression of love, and love as we all know can move mountains. The gradual transformation of ancient karma is like watching a saucepan of water slowly coming to a boil — for a considerable amount of time, nothing appears to be occurring. At a certain point, however, you sense a great pressure building within the saucepan, and you see the telltale signs that something big is about to happen — vapour begins to rise, a tangible heat begins to emanate, tiny bubbles prick the surface. When the explosion finally happens, it happens all at once and appears to be unstoppable. This describes the process of the dawning of the siddhis — the higher divine consciousness hiding in your DNA.

It is the same with human evolution. One day you reach an incarnation in which a great sense of something impending begins to surround you. Your life is littered with the signs and promises of a great and impossible sounding dream. You feel the pressure of this other reality more than you ever have before, and during these last stages your final tests are the most intense, as the most ancient vestiges of your karma are burned white-hot from your very DNA. When the supernova finally occurs in you, your identity and even your gifts are crushed as a new field of service emerges through you. This is the siddhic state. The physical vehicle and its chemical genetic coding remain to determine (and limit) the expression of your particular Divine archetype, but the sheer voltage of the light frequencies that has ended your sense of separation has literally washed your DNA clean of its memory. Only then do your true siddhis emerge.

The 32nd Siddhi is the state of Veneration. It is all that remains when fear has burned away. Veneration is the state that occurs when you see and know your place within the great living chain of life. You can look down the spiral of evolution and see those less evolved than you, and you can look up the ladder and see those far beyond you. Because you look at your place without a sense of personal identity, all you can do is experience wonder. You see the interdependence of all life, and you know the One Force that motivates and moves all life, including your own tiny form. Veneration is about feeling both tiny and huge at the same time!

In the siddhic state, the ring-pass-not of death has been breached so that you have a real experience of immortality (the 28th Siddhi), since the One Light is now entirely focused through your spirit. All you see is the beautiful spiralling arc of evolution through the myriad chains of reality. At this level, you are way beyond the personality that identifies with its individual vehicle. You are even beyond the soul or causal body, which continuously incarnates through multiple vehicles throughout time. The crystalline surface of the soul itself has shattered, which means that the consciousness within your genetic chain has transcended or outgrown its vehicle. There is no ounce of separateness left, so there is nothing to reincarnate anymore. For you, the entire notion of evolution has now come to an end. This great mystery is described in more depth in the 22nd Gene Key.

Every siddhic state leaves a Divine message before it leaves this plane of being and returns to the formless state. This message is rather like the autobiography of consciousness through all its lives and experiences in that particular chain of life. Thus the great sages each leave us something unique that builds on the wisdom of previous sages. Those who have attained the Siddhi of Veneration have always incarnated in some form of spiritual lineage. Consciousness in form always has to move through the limitation of a lineage. This lineage is an archetypal lineage rather than a uniquely genetic or social one. In the language of the Gene Keys, it is called your fractal lineage or fractal line. The fractal lineage of Christ for example, has nothing to with the religion of Christianity — it has to do with vibration. Thus by way of example, the Indian saint Ramana Maharshi is a descendent of the Christ lineage even though he knew little of Jesus' life or teachings. The great lineages of the Tibetan culture are examples of fractal lines that are both vibrational (the Buddha lineage) as well as social and genetic, since these lineages were for many centuries deliberately confined to a single culture. Now with the spreading of these powerful teachings, many evolved lamas and tulkus are incarnating into western bodies — even so, the fractal lineage remains.

Veneration is about being on a spiral ladder. You stand on the shoulders of those who have come before you, and you allow those who will go beyond you to stand upon your shoulders in turn. This is how higher consciousness extends and expands itself into humanity. What you venerate is the chain itself, as well as all the beings above and below you. You realise that even the humblest insect is allowing you to stand upon the shoulders of its consciousness. This gives rise to a fathomless feeling of reverence for all sentient life. Veneration is rooted in respect, reverence and gratitude for all aspects and forms of the One Consciousness behind creation. It is an intense aroma or perfume that is given off by consciousness itself. For the one experiencing this Siddhi, this perfume is detected everywhere, in everything, all the time.

The great secret to this 32nd Siddhi is to be found in the simplest of symbols — water. In human physiology, the 32nd Gene Key represents the regulation of liquid in your body. This is why it has such a deep connection with your genetic memory, since water holds memory. There is a peculiarity within the hydrogen atom that allows it to transfer consciousness as memory. Since our planet and our bodies are made up primarily of water, this is the medium through which our collective consciousness evolves. The water cycle on our planet is actually how we are awakening. Every form that dies releases its water content back into the water cycle, which means that every form releases a finite number of more evolved hydrogen atoms back into the world. Thus, the hydrogen atoms in every vegetable you eat pass through your body and leave through your sweat or urine more evolved than when they entered your body. This evolving chain of consciousness is present and occurring in every single life form on our planet through the medium of the water in the food chain. The key to evolution is digestion!

The more you contemplate this 32nd Siddhi, the more reverence you will inevitably feel for all life. As it penetrates deeper and deeper into your heart, you will begin to smell the divine perfume of consciousness as it moves between and behind all forms of life. Eventually, no matter what is before you, all you will see is the Self, the Atman, the Divine being, perfectly playing and weaving and evolving right before your very eyes.

The more you contemplate this 32nd Siddhi, the more reverence you will inevitably feel for all life.

ne Key

SIDDHI
REVELATION

GIFT
MINDFULNESS

SHADOW
FORGETTING

ogy: Throat/Thyroid

cid: None (Terminator Codon)

TTING

n this planet, there is only one thing they bring with
out ordinary memories as we understand them. There
cases, we do not retain memory of our past lives or of
we do not recall *the before*, we still carry the memory
e are cultures upon this earth that have devoted entire
ocess of incarnation — of how we enter and leave the
oists, to name a few, have left us reams of knowledge
d certainly the most ancient of these cultures is the
das.

buted to the *Rishis* — great teachers who attained liberation many
ago and left us a set of profound maps and teachings that would assist us in attaining liberation and,
in time, escape the so-called *maya* or world of illusion. One of the foundation stones of the Vedic tradition
is the notion of karma — the idea that every act, thought or intention carries a charge, which in time creates
a domino effect that returns to us at a later date. Most of us are familiar with this principle. Perhaps not so
well known, however, is the doctrine of the *sanskaras*. The sanskaras are the specific memories that you carry
from life to life according to the karma you have taken on in this lifetime. These are actually far more than
simple memories — they are charges of kinetic energy stored in the sheaths of your consciousness, which
over time determine the shape of your life and your destiny. Sanskaras give rise to all human desires, which
in turn create more sanskaras. Thus the ancients say you are caught in a wheel or a net of your own making
— paradoxically unable to remember who you truly are because of the memories you keep creating.

The 33rd Shadow is the great shadow of our forgetting. As we see with the 12th Shadow and the 56th
Shadow, this is one of the three great tests we have to face in our evolution on this planet.

There is no situation on earth that cannot be used as a means to raise your frequency and open your heart to your inner Divinity.

The 33rd, the 12th and the 56th Shadows are all related genetically to the three stop codons in the human genome. Since it is your memories that keep you in the state of forgetting, your greatest challenge in life is to see through your own illusion and thus slip through the net itself. How then, do we ever escape this mad paradoxical hall of mirrors when every desire we have serves only to tighten the net around us? Well, there is one desire that is the exception — the desire to remember yourself. This is the desire that the 33rd Shadow conceals. The desire to escape the maya begins your process of awakening and for the first time begins to unwind your sanskaras instead of feeding them.

Your sanskaras are truly wound opportunities. They are like karmic wormholes that pull you toward certain people and push you away from others. The Programming Partner to the 33rd Shadow is the 19th Shadow of Co-dependence. Thus, your relationships offer you the greatest opportunities to *unwind the wound*. The very word *wound* clearly indicates how we are bound by something that is literally wound around our genetic code. These deepest of memories or sanskaras are what give rise to your most challenging relationships in life. Co-dependence is about being energetically wound around someone else in a way that feels deeply uncomfortable and destructive. Even so, it is just such relationships that offer you the most direct way to pierce the veil of your forgetting. It is these difficult relationships in your life that eventually drive you to question yourself, your love and the underlying reason for your life.

To unlock the hidden secrets of your sanskaras, you must begin to fully face your own life and the pain that lies inside the various layers of your being. The three stop codons are genetic markers inside your physical DNA that, when activated, begin to unplug you from the illusory world of the outer senses. This codon ring is called the Ring of Trials for good reason. Every human awakening must arouse the hero or heroine inside you, and that inner being must face its mythical trials. It is only the intensity of your own dissatisfaction that has the capacity to jog your awareness out of the sleeping field of the maya. It all begins here with the 33rd Shadow, which seals the mass consciousness deep within the matrix of the senses. Despite many popular beliefs, the specific karma that you carry in your physical body has no direct connection to your past actions or your past lives. Your sanskaras are part of a collective energy field that is cleansed and reset each time you take an incarnation. To understand this mystery in depth, you can explore the 22nd Gene Key and its great mystical theme of redemption.

To really understand how your sanskaras work you have to contemplate your relationships in some depth. There are always certain people in your life to whom you are so strongly drawn that you sometimes feel you remember them from somewhere. This feeling of cellular memory between people is a sign of a karmic bond, and all karmic bonds are formed from sanskaras. Such relationships are always intense and can be very challenging. They are love/hate relationships. When you enter deeply into such a relationship and stay committed to its process you are courting the presence of grace. To accept the trial is to transform the co-dependent pattern of the relationship into a higher frequency and this takes great love and surrender. There is no situation on earth that cannot be used as a means to raise your frequency and open your heart to your inner Divinity.

The 33rd Shadow therefore governs all the endless cycles of existence and incarnation on this planet. It keeps you deeply asleep by concealing the past from you. Around the energetic and etheric boundaries of our planet lies a great veil called the *ring-pass-not*. The ring-pass-not is an energetic grid that connects the higher planes to the lower planes. Until the frequency passing through your DNA rises above the level of this great atmospheric veil, your true universal and eternal nature cannot be revealed to you. If you were to remember that you have lived this same old story in different nuances a billion times over, and that it still causes you the same suffering, you would immediately snap out of your age-old human patterns and transcend that suffering. The 33rd Shadow keeps us hidden from this greater life — it keeps you in retreat here on this planet of suffering. It keeps humanity locked into these material forms until such time as your consciousness begins to spontaneously awaken of its own accord. The 33rd Shadow ensures that we remain in the illusion that we are separate and alone, instead of being unified and *all one*.

REPRESSIVE NATURE – RESERVED

These are extremely hidden people. They find it very difficult to communicate with others and often cannot break out of their own silence. They may have the experience of being forgotten by the world. Their natural tendency is to simply hide from people and they usually find all kinds of ways to do this — in their work, through an enforced discipline or life choice, by living in isolation or far from others, or simply through a deep psychological insecurity that keeps them in a kind of torpor or numbness. These people find it very difficult to stay in relationships since intimacy demands that they come out of their shell. When they do finally begin to awaken, they can surprise both themselves and others with how much wisdom they have absorbed over the course of their life/lives.

REACTIVE NATURE – CENSORIOUS

Rooted in anger, the reactive side of the 33rd Shadow will express itself as an invasive behaviour pattern that unconsciously tries to provoke reactions in others in order not to feel alone. The nature of this pattern is censorship. These people can be adept at understanding the emotional locks that keep others in a state of forgetting. Thus they try to get attention by pointing out the negative patterns of others, sometimes under the guise of helping them. This almost always achieves the desired reaction — anger — from the other person. However, this way of getting attention is obviously extremely self-destructive and unsatisfying. It leads to a build up in resentment and creates more anger until an inevitable explosion ensues. When these people learn to break their pattern of projecting their rage onto others, they finally begin to remember what it is like to love and be loved.

THE 33ʳᵈ GIFT – MINDFULNESS

THE DEATH OF THE EGO

When you raise the frequency of the 33ʳᵈ Shadow to the 33ʳᵈ Gift, it becomes something rather magical — a quality called Mindfulness. Mindfulness is a word often used in Buddhism where it is understood as one of the great attributes of meditation. To be mindful means to be attentive, but it also suggests more than this: the person who displays the Gift of Mindfulness is someone who has loosened the grip of both extremes of the 33ʳᵈ Shadow — they are no longer held captive by their own unconscious desires, fears and reactions. At the level of Mindfulness, they may still hide or react, but now they can *see* themselves doing it. Once the process of self-remembering has begun, in a certain sense you enter a kind of purgatory. At least when you were asleep in the Shadow frequency, you were ignorant of your sanskaras and the cause and extent of your own suffering. Now, however, you begin to see the very patterns that cause your own suffering, and the more mindful you become, the more pain and futility you encounter. After a certain time however, mindfulness begins to transform the kinetic energy within your own sanskaras and a new view of life opens up before you.

It is also through mindfulness that you discover how to refine and purify your nature so that you no longer create negative karma.

It is also through mindfulness that you discover how to refine and purify your nature so that you no longer create negative karma. This process of purification is an integral part of the loosening of your sanskaras and it can be hugely accelerated through working within your relationships. Through the Ring of Trials you learn how to discriminate (the 12ᵗʰ Gift) between that which keeps you trapped and that which brings you freedom, and you do this by moderating your thoughts, feelings, speech and actions. Similarly, through the other allied stop codon, the 56ᵗʰ Gift of Enrichment, you see how easily you are distracted away from your true nature by your senses. Thus you begin to be drawn towards experiences that enrich your spirit rather than experiences and addictions that simply stimulate or over-stimulate your senses. Mindfulness as such is a very enriching process. The karma inside your very DNA is systematically transformed back into pure essence. This essence in turn is so wonderful that you find yourself wanting to experience more of it. You have found the secret of all human desire — once it is purified into its natural state of Divine longing, it becomes the very fuel that returns you to your true centre of being.

In Buddhism, there is a well known meditation technique called *vipassana*. Although it can be translated in many ways, it points towards a witnessing self-awareness. In one sense, every one of the 64 Gifts is a type of vipassana — a self-witnessing that is a prelude to the shaking off of the negative patterns of your own Shadows. True transcendence only occurs at the level of the Siddhi, but the bridge to these higher levels is the myriad of techniques and themes of the 64 Gifts. At the Gift frequency, technique is still extremely valid which is why meditation or contemplation is so helpful. For most people, it is the Gift that is the most accessible route to the higher dimensions. In fact, you might say that the Shadow is the seed, the Gift is the flower and the Siddhi is the fruit. One step automatically leads to the next step. Mindfulness is not really a technique, although it can be in the beginning. Mindfulness is the organic flowering of self-awareness or self-remembering. As you awaken within the maya, the Gift of Mindfulness springs up from within you as a natural aspect of your nature. As soon as you begin to witness your karmic patterns in action, your sanskaras begin to unwind.

To be mindful means to catch yourself in the middle of a heated argument and to realise that you have forgotten your true nature — in effect, it is to see your own victimhood. As you begin to see through these unconscious tendencies that have driven your nature for so long, a new awareness develops within you. It is important to understand that, at the frequency level of the Gifts, you are not yet free from your patterns but you *are* engaged in a process that is progressively weakening them. It is at this level that another interesting phenomenon occurs — the closer you come to the siddhic quantum leap, the more challenging and intense your patterns appear to become. They are not, in fact, becoming any harder, but you are becoming more and more aware that you are asleep, which in turn makes you feel more and more uncomfortable. However, since you have also learned by now that you are unable to *do* anything to speed up your awakening, you start to witness your illusion beginning to crack around the edges. This is what the mystics refer to as the death of the Ego. Your mindfulness is the only thing that you can rely upon at this stage — and eventually, even that too will have to dissolve.

Interestingly enough, mindfulness is the prerequisite for genius. The Gift level is the level of your genius. When you can objectively view your thoughts, your feelings, your passions and desires — your own subjectivity — only then can you create art from them. If you are a scientific genius, you need to be able to think with your feelings, or feel with your thoughts. This is how genius emerges into the world — it is witnessed. This is also why many geniuses are uninterested in recognition — they know they did not create their genius. It only happens when you are not there! Genius is a state of consciousness — a certain stage in the raising of our frequency. True mindfulness occurs when you become aware that something else is beginning to look through your eyes, think through your mind and live through your actions. This *something else* — whether we call it genius, meditation or prayer — is the larger reality remembering itself through us.

THE 33RD SIDDHI – REVELATION

DIVINE CLOSURE

The three great stages of human awakening are captured here within the Ring of Trials and the three stop codons in your DNA. The first stage, represented by this 33rd Shadow of Forgetting, is the realisation that you are in hell, or in the words of the Buddha's First Noble Truth — all life is suffering. The second stage, represented by the 56th Shadow of Distraction, is the process through which your suffering is gradually transcended. The third and final stage, represented by the 12th Shadow of Vanity, is the confrontation with the true nature of Self. As you rise into higher and purer frequencies, the secrets of the higher realms are released sporadically into your lower nature and, over time, your physical DNA begins to mutate to accommodate these higher frequency wavelengths. Once your DNA has mutated to a sufficient degree to handle these higher frequencies permanently, it is ready to receive the full force of the siddhic realms.

When the 33rd Siddhi dawns, all that you were is forgotten and all that you could be is remembered. The fullness of the siddhic state causes time to disintegrate and memory along with it. It is not that you lose your physical memory. Rather your mind can no longer interfere with the purity of consciousness, which flushes the sanskaras out of your whole being. With no direct connection to the past or future, there is now only infinite time — a concept that cannot possibly be grasped in words. Revelation is a flood. Since the beginning of human culture there have been beliefs and myths that once, long ago, a civilisation existed — a golden age — in which utter harmony ruled and there was peace upon earth. Those same myths also relate that, when this age came to an end, humanity forgot how to love and be in harmony with life. Thus our existence was erased by a great deluge or flood that swept the world clean, leaving but a handful of survivors.

Mythology holds many hidden secrets. Above all else, such myths conceal the hidden codes of life in pictorial form. Every myth has arisen directly out of the human genetic code, which contains all mysteries in its primal links. This is why the great sages have guided us to look within ourselves for the kingdom of heaven. The myth of the flood or deluge, whether partially based on historical fact or not, is a deep symbol within the human psyche — the symbol of Revelation. After the flood, there is always a sign — a dove, a rainbow — the symbol of a new world. The flood itself is a memory of our own future. Its existence testifies that we will each one day be washed clean by the overwhelming tsunami of consciousness — and that your existence as a separate individualised creature will one day end. Thus Revelation brings an end to endings. It ends all myths. It ends evolution and it ends humanity. When Revelation comes, the universe itself will disappear. You cannot really say that a new world will appear out of the old one because time itself will dissolve. Therefore all words, thoughts and symbols will also disappear, since they depend upon time for their existence.

True remembering is a kind of detonation inside you — a flash flood that wipes out your past and reveals the greatest mystery of all — the mystery of the eternal Now.

Out of Revelation comes true silence. The beings manifesting this Siddhi have nothing whatsoever to say — and if they do speak, it is to say that there is nothing whatsoever to say! (There are many amusing ways to say this). Of course, this Siddhi is also about remembering. It is literally as the word suggests — re-membering — to enter into all the separate members of your greater body until you realise there is only one of everything. True remembering is a kind of detonation inside you — a flash flood that wipes out your past and reveals the greatest mystery of all — the mystery of the eternal Now.

Apart from the obvious enlightenment that comes with every siddhic state of consciousness, there are other natural manifestations of this Siddhi. The one who is swept away by the 33rd Siddhi also brings a sense of closure to the world. The presence of this Siddhi in the world signifies the end of an era or epoch. True revelations always emerge at natural end-times in human history and they often leave behind strange sounding prophecies about the next age.

This Siddhi is also known as a great revealer of hidden secrets. It releases unique secrets through its vehicles into the collective consciousness. Because of its unusual status in this regard it can also be said to have an archetypal connection to evolutions beyond human consciousness, such as the angelic kingdoms. Every stop codon in your DNA precedes a start codon of another strand of interlinked code. Even though we may talk of evolution coming to an end, there are kingdoms that exist beyond our concept of evolution. The 33rd Siddhi is an access portal to these higher and wider realities that begin where you, the human being, come to an end.

Whenever we discuss the siddhic consciousness we find ourselves in a world riddled with paradox. From the point of view of the ultimate, there is no world, no evolution, and no point of view. However, the human vehicles that awaken to the siddhic level always release something unique into the collective consciousness of humanity before they die — and this something always contributes to the evolution of human consciousness. In other words, even though the story has ended for such beings, their awakening fits into and contributes to the story of humanity. They too are players in the great drama, and they must play their roles before they leave, even though they know there is no one to leave and nowhere to go. This is the testament of the 64 Siddhis — each Siddhi is still a limited expression of the infinite.

The 33rd Siddhi actually reveals the individual and collective mythology of enlightenment. Enlightenment itself evolves. Even though the state remains absolute, the means by which humanity awakens as a totality has its own storyline within the maya. By opening us up to higher realities and angelic frequencies that lie beyond our physical DNA, the 33rd Siddhi accelerates the evolution of human consciousness. This is such a paradox because the siddhic consciousness knows that higher consciousness and even notions such as frequency are themselves illusion! Human DNA *does* have miraculous qualities hidden inside that are activated by the siddhic frequencies, but it also has its limitations. Other vehicles exist beyond the binary code of DNA and they exist in higher dimensions, in what we call the subtle bodies.

Thus the great cosmic drama unfolds, and every individual moves through progressive revelations before arriving at the ultimate Revelation. The consciousness within form always has a storyline to follow. The trick is to fall in love with your own story and follow it without holding anything back. Two things are then assured — firstly, you *will* arrive at the story's end, and secondly, your own story will be utterly unique and unlike anyone else's. This is why it can be so distracting for you to hear someone else's story about how they attained an enlightened state, or what disciplines or teachings they followed in order to reach that state. The truth is that nothing you do or don't do changes when and how you reach the ultimate. You simply have to have faith in your own storyline. This is also why it is so rare for humans to attain these states — there is no one to follow, the path is virgin and wild and when your revelation finally dawns, it does so without your even being there!

≣ 34th Gene Key

The Beauty of the Beast

SIDDHI
MAJESTY

GIFT
STRENGTH

SHADOW
FORCE

Programming Partner: 20th Gene key

Codon Ring: The Ring of Destiny
(34, 43)

Physiology: Sacral Plexus

Amino Acid: Asparagine

THE 34ᵀᴴ SHADOW – FORCE

THE BANE OF TRYING

The 34th Shadow concerns the notion of individual human power. It represents an ancient part of our genetic matrix that is primarily based on individual survival — the survival of the fittest. This primal power source has its roots deep within our genetic past and began when the very first plants appeared on earth. It was also strongly apparent during the reptilian phase of evolution on our planet. The reign of the dinosaurs during the Mesozoic Era is an archetype of the power lying within the 34th Gene Key. In human evolution, it was this power determined to survive that literally forced the spines of our early hominid ancestors into progressively more upright stances. This is the force that made us different from all other mammals, because the moment we were upright our brains began to develop differently.

Despite our burgeoning intelligence, the 34th Shadow still lies within us as the urge to use force to make things happen in life, and it can be highly destructive when influenced by low frequency vibration. The 34th Shadow carries a primal quality about it — it is not even animal, but is far more ancient. It is simply a pure evolutionary force whose prime directive is survival, and its only prerogative is to maintain life within a particular organism. You would not even call it selfishness since that implies an awareness that others exist. This Shadow creates a kind of intense self-absorption that, when applied to a modern human, leads to brute power without self-awareness. This 34th Shadow therefore gives rise to a basic human law — at low frequencies, all human beings are destructive to the collective. The nature of modern society means that DNA, when behaving according to these archaic rules, inevitably destroys itself.

You can perhaps see how the ancient intelligence in this 34th Shadow was once absolutely necessary in order for humans to survive and evolve beyond other forms of life — especially other mammals. Through trial and error, the force within this Gene Key taught our bodies how to outwit other species. Yet in today's modern world this ferocious competitive power is actually the greatest threat to our collective survival, and not just to our own survival but to that of the entire planet. The programming partner of the 34th Shadow is the 20th Shadow of Superficiality whose repressive aspect is Absence, which denotes an absolute lack of what we call humaneness. When the 34th Shadow acts, it actually has no sense of what it is doing. It just acts without thinking or caring. Awareness may well creep in after the activity has finished, but during the activity there is only pure mechanised absorption.

In terms of modern human beings and their behaviour, the 34th Shadow can be best understood through the notion of *trying*. Trying implies forcing something to go a certain way when it won't go there easily on its own. Each time someone is caught in a space of trying to make something happen against the flow, they are under the influence of this Shadow frequency. Such people seem unable to stop moving in the direction in which they have set out, and any attempt at help or guidance from the outside falls on deaf ears. As an aspect of the codon Ring of Destiny, the 34th Shadow is linked chemically with the 43rd Shadow of Deafness. Under the influence of this Shadow you become totally lost to the force that is moving through you, even though it may be harmful to you or others. When operating at a low frequency, this 34th archetype is unavailable to outside influence. Consequently you can see how, at its extremes, this energy can give rise to the most horrendous sub-human acts.

If you have this 34th Gene Key as a major aspect of your Hologenetic Profile you will tend to be confronted by its Shadow aspect through the responses of others. This Shadow tends to infuriate other people since they cannot understand how you could be so blind to outside influence. It often seems to others as though you are behaving in an inhumane or foolish manner. This frequently results in some kind of challenge or complaint from the other person, which may take you by surprise. If you cannot maintain your frequency at a certain level, this kind of interference from others can create unwanted volatile situations in your life. If you are trying or forcing something, you will continually meet resistance from others. The 34th Gene Key is unaware of itself even at the highest levels so it is not a matter of changing yourself. It's simply about doing the right activity at the right time.

In a child the 34th Gene Key can be seen in its purest aspect where it manifests as an endless stream of activity without any awareness of boundaries or of anyone outside them. Obviously children need to learn the importance of boundaries, but it is the 34th Shadow that is often unconsciously imprinted in your character as a child. Children with the 34th Gene Key in their Hologenetic Profile need a great deal of space and freedom as well as proper boundaries, although they will tend to find their own boundaries as they go along. Such children cannot be compared with *normal* children and need to be allowed to develop naturally. They have within them the ability to learn to distinguish between force and strength, and they will discover this in their own way without excessive help or conditioning from outside. As we shall see at the higher frequencies, there are enormous benefits and gifts hiding in this Gene Key. No matter who you are or what your Hologenetic Profile says, every human being has the capacity to tap into this aspect of our DNA and discover an inexhaustible reserve of inner strength.

We can see from where this Gene Key relates physiologically that strength is to be found deep within the sacral plexus. It emerges from the area below and around your navel. This area of the body has long been known as a centre of great power in human beings, and its real power lies in the fact that there is no obvious awareness located there. However, between the sacral

*At low
frequencies,
all human
beings are
destructive to
the collective.*

and solar plexus there is a huge and complex neural network. Despite our modern emphasis on the primacy of the brain as our main centre of intelligence, the body's real nexus of intelligence is in fact centred in the belly. The 34th Shadow bypasses this belly-centred intelligence and tends to lead you into unconscious mental loops within the brain. Force is always rooted in the mind, whereas power comes from the belly. Power is natural, grounded and universally connected to all life, flowing as it does from the great umbilical centre within your being. The awareness within the belly is beyond self-awareness because it is beyond the self. As such it can be called pure awareness. Force happens when you forget to trust in this true centre of pure awareness.

REPRESSIVE NATURE – SELF-EFFACING

In those with a repressive nature, the 34th Shadow essentially hides from its own power. It is afraid of itself. These are people who put themselves down and accept other people as their authorities. The self-effacing nature is about letting others walk all over you. Such people really become slaves to other people's principles or to society in general. All the magnificent individual power within this archetype is held back, often because of a difficult childhood. Thus all the energy within these people becomes backed up, leading to tiredness and lack of energy rather than to the natural expression of the 34th Gift, which is tireless energy and strength. Ultimately, these are people who will have to free themselves from the situations in which they find themselves. As soon as they do this, they will find that their true strength returns.

REACTIVE NATURE – BULLISH

The reactive, angry side of this Shadow manifests as a bullish or bossy nature. These are people who use force to bully others around. Their main problem is communication and inappropriate behaviour. Because they are generally unaware that they are being so bullish, they become furious when challenged. Such people will continue to be bull-headed even after they are warned or have been challenged, and this usually ends badly for them. Even though they are often perceived as life's bullies, they are deeply misunderstood. They have been conditioned to behave in this way and are unaware of what they are really doing. What they need is to find a healthy activity into which they can channel their anger rather than constantly aggravating others.

THE 34TH GIFT – STRENGTH

THE OLYMPIANS

When the 34th Shadow gives way to the 34th Gift, a truly beautiful archetype emerges — human strength. What is fascinating about the connection between the Shadow and the Gift is the incredibly fine line between these two activations of the same essential energy code. The difference between the Shadow of Force and the Gift of Strength is both miniscule and enormous. The 34th Gift employs the same terrific primal vitality, but unlike the Shadow it does so through appropriate activities and correct timing. The result is a creative manifestation of aligned life force that always catches the attention and praise of others.

Individual strength is something absolutely natural to all human beings. Obviously we are not talking about physical strength here, although people with this gift often can be physically strong. We are talking about the ability to act in harmony with natural forces — the real definition of strength. When applied to physical activity, which is what this 34th Gift is all about, strength has no element of trying or forcefulness within it. It simply flows out of you and you become effortlessly at one with the activity. Effortlessness in this context does not mean that there is no exertion — there may be great exertion — but there is no resistance. This fluid efficiency is one of the main hallmarks of the 34th Gift.

Another key aspect of the 34th Gift is display. We already know that this Gene Key has ancient connections with the early cycles of life on the planet and its oldest connection is with the plant kingdom. Plants depend upon insects, birds and bees to generate and proliferate. In order to attract these other agencies, flowers bloom into all manner of beautiful shapes, colours and aromas. In humans, the 34th Gift shares this primal need for display, although not through any kind of ego. We need to remember that the 34th Gift has no awareness of itself at any level. Obviously, the tendency to draw attention draws negative attention at the Shadow frequency, but at higher frequencies it draws positive attention. It is out of the 34th Gift that all kinds of examples of human strength are born.

The 34th Gift is the gift of heroes and heroines. It is one of the greatest and oldest of human archetypes. True heroism occurs without awareness and is a wholly individual act. This is why every human being alive feels deep resonance to the heroes of myth or the heroes of contemporary culture. Heroism denotes strength. Ironically, though, true strength is quite unaware of itself. Many people who have committed incredible acts of bravery have afterwards described the experience as devoid of any intention on their part. This kind of heroism is entirely accidental (unlike the highest frequency of the 21st Siddhi, which is Valour, an entirely different archetype). Often heroes feel uncomfortable about being praised, because it seems to them that the whole thing was out of their hands. However, even if it is accidental, the heroic act is never interpreted as such by others. It is usually set upon a pedestal and glorified.

Being such a physical archetype, the 34th Gift has a deep connection to activities involving physical movement such as sport or dance. But this gift is not so much about team sports; rather it presents as individual flair. These are people whose very movements have a hypnotic quality evoking admiration from others. They are our sports heroes and Olympians who seem to convey and radiate an individual power and poise beyond the norm. Mastery of human movement actually emerges only when awareness has withdrawn, leaving simply an intense Self-Assurance, the programming partner of the 20th Gift. The 34th Gift can be demonstrated in myriad fields of human endeavour, that is, in any area in which an individual rises above the common man. These people become our icons and gurus in business, sport, war and sometimes government. There is a kind of primal power at work within such people, and it is impossible for others to miss it. In a world where the lower frequencies dominate, true inner strength never goes unnoticed.

Along with the 43rd Gene Key, the 34th Gene Key forms the Ring of Destiny — an unusual genetic configuration that has a huge effect on your external life or destiny. When people speak

In a world where the lower frequencies dominate, true inner strength never goes unnoticed.

of destiny they often refer to a force that lies outside the reach of humanity. Yet the secret to destiny has to do with frequency. It is through this codon that the two great forces of evolution and involution come together. There is a great mystery here. Do you raise the frequency of your own DNA and change your destiny? That is the view of evolution. Or does a higher force outside you make it possible for you to raise the frequency of your DNA? That is the view of involution. The paradox can only be solved through another paradox. Both are true and both are interdependent. Heroes and heroines are made both on earth and in heaven.

THE 34TH SIDDHI – MAJESTY

WHERE EPIPHANY MEETS MAJESTY

As mentioned, the 34th Gene Key has an unusual attribute that remains at all its levels of frequency — that of being unaware. Force is unaware of itself and leads to destruction; strength is unaware of itself and leads to admiration. Here at its peak, the Siddhi of Majesty blossoms. This is the majesty of the human form in motion, and the form itself *is* pure awareness, which is why it cannot be awareness *of* something. It quite simply reveals the true beauty of consciousness in form. Majesty is not really a state of being — it is the truth of all nature in motion. At the siddhic level of being, everything alive displays majesty. Even the struggles of the low frequencies are revealed as majestic in their own distorted ways. But this 34th Siddhi in actuality concerns humanity more than any other aspect in human DNA — it is the Siddhi of the naked ape, of Adam and Eve, of the Divine energy that constantly moves through the human form.

It is through the 34th Siddhi that we get the idea of man as a God. This notion has been immortalised in our myths in which the Gods appear on earth and take on physical shapes, or through man's attempts to convey Divine authority on certain chosen people throughout history. The great paradox is that if Gods *were* to assume the shape of humans, they would never know they were Gods! This is the true meaning of this 34th Siddhi: the Divine essence of creation can only move freely through the body when the identity has ceased to exist. Such people move in a way that is bewitching to the eye. Out of this Siddhi have arisen the physical practices that aim at higher consciousness; for example, the mudras and asanas of yoga, or the movements of Tai Chi, all of which are spontaneous expressions of Divinity moving through the form of someone in whom the 34th Siddhi once dawned.

These sacred movements, gestures and dances contain the codes of higher consciousness. However, such practises can be great tricks that keep you thinking that a path to higher consciousness exists. There is no such path. Higher consciousness is simply a sudden leap. The sacred movements can indeed gain humans access to higher states of consciousness, but they do not lead to the leap. The problem of the sacred movements is that they were originally spontaneous expressions of higher consciousness. When they are copied, they are no longer spontaneous, even though they may lead to glimpses of higher consciousness. None of this is to say that sacred movements and practises are worthless. They are a natural stage for many people on the path towards transcendence. One just has to realise their origins.

Out of the 34th Siddhi also comes the highest level of the martial arts with their notions that true strength lies in emptiness. The highest art of Chi Kung for example is known as *empty force,* in which the practitioner has dissolved all semblance of identity and has become a conduit for the intelligence of life itself. This is also where we get the concept of *no mind* in the highest art forms of Zen Buddhism. If this Siddhi dawns in a person, they do not communicate through words or language but through their actions and movements. Many of these people were the originators of the sacred arts such as calligraphy, music, dance and art. A work of art done by someone with the 34th Siddhi is always regarded as a work of unparalled genius and originality. Nevertheless, for the siddhic artist, the art itself is in the doing, whereas the results are meaningless. This truth is beautifully symbolised by Tibetan sand art, where incredibly intricate mandalas are created over months and, when finished, are left in the open wind to be blown away in a matter of hours.

The 34th Siddhi contains another great secret concerning awakening. Through its genetic connection to its programming partner — the 20th Siddhi of Presence — the 34th Siddhi requires that an individual transcend their genetic fear of survival by bringing their presence fully into every moment. The challenge for the individual is to let go of his or her individuality which feels like their very essence. Indeed, your survival appears threatened by dissolution into the ocean of Presence. In truth, this is the whole point — you cannot survive the leap into the siddhic level! The final surrender required of you before dying into the siddhic frequency is the surrender of your physical body and the vibration of its movement. Once you have entered the siddhic frequency, all fear is purged from your system and the pure awareness of your body's physical intelligence is revealed. It is as though the entire universe were moving through your body.

The 34th Siddhi demonstrates one of the key attributes of life itself — that of efficiency. In the Ring of Destiny the epiphany of the higher planes encounters and infuses the majesty of the body on the lower planes. The result is pure fusion as spirit enters matter and imbues it with divinity. When life is not interfered with in any way, it becomes highly fluid and efficient. The more you try to force life to flow where you would like it to go, as opposed to where *it* wants to go, the less efficient you become and the more energy you use. The most commonly used analogy to this aspect of the 34th Siddhi is the comparison of the life force to water. To echo the words of the great sage Lao Tzu:

"The very softest thing of all
can ride like a galloping horse
through the hardest of things.

Like water, like water penetrating rock.
And so the invisible enters in.

That is why I know it is wise
to act by doing nothing.
And how few, how very few understand this!"

If this Siddhi dawns in a person, they do not communicate through words or language but through their actions and movements.

䷢ 35th Gene Key

Wormholes and Miracles

SIDDHI
BOUNDLESSNESS

GIFT
ADVENTURE

SHADOW
HUNGER

Programming Partner: 5th Gene Key

Codon Ring: The Ring of Miracles
        (35)

Physiology: Thyroid/Parathyroid

Amino Acid: Tryptophan

THE 35ᵗʰ SHADOW – HUNGER

THE HUNGER OF THE SPECIES

There is a perpetual and innate hunger built into all human beings. This hunger operates on many different levels within your being and is caused by the chemistry generated by the 35ᵗʰ Shadow as it is refracted through your DNA. The classic translation for this 35ᵗʰ hexagram of the I Ching is the word *Progress*, which is very apt since the 35ᵗʰ Gene Key drives all human progress. At the Shadow frequency progress is expressed as outer evolutionary progress, embodied for example in humanity's recent technological revolution. True progress, however, has little to do with physical manifestations and everything to do with the progress of human awareness. In a nutshell, what the 35ᵗʰ Shadow does is divert true progress out into the world of form, thus sacrificing its potential to transform the inner structure of humanity itself. Thus, because of this Shadow, the outer world evolves at the expense of the inner.

Through its associated amino acid tryptophan, the 35ᵗʰ Gene Key is connected to the secretion of serotonin within the body. Serotonin is well known as a chemical that induces states of satiation and deep fulfilment. As a result of the interference from the Shadow frequency passing through this codon, serotonin production within the body is inhibited, leaving human beings with a perpetual feeling of hunger. This deep unrest within the human body drives human beings out into the world to try to find something that can bring an end to what is essentially genetic hunger. This hunger drives all human experience, from drugs, food and sex (all of which may increase serotonin levels briefly) to business, religion, science and even war. However, no matter how you try to satiate your hunger in the outer world, it will never be enough. Our fate is sealed by this 35ᵗʰ Shadow because no external crutch or method can ever replace the natural balanced chemistry of your own body.

The 5ᵗʰ Shadow of Impatience is the programming partner to this 35ᵗʰ Shadow and it operates alongside hunger, pulling you away from your inner natural rhythm with life.

It is your very impatience that fuels your hunger, and your hunger that fuels your impatience — a classic biofeedback loop that keeps human beings operating at the Shadow frequency.

All external human progress, particularly in modern times, has come about because of this 35th Shadow. We are a hungry and unfulfilled species — we don't know what we want, but we want it now. It is important to understand that this endless wave of hunger in human beings is not an individual programming function of your genetics. It is universal to our whole species, driving us outwards to explore and conquer the world around us without a thought to the consequences. This is what humanity is right now. When you are truly hungry, you cannot think about the consequences of eating — you are simply possessed by the urge to fill yourself.

One of the great destructive aspects of the 35th Shadow is seen each time you have temporarily filled yourself up. The moment you are full, you slowly begin to feel empty again, and so the cycle continues. And it is at this very stage that we make our greatest mistake — we either become attached to the external means of fulfilment and enter a cycle of addiction, or we blame the external means (often a person) for not fulfilling our hunger and move on to something or someone else. It is because of this basic pattern of disappointment and blame that the human spirit is unable to find lasting peace. The destructive tendency spoken of above comes about through your identification with the object or person that promises your fulfilment. If you become addicted to it, you destroy a part of yourself since you cease to grow. If you blame your disappointment on another, you tend to want to destroy that other, whether it is a person, a religion or a drug.

It is because of this basic pattern of disappointment and blame that the human spirit is unable to find lasting peace.

It is fascinating that so much of our modern society, with all its good points and its bad points, has been created due to a basic chemical imbalance within the human brain! Everything in life occurs for a reason. Modern science has now even identified that low levels of serotonin in the body lead to states of depression and craving. What science doesn't realise is that almost all human beings have low levels of serotonin because what scientists are taking as the norm is really the Shadow frequency. The great dilemma of the 35th Shadow is that any attempt you make to raise your level of serotonin production from the outside fuels the hunger even more. If, for example, you take a pill that increases serotonin, you may experience for a brief period what the higher frequencies feel like. But such an experience makes you hungrier still and thus all kinds of additional problems are created.

The only way to redress the basic chemical imbalance within the 35th Gene Key is to make the leap out of the Shadow frequency altogether. This means you must first understand your chemical dilemma — and then you must face it with every fibre of your being. It is awareness of the problem that will finally break the grip of the Shadow frequency. You must enter so deeply into your genetic craving that you break its power over you. You cannot discipline yourself out of this problem — you have to dive right into it. In effect, you have to come to realise that whether you starve yourself or stuff yourself, you are never going to solve the problem. As you turn to face your humanity and your hunger, you will see how helplessly you are held hostage by it. You will see how it subtly drives all of your actions in the world.

Only when you see how trapped you are by this chemistry does a new sense of freedom begin to rise up within you. You stop the energy from moving out into the world and drive it back inside. This is when the leap occurs — there is an inner crossing-over within your awareness — and the great adventure of the 35th Gift awakens you to a new level of awareness.

REPRESSIVE NATURE – BORED

The repressed version of the 35th Shadow is about boredom. This is when the deep hunger within someone is thrust back down into the unconscious. Such people are afraid of what their lives might become if they allow the hunger to dictate their direction, so they find ways to keep it beneath the surface. This inevitably leads to a profound sense of boredom as the hunger constantly bubbles up to the surface and tempts them in a new direction. Even though these people are bored, they may not know they are bored. It takes huge inner strength to hold down such a genetic force, and the usual result of such pressure is a great reduction in a person's vitality and love of life — their faces do not shine, their eyes are lifeless and their lives are unadventurous and empty.

REACTIVE NATURE – MANIC

The reactive 35th Shadow is too busy to be bored. These are people who are so afraid of their lives being empty that they continually fill them with activity. They constantly look for external stimuli to fulfil their hunger, thus they move from one flower to the next in a kind of lustful frenzy. Such people become the great blamers of others for their circumstances, making it very difficult for them to maintain any lasting sense of commitment. They will try many different things in life but will not manage to knit those things together to form continuity. Nor will they garner true wisdom from their many varied experiences since they are always disappointed in their expectations. The manic nature is driven by the deep urge to escape its own past and find the perfect place, person or situation to fulfil its dreams. In wishing for such a state these people miss the real gold, which is the adventure itself.

THE 35TH GIFT – ADVENTURE

THE ONLY PROGRESS IS LOVE

To break through the hunger barrier is to break into the heart. The hallmark of the Gift frequency is that it always involves love. Love is the only force that can bring an end to the hunger of the 35th Shadow. With love, seratonin production is increased and stabilised within the chemistry of the human brain. With love, the energy stored within the 35th Shadow is released into the human being instead of being squandered in the world. Such an implosion of energy finally allows the inner human being to evolve rather than only the world outside us. For humanity, progress means you have to evolve in your awareness, and your next step beyond mental awareness is to leap into the higher awareness of the heart.

*It is through
a higher
functioning
of your solar
plexus that you
can experience
what love
truly is.*

What we are referring to as the heart actually has its roots within the complex ganglia of the solar plexus system. It is through a higher functioning of your solar plexus that you can experience what love truly is. Love has nothing to do with another separate person but is the connective tissue underlying all creation. Once the solar plexus has evolved its higher genetic functioning (a process described in depth in the 55th Gene Key), love becomes the reality that you live inside. What most human beings currently view as love is the brief fulfilment of the hunger of the 35th Shadow by means of an outer stimulus or person — a distant hint of what love really means. Once the frequency of the 35th Gift has been raised then all human progress finally turns inwards instead of outwards. The final frontier of the human hunger for adventure does not lie outside us but within our very nature and being. It is not outer space we long to explore, but inner space.

The 35th Gene Key stands alone as an island within the human genome since it is formed by only one codon, which activates the amino acid tryptophan. In this sense the 35th Gene Key is similar to the 41st Gene Key, which represents the initiator codon that begins all human genetic processes. However, the 35th Gene Key does not hold such genetic importance as the 41st Gene Key. This makes it something of an anomaly, and an anomaly in genetics always conceals a mystery. The mystery of the 35th Gift is captured in its name — the Gift of Adventure. Like the 41st Gene Key, it is an aspect within humanity waiting for a specific time before it switches on. Yet, unlike the 41st Gene Key, the ability to switch certain genes on and off does not always lie outside your grasp. *The 35th Gift is the only place within all human DNA that human beings have a choice in how their reality is constructed.* As you might imagine, this is quite an important codon within human evolution!

There are many arguments among geneticists, evolutionaries, theologians and mystics about whether or not there is such a thing as free will. Everyone has their viewpoint within the duality of the *maya* or human mind. Until you enter a siddhic state, there can be no resolution to this conundrum, so you either have to choose a viewpoint or remain neutral. From the viewpoint of the 35th Gift then, humans can indeed influence their own destiny by directly influencing their DNA. The secret is adventure, and true adventure can only be based upon love. Adventure is what happens when the human spirit breaks free from the mind. Adventure is a state of being that comes from the realisation of the oneness of all being. Adventure is not yet immersion in that oneness, but it is what it feels like to swim within that oneness. Oneness cannot be known mentally, even though it can be logically deduced. Oneness can only be felt when enough serotonin is produced within our bodies.

It is not simply a matter of pumping more seratonin into your body that creates the experience of the higher frequencies. Seratonin is only one link in a whole chain of interconnected chemicals (such as pinoline and harmine) that are secreted within the laboratory of the body. Higher consciousness comes about due to the delicate balance of all these chemicals in highly specific quantities and cocktails. So exact and finely tuned is this balance inside each person that it can only come about through a natural inner organic process. This is the great adventure that you take as the 35th Gift arises in your inner being — you enter a process of alchemical co-creation within your own body.

As you raise your own consciousness through love, these delicate chemical matrices begin to spontaneously form within your physiology. It is a natural evolutionary process of refining and distilling your inner essences through the endocrine system.

To live with an open heart is to live in a perpetual state of adventure. Adventure means that there is still some fear inside you, but you have reached a frequency that you know is high enough to outwit that fear. Every human being has the choice to take a path of love. That is the mystery of the 35th Gift — progress can be made if you are ready to take the great leap and risk it all for the sake of love. Of course, for most people the fear of the Shadow frequency field is too great and they are caught in the endless cycle of trying to fulfil their cravings. However, every now and then a human being takes a momentous step into the 35th Gift where they discover a wormhole to another dimension. When you give to another unconditionally you actually stimulate the secretion of serotonin. This not only makes you feel happy but also induces a deep state of trust that you are in harmony with the entire universe.

In the world that we currently inhabit, giving unconditionally is such a radical move that it literally short-circuits other people's brains. It gives others a moment of pause where they are forced to look at themselves and their lives in the mirror of your actions. It *can* cause people to go into even deeper fear patterns. The 35th Gift is the only place in the entire 64 Gene Keys where a human being can actually do something consciously to raise their frequency — they can embark on the adventure of love and catalyse a genetic speeding up of their own evolution. This, by its very nature, will have enormous impact on humanity's evolution.

If you happen to be reading this 35th Gene Key, then you have stumbled upon the greatest single secret hidden within all human DNA. In a book filled with secrets, you have found the simplest and easiest. Indeed, certain translations of the original I Ching have named this 35th Hexagram *Easy Progress* as opposed to just *Progress*. The easiest and quickest way to change your life for the better is to give your love unconditionally in as many areas of your life as you dare. If you take on this adventure, you will actually affect the minute workings of your DNA. You will stimulate new chemical messages to pass from gene to gene and you will enter a whole new world of adventure, one that most people equate with the world of fantasy.

THE 35TH SIDDHI – BOUNDLESSNESS

A BACK DOOR INTO BLISS

If you read every one of the 64 Gene Keys, you will gradually begin to realise that written into the structure of human DNA is a great evolutionary plan in which humanity will eventually transcend the human form itself. All individuals, families, racial groups and gene pools are caught up within this master plan, which appears to have its own hidden timing. Yet, within this master genetic plan, the 35th Gene Key stands out as an anomaly, as though the gods who designed the machine put in a hidden short-cut through which any individual could bypass the laws of timing built into the machine itself.

This is precisely what the 35th Gene Key is — it is a hidden wormhole that leads to a vast realm beyond our comprehension. It leads right to the heart of the 35thSiddhi — the Siddhi of Boundlessness.

The interesting thing about the 35th Gene Key is that it breaks all genetic laws. It allows human beings to take charge of their own evolution and outwit the global genetic programme. Just as the 35th Gift breaks laws, so does the 35th Siddhi. It invites human beings to expand their consciousness so far that they leave the sphere of evolution altogether. As we saw, the 35th Gift still contains traces of fear, since fear is what gives the edge to the adventure of life. At the level of the 35th Siddhi the fear is gone because you are gone: the adventure is over. Boundless-ness is consciousness. There is nothing further one can say about this 35th Siddhi. However, one *can* say things about the 35th Siddhi from within the *maya* of the mind. The 35th Siddhi of Boundlessness has great pulling power — even as a concept. If you can entertain the concept of boundlessness in your daily life, then you are automatically drawing yourself away from the lower frequencies.

Very few human beings can handle the concept of boundlessness because it has a diffusing effect on your mind. It invites you to move out of your mind and into your heart. It is really the heart that knows no bounds, which is why consciousness is the same word as love. Along with its polarity, the 5th Siddhi of Timelessness, the 35th Siddhi forms the arc of the infinite. There are no boundaries to life and there can be no beginning and no end. Even though, through the 64 Gene Keys, we see that human evolution will one day come to an end, it will nonetheless progress to another form and then another and another. Within our modern scientific world-view, astrophysicists tell us that we can measure the size of the universe and that it has a finite age. Through the 35th Gift and its Siddhi, sitting there in our DNA, we know that these are half-truths because, somehow implanted deep within us, lies the concept of infinity.

No mind will ever solve the mysteries of the universe — of time and space — because mind is only a tiny facet within those mysteries. The heart, or what some people call the heart-mind, is the only aspect within a human being that can unlock the mystery of boundlessness. So what can the 35th Siddhi tell us? It can tell us only one thing — that pure unconditional love can break all the laws of the cosmos. Whenever you taste even a fraction of this kind of love, anything becomes possible. Because we inhabit a form and live in a world where magic seems impossible, more and more human beings are turned away from the truth that anything in the universe is possible. When the Siddhi of Boundlessness enters the world, it allows conscious-ness to penetrate right into the heart of matter, which bends the laws of matter. Such a siddhic being represents a shortcut through evolution itself, and many people can experience huge awakenings and miracles around such a being.

Pure unconditional love can break all the laws of the cosmos.

As the single heart of the Ring of Miracles, the 35th Siddhi brings the miraculous into the world. It opens up all kinds of possibilities to those who allow their minds to become like a child's. To the child, anything seems possible — this is the state of boundlessness. To activate such streams of miracles, you have to break free from the beliefs and limitations imposed on your mind by the world and those around you.

The 35th Siddhi dispenses with all boundaries and limitations imposed by the mind. Throughout history men have flown, ascended and dematerialised right before people's eyes. At the level of the 35th Siddhi, anything is possible. Of all the 64 Siddhis, perhaps with the exception of the 60th Siddhi, this 35th Siddhi has but one purpose — to break the laws of form and to manifest miracles so that people can expand their awareness and open their hearts to a greater mode of being.

It is because of the 35th Siddhi that the traditional meaning of the word *siddhi* as spiritual power has come into the world. In Eastern tradition, different pantheons and systems number different Siddhis. The Siddhis are generally seen as distractions along the path of liberation. But in the case of the 35th Siddhi, they are a natural manifestation of liberation itself, arising out of the state rather than preceding it. Simply to be in the presence of a true miracle is to open your eyes to a whole new horizon. It can change your life and open your heart in a single mind-bending, reality-shattering experience. In this respect, the 35th Siddhi also finds a strong connection with the 22nd Siddhi of Grace, since it is grace that decides when, where and to whom a miracle will occur. The very existence of the 35th Gift and its Siddhi is a testament to the power of grace. We can draw great comfort from the fact that deep within the DNA of each and every one of us the 35th Siddhi lives as a constant reminder of the infinite possibilities of a life lived in the name of love.

SIDDHI
COMPASSION
GIFT
HUMANITY
SHADOW
TURBULENCE

▦ 36th Gene Key

Becoming Human

Programming Partner: 6th Gene key   **Physiology:** Solar Plexus
Codon Ring: The Ring of Divinity   **Amino Acid:** Proline
(22, 36, 37, 63)

THE 36ᵀᴴ SHADOW – TURBULENCE

THE DARK NIGHT OF THE SOUL

The 64 Gene Key Shadows provide the grist for the mill of all human suffering. Each shadow provides a mythic challenge for humanity to move through which, at the individual level, is fought within the physical battlefield of the body. The 36ᵗʰ Shadow is an emotional battle that must be fought by every individual at certain points during his or her life. As a vibration felt across the length and breadth of our planet, this Shadow manifests as a collective emotional turbulence. This turbulence exists because uncertainty exists and every human being knows that disaster could strike at any moment. There is a great deal of heavy propaganda for this 36ᵗʰ Shadow, particularly through the modern mass media. The continual barrage of negative news via television and the media ensures that most human beings are programmed with an unconscious background field of nervousness and turbulence.

At a certain point in history, the 64 patterns or hexagrams of the Chinese I Ching were adapted for the purposes of divination and for predicting cycles within time. It is the modern use of the I Ching as an oracle that is so popular today. The ancient sages noticed that when certain hexagrams were drawn, they indicated periods of intense crisis or danger. The 36ᵗʰ hexagram was one such hexagram, and thus it was given the ominous name *The Darkening of the Light*. There is much truth within this name, although when understood at its most profound level, the 36ᵗʰ Shadow contains an enormous wealth of consciousness.

Extrapolated to our genetics, the 36ᵗʰ Gene Key represents a part of our chemistry that goads human beings to break through boundaries of experience. At a collective level this is a vital aspect of human survival — to learn what is dangerous and what is not. Evolution in this sense can be viewed in a much wider context. It *uses* us humans as experiential eyes to probe the outer limits of consciousness itself and demands that we look into the darkness, put our fears aside and plunge in regardless. Because of this Shadow, there can be no human life free of emotional turbulence. It is deeply engrained in our myths.

The modern phenomenon of the television soap opera is a prime example of our collective recognition of the emotional turmoil all humans have to go through over the course of their lives. One of the greatest determinants of our frequency is how we deal with emotionally challenging situations.

Nowhere does the 36th Shadow wreak more destruction than in the field of human relationships. Keep in mind that the programming partner of this Gene Key is the 6th Gene Key whose Shadow is Conflict — the breakdown of communication between individuals or groups. One of the deepest human yearnings is for a perfect loving relationship. It is a yearning that has been with us since the beginning of time, and the journey towards this ultimate dream of ours begins here in the 36th Shadow. The journey begins, as do so many of our myths, with the challenges of our sexuality and with the guilt that follows it.

One of the greatest determinants of our frequency is how we deal with emotionally challenging situations.

At a low level of frequency the 36th Shadow manifests most strongly through our sexuality where it becomes sexual lust. Sexual lust is simply a low frequency expression of the evolutionary urge to break through another boundary of experience. Lust itself is an incredible chemical turbulence felt within the body. When it presents itself, it can create all manner of emotional disturbance in your life. Seen without a moral connotation, lust is a very pure energy. Whether you try to resist it or yield to it, one thing it does is make your life more interesting in terms of your experience! This is exactly the purpose of the 36th Shadow — to create an interesting storyline to your life thereby pushing you forward. However, at the lower levels of this shadow frequency, human beings can go through suffering after suffering until they finally begin to evolve.

The problem at all low levels of frequency is that the patterns perpetuate themselves. With the 36th Shadow lust can be suppressed, causing illness, or it can be expressed, causing guilt and/or deception. Lust in itself is not a negative energy, but is dirtied through entanglement in human morality. Sooner or later, all human beings must raise their frequency in order to escape being victims of these Shadows. When this happens, problems such as sexual lust cease to be problems. They are dealt with cleanly and honestly, with openhearted communication and freedom from guilt. No feeling in us is ever *wrong* or *bad*. Every human being is innocent of their own chemistry and the way it makes them feel. Difficulties emerge when we react or repress our feelings out of fear or anger, creating enormous turbulence within our own lives and those closest to us.

Among mystics, the 36th Shadow has long been known as the *Dark Night of the Soul*. It is an archetype within human beings that pulls us towards the unknown, and that draws the unknown towards us. In so doing, this Shadow tests your levels of frequency and offers you the opportunity to transcend your own suffering. It tends to draw crisis towards itself like a magnet. If you do not break out of your own victim patterns but continue to resist your true nature, you will keep being shown the same lessons again and again until you learn from them.

The potential suffering that exists on account of this Gene Key is part of a genetic family known as the Ring of Divinity. This is a quite beautiful revelation. Each of the four Gene Keys within this codon group has the capacity to awaken the highest consciousness inside you.

Being allied to the Grace of the 22nd Gene Key, these four archetypes bring intense emotional experiences and challenges to your life. They also tend to plunge you into a deep dark night in which you are riddled with self-doubt (the 63rd Shadow). Ultimately however, the theme of this codon group is redemption. As you allow yourself to be purged by the power of your own suffering, you will realise the utter bliss of the higher states and will recognise your own huge good fortune. As we shall see, this codon group in many respects is deeply connected to the true meaning of the Christ consciousness. If you feel a strong connection to it, or have it as a key aspect of your Hologenetic Profile, you are indeed a fortunate human being and your journey will inevitably lead you to this realisation.

REPRESSIVE NATURE — NERVOUSNESS

When emotional turbulence is resisted, it becomes nervousness that ripples through one's entire body and aura. These are natures that are terrified of change and that attempt to maintain outer calm at all costs. The cost is that the turbulence becomes internalised and gets trapped in the nervous system. Such people are unable to relax and send nervous waves into their immediate environment as well as into others, destabilising the very things that they are trying to keep stable. These are also people who tend to repress their sexuality out of their deep-seated fear of change. They can be very hard to approach or get to know. Nevertheless, the terrific primal force of our sexuality cannot be indefinitely repressed, and the usual result is a nervous breakdown or cancerous disease of some kind.

REACTIVE NATURE — CRISIS PRONE

When emotional turbulence is expressed without clarity or honesty, it results in repetitive destructive emotional situations. These are people whose lives run like the soap operas. The reactive nature is more likely to have sexual affairs and then conceal them through guilt. However, as with the repressive nature, non-acceptance of your true nature always comes back to haunt you, and thus life will always tend to draw an emotional crisis into such a person's life from a different and usually unexpected direction. These people do not realise that their inability to deal with their emotions and sexuality cleanly and honestly creates further turmoil in other seemingly unrelated areas of their life.

THE 36TH GIFT — HUMANITY

THE SPIRIT DESCENDING

When your struggles with sexuality and emotional turbulence are finally embraced openly and honestly, a remarkable Gift is born — you finally graduate as a human being! The 36th Gift is the Gift of Humanity, and this is what human suffering is really all about. It is our suffering that connects us all together. It opens up your eyes beyond your self-obsessions and forces you to evolve beyond your selfishness. To be a fully integrated human being means to begin to transform your own suffering and open your heart to life.

A person with the Gift of Humanity is a person who truly understands human emotions and who consequently understands all people. You can see here how the programming partner of this gift, the 6th Gift of Diplomacy, matures simultaneously as this 36th Gift matures. The moment you let go of the notion that you are somehow a victim of the fates, you begin to finally communicate cleanly with others.

The Gift of Humanity is truly a Gift that can only be earned. These are people who have looked deeply into their own Shadows and who have grappled with the challenges of their sexuality and their emotions. These are people who are in the process of accepting their own suffering and thus experience it at a different level of frequency. Only at the highest levels is our suffering instantaneously transformed into rapture. At the Gift level however, suffering is still suffering, but it brings human beings together rather than pushing them apart. This Gift is about humaneness — it is about working from your heart. When you begin to live from your heart, you suddenly have an antidote to fear.

The 36th Gift is driven by the same urge as the 36th Shadow — the evolutionary impulse to experience new feelings and new situations in order to learn from them. Here, with an open heart, the 36th Gift can negotiate challenging and potentially turbulent emotional situations with maturity and diplomacy, taking one's own and others' feelings into consideration. People with the 36th Gift are thus the kind of people that others turn to when in distress, since their aura resonates strongly with the common human theme of suffering. At this level these people are no longer overwhelmed by emotions as they tend to be at the Shadow level, but have opened and expanded themselves to life through harsh experience. This makes them emotionally capable of handling all manner of traumatic experiences.

As a living transmission, the 36th Gift carries a great teaching into the world — the celebration of humanity. It contains a natural spirituality that is more humanist than Divine. In this sense it is one of the most grounded of all the Gene Keys. It is a bridge for the higher bodies of humanity to move down into the lower planes and transform them. It conveys the image of Christ descending into hell and absorbing its frequencies fully into his own being. All human beings partake of this Gene Key, whether it is prominent in your Hologenetic Profile or not. Every emotional struggle that comes your way talks directly to this aspect within your DNA. At the Gift level, the 36th Gene Key has learned to accept the pain rather than run from it. In embracing pain, you are actually trusting in your own strength and in life. If you are in pain, then that pain is there to teach you something. It is an invitation to accept your humanity in more depth and to feel the humility of living within a mortal form.

The deepest role of the 36th Gift is to help humans to become human.

The presence of the 36th Gift in someone's genetic makeup gives a deep insight into what kind of life they will live. People with the 36th Gift prominent in their Hologenetic Profile will have a strong theme of emotional healing running through their lives, and will tend to draw experiences and people into their lives who reflect this. The higher they are able to raise their frequency, the clearer and more profound their emotional experiences will become. The evolutionary urge to go on exploring new territory ensures that these people will never lead dull lives, and at the higher levels they can experience the most profound, heart-opening experiences.

The 36th Gift is an archetype of the true purpose of all human suffering. These people are here to teach us that life is full of flavour and that pleasure and pain always come together. The deepest role of the 36th Gift is to help humans to become humane by respecting others and by embracing your own suffering whatever it looks like, rather than being dragged down into the depths of victimhood. It is this courage and deep acceptance of the life process that will eventually give rise to the highest of all human expression — the expression of compassion.

THE 36TH SIDDHI – COMPASSION

THE BEATITUDES

There is perhaps no better symbol for the entire journey from the 36th Shadow to the 36th Siddhi than the life of Christ. In examining the 64 Siddhis, one can see that certain Gene Keys appear to have very strong connections with certain mystical lineages or figures. The myth of Christ is one of these key repeating lineages, but the lineage does not so much concern the historical personage of Jesus himself as the frequency represented by him. The symbol of Christ on the cross is a powerful reminder of the truths inherent within this 36th Gene Key. It reminds us that we are each mortal and that there is great beauty in our humanity, echoed in Christ's assertion that he was *the son of man*.

People who do not understand the true relevance of the symbol of Christ on the cross may see this as an inglorious representation of a god. In most religions, the prophets or gods are seen as powerful, beautiful or awesome, whereas the figure of a man dying helplessly on a cross appears to be a deep representation of a victim state of consciousness. However, it is this aspect of Christ that makes him such an approachable figure for so many people. The very fact that he is suffering makes him more human. This approachability is what sets the 36th Siddhi so far apart from many of the other manifestations of the siddhic level of consciousness. This is a Siddhi that spans the chasm between man and god, between being a victim and being enlightened. It speaks to every single human being, and it asks each of us to ask ourselves a single question: "Why must I suffer?"

The answer to the question of suffering lies here within the 36th Siddhi. It is the same answer given to us by the 25th Siddhi — it is all about Love. However, the Universal Love manifested through the 25th Siddhi is very different from the Compassion expressed through this 36th Siddhi. Universal Love almost has a distant quality to it, as though it came from the gods themselves. The 36th Siddhi, on the other hand, speaks in the language of man, since it can only flower through a journey of intense suffering. Such is the path of the 36th Gene Key. This is the flowering of suffering itself, and as such it can no longer be understood through the word *suffering*. It is the fragrance released after the storm is over, and just like the 36th Gift of Humanity, this is a Siddhi that has to be earned.

Beings like Jesus, who attain final liberation through this Siddhi, are salt of the earth people. They are burnished over and over again through the trials their life brings to them, and

*These are
the most
heart-melting
people ever
to walk the
face of the
earth.*

once they have made that magical step away from seeing suffering as negative, it stops being so. It actually becomes their muse. At the siddhic level, your own suffering becomes so universalised that it encompasses all humanity. All boundaries of self dissolve. Suffering and Compassion merge into one as your human heart explodes with the feelings of all human beings — their pleasures and their pain, their aches and longings, their vices and their virtues. At this siddhic level, as the programming partner of the 6th Siddhi testifies, Peace reigns supreme.

In every mythic journey there must always be a final trial — a moment of utter hopelessness and helplessness, like Jesus hanging bereft on the cross crying out in agony to his God for understanding. These moments — these *darkenings of the light* — come to us in life not so much to test us, but to cut us so deeply with our own humanity that we are reminded of the awesome power of our own compassion for all our fellow men. Such trials give us the opportunity to make huge evolutionary leaps out of our Shadow states and into the higher frequencies. There are many traditions which speak of this Siddhi in terms of miraculous transformations made by truly evil beings who in a moment of utter horror and complete surrender, make the leap straight from the shadow state into the siddhic state. This is one of the very few Gene Keys that allows such miraculous leaps of consciousness.

Those in whom this Siddhi has awakened tend to keep moving into those areas of life where the light is most darkened. They usually flourish during periods of crisis or war, or live and work among the poor and wretched. The 36th Siddhi goes on longing to explore the frontiers of consciousness within form, and after a certain point it ceases to care or identify with its own security at all. Anyone manifesting the 36th Siddhi has at some point sunk into such a dark space of helplessness themselves that it is impossible for them to experience anything as intense again. This has the effect of annihilating all fear from their system, so that true peace emanates from within them. The simple presence of such people and the look in their eyes or the softness of their voice can trigger all manner of so-called miracles in those whom they meet. These are the most heart-melting people ever to walk the face of the earth. Wherever they go, the true frequency latent within human suffering is immediately released and compassion tears open people's chests and explodes inside their hearts. Even the darkest of natures can be instantaneously brought to tears through the presence of this 36th Siddhi.

With the above in mind and held softly in your heart, perhaps you may come to a deeper intuitive understanding of the true meaning of the Beatitudes, those famous words of Jesus uttered in the Sermon on the Mount:

> *"Blessed are the poor in spirit, for theirs is the kingdom of heaven.*
> *Blessed are those who mourn, for they will be comforted.*
> *Blessed are the meek, for they will inherit the earth.*
> *Blessed are those who hunger and thirst for righteousness, for they will be filled.*
> *Blessed are the merciful, for they will be shown mercy.*
> *Blessed are the pure in heart, for they will see God.*
> *Blessed are the peacemakers, for they will be called the sons of God.*
> *Blessed are those who are persecuted for righteousness, for theirs is the kingdom of heaven."*

SIDDHI
TENDERNESS

GIFT
EQUALITY

SHADOW
WEAKNESS

▤ 37th Gene Key

Family Alchemy

Programming Partner: 40th Gene Key

Codon Ring: The Ring of Divinity
(36, 37, 22, 63)

Physiology: Solar Plexus (Dorsal Ganglia)

Amino Acid: Proline

THE 37TH SHADOW – WEAKNESS

THE REAL POLE SHIFT

As humanity gradually begins to move out of the astrological age of Pisces and enter the new Age of Aquarius, we are experiencing a shift in the emphasis of the global archetypes that imprint us as a species. The 37th Gene Key and particularly its Shadow frequency have marked the age that we are now leaving. Those with an eye for occult information will perhaps recall the number 37 as being the number of Christ taken from Gematria, the ancient Hebrew system of alphabetical numerology. We are moving out of the sphere of the specific mythology surrounding Christ and into the sphere of the third aspect of the sacred Trinity whose new mythology is based upon the great feminine archetypes of synthesis. This does not mean that the power behind the Christ myth will in any way diminish as we enter this new phase of evolution. What will change is our understanding and interpretation of the myth as its hidden aspects finally reveal themselves. This process ushers a flood of new mythic feminine images and archetypes into the collective consciousness.

The 37th Shadow is the Shadow of Weakness. Weakness as we shall see, is nothing but a projection of the male psyche onto the female psyche as, in the West, women have been viewed up until recently as *the weaker sex*. However, what appears to humans as weak is actually something that we cannot yet understand. We may see this enacted even in the externals of the Christ myth itself, in which one man surrenders to forces that finally overpower him. Naturally, it is the inner meaning of the myth that holds the real key, for it is only in the sacrifice and the subsequent resurrection that one finally comes to understand the hidden nature of Jesus' actions. Much about this 37th Gene Key therefore concerns the nature of a force that is perceived as weak but that inevitably shows itself to be the opposite.

The 37th Shadow represents the inequality between yin and yang forces on our planet. The natural tendency of evolution has been to favour the physically stronger. At low frequencies, our genes know no other reality than the survival of the fittest.

However, we have now reached a stage in evolution in which physical strength no longer governs our future direction. That future direction lies wide open for anyone, regardless of sex or strength. What we used to perceive as weak or strong is totally changing, even reversing. Those who thrust themselves to the top of the hierarchy through brute strength and subversion are losing their power and those who hold a vision of synthesis based on self-transcendence are gaining more power. In this way the world is transforming.

The 37th Shadow is undergoing a major mutation throughout humanity that will only build in speed and frequency as we move beyond its old boundaries and definitions. It is because of this mutation that the entire social structure of our civilisation is crumbling. Here in the West the traditional family unit is experiencing a natural demise, and at such times great social change and unrest is much in evidence. Because of this inner rebalancing within the human psyche, the fabric of our society is being forced to change. The repressed feminine side of humanity is rising up to the surface once again and this is changing the basic patterns of the traditional roles for men and women. This uprising of the yin force is actually beyond gender, but it is also confused by gender. Many women today are under the impression that the time for women is coming and the time for men is over. This attitude is simply another form of the 37th Shadow, which always over-emphasises one aspect of the polarity.

The true definition of Weakness can be seen when the feminine principle is witnessed serving the masculine principle, which is how the world has evolved and survived up until this point. Wherever you see the submission of the female to the male you are seeing the 37th Shadow is action, and it can result in only one phenomenon — Exhaustion, the 40th Shadow and programming partner of this 37th Gene Key. In the way our societies are currently structured many women may feel impotent, financially dependent on men on account of having to stay at home and bring up children. Of course there are always exceptions to this, but it is a general truism. The pole shift that is upon us will see the reversal of this tendency, so that the masculine serves the feminine. One of the modern reactions to this in developed societies can be seen in the large numbers of women having to go out of the home to work, usually leaving the children in some form of care or institution.

Such a phenomenon can only create more division in the world, since the ones who suffer in this are always the children. The 37th Shadow of Weakness manifests both through men who cannot see holistically and women who try to redress the problem in the only way they know how — through entering the male world of hierarchy and competition. All such social issues as these invoke strong emotional reactions from men and women, as do the same issues when connected to religion. There can be no answers to these imbalances from the human mind. They are simply a sign of the transitional times that we are living through. Because of the emotional charge between the yin and yang — between men and women — to even discuss such issues arouses huge hidden tensions and opinions. In this respect one of the great signs of the 37th Shadow is your identification with your gender in the first place. The real issue is one of yin and yang — of an imbalance between archetypal forces — not between their manifestations. There are always exceptions to the general rules, so that there is always room for individual uniqueness.

One of the
great signs
of the 37th
Shadow is your
identification
with your
gender.

The mutative energy moving through this 37th Gene Key is also bringing an end to many other structures that have been a part of human civilisation for so long. One of these structures is organised religion, which is essentially a male-oriented phenomenon based upon division and worship. The masculine left hemisphere of the brain divides in order to understand, whereas the feminine right hemisphere grasps the fundamental unity that is always and already present. This movement from worship to embodiment is such a profound transition for humanity because it cannot easily be understood through the mind. The Shadow of Weakness is actually a complete fabrication of the mind. We need these external crutches only as long as we perceive ourselves as weak without them. The moment we drop the crutches, we will come to see that we had strong legs all along.

The essence of the dilemma of the 37th Shadow is seen most clearly through this imbalance within the individual. What we fear most deeply through the 37th Shadow is lack of support. Because our frequency does not allow us to feel a part of the whole, and therefore supported by the whole, we only give with the assurance that we will receive something back. This is the foundation of social economics, and its root is fear. What we don't have access to at this Shadow frequency is the feeling of deep trust in the collective. We don't know that when we give from our heart, we will receive far more in return from somewhere in the collective energy field. Because we cannot anticipate *how* this energy might return to us, we do not trust it. Consequently you can see that when you allow your mind to lead your life, you create weakness in the world because you sever the chain of support that exists naturally between all beings. Thus, it is in the human heart and its ability to grasp the holistic essence of life that our true collective future lies. In an individual human being this change translates as the mind serving the heart.

REPRESSIVE NATURE – OVER-SENTIMENTAL

When the 37th Shadow is repressed it tends to manifest through inauthentic giving. These are people who are ruled by their emotions rather than their hearts, and they often believe that the two are the same. They repress their own fears through a dreamy mentality that may speak great words about humanity but that lacks the inner strength to take a stand for itself. These are the people who, as victims of their own emotions, give the human heart a bad name. Whereas the heart is strong and fearless, the emotions can be fickle and overwhelming. Such people hide behind and within their emotions which blind them to the deeper truths inside them. They have a tendency to *over-process* at the emotional level, which makes them draining for others to be around. The deepest irony is that these types of people are actually addicted to emotional states as a means of repressing their deeper fears.

REACTIVE NATURE – CRUEL

The reactive nature projects its own fears out into the world where it sees only inequality. Out of this core belief, these people become hardened around their heart until they entirely forget what love feels like. As the polarity to the repressive nature, these people tend to see others as over-sentimental and will take advantage of their good nature. Out of these two patterns —

the repressive and reactive — our current civilisation is created. Only those who are no longer victims of their own or others' emotions can break the victim/abuser patterns lying at the heart of our communities. The people who we perceive as cruel need willing victims in order to rise up in the community. Once someone takes a stand anchored in their heart, these abusers are the first people to crumble.

THE 37ᵀᴴ GIFT – EQUALITY

THE ASCENT OF THE FAMILY

It is in the 37th Gift of Equality that a great hope is presently arising for humanity. Equality is such an easily misunderstood term. All humans are born equal since we all share the same genetic code, but as we grow and enter society it becomes apparent that we are not so equal. In reality, equality is a matter of perception and it depends upon frequency. At the level of the 37th Shadow, which sees humans in terms of strength and weakness, equality is nothing but a dream. Yet, at the Gift level of consciousness, equality is a living ideal that lifts people out of the game of victims and winners. When you operate out of the 37th Gift, you are living directly out of your heart and all else is secondary.

From the point of view of the human heart, all humanity is one family. This is not a mere sentimental dream but an all-powerful truth rooted in a vast inner current of strength and love. Without love, you are always operating from the Shadow frequency and all you see and create is inequality. As the great mutation triggered by the 55th Gene Key sweeps through humanity, the very core structure of our civilisation will readjust itself to this shift in consciousness. The flaws in the current social fabric of all cultures cannot be ironed out but only phased out. A new world will have to grow within the ashes of the old, and it will be built upon the rock of the 37th Gift of Equality. When the balance within the human psyche is redressed, there will be no need of external religion. Neither will we have a patriarchy nor even a matriarchy — we will have a single stream of consciousness experiencing the true meaning of the word family.

Out of this 37th Gift, a new vision of the family will be born. The family unit is the most powerful crucible of love through which humanity will be transformed. The sheer force of love contained in the family is second to none. The love of a child for its parent and vice versa is as potent a force as any in the universe. As we create a new society based around the liberation of this love, our world will change very quickly indeed. The problem to date has been that the family has always been restricted to its gene pool or tribe, and it has always been contained within a hierarchical structure. The current social, governmental and educational structures on our planet do not serve the family first. By over-emphasising individual achievement, they encourage competition between families, which also creates competition *within* families. Through the 37th Gift, humanity will come to see that it is the family that makes all human beings equal. We will find a new vision of family, both locally and globally, and we will in time extend this vision to encompass the entire human family.

The family unit is the most powerful crucible of love through which humanity will be transformed.

The great mutation that is moving through the 55th Gene Key will first manifest as a change in our children as they are born with a new higher activation of their DNA. These children in turn will need a new family structure to support their collective awareness. Their very impact at an auric level will cause this structure to manifest. Even now, communities with higher awareness are beginning to understand new ways of conceiving and implementing this model. Families are the life blood of humanity, and a healthy family is a vehicle for such a potent force of creativity and love. The new family will become a vessel for collective ascension, as we recognise the power of the new awareness carried by the coming generation. When we begin to build a civilisation based around the view of an incarnating child, we will create a truly magical world.

The love between a child and a parent is an archetype of unconditional love. At the Shadow frequency however, all love is conditional, which means that essentially it is not love. True love is based upon giving without the need for a return. The entire Shadow frequency is thus based upon the principle of the bargain — of giving in order to receive. This *bartering consciousness* is what lies at the heart of modern civilisation. It is seen through economics, government and especially in relationships. It is here in the dynamics of all human relationships that bartering was born and it is based upon the fear of not having enough. Through the 37th Gift you will discover a great economic secret — the more you give, the more you receive. Many people have misinterpreted this truth down the ages and have wondered why it does not appear to be true. Giving cannot be faked. You can give from your mind and you can give from your heart. Giving from your mind is always conditional, no matter how subtle, because it always has a secret expectation or hope attached to it. True giving is an act of insanity when viewed from the mind!

The giving that comes from being in your heart leads to equality because it is *based* upon equality. The human heart carries this equalising force wherever it moves — and it illuminates divisions in others by treating everyone with respect. Despite the mind's fears that equality will create a homogenised world in which everyone is the same, true equality is based upon respect for individual uniqueness; indeed, it thrives on it. This kind of equality means starting with yourself. Giving is another place that the Shadow frequency can hide and pull you down — sometimes you may give to others out of a need for personal recognition or as a diversion from your own issues and fears. First of all, then, you must learn to give to yourself. As you will discover, it is through the love of your own self that you will find your love of others. In order to support another, you must first ask your heart how you can best support yourself, and out of that will come the answer of how to give so that both giver and receiver are fulfilled. To give to yourself in this way means to trust completely in the wisdom of your heart — it really is as simple as that.

Experienced as friendship, the 37th Gift is the glue that holds humanity together. In fact, true friendliness is the universal mark of the Gift frequency. Our current age is truly an exciting time to be alive because this 37th Gift is giving birth to a new social paradigm all across our planet. It will create out of itself a new social framework that synthesises business, education and family rather than setting them apart from each other.

It will also bring an end to isolated family groupings and communities by networking, both geographically and globally, to create a new model for community that is based upon giving rather than controlling. This energetic type of friendliness will also bring an end to the notion of separate communities based on the worship of an external God or authority. Through the awakening of the 37th Gift, each community will find that it holds the same core principle as all others — namely support of the earth, and empowerment and nourishment of humanity as a whole.

THE 37TH SIDDHI – TENDERNESS

THE SACRIFICIAL LAMB

To enter fully into the mystery of the 37th Siddhi, we will have to move more deeply into the mythic figure of Christ. Although there have been other figures with a similar mythology to Christ, such as the Norse Odin, who was hung from a tree, or even the Egyptian Osiris whose body parts were scattered far and wide, the figure of Christ and his exact mythology has entered so deeply into the world psyche that it has a particular resonance with the times in which we are living. One needs to be clear from the beginning that we are not talking here of the Christ as he is generally understood by the Christian religion. We are looking at this man's life as a symbol of a deep archetypal process that governs all human beings, regardless of their beliefs or culture. As the second aspect of the mysterious Trinity, the Christ unites the two polarities of the masculine Father (prime yang) with the feminine Holy Spirit (prime yin). This is why Jesus is said to be both the Son of God and the Son of Man. He is the offspring of both yin and yang and is therefore the great equalising force between them.

We have seen that the 37th Gene Key represents an evolutionary phase that is now coming to an end, which means that something new must also be born from it. The Christ myth itself must give birth to another dimension, and this dimension concerns the role of the feminine. As the story of Christ is currently understood, Jesus is seen as an embodiment of the Divine man, born of a virgin womb and untarnished by sexuality. In the popular story, Jesus has no sexual encounter or attachment to women whatsoever. He stands alone as a proud and potent symbol of the patriarchy that sculpted this understanding of his life. However, without wishing to offend exponents of the popular version of the Jesus story, this interpretation of Jesus is somewhat sterile in terms of its connection to human sexuality. In the Christ myth, Jesus is pitted against sexuality in the form of the devil, and the devil has obvious connections to the feminine through the earlier myth of Eve and the serpent in the Garden of Eden.

The 37th Siddhi is the Siddhi of Tenderness, and tenderness is a quality most often associated with the exchange of love between a parent and child. Tenderness in particular is an archetypal motherly essence. Jesus himself is universally connected with the figure of the lamb, whose essence is represented by this 37th Siddhi. This lamb has a deep symbolism with our coming age because it represents the spirit of surrender to an overwhelming inner power.

The sacrifice of the lamb corresponds to the giving up of identification with outmoded outlooks and the embrace of a wider reality that lies beyond our current view. We, humanity, are the lamb and the universal spirit holds us in tenderness, as a mother cradles her child. As individuals we each have to bow our heads and accept the devil within us instead of pushing it down deeper into our psyche. The very thing that terrifies us the most will utterly transform us if we can but trust it — it is the source of our sexuality, the dark feminine shadow place within every human being that contains the greatest mystery of who we are and what we can one day become.

The 37th Gene Key has emerged into our current world through the sacred sacrament of marriage. It became sullied and stagnant through its manifestation as a social contractual convention. The marriage we see in the world today is but a shadow of the true inner ideal of marriage. On an individual level, marriage concerns the inner balance and union of yin and yang within each human being. The Christ myth must therefore reflect this inner union through the marriage of Christ to his sacred bride, represented by Mary Magdalene. The modern resurgence of ideas that explore Christ's connections to the figure of the Magdalene can be seen as a reflection of the resurgence of the repressed Divine feminine. The Siddhi of Tenderness is the awakened aura of this mystical union since tenderness is an energy devoid of tension or sexuality. It is in fact the transcendence of sexuality.

A great paradox is emerging through this 37th Siddhi — the Divine feminine will primarily enter the world through men, via the power of the mother. The collective imbalance of the feminine archetype is not to be found in women but in men. The counterbalance to this is the rising of the masculine force of independence burgeoning in women. It is through women's independence that a new social structure will be born in the world that will place the early upbringing of children before all other factors. When children are nurtured by both mothers and fathers, and mothers and fathers are properly supported rather than torn apart due to social tensions, then children will come into the world emotionally balanced. This 37th Siddhi of Tenderness is the natural climate that surrounds all parents and children. When boys are brought into the world surrounded by this support and tenderness, their feminine side will no longer be compromised and the world as we see it will begin to change as these boys grow into emotionally balanced men.

Conversely, it is through future young girls that great social change will be brought into the world. Energetically speaking, it is the feminine that carries the vision of synthesis and the masculine that will build it. This is the generalised blueprint of future society — it will be a world in which the masculine principle serves the feminine principle. So what we are now seeing is not a decline of the family unit but a complete redefining of what family really means. Family is the structure that creates an aura of tenderness across the planet — empowering individuals and supporting communities simultaneously. In the future, family will no longer be an isolated tribal phenomenon but a collective breath pattern that unites all human beings.

In the future, family will no longer be an isolated tribal phenomenon but a collective breath pattern that unites all human beings. This is the future that is coming. It is the same vision that Christ held as the natural state of humanity. It is what he referred to as the kingdom of heaven.

This is the future that is coming. It is the same vision that Christ held as the natural state of humanity. It is what he referred to as the kingdom of heaven.

The 37th Siddhi is unique among the Siddhis in that it does not manifest through a single individual. It is the heart of the human family, drawing all individuals into its infinitely soft and loving embrace. It is not a force outside of humanity, as we have assumed for so many aeons, but is our collective inner nature. As a vital aspect of the genetic codon family known as the Ring of Divinity, the 37th Gene Key has some very powerful genetic programming companions — the 22nd Siddhi of Grace, the 36th Siddhi of Compassion and the 63rd Siddhi of Truth. Each of these codes brings the eventual manifestation of the full Divine being into the world. As we move beyond the tribal and blood definitions that have dictated our traditional lifestyle for so long, we will see how limiting our current definition of family really is. The 37th Siddhi is an upsurge of consciousness that will slowly unite human beings as the 55th Gene Key frees us from our victim consciousness. Tenderness is a force to be reckoned with, and as its programming partner — the 40th Siddhi of Divine Will — testifies, it is a force that has us all set in its sights. In its very softness it is insurmountable, invincible and inevitable.

38th Gene Key

The Warrior of Light

SIDDHI
HONOUR

GIFT
PERSEVERANCE

SHADOW
STRUGGLE

Programming Partner: 39th Gene Key
Codon Ring: The Ring of Humanity
  (10, 17, 21, 25, 38, 51)

Physiology: Adrenals
Amino Acid: Arginine

THE 38TH SHADOW – STRUGGLE

A FIGHT WITHOUT PURPOSE

The 38th Shadow joins its programming partner, the 39th Shadow, to form a very heavy duo. This is ancient genetic programming based on individual survival. Both these Shadows have strong connections with the animal kingdom through their role within early hominid history. Before we begin an exploration of the darker side of this 38th Gene Key, one should hold the following perspective: without either of these Shadows, there would probably be no humans walking the earth today. This can be a very dark place within the genetic matrix because it represents a primal energy whose main instinct when threatened is aggression. In animals, this natural ferocity is embodied in the behaviour of a mother when her cubs are threatened. As this aspect of DNA developed in human beings it became the foundation of the evolutionary law of the survival of the fittest. This Gene Key is also deeply connected to individual health and well being.

This same genetic wiring still influences all human beings today, regardless of how closely associated you are to this Shadow. It is there in the collective energy field of humanity, and it is particularly strong when certain groups of human beings are threatened by other groups, manifesting strongly through its aggressive reactive side. This 38th Shadow likes and needs a fight, and the nature of this fight is dependent upon its frequency. At a low frequency, this Shadow fights with others, with itself or with life itself. It is after all called the Shadow of Struggle. This kind of struggle is very disempowering, both for the struggler and for anyone else who happens to get locked into the combat. The word *locked* here is very appropriate for that is the nature of this 38th Gene Key. You can see that the Gift level is called Perseverance, but at a lower frequency this simply manifests as a stubborn persistence to fight, no matter what the outcome. When combined with its programming partner, the 39th Shadow of Provocation, the 38th Shadow simply digs its teeth into someone or something and doesn't let go until that someone or something gives way. At a low frequency this always causes some kind of destruction.

*If there is
nothing left
to fight for,
you may cease
to exist.*

More often than not, this Shadow manifests in relationships when one or both people are not feeling fulfilled in their lives. The 38th Shadow is about the battle to find a sense of purpose in life, and in this sense it also has a strong magnetic connection to the 28th Shadow of Purposelessness. If there is no real alignment of purpose within a being, then the provocation carries a frequency that results in a very destructive dynamic in the relationship. These kinds of dynamics, which are very common in the world, gradually grind down the love within the relationship and the fighting actually becomes an addictive need. Every relationship has its own inner purpose, but if the purpose of two individuals is not initially aligned, then the true nature of the relationship never has a chance to reveal itself. There is a great deal of sadness within this 38th Shadow. If there is no real sense of purpose in your life, one of the chief vents for this sadness is to channel it aggressively at those closest to you, usually members of your own family.

Other than the absolute necessity of living out your true purpose, there is a great secret to breaking the addictive patterns of struggle embodied in this 38th Gene Key: Simply take a breath. Struggle is a pattern that locks you into a certain breath pattern that makes you totally forget yourself, as well as rendering you deaf to all outside influence. It is as though an outside force has you in its possession and forces you to keep banging your head (or someone else's) against a wall. The moment you take a breath and pause, the pattern is broken and a space arises in which you can reorient your energies. At low frequencies, the tighter you hang on to things, the more resistance you create. By providing gaps in the pattern, you enable transformations to occur, and it is usually in these gaps that the answer to the struggle is resolved. You can actually see from these three Shadows — the 28th, 38th and 39th — how struggle and resistance allow you to diagnose whether you are fulfilling the true purpose of your life. In that sense, they can be extremely useful.

At a collective level this 38th Shadow has other manifestations in the world. The human addiction to struggle and violence is compounded through the pairing of the 38th and 39th Shadows. These Shadows see everything at a personal level, which means that they have very little perspective. History has shown that many great wars have been fought because certain individuals in positions of power reacted aggressively to another individual or tribe based on a personal agenda. The 38th Shadow is not an archetype that thinks. If it feels in any way threatened, it reacts in an extremely aggressive manner without consideration of who might be hurt. Many innocents have been slaughtered because of the 38th Shadow acting without taking pause to consider the logic or justice of its action. The core of this individual need to struggle can only really be understood when it is viewed at an unconscious level. At the deepest level, struggle maintains the illusion of your separate identity. As long as you can fight, you can remain in control of your environment. This reflects the greatest human fear — if there is nothing left to fight for, you may cease to exist.

It is the unconscious fear of death that prevents human beings from transcending their individuality. The 38th Gift invites you to transcend your separateness *through* your individuality by means of serving a goal beyond yourself. But at the fear level we never stop to take a breath, so it never enters into our awareness to use this terrific life force in a creative way.

Thus the whole energy of stubbornness and deafness within this Gene Key results in a planet where most of the world still has to struggle to survive. At the same time, the minority of people who do not outwardly struggle — the developed countries — continue to struggle internally through the stresses of living lives without a strong sense of purpose. The repressive nature of this 38th Shadow is defeatist, and the majority of the western or developed world also holds this attitude. People simply do not have the perseverance to tackle the great issues (for example, world poverty) and this collective defeatism means that we do not even try.

REPRESSIVE NATURE — DEFEATIST

The defeatist attitude is one pole of this archetype. It is what happens when the energy within this Gene Key collapses. Those with this kind of nature store an enormous amount of tension in their bodies. This tension results from their failure to direct their huge life force in a direction that serves a higher purpose than their own. These are people who at some level deep inside themselves have given up, resulting in a dampening of all enthusiasm for life. At its extreme this becomes deep depression. These people tend not to project their misery outward but internally blame themselves, thus locking the energy even deeper in their body. Unfortunately, there is no way that an outside influence can pull someone out of this state of shutdown. It takes a huge inner motivation within the individual to lift free of their own inner demons. However, once they find a cause worth fighting for, suddenly all that latent energy is released into the world and the tension leaves their being.

REACTIVE NATURE — AGGRESSIVE

Just as the repressive nature lacks the stomach for a fight, so the reactive nature cannot help fighting, yet always ends up fighting the wrong things and the wrong people. The reactive nature is about projection. These are people who project all their anger and aggression onto others. They are the people who embody the very definition of struggle — to fight for no purpose. Having no sense of purpose, they are constantly locked in combat with other people in order to release the pent up tension in their bodies. They actually become addicted to fighting as a means to discharge this tension they feel inside. Obviously, these people do not have successful and loving relationships. They can be tyrannical and controlling. If, on the other hand, they are able to channel this aggression into some kind of higher purpose, then they immediately drop their aggressiveness and transform the Shadow into the Gift of Perseverance.

Here you learn the difference between meeting obstacles and fighting resistance.

THE 38TH GIFT — PERSEVERANCE

THE INDOMITABLE SPIRIT OF THE UNDERDOG

The only difference between the 38th Shadow and the 38th Gift is the nature of the fight. You have only to find the right fight, and your whole experience of this Gene Key changes. As you pour your heart, body and soul into a fulfilling fight, it ceases to be a struggle.

Here you learn the difference between meeting obstacles and fighting resistance. Resistance is what occurs when you are pushing against the universal flow and is a hallmark of all the Shadow states. Obstacles, however, are natural to the rhythm of life. Obstacles test your commitment and surrender, allowing you to forge new skills and hone your excellence. Obstacles are always gifts in disguise. The 38th Gift is designed for obstacles — in fact it loves them. If you are someone with the 38th Gift in your genetic profile, every obstacle is a wonderful and vital opportunity to feel more alive and to fulfil your higher destiny.

The 38th Gift is the Gift of Perseverance. This is a Gift that really thrives when you are up against the odds. People with this Gift make the impossible look effortless, even whilst exerting themselves fully. These people are extremely active and physical. They have a genetic need to push their bodies and love being right in the thick of the action. As we saw with the 38th Shadow, this is not a thinking archetype. These are people of action. The only trick for them is to know when to act and when not to act and this is where the Gift of Perseverance comes into play. It knows when to hold its energy back. In other words, these are people who have learned to draw a breath before they rush into some new action. This breath is not so that they can think — it is simply to make sure they are responding from truth rather than reacting from fear or anger.

As with the 39th Gift, the 38th Gift has a strong link with the archetype of the warrior. In the modern world, the path of the warrior is not what it used to be. All kinds of new domains have appeared where this Gene Key is now played out. There are warriors in the business world, warriors in government and education as well as in the sciences and arts. Wherever there is a fight for a higher purpose especially against insurmountable odds, there you will find the 38th Gift stretching itself to its limits. Where the 38th Shadow is always fighting out of fear, and usually for survival, the 38th Gift fights for love. In our modern world, the only domain for the 38th Gift is the battle against the Shadow frequency, which is the battle against the collective frequency of fear. The 38th Gift does not think about what it is doing. It does not stop to consider how foolish its behaviour may appear. It simply knows in its heart when something is right and once it has committed its full energy to its appointed task it will never, never back down. Even so, this does not mean that such people are beyond fear, because they are not. However, the frequency of the Gift level insures that love always wins over fear.

The 38th Gift is part of a genetic codon family known as The Ring of Humanity. Each of the six Gene Keys in this group represents an archetypal aspect of the human story. In this story, the 38th Gene Key sets the pattern of all human struggle — the struggle of all form to reach spirit. In human beings it is represented by the quest of the inner warrior of light against the dark forces of your lower nature. It is this quest or battle that stands behind all external human conflict. Perseverance is the most essential human attribute of the inner warrior because the darker forces of human instinct are so deeply embedded in our nature. Thus it is often through defeat that human beings learn to be stronger. Over time, through perseverance, love and trust, you will eventually attain victory and experience your own divinity.

People with the 38th Gift become our heroes. These people take a stand, absorb all the obstacles along their path and eventually win their battles despite appearing to be the underdog. With the 38th Gift comes the certainty of final victory. The only necessity is to find the right battles. For them, the right battles are those that empower others to stand up for themselves rather than remain victims of mass propaganda. There are two huge benefits to the collective that emerge from the manifestation of this 38th Gift. The first is a great challenge to the collective fear — it says that nothing is impossible if you stand up for what you believe in. The lives of people with the 38th Gift are living proof of this. The second benefit is that the 38th Gift also proves that you can fight a battle and win without resorting to violence and corruption. This does not mean that the 38th Gift is in any way *soft* — on the contrary, it can be fiercely aggressive in its service of a higher goal.

Many of the qualities that may be perceived as negative in the 38th Shadow become incredibly potent when they are put in service of a higher aspiration. Stubbornness can be an awesome quality when it is used to wear down a more powerful enemy. Aggression can be channelled into an appropriate force or used as a specific tactic at specific times. Even deafness can be wonderful when you need to edit out the negative propaganda of others. All of these qualities are fused together into this single word *perseverance*. What's more, a further aspect comes into play at the Gift level that is lacking at the Shadow level — thrill. The experience of fighting a battle against the odds, but for a higher purpose, fills you with an indomitable spirit that feeds on itself and becomes stronger and stronger. At a certain level, this spirit enters into the realm of an extremely pure frequency, one that gives its name to the 38th Siddhi — the realm of Honour.

THE 38TH SIDDHI – HONOUR

"GREATER LOVE HATH NO MAN..."

Honour is a dynamic, living energy field in itself. It is the Siddhi that dawns when the warrior archetype reaches its ultimate potential. Honour is the energy field of every single human being when they are living their individual truth. In this sense honour is something of a paradox. Even though it seems to be the peak of human individuality, it also binds all human beings together at the level of the higher self. In the field of honour, all human beings become one. This is the true meaning of the highest warrior code. Honour turns combat into a dance. It is the swordplay of love itself. With Honour, death is transcended and any human action can become immortalised. Even the act of killing another can become honourable if it is performed from within the energy field of true Honour. In such extreme cases, the one who is killed must always give permission to the conqueror in order for the true contract of honour to unite them.

To honour someone is to hold them to their highest frequency no matter what their current frequency may be.

What then constitutes an act of Honour? Honour is rooted in a love so pure that it sacrifices itself without a thought to serve a higher aspiration.

Every act of honour carries an equalising force into the world rather than a divisive force. This is how you can tell whether it is true or not. If the act brings people together into the high frequency field, it is an act of honour. Honour always carries the frequency of mercy and surrender within it. Many appalling acts are committed in the name of honour. Only if they have the effect of uniting all people involved and if they contain mercy are they honourable. In this sense, the whole essence of honour is contained in the expression of *honouring another*. To honour someone is to hold them to their highest frequency no matter what their current frequency may be. Thus a person manifesting honour may be subjected to all manner of victimisation at the hands of dishonourable people, but he or she will still never sink to their level. On the contrary, he or she will continue to honour them and hold them in the highest regard.

The Siddhi of Honour carries a certain type of mythology within it. We need to remember the journey consciousness takes through this Gene Key — from Struggle to Perseverance to Honour. The final accolade of Honour represents a complete transcendence of all struggle, but at the same time it still speaks the language of battle. This means that the 38th Siddhi retains a deep mythic connection to struggle, aggression and all the darkest sides of human nature. People in whom this Siddhi manifests are the highest form of warrior. They become our holy warriors — those people who cannot be tarnished by the world and who surmount the most awesome of obstacles. The 38th Gene Key, as we have repeatedly seen, loves a fight. These people have raised the frequency of that fight to a dance, to sacred theatre, and upon that stage they play their role with utter surrender. When the siddhic state occurs within a human being, that being loses all sense of a separate self. In many ways, they are simply shells through which their genetics continue to function. You might call them empty, or more accurately, you might say they were filled with the essence of life itself or divinity. Every siddhic state carries such a powerful vibration that it either strikes fear in someone's heart or it inspires that one to greatness. Either way, honour always reveals a person's true nature.

The 38th Siddhi is closely allied to the theme of death. Honour and death have always been closely associated. This is because the ultimate honour is to give your life for the sake of others. As Christ himself spoke: "Greater Love hath no man than this, that a man lay down his life for his friends". In our ancestral past, this Siddhi has long been deeply associated with conflict and war. However, at the siddhic level this association is symbolic rather than literal, for the 38th Siddhi fights to bring an end to all human conflict. At times, it may enter into human conflict and sacrifice itself to bring about that end. Nonetheless, in truth this Siddhi, like all other Siddhis, is based upon the utter respect for all sentient life. The ultimate fight then is the fight against fear. This has been symbolised forever by the forces of light fighting against the forces of darkness. As the true warrior of light, the 38th Siddhi does not fight the lower nature; rather it absorbs the forces of darkness deep inside itself through its actions, and in doing so shines out with immense purity in the world.

Honour is not about winning or losing. In a sense, the role of honour is to make a mockery of the entire concept of winning and losing. More often than not honour is embodied when a stronger force deliberately surrenders to a weaker force.

In other words, the being of light surrenders itself into the powers of darkness. This reversal of the human norm tends to highlight the futility of all human conflict and often results in the weaker force being transformed or *converted*. Such is the paradoxical power of an act of true honour. It always turns surrender into victory. This surrender is symbolised by death itself, since any true surrender signifies death to the separate self. After death, the honourable act only grows more and more in power until it becomes mythic. One of the most obvious examples of this is the surrender of Christ on the cross.

Whenever a Siddhi dawns in a human being, its programming partner is also activated simultaneously. Thus with the 38th Siddhi, we also see the 39th Siddhi of Liberation manifesting. This shows the true nature of honour — to unleash a chain-reaction resulting in a stream of dynamic liberating energy lasting for generations and generations. At the Shadow level we also saw how the 38th Gene Key is connected to the 28th Gene Key through our genetics. Thus we can see the interlinking of the themes of Honour and Immortality (the 28th Siddhi). Every true act of honour endures forever. People never forget the truly great lives, and as their stories are told and retold down the ages, the siddhic energy of the original being simply continues inspiring and liberating others. This is because the highest potential within every human being resonates to the field of Honour. When we hear tales of a great act or a great life, whether it be contemporary or a myth from the past, it reminds us of who we truly are — warriors fighting to overcome and dispel our personal and collective fears.

▤▤ 39th Gene Key

The Tension of Transcendence

SIDDHI
LIBERATION

GIFT
DYNAMISM

SHADOW
PROVOCATION

Programming Partner: 38th Gene Key
Codon Ring: The Ring of Seeking
(15, 39, 52, 53, 54, 58)

Physiology: Adrenals
Amino Acid: Serine

THE 39ᵀᴴ SHADOW – PROVOCATION

ATTITUDE AND ALTITUDE

Even though there are so many incredibly different flavours and possibilities within the human genetic matrix, there are relatively few *themes* governing humanity. The 21 Codon Rings program us collectively and individually to move along certain easily recognisable pathways. All the archetypes that appear on our television screens and in our novels and myths are therefore represented in our DNA. The 39th Gene Key and its programming partner the 38th Gene Key hold a truly unique place in our collective cosmology. Above all other archetypes, these two represent the myth of the warrior.

Legends and myths have extraordinary power over our different cultures. Even today, when campfire stories have been replaced by the silver screens in our living rooms, the myth of the warrior still holds us captive. We shall see in due course why we aspire so strongly to attain the highest frequencies of this 39th Gene Key. Meanwhile, the equally potent lower frequencies of the 39th Shadow are extremely sinister, as they currently hold the majority of humanity within their thrall. The 39th Shadow is responsible for maintaining a planetary vibration dominated by a single quality — violence.

Rooted in your adrenaline system, this 39th Gene Key is a highly dynamic code that is all about action. It has a gutsy, explosive and primal quality to it. People with these Gifts and Shadows are not led by their heads but by the primitive urge to act, whether wisely or foolishly. You can see from both the repressive and reactive sides of the spectrum, that this is a dangerous Shadow with which to cross paths. It is rooted in the fear of being trapped, which is essentially the fear of losing your individual freedom to act. We all know how dangerous animals become when trapped, and we humans carry that same fear in our ancestral DNA. There are many reasons why humans can be incited to violence but here we find one of the oldest reasons of all — the threat to individual freedom.

The 39th Shadow is like a cobra waiting to strike. This violence is personal, and when aimed precisely at its target seldom misses. We most often see its manifestation in our personal relationships.

This is the Shadow of Provocation or, in everyday language, *button-pushing*. Have you ever wondered how it is that your partner, parent or child knows the precise code, tone and word that really provokes your anger? Violence comes in many forms, and it does not have to be physical. Emotional violence is acoustic, using intonation to achieve its result. It isn't what they say — it's in the tone they use! Tone never lies, although it is usually totally unconscious. Like a child whining at the exact pitch that irritates you, every provocation has its tone. When someone is trying to make you feel guilty for example, they unconsciously use a self-pitying tone that touches the exact location of your guilt. It's all rooted in sound.

As we have seen throughout this book genes can be adapted, even though they cannot easily be altered. It is a matter of frequency, and frequency is about sound. As your attitude shifts, so your frequency shifts. If you focus only on the negative aspects of a situation, then your frequency drops. If you focus on the positive, then it lifts. It depends on which octave you are tuned to. Attitude in this sense is always acoustic. What kind of process does information go through as it enters your ears? How have you wired your own thinking and feeling to respond to the myriad intonations from the environment? Until you create an inner transformer to tune out lower frequencies and tune in higher ones, the lower frequencies will hit your DNA and bounce back as a reaction rooted in a similar low frequency.

The 39th Shadow has a single goal — to provoke you. If you are easily provoked by a person, any person, then you are under the influence of the lower frequency of this Shadow. You are trapped as soon as you react to the provocation. The provoker and the provoked are simply playing out an age-old chess game of abuser and victim. Most of us play both roles at different times and with different people. The programming partner of the 39th Shadow is the 38th Shadow of Struggle. It is a genetic law that the moment you react violently (emotionally or physically) to an emotional provocation, you are engaging with the energy of struggle and effort. In this instance the violence is aimed personally and directly at a single person. This is different from anger that is simply released in a more healthy way without being directed towards anyone in particular.

The power of the 39th Shadow over human beings lies in our tendency to take things personally. The astonishing reason for this is that we think we exist as individuals, whereas in fact we are quantum patterns in an ever-changing web of energy. The next time someone provokes you emotionally, try to understand what really just happened to you. If you reduce it down to essentials, all that happened was that someone resonated some speech codes by means of their vocal cords, and this simple collection of tones caused you to react in a violent unconscious emotional manner. In the old days they would have called that other person a wizard to exert such control over you.

In light of the above, all violence can be understood as an acoustic field in which individuals react automatically and personally to speech codes according to pre-conditioned patterns. In other words, violence is an addiction for most people on our planet, which is why we endlessly watch it on our television screens. Even so, most of us feel that we have very little to do with the world news or the many atrocities going on in the world around us.

Violence is an addiction for most people on our planet, which is why we endlessly watch it on our television screens.

We tend to think that these situations are the fault of other people. However, the deepest and most shocking truth is that every time we either provoke or are provoked by someone else in an emotionally violent manner, we are propagating the world energy field of violence and co-creating these very news headlines.

All patterns held at a low frequency trap the human life force deep within your body. The 39th Shadow ensures that you do not breathe deeply. When you do not breathe fully, you deplete your energy and reduce your capacities. One of the sure signs of the 39th Shadow is fatigue. People who lack dynamism in life are people who have allowed themselves to feel trapped in some way. The irony is that the solution does not lie in changing their outer lives, but in changing their attitude. When your attitude changes, then your outer life begins to reflect this. If you just change your outer circumstances without changing your attitude, you will simply end up feeling trapped again.

One final intriguing insight about this 39th Shadow is that it influences the shape of the human body. The shape of your body is primarily determined by the way you metabolise the food you eat. When you operate at a low frequency you cannot metabolise your food properly, so you feel hungrier than you actually are. However, it is your spirit that is hungry rather than your body, and no matter how much food you eat, you will never quite feel sated. Your spirit is actually hungry for creativity — it has an urgent need to express its inner beauty. Thus the world reflects the 39th Shadow through global eating patterns. In the West, one of the great new health problems is obesity, whereas in the East it is malnutrition. These are both outward reflections of the power of the 39th Shadow operating at a collective genetic level. Until we humans release our own unique creativity into the world, we will be caught in one of these two uncomfortable extremes.

REPRESSIVE NATURE – TRAPPED

The foundation of each of the 64 Shadows is to be found in its repressive nature — the state of fear. Fear lies even deeper than anger, which lies at the source of the reactive nature. When fear is expressed it becomes anger. However, when fear is not expressed it traps us. All repression is rooted in being trapped. The most perfect trap is the one that we are unaware of and this is the case with the 39th Shadow. What is really trapped is our unlimited reservoir of life force and creativity. The majority of humanity is in fact trapped by the lowest frequencies of the 64 Gene Keys. It is fear that keeps us asleep. Those with the 39th Shadow in their nature are the ones whose life energy is often most frozen. Locked into habitual emotional patterns, their awesome human potential lies completely inert, as though asleep. This is why human awakening can be such a powerful event because as we awaken, our life force is released and our latent creativity is born.

REACTIVE NATURE – PROVOCATIVE

The first stage in awakening is in fact to get angry! It takes the energy of anger to crack the layers and layers of inertia and repressed fear. As we begin to feel our fear, it transforms into anger and we begin to act it out or project it onto others. This is the energy of provocation. It is actually fear made dynamic, but there is not yet freedom within it. We are still trapped by our anger and by our need to externalise our fear through causing others pain. People who are provocative by nature are usually unconscious of the fact. By provoking others they are simply heaping misery back on themselves. Such people are really seeking love at a deep level but are doing so through negative attention, which is abusive. Most human beings are caught by the lower frequencies of this 39th Gene Key, witnessed by every time we speak words whose underlying intention is to cause pain to another.

THE 39TH GIFT – DYNAMISM

THE PRESSURE OF CREATIVITY

As you activate the higher frequency of this Gene Key, you begin to have a notion of the nature of the energy that drives all children. As we know, children seem to have endless energy that just crackles in the very air around them. Many parents are literally dumbfounded by how active children are every moment they are awake. This is the raw energy of the 39th Gift — the Gift of Dynamism.

It is interesting to note how, as you make the transition from child to adult and your personal and cultural conditioning gradually take you over, you also experience a decline in your natural dynamism. By the age of seven years old, most of your inherent patterns are already set in stone. It is therefore essential that children be given enormous amounts of time and space within their first seven years, as their natural dynamism needs to flow in an unimpeded manner. Of course children need boundaries, but primarily, in their first seven years all they really need to do is play. Play is dynamism exemplified. It is the unbridled expression of the basic energy of life. Sadly, in many cultures, mostly western, children are sent to school at some point within their first seven years and their natural seven-year cycle of unstructured play is interrupted. This early emphasis on mental learning interferes with the child's basic genetic need to express its dynamism through its body, rather than through its mind.

One of the reasons for modern society's insistence on early learning is our misunderstanding of the difference between knowledge and genius. There is a collective belief that genius is something of a freak occurrence — that it is somehow not the norm. There is also a collective belief that genius has to do with knowledge, rather than intelligence. But true intelligence has nothing to do with the mind, even though it can manifest through the mind. Genius is simply the fruit of a natural intelligence that has not been interfered with. Whereas knowledge can be forced, genius needs to be given a great deal of time and space to develop organically. The first seven years of a child's life is the natural seeding time of their later genius.

The first seven years of a child's life is the natural seeding time of their later genius.

Having said all of that, the news is not all bad. No matter how good or bad your upbringing may have been, all conditioning can be reversed. However, this deconditioning usually involves a journey of awakening in which you have to raise your own frequency back to its original state. When this occurs, your natural dynamism returns, and it returns as creativity. The same energy that is play for the young child matures into creativity and genius in adults. One of the simplest and quickest ways to raise your frequency is to do what you love in life. If you do what you truly love you will unleash your creative dynamism, and the more creative you are the more energy becomes available to you. This is the simple equation that is missed by so many people. As with each of the Gene Keys that are a part of this Codon Ring — the Ring of Seeking — the 39th Gift creates enormous pressure in the individual and their immediate environment. It is this pressure that gives rise to so much creativity.

The same pressure that was provocative at a low frequency is also provocative at a higher frequency, but it is a matter of what you provoke. At the low frequency, fear provokes fear and anger provokes anger. At the higher frequency level of the 39th Gift, Dynamism can be contagious. Whenever you meet someone expressing their latent genius, it draws you in. In such a person's presence you feel capable of so much more, and your horizons expand as you begin to breathe fully once again. You will be amazed at how prolific people of genius can be in terms of their creativity. They never seem to run out of energy, as though they are being driven by a force beyond themselves. This is the 39th Gift in action; it is about action and energy — it doesn't wait but initiates, pushes, sparks and catalyses. A person living out the 39th Gift unleashes a tidal wave of creative activity wherever they go.

The other aspect of the 39th Gift is that it knows no fear. This is the true spirit of the warrior. There are many grades of warrior. There are those who simply fight unconsciously because they cannot help it — these are the people who cannot resist whatever it is that has provoked them to action. But there are higher types of warrior too, those who are provoked but who manage to control and direct their anger. Because these people are operating out of a higher frequency of awareness, they will defeat those who cannot control their anger. Beyond even this, at the siddhic level, are those rare warriors who cannot be provoked at all. They have seen the nature of reality: there are no individuals in the first place, which makes it impossible to play the game of conflict anymore.

In conclusion we can see that at the Gift level of frequency, the same energy that gives rise to violence becomes creative action. The provocative energy of this Gene Key is no longer used in reaction, but is now used in the service of provoking creativity and freedom in others. The energy of the 39th Gift is not afraid of hurting another person's feelings if it means breaking them out of a lower frequency energy pattern. The pressure of Dynamism cannot help but create more dynamism. The real essence of this 39th Gift is thus to touch other people's spirits in a way that helps them break out of their traps and into higher levels of freedom and energy.

THE 39TH SIDDHI – LIBERATION

THE CRUX POINT

Perhaps from reading and contemplating this 39th Gene Key, you can begin to have a sense of what happens as this same energy hits the highest planes of frequency. There is always a point where the Gift suddenly makes the quantum leap to the Siddhi and the blossom becomes the fruit. In the case of the 39th Gift, Dynamism becomes Liberation when it finally forgets itself entirely. The 39th Gene Key is highly individualistic — it has to be in order not to be taken in by the status quo. Unique creativity has to break out of all conditioned patterns in order to rise to the heights of genius. The Ring of Seeking provides the pressure that drives human beings to eventually transcend themselves.

At the siddhic level, an interesting thing occurs when all this genetic pressure breaks through to a higher dimension. Consciousness merges back into the totality, which means that it has to give up its greatest gift — individual freedom. There is a great irony here. Only in giving up the illusion of your individual freedom can you receive an even greater gift — the liberation of being. Few people are able to make the transition from creative genius to divinity because you must give up the very thing that appears to fuel your creativity — your individuality. To the genius, the giving up of one's individuality seems akin to death. But in order to attain a siddhic state, you must pass through the Gift level of consciousness and give up your seeking. This means that with every one of the 64 Gene Keys, you will have to give up your hard-earned Gift in all its glory in order to make the leap to the final level.

Only in giving up the illusion of your individual freedom can you receive an even greater gift — the liberation of being.

Dynamism as we have seen, feeds upon itself. To be dynamic is to create more and more possibilities through action. The by-product of Dynamism is that it serves many people, which in turn raises its vibration. The more people you serve, the more rarefied your frequency becomes until one day you reach the crux point. The crux point is not easy to describe. It cannot be predicted and it cannot be prepared for. Mystics have usually described it as a death and a subsequent rebirth. In the case of the 39th Siddhi, it feels like a physical death because this is such a physical, dynamic Gene Key. You have to put your entire being on the line. All your plans, creativity, work, intent and service must be surrendered into a single vanishing point. When this event arrives, there is no mistaking it. It comes with the guarantee that you will transcend. There are no failures in this sense. You can only move through the tunnel of agony until it turns into ecstasy.

Liberation is one of the strangest of phenomena. It is perceived by the un-liberated as an experience, even though it is not an experience since there is no one to experience it. Even though liberation is often described as an event, it is not an event. Even though it appears to happen within time, it does not. It occurs outside of time. There are no sounds that can be vibrated that resonate with this phenomenon. There are no words to describe liberation because there is no process to describe. It is perhaps best understood as analogous to dying.

Liberation is beautifully represented through the metaphor of the warrior's death. The warrior's life is one of preparation, adversity, energy and finally, death. The perfect death for the warrior is to die in battle. In ancient society, it was a great honour for the warrior to die in battle (the 38th Siddhi is the Siddhi of Honour). All of this is a metaphor for something that exists at a much higher level than we generally realise. Of course, the warrior is you — the independent human being. Your battlefield is the world, and the war is life. The true warrior is the one who is willing to die for the highest cause, whether that is represented by your country, your cause, your brothers and sisters or your children. The important part of the metaphor is that the warrior has to give his or her life for another, just as Christ died for humanity upon the cross. As in all of the great legends and stories, there is always a rebirth that takes place after the warrior has given his life for a higher cause. This rebirth is liberation, and it occurs only after great effort and many trials. This is the way of the 39th Gene Key, and the reward is the annihilation of fear through the death of the smaller self. This is the true meaning and symbolism of the warrior and is why we all aspire to it.

Liberation may sound similar to freedom, but there are subtle and important differences. Freedom represents the 55th Gene Key. There is in fact a deep connection between the 55th and 39th Gene Keys. At the siddhic level, the resonance of each word is precise and specific. Liberation is a provocative energy — it tests and challenges those who are drawn to it. The person manifesting the Siddhi of Liberation is profoundly dangerous to be around — not in the sense of physical danger, but in the sense of the Shadow patterns. This is not a person that treads lightly around you. This is a force that bores right into the heart of your low frequency patterns and strips you down to the bones of your true being. These are the teachers who direct the force of their love right at the weakest chink in your armour. Just as the lower frequency of this Gene Key pushes your buttons out of reaction, the sage displaying the 39th Siddhi will use any technique at their disposal to test the limits of your surrender. If you take anything personally, then you still have limits to your surrender.

The 39th Gene Key is related to the 55th Gene Key in the same way in which a spring mechanism releases an explosive energy. When all human seeking has exhausted itself a great need to act arises, and this pressure is currently building within all human DNA. Life itself is struggling to transcend its struggle through human beings, and the 39th Shadow builds the tension for change until it is released through the human genome as the Siddhi of Liberation. Even before the mutation of the 55th Gene Key therefore, we will hear the dynamic explosion of this energy, which provokes the mutation itself. Thus we will see how the energy of liberation ultimately provokes the realisation of Freedom.

40th Gene Key

The Will to Surrender

SIDDHI
DIVINE WILL

GIFT
RESOLVE

SHADOW
EXHAUSTION

Programming Partner: 37th Gene Key

Codon Ring: The Ring of Alchemy
      (6, 40, 47, 64)

Physiology: Stomach

Amino Acid: Glycine

THE 40ᵀᴴ SHADOW – EXHAUSTION

THE ENERGETICS OF FORCE AND WILL

The 40th Gene Key and its Shadow concern the correct or incorrect use of the power of the human will. The secret of this Gene Key is found in the difference between the words *energy* and *force*. Energy in this context refers to the natural vitality that flows into your actions in the world. Whenever your actions are in alignment with the universe the requisite energy is supplied from deep within your being. However, when your actions do not flow from your true source but are forced, your energy will be depleted.

The malfunctioning of the 40th Shadow is deeply connected to the conversion of food and liquid into energy through the medium of the stomach. In the age-old traditions of oriental medicine, the overall health of the human body is seen in terms of its vitality or *Chi*. According to this tradition, there are two forms of chi: pre-natal chi, which is the inherited vitality that you are born with and which governs your eventual age, and post-natal chi, which you extract from food and nature. The oriental approach to health is based upon conserving your pre-natal chi wherever possible, while at the same time boosting your post-natal chi. The 40th Shadow of Exhaustion comes about due to ineffective energy conversion of food and liquid into post-natal chi, resulting in the body having to draw upon its precious reserves of pre-natal chi. Coupled with its programming partner, the 37th Shadow of Weakness, these two low frequency energy patterns gradually wear human beings down to the bone. Like all thirty-two of the Shadow pairings, this process is a vicious cycle.

At the higher levels of frequency, the 40th Gene Key is actually responsible for transforming our civilisation and society through fruitful alliances, solid boundaries and mutually beneficial exchanges between individuals, communities and even entire nations. However, the 40th Gene Key has the Gift of Resolve, which means that people with this Gene Key in their Hologenetic Profile are born with great willpower, a greatly misunderstood human faculty. We usually see willpower as something that all human beings can access as long as they have the inner strength.

There is a credo in the West that says if you want something badly enough, you can always make it happen — it's simply a matter of willpower. This is the kind of conditioning that the 40th Shadow feeds on.

Willpower used in the wrong direction becomes force, and even though it may still succeed in its endeavour, the results to the body will be catastrophic and irreversible. When someone begins to force their life away from its natural direction, problems related to the stomach and to digestion crop up. Because the energy from food is not being properly converted, there is usually a build up in stomach acid that over time can lead to all kinds of far more serious problems, ranging from ulcers to full blown cancer. The incorrect use of your will also puts an enormous strain on your kidneys and adrenals as the body struggles to supply the energy to push against the universal flow. The result in the long run is premature ageing, disease and exhaustion. For most human beings, this is the normal way. Even so, the human body is an incredibly tough organism and can sustain an enormous amount of punishment.

There are two ways in which the 40th Shadow can overwhelm you — either through attempting to force your own willpower without having adequate support from others, or from allowing others to take advantage of your weak willpower and giving in to compromise. This latter scenario is very common among the business community, where people work in jobs with little or no resolve, and for fees that are far too low. The problem is that when you compromise and work in a job that doesn't allow your spirit room to breathe, your low frequency actually reinforces your low self-esteem, so you accept whatever you are paid.

Another way in which the 40th Shadow works is through individual force of will. These are the people on the other end of the same dynamic — the slave drivers who take advantage of the people with the weak willpower. Such people isolate themselves from others, becoming lost in their own ambitions and their addiction to work. The steady energy drain on their core vitality lowers their frequency and closes their hearts to others. Despite this, such people often maintain their direction through sheer force of will, but at the cost of their very humanity. As you will see, the most important secret to this Gene Key is the art of relaxation. For those under the influence of the 40th Shadow, relaxation is virtually impossible. In point of fact, true relaxation is one of the most absent qualities in civilisation today.

The 40th Shadow is partly responsible for another low frequency human state — that of loneliness or isolation. When you work at something you love, using your energy harmoniously, you automatically activate support from others and therefore do not feel lonely. However, when you push against the flow or give in to someone else's flow, loneliness always rears its head. When you allow your energy to be misused by another, which is the repressive nature of this Gene Key, your natural support network is withdrawn, giving you the feeling of being cut off from life when, in fact, it is your own action or lack of action that has caused this state.

Yet another manifestation of this Shadow frequency is a type of loneliness that comes when individuals isolate themselves from others through their refusal of support or by constantly biting the hand that feeds them. This is a typical behaviour pattern of the reactive nature of this Shadow.

This latter brand of loneliness is less obvious because such willful people often appear to have great strength and independence, whereas in reality at a more unconscious level they are just as weak and lonely as the repressive side. The 40th Shadow is adept at denial, and its prime denial concerns feelings — people with this Shadow active often deny the fact that they have feelings, and it is this denial that ultimately leads to their downfall.

Whenever human beings emotionally isolate themselves from others they put themselves in great danger. The subtle quantum field known as the astral plane simply makes it impossible to separate yourself from other human beings. This kind of denial turns the negative frequencies in on themselves, driving them deep into your body where they will gradually become more and more acidic, eating away at you. The 40th Shadow is one of the most deep-seated causes of cancerous diseases on our planet. It takes root whenever someone is unable or unwilling to face and feel the depth of their emotional pain. It is through this Gene Key that you have to deal with the fact that you alone are responsible for your health on all levels. Although others can offer you support and at times hold your hand, it is you alone who must face your life and all that it brings you.

REPRESSIVE NATURE – ACQUIESCENT

These are people who lack firm boundaries in life. People with this nature easily acquiesce to manipulation by others, lacking the will to stand up for themselves. This permissive habit is rooted in denial of one's own needs, typically arising from patterns and coping strategies that developed in childhood. These people will give of themselves tirelessly and usually to people or organisations that never really care about them. This 40th Shadow drains people who do not value themselves and their energy enough. When such people come out of their denial and make a stand for themselves, and themselves alone, their life can dramatically improve.

REACTIVE NATURE – CONTEMPTUOUS

The reactive side of this Shadow is about the denial of rage. Like the fear of the repressive nature, this rage has its roots in a difficult childhood. This kind of denial distorts the rage into contempt for others. These people can be extremely arrogant. They prey on the weaknesses of others for their own benefit. Such a sustained, disdainful attitude means that these people cannot afford to allow anyone to get too close to them. This reactive denial feeds on disrespect for others, and thus people with this reactive nature gradually erode their energy by cutting away all their support. Because their strong will is fuelled by unconscious rage, they do not often appear to be exhausted, but inside, their constant need to isolate themselves from their community takes its toll in the end.

The 40th Shadow is one of the most deep-seated causes of cancerous diseases on our planet.

THE 40TH GIFT – RESOLVE

THE LOST ART OF DOING NOTHING

When you activate the 40th Gene Key at a higher frequency, exhaustion becomes a thing of the past. When your life force is aligned correctly, you will discover that there is a vast amount of energy available to you. The same forces that drive the Shadow also drive the Gift, but the result is completely different. Everything about this 40th Gene Key is about having boundaries, and in order to create boundaries you have to be able to deny other people access to your energy. You have to be able to say no. The correct use of this kind of isolation ensures that neither your energy nor your resources are ever depleted. The energy that goes into isolation can be a wonderful ally in life if used correctly. It is this ability to set personal boundaries around your energy and your time that gives rise to the Gift of Resolve.

The 40th Gift of Resolve is about becoming adept at giving to yourself.

The 40th Gift of Resolve is about becoming adept at giving to yourself. Ultimately it is about deep physical relaxation. It is that wonderful balance between serving the world and serving your own enjoyment. True resolve is not possible unless you know how to relax. Many people in the hectic modern world mistake the word rest with the word relaxation. We all need rest, yet perhaps even more, we need relaxation. Rest allows the physical body to recharge, but relaxation allows all of our subtle bodies to recharge as well. When we relax fully, our emotional and mental health is assured as well as our physical health. The 40th Gift is a genetic reminder inside each of us of the great importance of relaxation in life. Life was never designed to be as hard as we humans sometimes make it. We have even designed a world that forces us to expend our willpower beyond what is comfortable and natural. The 40th Gift then is a Gift that knows how to save energy at all levels. It knows the great mystical importance of *wu wei*, the timeless art of doing nothing.

For a person with the Gift of Resolve nothing is really an effort because nothing has to be forced. These people may work extremely hard and use a great deal of energy, but unlike the Shadow, they do not exhaust themselves. They know when to stop, and more importantly they know when to say no. Work that is in harmony with your true nature is not really work at all. In this sense it doesn't require willpower. Resolve means that willpower is built into the activity from the outset, so it flows effortlessly out of you rather than having to be forced. In addition to this sense of effortlessness, the 40th Gift engenders a great deal of respect and support from others. Unlike the Shadow frequency, people living at the Gift level do not feel a distrust of support, even though they will always maintain the integrity of their own space. These people actually inspire support and help from others. They often become the backbone of teams or projects because of the sheer power of their resolve.

The other aspect of this 40th Gift is a truly magical insight that is applicable to everyone, regardless of their genetic predispositions: sometimes saying no to the right thing brings as much power as saying yes, and in some cases it brings even more power. People with the 40th Gift act as agents of nature — if their resolve does not allow them to do something or offer someone their resources, then that person belongs somewhere else in the grand scheme of things.

Even though the other may be disappointed or resentful at first, in the end they will see that the flow was correct for both parties. To say a firm and resolute *no* to the right person or thing is to stand rooted in the ground of your true being.

Resolve is more than just willpower or determination: it is the flowering of your aloneness. This 40th Gift thrives on solitude. The source of all your power lies in your love of being alone. This does not mean to say that people with this Gene Key in their Hologenetic Profile always have to be alone. It means is they do not suffer from loneliness because their life force is constantly in the process of giving birth. This inner abundance gives terrific strength to your aura and is likely to make you much sought after by others. The 40th Gift is a genetic balance to its programming partner, the 37th Gift of Equality. Whereas the nature of the 37th Gift is to endlessly offer support and succour to others, the 40th Gift balances this through ensuring that you also always have enough time, space and enjoyment for yourself.

Every human being has to eventually learn from this Gift. It brings balance to our lives by reminding us of the true power of our aloneness. In the words of the poet Rilke:

"The individual who senses his aloneness, and only he,
is like a thing subject to the deep laws, the cosmic laws.
If a person goes out into the dawn or gazes out into the evening filled with happenings,
if he senses what happens there, then all situations fall away from him as from someone dead,
even though he stands in the midst of life."

THE 40TH SIDDHI – DIVINE WILL

COMPLETE PHYSICAL RELAXATION

At the highest frequencies, the Gift of Resolve becomes transformed into pure Divine Will. In many mystical pantheons, the qualities of the divine are divided into three main cosmic organs — the Divine Mind, the Divine Heart and the Divine Will. Of these three, Divine Will is usually seen as the primary faculty out of which the other two emerge. The entire notion of the existence of a Divine Will is an aspect of humanity's need to know that there is some kind of beneficial force that controls the universe. In other words, this 40th Siddhi actually concerns our perception of the existence of God.

Two polarities within our genetics — the 37th and the 40th Gene Keys — form the bedrock of our beliefs and experience concerning the existence of a higher power. If we look at the 37th Siddhi of Tenderness, we can see that all those who have attained realisation through the 37th Siddhi have left an imprint within the collective psyche of humanity concerning the nature of the divine, and this imprint reflects a deeply tender and loving force underlying all creation. This tenderness is reflected in all mythologies and world religions that see the divine force as either a mother or father figure. In this view we humans are seen as the children of God. However, when you look through the 40th Siddhi you see an entirely different picture, one that has caused great confusion amongst mystical seekers for millennia.

Masters who have attained the 40th Siddhi are the great mystical deniers of God. Just as the 40th Shadow denies its own or other's needs, the 40th Siddhi denies humanity's need for a God in the first place. This is an extremely powerful expression of the siddhic state because whenever it appears in the world it essentially throws human seekers into a panic!

When a man or woman has attained the ultimate siddhic state known as enlightenment, an undeniable energy emerges from them — the pure energy of consciousness itself. Such people speak with great power. Humanity's collective need to know that God exists is actually a need based on our deepest fear — the fear that we are alone in a Godless universe. When the 40th Siddhi expresses its divinity it ironically does so by denying the existence of any separation between humanity and divinity. In doing this, the 40th Siddhi forces humans to confront one of the great problems of seekers — that the seeking itself gets in the way of ultimate realisation. If someone attains the siddhic state through this 40th Gene Key they will have done it in spite of God. This is the path of the mystical denial of the need for help from a God. Such people will follow no other teaching or master but will travel their path absolutely alone. When they attain to the highest state, they will often speak about it in the terms of their denial.

When the 40th Siddhi speaks, it will say that there is no path to God because there is no God outside of your aloneness. They may say that all holy practises and methods of seeking God are futile. They may not even talk about their state as *mystical* or spiritual. These are people who often refute the whole notion of spirituality and holiness. Because of their radical position such masters are not generally popular with the masses, or even with most seekers, but the vibration behind their words is undeniable to anyone who comes near them. The 40th Siddhi exudes an aura of splendid aloneness and displays an exquisite independence from the normal needs of human beings. Their words are simple, logical, penetrating and sometimes deeply shocking. This is how consciousness uses denial as a means to bring others into deeper authenticity, by destroying all your human hopes of attaining a God-like state. Ironically, it is only when your hopes have been dashed that you are empty enough to experience such a state. These are deeply paradoxical teachings.

As you can see, the Siddhi of Divine Will is riddled with paradox. To those outside the siddhic state, Divine Will seems like such a powerful concept — there is a reason behind everything and that ultimately everything remains in the hands of a higher force. To the one immersed in this 40th Siddhi, the paradox is wonderful — that God can only visit you when you are not there. When the experience does happen, the paradox is revealed in all its glory; every human being is a point of Divine Will, and yet nothing exists outside of humanity. From this viewpoint you are absolutely free to do as you wish in life and yet at the same time nothing you do is really in your hands.

You are absolutely free to do as you wish in life and yet at the same time nothing you do is really in your hands.

Within this Siddhi lies the great mystery of Free Will and the state that the mystics refer to as *choiceless awareness*. Neither concept has any meaning at the siddhic level because in this state of awareness there is no *one* to make choices or not make choices. Such are the games that the 40th Siddhi plays and such are the horrors with which the 40th Shadow struggles.

The 40th Siddhi has been responsible for what is generally termed *the negative approach* to realisation, whereas its programming partner, the 37th Siddhi of Tenderness, represents the path of the true seeker of God. Both paths are wound deeply into the human genetic storyline, and both paths will eventually have to be transcended and left behind in order for the true state of realisation to bloom.

Despite its paradoxes, the 40th Siddhi remains one of the great mystical chess pieces in human evolution. In the teaching known as the Seven Seals (detailed in the 22nd Gene Key), the 40th Siddhi represents a code in human DNA that will transform humanity on the physical plane. At an individual level, the 40th Siddhi holds the key to complete physical relaxation in which every molecule of DNA inside your body is functioning at its optimal frequency. This requires a huge process of karmic release, which is the job of the codon ring known as the Ring of Alchemy. The four Gene Keys in the Ring of Alchemy will transmute your DNA so that there is no longer any interference in your physical body. As you may understand from reading the 22nd Gene Key, this means that your highest subtle bodies can manifest directly through your physical body. This is the real meaning of complete physical relaxation; it is a state beyond the comprehension of your mind. It is the direct manifestation of Divine Will.

SIDDHI
EMANATION

GIFT
ANTICIPATION

SHADOW
FANTASY

▤ 41st Gene Key

The Prime Emanation

Programming Partner: 31st Gene Key

Codon Ring: The Ring of Origin
(41)

Physiology: Adrenals

Amino Acid: Methionine (Initiator)

THE 41ˢᵀ SHADOW – FANTASY

THE GENETIC WHEEL OF SAMSARA

The 41ˢᵗ Gene Key and its various frequencies comprise a truly remarkable archetype. It stands alone within the human genetic matrix with a very important function. It relates to what is known in genetics as the *start codon*. Since this is such an important Gene Key it may help to understand exactly what it means.

Below is an example of a section of genetic code transcribed into letters. The genetic code is made up of combinations of only four letters A, T, C and G. These letters are called *bases* and represent the basic building blocks of the entire code. Hidden within these billions of letters are specific instructions for the body to follow. In deciphering the code of life, scientists discovered that there were places within the sequence where the body always seemed to know to begin building. They found that whenever the body sees the letters **atg** in a sequence, it always acts on the instructions that follow. Thus they called this the *start codon* because it operates like a front door key into the code itself.

caattgtcatacgacttgcagtgagcgtcaggagcacgtccaggaactcc
tcagcagcgcctccttcagctccacagccagacgccctcagacagcaaag
cctaccccgcgccgcgccctgcccgccgctgcg**atg**ctcgcccgcgccc
tgctgctgtgcgcggtcctggcgctcagccatacaggtgagtacctggcg
ccgcgcaccggggactccggttccacgcacccgggcagagtttccgctct

From the above description, you should see how important this 41ˢᵗ Gene Key is. As a genetic archetype of functioning within human consciousness, its message is of huge import to us all.

At the Shadow level of consciousness, the 41ˢᵗ Gene Key centres around the issue of fantasy and dreams. Being in the thrall of the 41ˢᵗ Shadow of Fantasy is like holding the key to your dreams in your hand, but never turning it in the lock. Whether you have this 41ˢᵗ Shadow in your Hologenetic Profile or not, you are undoubtedly available for its influence because like all the Shadows, it exerts its greatest power through the

*Over half of
your genetic
code derives
from other
organisms
from an
earlier
evolution.*

collective frequency of the planet. It is because of this 41st Shadow that our planet is populated by people who dream of a better life but who, for one reason or another, are unable to bring these dreams into reality.

The 41st Shadow creates a continual pressure within humans; it is the pressure to evolve. When this pressure is distorted by a low frequency field, as is the current state of humanity, it becomes distorted into the pressure to feel happy. Thus begins what the ancients named the *Wheel of Samsara* — an endless cycle of suffering in which humans become trapped by the need to satisfy their desires. With the mass distortion of the 41st Shadow, it is as though the body of humanity has misread the instructions left within its collective DNA. It all begins here, in this Gene Key. It is not desire itself that is the problem, because desire (the 30th Shadow) comes after Fantasy. Fantasy is the spark that ignites the fuel of the desire.

So how is it that we have misread such a vital code within our nature? And is there anything we can do about it? There are many answers to be found within the higher frequencies of this Gene Key as you shall see. For the moment, let us try and see how this Shadow works, and how effectively it keeps human consciousness from flowering. The key to the problem, as ever, lies within your mind. The evolutionary pressure that surges up in human beings carries our entire genetic history within it; that is, our evolution all the way from amoeba to Homo sapiens. Over half of your genetic code derives from other organisms from an earlier evolution. All this history carries an enormous weight inside you. In one sense it drags you down, but in another it urges you to want to break free. To face the pressure of all this past inside you takes enormous courage and, as you let it in, you want to run and distract yourself away from it in any way possible.

Because of the *ancestral weight* carried by this Shadow, it has a deep connection to human appetite and energy. This deeply uncomfortable pressure coming straight out of the 41st Shadow causes us to literally hunger for a better future. By the same token it has connections with eating, not eating and depression. The distortion of the 41st Shadow can lead to all manner of weight problems and energy problems — from chronic fatigue to hyperactivity. All of these issues are ultimately rooted in the mind and its ability or inability to fantasise about the future. This Shadow constantly shuffles between the dream of being full and the urge to be empty. When the gauge reads *empty*, you fantasise about being full. Whether that means dreaming of meeting your perfect soul mate, having lots of money or devouring a bar of chocolate depends upon your individual mind and its conditioning. Conversely, when the gauge reads *full*, you can sink into heaviness in which you feel weighed down by your past and you experience the urge to purge.

The Shadow of Fantasy prevents you from feeling complete because the mind doesn't rest in the now, but swings between dreaming of the future and rehashing the past. However, the greatest problem with this Shadow is that it prevents you from actually fulfilling your dreams. You become addicted to the hope that the dream brings to your mind, rather than actually launching off in the direction of the dream. This is a major reason why people live off the fantasies provided by their culture — the movies or alternative realities created through technology and the internet which have become humanity's latest great fantasy addiction.

The 41st Gene Key is also always involved with issues of leadership. The leader that knows how to manipulate fantasy can have enormous impact over others. The programming partner to the 41st Gene Key is the 31st Gene key with its Shadow of Arrogance. This 31st Gene Key's theme of leadership is marred at the shadow level by false humility, which is really the same thing as arrogance. True leaders have to overcome a deep fear of humiliation, because when the 41st Gene Key declares its dreams aloud and stands behind them, they risk being very easily misunderstood. Those who bring the new into the world have always faced this challenge.

In conclusion then, the 41st Shadow of Fantasy sets up the great *maya* or illusion in which most human beings live. As a primary evolutionary pressure it triggers a *false start* into a mental representation of your life, thereby short-circuiting your ability to live in the present moment. Fantasy itself is a beautiful thing, but if it prevents you from really living it becomes an escape from life into the mind. Once you have become trapped in these addictive mental patterns, it is very difficult to recognise them and break out of them. However, in every turn of the wheel of fantasy, as the cycles of your life periodically reset themselves, there is that chance that the next time round, the code will be correctly interpreted. When this happens to you, you will engage the true start codon rather than its shadow. Instead of escaping back into the vicious cycle of fantasy and hope, your higher purpose will be unleashed and will manifest in the world.

REPRESSIVE NATURE – DREAMY

The repressive nature of the 41st Shadow is about escaping life through dreaminess. These people have an agenda that is dictated by fantasy. They really do not live in the real world. Whatever you say to such people, they will interpret it through their dream. Such people do not actually want to fulfil their dreams, but are addicted to the inner worlds created by their minds. They are governed by deep-seated fear that doesn't allow them to have proper human relationships. This dreaminess often presents as a kind of lethargy that gives rise to the gradual break-down of the energetic systems within the body — in particular the vascular and the digestive systems. The only way for these people to break out of their mental cycle is to begin manifesting their dreams on the material plane.

REACTIVE NATURE – HYPERACTIVE

The reactive nature of this Shadow can be a sheer bundle of nervous energy. These people are always ahead of themselves, being absolutely carried away on the fuel of their dreams. This kind of nature inevitably leads to burnout as the power of the fantasy they are trying to bring into the world outstrips the limitations of the material plane. Their insatiable hunger drives them into deeper and deeper problems, whilst placing enormous pressure on their nervous system. The inevitable breakdown of such a nature is often highly dramatic and destructive for anyone else involved. The hope for these people pivots upon their ability to allow others into their inner life and relinquishing their one-pointed obsession with manifesting their dream exactly as it appears inside their minds.

THE 41ST GIFT – ANTICIPATION

THE RING OF ORIGIN

As we saw with the 41st Shadow, this Gene Key represents the basic pressure to evolve — to seek out new pastures and new experiences. This evolutionary impulse is the secret to the 41st Gift — the Gift of Anticipation. There is an interesting phenomenon that occurs the more you raise the frequency passing in and out of your DNA. You become more and more sensitive to the hidden properties in the world around you. One of the things you become particularly aware of is the presence of morphogenetic fields. Morphogenetic fields were first postulated by the scientist Rupert Sheldrake, but many older cultures have spoken of such energy fields throughout history. Essentially, a morphogenetic field is an invisible energy grid that communicates specific information across time and space. Depending on your sensitivity, you can pick up information about the past or the future from the field.

The 41st Gift is a particularly special Gift. As higher frequencies begin to operate through this Gene Key, it picks up very specific information from the morphogenetic field. Every one of the 64 Gene Keys operates out of a particular morphogenetic field. For example, the programming partner of this Gift, the 31st Gift of Leadership, attunes itself to all higher frequency leaders on this planet, even inspirational leaders from our collective past. Each Gene Key draws strength and power from those who have come before and, at the same time, also attunes to those who have not yet been born. It may explain why certain people seem to have premonitions about the future. Both of these Gifts — Anticipation (41) and Leadership (31) — operate together, one reinforcing the other. The greatest leaders are those who build upon the past and anticipate what is coming. Likewise, people with the Gift of Anticipation are naturally held up as leaders.

The 41st Gift has a single genetic agenda — it is always attuned to the next evolutionary energetic grid that is waiting to descend into form. Behind it are hidden the blueprints of evolution itself. Which blueprint it picks up depends upon the level of frequency and the person's cultural conditioning and geography. We pick up different blueprints in different places, and as our frequency rises, the more details we see. Premonitions sometimes occur to people when their body receives a sudden surge of frequency through a shock or through being in a particular place with a very strong morphogenetic field. Most occult phenomena are the result of sudden electromagnetic surges through the 41st Gene Key. These impulses are often misinterpreted when the surge quickly dies back and the Shadow frequency regains its grip on the vehicle. The mind paints its own fantasy of what it just experienced and people can interpret these sensations and impressions in very diverse ways, from ghosts to past lives.

When you are able to maintain a high frequency through this 41st Gift, you will be able to literally *download* all manner of beautiful things from the morphogenetic field. All works of genius emerge from this field. Mozart is a fine example of a man with a strongly activated 41st Gift. His ability to *read* the completed scores of all his finest works straight from the

It is the difference between genius and non-genius. Genius manifests whilst non-genius dreams.

morphogenetic field is well documented, although it may not have been described in this kind of language before. Mozart was born into a world of music and was intensively tutored by his own father, one of Europe's leading musical masters. It is unsurprising then that Mozart picked up on the burgeoning musical morphogenetic field of his era, now known as the classical style. He anticipated its maturation and manifested it in the material plane. This is the real difference between the 41st Gift and the 41st Shadow. It is the difference between genius and non-genius. Genius manifests whilst non-genius dreams.

Since the start codon and its amino acid methionine are only coded by this single 41st Gene Key, it has a remarkable place within the system of chemical rings known as the Codon Rings. The 21 Codon Rings are threaded together as chemical chains that relay biological information between each other in a fractal manner. This method of information relay means that the body can essentially operate like a quantum biocomputer, organising multiple levels of information in response to stimulus from the inner and outer environment. Within this interlinked ring system, the 20 amino acids can be combined and recombined to create all manner of chemistries within the body. Nested in the heart of the ring matrix sits the 41st Gene Key — the Ring of Origin — pulsing with the source code that it then transmits to the networks of codon rings. Even though we may look at DNA as a double helix with the four bases strung out along a line of code, its actual functioning within the body as a whole is in no way linear.

The implications of the above are astounding. Every single start codon inside every single cell of your body is electromagnetically linked to each other. This is the foundation of the holographic body, which in turn is linked electromagnetically to the holographic universe. Every impulse inside you is therefore communicated to the whole, just as every impulse within the whole is chemically mirrored in your DNA. Once you have discovered the knack of influencing your own DNA, you can literally reprogram every cell within your body. However, in order to activate secret chemical formulae inside your body that accompany the highest states of consciousness, you must learn to overcome the inertia of humanity as a whole. This is the domain of the 41st Siddhi.

THE 41ST SIDDHI – EMANATION

THE OUROBOROS

As we have examined the 41st Gene Key, we have risen to the awareness wherein the physical body can be seen as a holographic mirror of the universe, the source of which is this very Gene Key. When we extend this model at a macrocosmic level to its ultimate frequency, we reach something both spectacular and terrifying. The question is: What does the cosmic start codon look like, and what does it mean to arrive at the source of all being? As the source of being, the 41st Siddhi is the Siddhi of Emanation. Emanation is a word that has been used by many ancient metaphysical and mystical systems as a means for understanding something that is really unapproachable. One of the best known of these systems is perhaps the Kabbalah, whose central tenet is a map known as the *doctrine of emanation*.

In a nutshell, the doctrine of emanation points towards a fractal model of the universe in which everything emerges infinitely as a mirror image of itself. The cabbalists used the symbol of ten spheres or *sephira*, one emanating from another, to represent the various levels of spirit descending into matter. The ultimate source, known as the *ain soph*, represents the unnameable, inconceivable, limitless light from which all else emanates. Although the Kabbalah is an incredible model of the universe, with many different uses and dimensions, it remains a flawed model because all models are flawed by the limitation of language. No language is able to approach this word *emanation*. It is a word that contains the concept of infinity.

The ancient Chinese named this 41st hexagram of the I Ching *Decrease*, which is rather profound because the 41st Siddhi is nothing less than a black hole. As you approach it, it begins to pull you into pieces. It sucks in language, it sucks in time and it even sucks in space. It reduces everything down to nothing. It represents the source of everything. This source has been given so many names by so many different traditions as man has tried to understand it. What is so fascinating about this 41st Siddhi is that it is the source underlying all true siddhic states. No matter what manifestation comes through our genetics, they are all ultimately grounded through this Siddhi that is not a Siddhi. In many ways, this is the nameless Siddhi, but we call it emanation because we have no other way of approaching it.

There is another interesting point that emerges from deep contemplation of the 41st Siddhi. Somewhere deep within your DNA lies a code whose sole purpose is to trigger the state we call enlightenment. In certain vehicles the body picks up a specific set of secret genetic instructions and triggers the neurological process of enlightenment. It may come as a shock to many mystical seekers that this process cannot be triggered by what we do up on the surface. Within the hologram of the universe, enlightenment is a spontaneous *acausal* non-event. You could say that Grace triggers this process, but the point is that it is a chemical process and no one knows why or when or to whom it will happen. It is an emanation that simply arises from an unknown source.

Within the hologram of the universe, enlightenment is a spontaneous "acausal" non-event.

The one whose body has undergone the process of true enlightenment cannot know how it has happened. Neither can they know to whom it has happened because no one remains present to experience anything. This Siddhi triggers the death of experience itself, which is impossible to comprehend from outside this state. From within this state, there is no underlying agenda — any manifestations of so-called heightened states of consciousness are meaningless. Bliss is meaningless, ecstasy is meaningless and God is meaningless. All the other 63 siddhic states are meaningless since they are all subtle expressions of the prime emanation. From the viewpoint of the 41st Siddhi everything is reduced down to mere physiology, which is why this Gene Key can be so terrifying for human beings. It makes a mockery of all your efforts, meditation, morality and systems for understanding life. There is no *how* within this Siddhi. This is the terrifying sentence for us because it means we can do absolutely nothing to trigger this state, which in turn means that we are utterly helpless.

The one in whom the 41st Siddhi is made manifest faces a terrible dilemma. If they speak, they know they will speak with an agenda. Even if they do not speak, people will interpret this as another agenda. Such people know that they are trapped. They know that their state can never be communicated and there is nothing that they say that can help anyone. In fact, whatever they do say will probably be misinterpreted. One thing that can be said about such people is that their *state* triggers great hunger in others. Whether that is a good thing or not is unknown! All they know is that for them there is no more hunger. They have somehow fallen off the wheel of samsara and in doing so, there is no longer any pressure to evolve. There is no evolution of consciousness. All evolution is a figment of our imagination and belongs in the domain of samsara itself. In a great final twist, even the 41st Siddhi is sucked into the very void from which it emanates, just as the eternal serpent of consciousness — the ouroboros — is seen endlessly devouring its own tail.

▤ 42nd Gene Key

Letting go of Living and Dying

SIDDHI
CELEBRATION
GIFT
DETACHMENT
SHADOW
EXPECTATION

Programming Partner: 32nd Gene Key

Codon Ring: The Ring of Life and Death
(3, 20, 23, 24, 27, 42)

Physiology: Sacral Plexus

Amino Acid: Leucine

THE 42ND SHADOW – EXPECTATION

WAITING AT THE EXPECTATION STATION

Almost all human beings are waiting at the Expectation Station. Expectation is the most profound mark of humanity. It is the dream that the future holds a greater promise than the now. Somewhere inside us, we are all waiting for our lives to improve — for a great day somewhere in the future when finally all will be exactly as we dream it can be. Somehow we will receive all the money we need, our love lives will be perfect and we will have absolute freedom to do all the things we have always wanted to do. But not now — first we have a few things to clear up before we can be truly happy. And so we go on, postponing and postponing... until we are old, and then of course it's almost too late. In fact it's never too late to be in the now. You have only to see through the futility of your personal expectations. If you can do this totally and with absolute honesty, you will discover one of the secrets of being human and being happy.

The 42nd Shadow is that aspect within your genetics responsible for keeping you always sitting at the expectation station. It does this by entangling your desires with your mind. When you read the 30th Gene Key and learn more about the nature of desire, you will see that desire in and of itself is not your enemy and as long as it is pure it can take you to great heights. However, the moment desire becomes entangled with the projections of your mind, it can only result in disappointment.

The 42nd Shadow is actually one of the Shadows at the source of all human mental fear, because the 42nd Shadow concerns death. This is the Gene Key that literally programmes us to die and it exists in every living cell as a built-in design for decay. Because our neocortex perceives life as time flowing by, it is the source of our anxiety about time running out. The notion of time and your response or reaction to its passing is rooted profoundly in this 42nd Gene Key which closes all natural life cycles, and in particular the seven-year cycle of growth and decay. Seven-year cycles have been noted by cultures all around the world for millennia. However, the most powerful seven-year cycles are the ones that appear to govern cellular generation within the human body.

There is an old adage we learned at school that every seven years we have a new body, since most of its cells have been totally replaced. This means that every seven years we arrive at a portal of some kind, and something must die in order for a new cycle to begin. How that transition occurs is the domain of the 42nd Gift or the 42nd Shadow.

The 42nd Gene Key is one of six Gene Keys that form a complex codon group in your DNA known as the Ring of Life and Death. In all cycles of physical cellular mutation, life is programmed to follow the archetypal processes represented by these six Gene Keys. All cellular life begins with the 3rd Gene Key and ends with this 42nd Gene Key. The 3rd Gene Key catches the essence of the beginning of life, through the Gift of Innovation and its Siddhi of Innocence. All life must innovate and adapt, beginning in innocence and gaining in experience. Through the 42nd Gene Key, life ends with the Gift of Detachment and the Siddhi of Celebration. Thus spiritual essence detaches itself from form, just as we human beings must detach ourselves from those who have passed away. At the highest level all death is really a cause for celebration, as we shall see when we come to explore the highest frequencies of this Gene Key.

Through this Shadow, human life is programmed to revolve around your expectations about your life and those surrounding you. Expectation itself should not be seen as a bad thing. It depends on how you react to your own expectations and is a measure of how much trust you have in life. Whenever you feel circumstances moving out of your control, you can immediately see how attached you are or how detached you are. Every time you identify with an expectation, you set yourself up for disappointment. It is actually possible to hold an expectation without being attached to it, which occurs quite naturally with the 42nd Gift of Detachment. If you are able to expand your consciousness and raise the frequency of this Shadow you will remember that you are merely a part of a greater natural cycle and that all events fit into a far wider picture than you can usually comprehend. Expectation and disappointment only plague you when your vision narrows to the event that you are caught up in, rather than seeing its place in a wider picture. As we human beings so often learn in life, many disappointments actually turn out to be enormous blessings.

The programming partner of the 42nd Shadow is the 32nd Shadow of Failure. It is easy to see how these two Shadows genetically reinforce one another to set you up for what you perceive as failure. In fact, the moment you think in terms of success and failure, you have already failed because you have become embroiled in expectation.

For such an innocent sounding word, expectation actually plays havoc with our lives. The mind only sees what you programme it to see which means that, at one level, it is actually co-creating your reality and influencing the flow of events around you. If you are expecting something bad to happen you may not actually notice the good around you, which means you cannot take advantage of it or even enjoy it. By the same token, if you are expecting something wonderful to happen and it appears not to, you miss out on the potential contained in the event before you. Your expectation takes you out of the present moment so effectively that you lose your place in the greater flow of the cosmos. Whether your expectation is optimistic or pessimistic, it narrows your field of vision and closes down the limitless potential that exists in each present moment.

Whether your expectation is optimistic or pessimistic, it narrows your field of vision and closes down the limitless potential that exists in each present moment.

REPRESSIVE NATURE – GRASPING

When expectation manifests through a repressive nature, it becomes an inability to let go and a continual grasping at life. These people simply do not want things to end, and do anything to keep everything the way it is. This is a deeply rooted fear of change. Obviously, everything in life does come to an end in order that something new can be born. There are many ways in which people resist change — through trying to remain young, through not allowing our loved ones to go their own way, through clinging in some way to the past. However, this clinging actually makes these people gradually wither away. When we do not embrace change and allow things to die and decay naturally we prevent life from renewing itself, thus sapping our own healthy energy and vigour.

REACTIVE NATURE – FLAKY

The reactive nature finds itself unable to fully complete anything in life. This is another way in which expectation subtly undermines us. These people try to avoid disappointment by not really making any form of commitment in the first place. They move from one thing to another without allowing the natural life cycles to complete themselves. The dilemma is that they continually get stuck in the same old patterns. By breaking a cycle before it reaches a natural conclusion, they simply have to begin the same cycle again in a different form. No matter what they do to escape the negative patterns, they keep reappearing in their lives. This can be a disaster in relationships and in their finances. People who are flaky have a deep-seated inability to follow through with things. They are unconscious victims of their own expectations regardless of whether they are pessimistic or optimistic.

THE 42ND GIFT – DETACHMENT

THE READER AND THE WRITER

Detachment is a concept that the western world finds quite challenging to understand. As one of the great aims of Buddhism, detachment appears at first to be an anti-materialistic philosophy that denies any sense of true immersion in the world of the senses. However, the Gift of Detachment when fully understood allows you the greatest freedom to explore life in its many facets. Despite how the word may sound, to be detached does not mean to *not* feel. In fact, if you are truly detached, you will feel more intensely than others because you will not have allowed your expectation, positive or negative, to constrict the event you actually experience. With detachment, you can even enjoy the experience of being disappointed!

Detachment occurs in the opposite way than one might think. True detachment cannot be forced through discipline. Any attempt to force detachment simply creates more attachment. This is often the reason why long-time celibates or monks *fall* from grace through their sexuality. You cannot hold down a natural impulse indefinitely. With true detachment one yields to every feeling absolutely.

This does not mean that it has to be acted upon but that it must be embraced, allowed and accepted. Oscar Wilde said that he could resist anything but temptation, and this is the essence of detachment — not to resist temptation but to dive into it with vigour, whether that be an internalised experience or an outer manifestation.

Detachment presupposes a great love and trust in life. It is about working *with* your expectation rather than being a victim of it. Detachment arises quite naturally of its own accord when you yield and trust in life. The 42ⁿᵈ Gift allows you to trust in the ebb and flow of events around you and to accept the growth and decay of our bodies and lives. This Gift sees life as a series of stories or tapestries intricately interwoven with each other. Even though you cannot immediately see the outcome of a given situation, you know that it is connected to a prior situation and bridges to another situation ahead in the future. With this detached perspective, you may begin to feel rather like your life is a book, and although you are the hero or heroine writing the story as you go along, you are also the reader, fascinated and absorbed, but never lost in the words or details.

The detached perspective of the 42ⁿᵈ Gift allows you to let go of your expectations very quickly, as life continues along its own course in spite of you. It also means that you can see clearly the places in the storyline where you still wish to exert some control. You realise that if you do not let go quickly, the emotional and mental waters will close over your head and drown you in torrents of your own regret, anxiety and self-pity. The Gift of Detachment represents the separating-out process between the emotions and the mind, and it takes place naturally as the frequency passing through this Gene Key rises. With detachment comes both understanding and healing. There is a great understanding and freedom that comes to every human being who surrenders deeply enough to their life. You learn to accept your own mortality and the intricate flow of events around you. With detachment, you become truly centred in your own being.

Detachment represents the process of letting go of control over your life, physically, mentally and emotionally.

You discover the Gift of Detachment through succumbing to your own humanity and the ordinariness of life. Then you appreciate life through your senses, you breathe the air deeply into your lungs, and you do not baulk at suffering or adversity. The 42ⁿᵈ Gift of Detachment represents the process of letting go of control over your life, physically, mentally and emotionally. It is a wonderful letting-go that travels deeper and deeper into your DNA as you gradually allow your expectations of life to fall into the background. The more you accept yourself the way you are, the more detached you become and the simpler life becomes.

THE 42ⁿᵈ SIDDHI – CELEBRATION

THE PUNCH LINE

The 42ⁿᵈ Siddhi is about the transcendence of death. These are the sages who teach humanity about the illusion of death. They do not teach us through knowledge but through their attitudes. This is a state beyond detachment. Detachment can never reach right to the core of existence because it feeds on the duality of the observer and the observed. In detachment, the observer watches the observed. When the 42nd Siddhi explodes — and it does explode — the observer *becomes* the observed. The indwelling awareness merges back into its source — pure consciousness.

With the 42nd Siddhi there is only one manifestation occurring and this is the state of Celebration. Everything is experienced as dying, just as it is with detachment, but the difference now is that something, somewhere deep within the vehicle begins to laugh. This laughter rocks the whole vehicle, since it emerges from the source of creation. The 42nd Siddhi sees the punch line of what it means to be human. When someone has entered the doorway of death deeply enough within themselves, they discover the profound meaninglessness of existence, and as this great truth begins to permeate them, from the inside out, it literally kills off every aspect of their own identification with what they are.

The joke at the heart of all creation can only be experienced directly. It lives within the heart of every cell of your being. You are constantly in a state of dying — every cell within your body is dying from the moment it is created — so in essence you are not really alive at all — you are simply a passage through which matter is endlessly converted into energy. When your awareness stops identifying with all this movement and finally comes to rest, you actually disappear. It is rather like sitting by a river and following the currents and eddies with your eyes as they continually disappear from view. If you stop watching the river and simply gaze at it with your peripheral vision, you often have a curious experience of timelessness. The river no longer has a direction — it simply comes at you and out of you from every part of your being. This movement of life that isn't really a movement is the sound of Celebration. You do not know whether something is being born or something is dying. There is simply the endless pulse — as the river laughs at the meaninglessness of it all.

The transition from the Shadow to the Siddhi, from Expectation to Celebration, does not occur within time. The process of detachment emerging over time as you accept your Shadow patterns is in fact an illusion. Even though it may feel to you as though you are making progress from one state to another, nothing has really occurred. Awareness is simply playing with itself. In a sense, it is playing at being serious. Awareness enjoys the seriousness of being on a quest, of being a hero or heroine within the drama of your life. However, when the siddhic state is experienced a great realisation dawns: you have always been asleep. Even as you thought you were evolving and that you were doing really well at being detached, you were still asleep. You cannot be half-awake! You are either asleep or you are awake. In spiritual circles, we often use the word *awakening*, as though this were a steady progressive occurrence from one state to another. In fact it is a sudden occurrence. It is not even an event. It is the end of all events!

The 42nd Siddhi really is akin to someone suddenly *getting* the punch line of a joke. Up until that point the whole storyline was a set-up that was leading to this final insight. And as with all great jokes, you can only laugh because in a sense you have allowed yourself to be tricked. The physical mirror of these truths can be found in human DNA. The DNA helix uses human forms to continually reshape and reinvent itself. It is like some cosmic serpent that endlessly keeps shedding its skins. Our bodies are those skins. Neither can you break through to this truth — the chains of DNA cannot be broken. It is more that you somehow slip through the net, almost as though it were a mistake, and suddenly you see the joke — that we are not those skins! We are the serpent — timeless, unpredictable, fluid and playful.

The one in whom the realisation of this Siddhi has dawned is filled with laughter, and all they wish to do is share this laughter with others. They do not know how or why this has occurred to them, and they do not know whether it will occur to anyone else. The siddhic state occurs outside of the human mind, silencing all the noise. There are no practises that assure it will happen, there is nothing at all anyone can do to trigger this state. It simply happens according to fate, which enrages the minds of those to whom it hasn't occurred! This is why there is nothing to do except celebrate — what else is there left to do? Celebration is the direct manifestation of true awakening, rooted as it is in the utter helplessness of what it means to be alive and dying inside a human body.

▤ **43rd** Gene Key

Breakthrough

SIDDHI
EPIPHANY

GIFT
INSIGHT

SHADOW
DEAFNESS

Programming Partner: 23rd Gene Key
Codon Ring: The Ring of Destiny
(34, 43)

Physiology: Inner Ear
Amino Acid: Asparagine

THE 43ʳᴰ SHADOW – DEAFNESS

SURVIVAL – SERVICE – SURRENDER

There is a natural process built into the lives of human beings, and this process essentially has three stages — survival, service and surrender. These three archetypal stages roughly represent the three levels or bands that form the spectrum of human consciousness upon which the 64 Gene Keys are built. At the Shadow level you are always operating out of your genetic survival mode, which is keyed into the frequency of fear. At the Gift level you enter into service mode, in which a transformation has taken place inside that enables you to operate more efficiently within the whole. The final stage involves a complete transformation as your individuated awareness is surrendered and your DNA is finally allowed to function with absolutely no interference from within. What is common to each of these three stages is the transformation process itself. At each level, it occurs in sudden leaps. We tend to move through long periods in which it appears that very little about us alters, and then suddenly, every now and again, an experience or event occurs which utterly changes our lives.

These transformational events can never be predicted in life; neither can the effects they have on us. As the ancient texts have always attested — the only certainty in life is change. At the Shadow level of awareness change is all you really want, but true change is denied you. It is because you *want* to change that you cannot change. The 43ʳᵈ Shadow of Deafness is what prevents you from hearing this simplest of truths. In this sense, deafness is the inability to hear what is going on inside you. Because the Shadow frequency is a field of fear created by human beings, you cannot hear the truth inside you due to the white noise generated by the sub-acoustic aura of humanity. The 43ʳᵈ Gene Key is about acoustics and inner listening. The Gift of Insight requires a calm inner environment, and yet the lower frequencies are so deafening that it is rare to find a human being with a clear, still aura. One clear insight from a higher frequency can completely change your life because true insight is a mutation within the structure of your DNA. The 43ʳᵈ Shadow blocks such inner events by tuning you into the lower planetary frequencies of pure noise.

The 43rd Gene Key is about surprises, and it will not reveal its secrets to a taut mind. The pulse of life follows a deeply focused rhythm that periodically undergoes unexpected fluctuations. This is true from the movements of the galaxies right down to the minute workings of the subatomic world. We can see these rhythms everywhere but if we look long enough we can also see the anomalies that break the pattern. This is why science for example has been unable to explain everything in the universe — because the goalposts keep changing! The human mind is unhappy living in such an uncertain universe. As an aspect of the universe, our very bodies are subject to these same fluctuations and we call them our moods. At the Shadow frequency, individuals do not want to accept that they cannot control their inner workings and their moods. We spend our entire lives trying to create the illusion of security and stability. Much of the world still exists at a level of subsistence that cannot afford to be concerned about how they feel — they are simply engaged in survival. But for those societies that have pulled free from the primal survival frequency, fear still persists. As the saying goes, we move from being warriors to being worriers.

The 43rd Gene Key is about the progression of consciousness in form, and this progression is measured through a heightening of efficiency in the functioning of that form. In wealthy societies for example, it is technological breakthroughs that allow a greater level of efficiency and autonomy throughout the society. Eventually these breakthroughs will doubtless spread to the whole of human civilisation, but even that will not make us happier. In many respects, once you have more individual freedom, you realise the depth of your own dissatisfaction. This realisation constitutes a vital stage in human evolution.

The fear that governs the Shadow frequency specifically attaches itself to the way your body feels. This sets up a great dilemma inside you because it makes you search for happiness. What we humans call happiness is simply one aspect of our natural chemistry; sometimes you experience it and sometimes you don't. A great deal of the time we do not feel happy because we spend so much of our time trying to create happiness in the future. Modern society is locked into trying to create external security. We assume that happiness comes with financial or marital security whereas in fact it is not dependent on anything external. The true purpose of the 43rd Shadow at a collective genetic level is to create a more efficient society so that we can arrive at the breakthrough that nothing external can buy us happiness! Our deafness to our fear eventually leads us to confront it. This is the power of the Shadow frequency as a driving force of evolution.

This 43rd Shadow impacts you on a purely individual level — you are really only interested in taking away your *own* uncertainty, so you try to find the perfect relationship and you try to make enough money so that you can relax, or you try to change your body or your lifestyle so that you will feel good about yourself. Modern society is built upon the individual's desperation to escape the way we feel. In truth we are not really deaf at all. We are simply too busy worrying to hear, or we are too busy *knowing what we are doing* to listen. Our dilemma is that we try to escape the loop of not feeling fulfilled and in doing so we in fact ensure that we stay in that loop. Everything you do to try to reach fulfilment actually has the effect of making your life more complicated (the programming partner here is the 23rd Shadow of Complexity).

Your hearing is really fine — it is simply drowned out by the internal noise inside your head! In our modern world, everyone is expected to know what to do in life, as though everything were absolutely sure. The truth is that nothing is sure, and deep down inside, no matter how you adjust or improve your lifestyle, your body always comes back and reminds you of this.

The original Chinese name for this 43rd hexagram is *Breakthrough*, and as we have seen, life is about sporadic breakthroughs. As you rise up in the planetary hierarchy, you realise that no amount of individual freedom or wealth can fill the void inside you. Only when you finally realise this does the first great breakthrough come. This breakthrough occurs when you begin to face the fact that you simply don't feel comfortable in the world. No individual can really feel comfortable in society because where society tends to be consistent and relatively stable, the individual is neither of these things. To precipitate this great breakthrough in awareness you have to look honestly into one of your deepest fears — the fear of being rejected by society. Inside every individual is a pulsing rebellious spirit, and it is this spirit that finally begins to awaken when you realise how deafened you have been by that which society, culture and even history expects of you.

REPRESSIVE NATURE – WORRIED

The two aspects of the 43rd Shadow both concern noise. This repressed aspect is rooted in internal noise, or the human tendency to worry. Worry is based upon the mind going round and round in circles trying to figure out how to escape worry. It pressures us into all kinds of activities that we hope will bring an end to the worry. Of course, once one activity has finished, another worry immediately leaps into the resulting space, thereby perpetuating a very uncomfortable mental loop. All worry is based upon fear. In the case of the repressed 43rd Shadow, it is the fear of not fitting into the world and of being an outcast. In essence this is a pure fear, but when repressed it becomes a monster that drives us again and again out into the world to try to end the feeling through some kind of external achievement. Only when we confront such fear directly do we finally realise how much creative power lies in our individuality.

REACTIVE NATURE – NOISY

The reactive aspect of this Shadow manifests as external noise, or the human tendency to simply keep talking. These are people who do not talk in order to communicate something — they are not in the least bit interested in what anyone else has to say. They are unconsciously trying to deafen themselves from hearing the way they really feel, which is miserable. Added to this, these people have a deep need to be accepted and understood by others, and because they are not listening to themselves they often speak inappropriately or at the wrong time. Thus instead of being accepted, such people generally feel misunderstood and are often outright rejected. This dynamic may further increase their fixed viewpoint and make them even more paranoid and angry at being so misunderstood. When taken to the extreme, such natures can end up feeling so ostracised that they take out their anger either on those closest to them or on society in general.

You realise that no amount of individual freedom or wealth can fill the void inside you.

THE 43RD GIFT – INSIGHT

THE CREATIVE REBEL

It is through the 43rd Gift that the spirit of the rebel is born. Every human being is born to be a rebel in the sense of filling a space in the world that cannot be duplicated by another. The wonder of the human being is this unpredictable, spontaneous liquid genius. As individuals, when we awaken to our true creative potential, that which is hidden in our DNA begins to rise up and pull us out of the survival frequency. It is then that we enter into the stage of service. When we think of this word *service* we may have a preconceived notion of what it entails. However, service to the whole is not the same as service to society. The service that the 43rd Gift brings into the world is rebellion. Without the creative passion and insight of the individual, life would not only be dull but would probably cease altogether. Without the spontaneous, the unexpected, the dangerous, we simply could not evolve.

A society caught in the Shadow frequency does not like the rebellious spirit at all. Such people cannot be controlled and appear to threaten the very security that human beings have spent so much time and energy trying to create. The nature of the individual human spirit is one of romance — we humans are all really poets, pirates and lovers. We refuse to be categorised or stereotyped into a fixed role in society. It is therefore hugely ironic that even though every single human being alive is potentially just such a hero or heroine, we have created a world in which such people are labelled outsiders. We worship these heroes but we do so from a safe distance. Therefore, the creative contribution of the awakened individual does not serve our society directly. It serves other individuals by inspiring them to take the same risk and discover the latent genius that lies inside them. Such service ultimately comes back to the whole because it constantly shakes up the system, and all systems need regular shaking, otherwise they become frozen and stagnant.

At the Gift level of frequency the very handicap of the lower frequency, deafness, actually becomes a great ally. Instead of not listening to yourself, you now stop listening to the status quo. This is the first law of the rebel — to trust in your inner voice no matter what the consequences. This is the true meaning of the Gift of Insight. Such insight does not make the rebel a destructive force. He or she is not a reactionary who takes out their frustration on the external, or wastes time blaming and condemning others — that is the game of the Shadow frequency. The rebel awakening through the 43rd Gift is immensely creative — he or she simply cuts a new pathway without worrying about where it is going. This is the deafness of genius — to forget the future altogether and simply be the conduit for creative insight. True Insight delights in Simplicity (the 23rd Gift), which in turn brings about efficiency.

The nature of the individual human spirit is one of romance — we humans are all really poets, pirates and lovers.

The 43rd Gift reflects everything back into you. You cannot afford to trust any external source that claims wisdom or knowledge. It is not that you stop listening to such things but that you stop being influenced by them. It is only when you trust your inner core that the breakthrough comes from deep within your cells — from the DNA itself.

As this 43rd Gift is rooted in the acoustic field, the insight is not seen but heard sub-acous-tically, and it is experienced as a flood of knowing inside the cells of your body. The process of insight is a very transformational experience as it touches you right to your core. You cannot make it happen. The only way you can experience insight is unexpectedly, so you have to give up looking for it. It is rather like the process of writing a song or a piece of poetry — the more you try, the harder it becomes. It has to be allowed to happen in its own time and in its own way. Once its does happen, you experience great inner freedom because every insight opens you to wider and wider experiences of human awareness.

In Japanese culture, one of the key words for insight is *satori*, which implies a sudden break-through from within. No teacher or system can lead you to the experience of satori but these moments occur more frequently as you relax more deeply into your inner nature. As they become more frequent, a certain deep spirit of trust begins to pervade and suffuse your being, as though life's mystery were awakening directly inside you. You begin to realise that although you cannot create these breakthroughs, you can expand the environment in which they seem to occur. This environment is one of deep relaxedness and solitude. Such solitude does not mean you have to withdraw from the world but that you use the same deafness from the Shadow to filter out all but the essential and beautiful from your life. You will become a more contempla-tive being, and your gait through the world will become more sauntering than directed. In this way your awareness turns inward without losing any of its worldliness, making you unique but integrated, profound yet accessible.

The 43rd Gene Key is bonded in your DNA to the 34th Gene Key. Together, this chemical coupling creates a codon ring called the Ring of Destiny. Human destiny is caught between the polarity of these two Gene Keys, with the 34th representing the human urge to evolve and become more, and the 43rd representing the Divine urge to involve and dive down into the world of form. The secret of this codon ring is the secret of timing, and through the ages it has whispered a great truth into the ears of those whose insight is developed enough to hear it — life is a mystery in which chance and love are dancing together. The more you allow love, the more chance appears to work in your favour. When you love totally, unconditionally, even chance is shown to be an illusion and the underlying cosmic geometry or *cosmometry* is revealed behind all things. It is only then that your timing becomes perfect and that which appears to be random is understood as an aspect of the holographic universe unfolding its myth through your life. Destiny is governed by the combination of human strength and insight. Only at the highest siddhic level do you transcend the force of destiny altogether.

Life is a mystery in which chance and love are dancing together.

THE 43RD SIDDHI – EPIPHANY

THE STAR AND THE MAGI

We can see how evolution on our planet is moving us steadily toward a place of increased individual freedom. Our technological advances have made us so efficient that more and more people have time in their lives to actually ponder the meaning behind life.

This stage in our evolution can only deepen, thus preparing our species for breakthroughs of a more spiritual nature as the Gift of Insight becomes commonplace in people's lives. As the human rebellious spirit comes alive, your DNA attunes to higher and more refined frequencies. This means that the age of individual truth will take over from the need for collective pseudo-truths based on insecurity. This work on the 64 Gene Keys is a prime example of the dawning of the coming age since it describes each of the codes of consciousness but comes without specific instructions. Therefore, it cannot claim to hold any truth. The truth lies inside each of us. When it is seen, read or heard it may indeed trigger a process of recognition within your DNA. True insight only comes in this way, when all outer instructions have been removed.

As we learned at the beginning of this Gene Key, all human beings move through three natural stages of evolution — survival at the Shadow level, service at the Gift level and finally surrender here at the level of the Siddhi. The 43rd Siddhi is that of Epiphany. It represents the final breakthrough of consciousness possible within a human being. Despite its obvious Christian heritage, the word epiphany is derived from a Greek word meaning *inner manifestation*. One of the meanings of epiphany has to do with the manifestation of a divine being, which is why it has become associated with the recognition of the baby Jesus' divine nature. The Siddhi of Epiphany is an inner surrendering or *giving up* that takes place in certain human beings. When it occurs, those human beings cease to identify with themselves and become manifestations of divine being. There is a great difference here between being a manifestation of *a* divine being and a manifestation of divine being. The epiphany is that there are no special exalted beings — it is the epiphany that consciousness is all there is.

In many ways, the epiphany of the 43rd Siddhi would be a great disappointment if you could survive it to be disappointed. It is the sudden revelation from within that you have been wasting your time looking for a God, even a God inside yourself. Epiphany is expressed beautifully through the Buddha's enlightenment, when after so many years of searching and meditation, he just became so tired that he gave up. It was only at this moment of deep surrender that he experienced his sudden epiphany — that he was just an ordinary man searching for something that he already was. Even though you hear this and may sense the truth of it, you still can do nothing to hasten your own epiphany. It can only come unexpectedly. Everyone who experiences the Siddhi of Epiphany does so differently — the Buddha had to be disappointed, but other people have experienced it in many unique ways. There is a story about an Eskimo shaman named Uvavnuk who experienced her epiphany while having a pee beneath the stars one night! No one can ever know how, when and to whom the epiphany will occur, and that is its wonderful mystery.

Through its associated Christian symbolism, the epiphany is also connected to the recognition of Christ through the medium of the three Magi and the giving of their gifts. This colourful story of the three wise men that came from the East and followed the star to Jesus' birth is pregnant with archetypal mystery. If one understands the legend of the Christian epiphany as an inner mythic code one will see that it contains a great hidden truth concerning humanity's final evolution.

The three magi and their gifts can be interpreted as many different things because so many esoteric myths are founded upon the triple nature of divinity. What we see is a uniting of this triple nature around the figure of the Christ child, drawn together by a celestial force, the star. The star can be seen as an emblem of fate or the movement of the spheres that lies beyond our human grasp. The magi and their gifts can be seen as three disparate aspects or levels of our self, drawn together and unified by the figure of the Christ.

The 43rd Siddhi is special in another way in that it is one of seven Gene Keys that provides a portal for Grace to move into the world of form. In the 22nd Gene Key we learn about the mystical opening of the Seven Sacred Seals, specific codes in human DNA whose primary purpose is to heal a vital and specific aspect of the human wound. The 43rd Siddhi represents the opening of the 4th Seal, which releases the profound human fear of rejection and opens the hearts of all individuals. This Siddhi has a powerful destiny within the human genome — to create mass breakthroughs across entire gene pools and thus open up the borders and boundaries that separate our communities and countries. Ultimately, the Epiphany is an explosion in the human heart, an openness and acceptance of others, and a Divine spirit of friendliness that is at the core of all human beings.

At an individual level, the 43rd Siddhi is and always must remain unique to each person in whom it appears. It allows you to see behind the workings of the universe and the human mind. It is the one breakthrough that you cannot anticipate or imagine because it brings an end to everything you think you know about your world. There is enormous laughter stored up in this Siddhi, as the epiphany lands you squarely in the middle of the greatest paradox of all time — that you are what you are and nothing can ever be added or taken away from your nature, regardless of your experiences. When the 43rd Siddhi manifests in the world, it often plays the fool in some deep mystical way. These are people who know that nothing they do or say is of any importance or relevance whatsoever, so they simply act as they like and say what they like. As a consequence of all this, such deeply realised people inevitably break all the rules of how an enlightened being is supposed to behave. It is as though they are constantly sharing a private joke with themselves. With the 43rd Siddhi, the only thing you know is that you know absolutely nothing. This knowing is so delightful to you that you will simply never be able to contain the wonder and beauty of it. You have become deaf to everything but the Divine!

With the 43rd Siddhi, the only thing you know is that you know absolutely nothing.

≣ **44th** Gene Key

Karmic Relationships

SIDDHI
SYNARCHY

GIFT
TEAMWORK

SHADOW
INTERFERENCE

Programming Partner: 24th Gene Key

Codon Ring: The Ring of Illuminati
(44, 50)

Physiology: Immune System

Amino Acid: Glutamic Acid

THE 44ᵀᴴ SHADOW – INTERFERENCE

HUMAN FRACTALS

The 44th Gene Key and its entire spectrum of frequencies concern a little understood subject that actually underpins the structure of human societies, as well as the science of incarnation. This subject, which will be briefly introduced through this Gene Key, is the existence and nature of *human fractals*. The term fractal refers to a phenomenon found in natural systems whereby patterns are found to endlessly repeat themselves holographically. The more you magnify a fractal image the more self-similar patterns you will find hiding within it. The term *human fractal* is an extension of this notion to human relationships, referring to the web of invisible patterns that bind certain groups of people together. In the classical Indian teachings, such ties between people are known as *karmic ties*. In the language of human fractals, every person you meet in your life is a part of the overall fractal pattern of your destiny.

When you look at fractal images, particularly those generated by a computer, you will notice how fractal geometry follows certain holographic lines and patterns. No matter where you go in the fractal, you find the same patterns repeating infinitely into inner space. As the genetic counterpart of cosmic geometry, human relationship fractals follow similar patterns. You will always draw towards you those relationship geometries that teach you exactly what you need to know in order to evolve your current awareness. Through HoloGenetics, the astrological profiling system behind the 64 Gene Keys, you can track these fractal relationships throughout your life. Wherever you look into your Hologenetic Profile, you will uncover the same repeating themes. What all of this means is that the people in your life — those closest to you — really hold the secret to your destiny and your higher life purpose.

As you learn from each relationship in your life and over time come to master the lessons each affords you, then the frequency of your entire Hologenetic Profile heightens in pitch and you begin to attract higher frequency fractals. A higher frequency human fractal brings new people into your life, most of whom are operating out of a far more transcendent awareness.

A sure sign that you are attracting a higher frequency fractal into your life is devotion. Your relationships will begin to carry a quality of devotion within them. If you do not learn the lesson of a certain relationship — even if you leave that relationship — the pattern will come right back into your life through another person. It is through the 44th Gene Key that all relationships are called towards each other and why it is so powerfully linked to the theme of ancestral karma and incarnation.

The ancient Chinese called the 44th hexagram *Coming to Meet*, which is of course highly appropriate since this 44th Gene Key concerns how, why and when people meet each other, as well as the results that emerge from group or family dynamics. Human fractals seem to operate at different levels — in relationships, families, local communities and even whole tribal gene pools. To really understand human fractals you have to look at life both holistically and holographically. If we view separate human beings like cells within a greater organism, we could say that certain cells bind together at different places within the body at different times for different purposes. In other words, there is a kind of master program that choreographs the movement and migration of human beings. Even though it appears that we as individuals are making free choices, it is really the master program operating through us.

The 44th Shadow works like a virus within the operating system of this master program, throwing the collective choreography out of synchronisation. This creates both local and universal interference. You end up with dysfunctional families, businesses that create more imbalance than balance, governments with inappropriate leadership teams and relationships that are never easy. The overall result is global chaos. It is exactly what we see today. However, there is one vital thing to understand — the master program cannot be wrong. If it appears to be doing something chaotic, it is in order to diagnose the problem, locate the virus and reboot the entire system.

Interference patterns that are transferred through your DNA must be resolved in order for a fractal to be "cleaned."

This is exactly what is beginning to occur on a planetary level. The master program is beginning to assess and eliminate interference patterns, beginning with the smallest components — individuals. The current state of our planet cannot be fixed from the top down. If a program is misfiring in such a huge and complex manner, you have to fix it at the grass-roots level — you have to reset the basic components, which are the individuals, followed next by relationships. The moment we begin to see relationships appearing in the world without interference patterns, we will know that the core human fractal is rebuilding itself from the bottom up. Once you have a clear relationship, assembling the rest of the fractal is a relatively easy task. Every clean human fractal begins as a binary — a relationship. This is a universal human blueprint, symbolised by the mother and father of the family, although the foundation relationship of this fractal does not require that these two be of the opposite sex.

Along with its programming partner the 24th Shadow of Addiction, the 44th Shadow of Interference represents an incredibly powerful virus that keep people from finding relationships that are healthy and loving. This 44th Shadow brings a great deal of genetic baggage from the past. Many people interpret such relationships as karmic, since the bonds seem so strong and the lessons so challenging. It may be that they are not so much karmic as trans-genetic.

In other words, these relationships span two separate fractals or gene pools. The Shadow of Addiction ensures that even if you leave such a relationship you are immediately drawn into another, somewhere a little further along the same fractal network. The real reason for such relationships is that they are clearing houses for ancestral memory. Interference patterns that are transferred through your DNA must be resolved in order for a fractal to be *cleaned*. Couples or partners who are actively involved in clearing out this collective genetic dysfunction are the pioneers of a coming new age. However, for the majority of the planet, dysfunction is the norm, and until two individuals in a relationship are ready to take on the ancestral interference pattern of their entire fractal, that fractal will remain dysfunctional.

REPRESSIVE NATURE – DISTRUSTFUL

The individual repressive nature of the 44th Shadow is based upon distrust. This distrust is both an inherited trait and a conditioned one. It is a fear-based response to one's early experience in childhood and it infects all of one's relationships. These are the people who have a single disastrous relationship and then shut down out of fear of the same thing happening again. Although they may not show that on the surface, this fear causes them to unconsciously distrust all relationships. Such people often find themselves working and living together with others, but at a subtle level they keep them at arm's length. They are haunted by their own past and rigidly defend themselves from ever experiencing love or pain again.

REACTIVE NATURE – MISJUDGING

The reactive side of the 44th Shadow is a master of misjudgement. These people do not shut down like the repressive side, but keep making the same mistakes in their relationships. The 44th Gift has a sharp instinct for people, but if the basic frequency of your vehicle is operating below optimum, your instinct misfires. Accordingly, individuals with this reactive nature ally themselves to those who really don't respect them, who turn against them or, in business, who are a drain on resources and often just plain incompetent. The classic example is someone who keeps breaking ties with certain people only to rediscover a similar pattern with the next person. What happens here is that their instinct cannot operate outside its own fractal, so they make alliances that are fraught with difficulties.

THE 44TH GIFT – TEAMWORK

BODIES AND BLOODLINES

As the frequency of this 44th Gene Key rises the Gift of Teamwork is born. This is actually a fascinating gift because at a very deep level it concerns the human sense of smell. Much research has gone into the incredibly sensitive olfactory abilities of certain mammals, but most people are unaware of how this sense operates at a higher frequency within humans. These are people whose genius is to read other people. They can literally size you up with a single handshake.

These are people whose genius is to read other people.

343

This is not an auditory Gift like the 57th Gift of Intuition that can, for example, read you from the tone of your voice over the telephone. The 44th Gift requires close personal contact in order for the higher aspects of your olfactory nature to work effectively. You have to be able to smell someone in order to read their true nature.

People with the 44th Gift use their sense of smell at a wider level than simply smelling through the nose. In fact, these people smell with their entire immune system and through every pore within their skin; they can pick up undetectable odours such as pheromones and even subtler hormonal signals from those whom they meet. At a deeper level, this Gift is about someone being in alignment with his or her own fractal. Once you have a high enough frequency, you can overcome the collective interference of the 44th Shadow and begin to pick up the scent of your true allies in life. Not only will you recognise the right people for you, but you will also begin to move differently through the world as your higher instinct begins to function accurately. For the 44th Gift, life is about picking up subtle scents. As you follow each scent in your life, you are following your higher fractal, and that is when the miracle of true teamwork can be experienced.

To really understand this Gift of Teamwork you also need to understand something of the mechanics of incarnation. The 44th Gift can smell its own higher incarnation and life purpose in those around it, which is why it knows so much about group dynamics. For many millennia, different cultures have held beliefs about reincarnation and these beliefs are very much in vogue in the modern new age revolution. However, the science of human fractals contains some revelations that believers in reincarnation might find interesting and possibly challenging. The first thing that needs to be said on this subject concerns the master genetic program that was mentioned earlier concerning the 44th Shadow. This master program operates through time as well as space. Since the program orchestrates human interactions through the medium of collective gene pools, at the purest level of consciousness, it is true to say that nothing reincarnates except the fractal line itself.

It is actually possible to view incarnation from a number of different dimensions. When you study and contemplate the profundities of the 22nd Gene Key, you will learn about the subtle layers of the human aura known as the Corpus Christi — the rainbow body. When you view incarnation from the view of the causal body, which is what many traditions refer to as the soul, you can see how this very subtle aspect of your consciousness returns again and again and it gradually becomes more and more lucid and luminous. Your causal body, which vibrates at a very rarefied frequency, also travels with the causal bodies of others across space and time in a kind of cosmic relationship dance. Such fractal lines of evolution have always been understood through the ancient laws of karma. However, when seen from a higher dimension, there is only one single consciousness that underlies all creation, and it cannot be divided. Therefore, those causal bodies that travel together across the aeons — your close friends, your family, your husbands, wives, lovers and even your enemies — are all really aspects of a single body that keeps dividing in order to play out a wonderful evolutionary story.

When seen from this higher dimension, the notion of an individual soul that survives death and travels onwards is a simplification of how consciousness uses fractal lines to incarnate. Only consciousness itself survives death. But the ancestral memory is stored in the blood and transferred through DNA down the fractal bloodlines. When certain people recall their past lives, they are really reading their fractal lineage and identifying with a single archetypal aspect of that fractal. Certain people have been able to recall incredible details about past lives, but again this is an aspect of the 44th Gift whose whole purpose is to recall those people who resonate with its fractal line. The information of your entire fractal lineage can actually be traced all the way back to the source of all fractals, but this takes an extremely clear vehicle resonating at the level of siddhic consciousness.

The Gift of Teamwork is therefore about recognising who belongs in your life. When you meet someone for the first time and they seem familiar to you, it is because their vehicle belongs on your fractal line. Within the greater body, your cells have business together. Taking this further, if you were able to assemble your own true fractal around you, then the dynamics of this team would be nothing short of awesome. There would be total trust within the group, or total love within the family. The world has not yet seen much of the archetype of the 44th Gift. People with this Gift have a knack for group mechanics and a gift for knowing people, but they still have to work with the destructive interference patterns of the totality. However, as we shall see with the 44th Siddhi, all that is on the cusp of changing.

THE 44TH SIDDHI – SYNARCHY

THE COMING OF THE QUEENDOM

The 44th Siddhi is truly remarkable. It concerns the complete understanding of the mechanics of human destiny and the entire story of humanity. The programming partner of this Siddhi is the 24th Siddhi of Silence, which is a key to understanding how this all works. As a concept, synarchy is the opposite of anarchy. The prefix *syn* means to act in concert, and *archy* means to govern. Therefore the literal meaning concerns the concept of collective rulership. Historically the notion of synarchy has been greatly abused politically — communism claimed to be synarchic, as did Hitler's fascist regime. We begin to move a little closer to its true implications through its use within various occult schools of thought. Synarchy was seen by various occult writers as representing a world led by a secret society of masters. This 44th Gene Key together with the 50th Gene Key form the genetic Codon Ring called the Ring of the Illuminati, and given that the 50th Siddhi activates the quality of higher harmony, we can see how deeply woven into human mythology this archetype truly is.

The myth of secret societies has been much in and out of vogue in the last hundred years. With the rise of the New Age movement it is currently much back in vogue, with numerous books postulating the existence of the hidden illuminati who are said to meet in secret and manipulate world events. Such conspiracy theories may have their roots in esoteric tradition which speaks of a hidden centre of the world, variously known as Shambhala, Mt. Meru or Agartha, and from which a secret circle of ascended masters or celestial beings govern the world from a higher plane. All such myths and stories have their place but are essentially innocent distortions of the 44th Siddhi of Synarchy. True Synarchy requires an understanding of human fractals in order to fully comprehend its vast vision.

Perhaps the best place to begin an exploration of the concept of Synarchy is the insect kingdom. There are two groups of insects that are known for their synarchic systems of government — ants and bees. Of these, bees are perhaps the better example. Many of the old esoteric traditions speak of a great being — *the King of the World*, sometimes also known as *Melchizedek* or *Sanat Kumara* — who sits at the centre of all creation and governs all life in our planetary system. This is akin to the power of the queen bee in a synarchy of bees. In a hive, all bees serve the queen and their synarchy is divided into different levels of *workers* and *drones*. Within the hive a single spirit appears to pervade the whole society, and the queen symbolically and chemically maintains the focus and direction of every single member. If the queen dies, the hive falls into chaos and also dies.

When the 44th Siddhi is present in a human being, then they see the entire tapestry of human interaction through time and space. They not only see it, but they dissolve into it. The consciousness within such a person is able to travel down every single fractal line within the cosmos. Because there is no resistance within their vehicle, their consciousness ripples down the fractal arms of both the past and the future of the universe. The secret to this is Silence. Such a being has to fall into complete silence in order to hear the movement of every cell within the great hive of being. However, this 44th Siddhi is about far more than just understanding the secrets of human destiny and the fractal patterns of time. To awaken through the 44th Siddhi presupposes that you occupy what is known as a *core fractal*.

If you were able to follow all human patterns back to their source at the Big Bang, you would arrive at what are called the *Three Source Codes*. When the Big Bang exploded, consciousness was seeded into matter in a basic trinary pattern. In other words, the compressed energy exploded into three primary fractal lines or arms known as the Three Source Codes. This trinary fractal branching throughout the universe is well known in modern chaos theory. These three octopus-like arms began to spiral outwards in ever more complex fractal patterns and gradually coalesced to form the rudiments of our material universe. Every human being alive today has within them a fractal shard belonging or resonating to one of these three original codes. This original trinary pattern or *trinity* over time has become the foundation of almost all the major world religions and mystical systems.

Every computer program has what is termed a source code. The source code refers to a hidden code written by the program that allows access to the main programming matrix. The only way that you can influence the original program is if you have access to this code. In the case of our universe there are three source codes and the 44th Siddhi allows access to any one of them. Moreover, written into the master program of the overall fractal pattern of our universe are specific points — in what we call human history — where the program will mutate.

In other words, it has a built-in design to evolve beyond itself. If we consider once again the synarchy of the bees, we see the existence of a kind of hierarchy, but one that is circular rather than linear. No one in a synarchy is above anyone else because at all times there is an awareness of the group's inherent oneness. Each unit has a perfect fit in the overall geometry and, if there is no resistance through the individual forms, the totality can function as one. Within this vast tapestry are centres of force known as *core fractals*. Core fractals are what we term enlightened beings or masters. Every time a human core fractal is awakened, the entire arm of that fractal slowly begins to awaken. Thus it is said that certain beings or avatars come to earth to take on the sins of the masses. This refers to the awakening of core fractals and the domino effect through their ancestral genetic chain.

The being expressing the 44th Siddhi therefore occupies a specific genetic vehicle whose mutation causes, in time, a chain reaction through humanity. If you reduce this down to its essentials there is actually no glamour in the synarchy. The awakening of a core human fractal simply represents the activation of a previously dormant code within the overall program. As the core fractal is activated, it removes the virus that is wound around its genetic lineage, the virus that represents the Shadow frequency. The awakening of a core fractal essentially involves the death of the illusion of its individuality. The moment it gives up its independence, it becomes a clear conduit for consciousness to continue along its lineage. There are a finite number of core fractals — 144,000 to be precise — which is why this number has long been associated with a secret society of beings whose role is to awaken the planet.

One final revelation concerning the Synarchy is that it has existed since the dawn of time. It is simply distorted by an interference pattern that has also always been there from the beginning. This distortion or *sacred wound* is literally *wound* around the arms of every single fractal line. It is the reason for human suffering, but it is also the reason for what we call evolution. As the universe awakens, the interference is gradually cleared and the synarchy that lies beneath emerges. Interestingly enough, humanity has always sensed the existence of the synarchy — it lies in all our myths of a past golden age or a future paradise or heaven on earth. It is written into our destiny to realise the synarchic nature of our species, although the greatest irony of all is that by the time *it* arrives, *we* as individuals will all have disappeared!

THE 44TH GENE KEY

As the universe awakens, the interference is gradually cleared and the synarchy that lies beneath emerges.

45th Gene Key

Cosmic Communion

SIDDHI
COMMUNION

GIFT
SYNERGY

SHADOW
DOMINANCE

Programming Partner: 26th Gene Key
Codon Ring: The Ring of Prosperity
      (16, 45)

Physiology: Thyroid
Amino Acid: Cysteine

THE 45TH SHADOW – DOMINANCE

THE HOUSE OF DOMINOES

When you look into the core reason for the major problems on this planet, you will find that they all centre around a single issue — food. Because human beings have to eat in order to survive, food is symbolic of real power. Everything else springs from this single fact. In today's modern world, food is symbolised by money. As long as you have enough money you can survive. However, if you were immediately transported to the wilderness of northern Alaska your money would mean nothing, which shows that money is really only a veneer placed across a deeper issue — whoever controls the food resources has the power. The old Chinese name for this 45th hexagram in the I Ching is *Gathering Together*, which like all these ancient names contains many levels of wisdom. Very early in evolution, humans discovered that in staying together they greatly increased their chances of survival. When hunters worked together in teams they had a better chance of making a kill. It was out of this gathering together around food that our entire civilisation was born.

One of the earliest phases of human evolution was the hunter-gatherer. These early people were purely nomadic, moving from place to place, gathering and foraging for food as well as killing wild animals as they went. An evolutionary leap was made as early humans discovered that by staying in one place, cultivating crops and domesticating animals, survival was even more ensured. This move to a more efficient agrarian lifestyle based on the control of territory as a means of controlling the production and distribution of food was the crucible of our modern societies and nations. Along with this change came another human development — that of civilised hierarchy. As tribal groupings multiplied, natural hierarchies developed, either around a network of elders or around a single *alpha* leader. In a hierarchical society individual members were no longer equals. Power was distributed according to other needs, such as skills and usefulness, or was inherited through patrimony. This basic model of hierarchy is still in place today, although in the West, food and territory have been replaced by money and wealth.

*Ultimately
any system
that has been
built on a
foundation of
fear will by its
very nature
crumble.*

The 45ᵗʰ Shadow of Dominance lies at the heart of this whole evolutionary tendency to operate out of hierarchies. It is the way in which our brain is programmed, even though it is now showing itself to be hugely out of date. The fact is that our modern technologies and infrastructure would allow us to live in a world without hierarchy. That is our utopian dream. However, human DNA has not yet been updated, so the old fear-based thinking still prevails. We remain victims of the Shadow of Dominance because we still live for ourselves rather than for each other. Survival is limited to the immediate or extended family, if there is one. As we can learn from the 7ᵗʰ Gene Key, humanity actually has a predisposition to create hierarchy because some of us are genetically programmed to become leaders and be recognised as such by others. This is the basis of the concept of the *alpha* male in the animal kingdom. However, this isolation into blood families is the very boundary that human beings will have to break through if a new mode of thinking and living is to emerge into the world. Even in modern day business with all its emphasis on teamwork and corporate culture, your primary commitment is to feeding yourself and your family. Tribal thinking still dominates over collective thinking and family loyalties far outweigh corporate commitment.

This is where the fear lies — within the 45ᵗʰ Shadow. This Shadow does not allow human beings to transcend family blood ties. Our loyalty is primarily confined to our immediate family. Naturally this is not always the case, as a global mutation is currently underway and is opening up our awareness to new paradigms beyond that of family loyalties. In the West especially, we are witnessing the gradual demise of the traditional family model, with its built-in patriarchal hierarchy. For many people this is a very frightening time since the family model of living has been the norm for so many generations.

The human need to find a place of security in the hierarchical chain is based upon dominance. That is to say, you always have to push someone else down in order that you can rise up. It is a fear-based system focused upon the oppression and dishonouring of other human beings. This is also the essence of the modern day business model even though the majority of those who work in business participate in this model unconsciously. The programming partner to the 45ᵗʰ Shadow is the 26ᵗʰ Shadow of Pride, and those who enjoy feeling important do not wish to give up the illusion of their authority.

The system we have created keeps us working within the hierarchy because the hierarchy feeds us. You have to pay taxes and mortgages *as well as* work for the system! It is important to understand that no individuals are to blame, not even those few multi-billionaires who appear to run the world through the big multinational corporations. Neither are the politicians to blame as we would like to think. It is our genes that are really to blame because it is out of our most ancient fear of not having enough that we have created this whole civilisation. Every single human being is an unconscious participant in the world we see around us. No one escapes the hierarchy unless they decide to reverse evolution and go right back to their hunter-gatherer roots. Where, then, do we go from here?

We live during an uncomfortable transitionary era. As the old family-based tribal system is beginning to crumble around our ears and the old values cease to have power or meaning, we are left with a feeling of profound social instability.

At the same time the ancestral urge to dominate persists and has found a whole new arena — global finance. Now our hierarchical struggle revolves around money. Everything — health, education, food, and government — is all based around money, and there is nothing we can do about it. This does not mean to say that we cannot bring about a revolution and improve the way the old system works. This is precisely what is happening now and what is needed. However, ultimately any system that has been built on a foundation of fear will by its very nature crumble. We are living in a house of cards. The awareness that is rapidly coming to human beings will eventually outgrow the need for any system devised by the human brain. By the time that happens all concept of hierarchy will have long since dissolved.

REPRESSIVE NATURE – TIMID

The repressive side of the 45th Shadow is made up of all those people who bow their heads to authority. These people allow their spirits to be dominated by those higher than themselves within the hierarchy. Such timidity compromises its freedom for the sake of not rocking the boat. Since it is rooted in an unconscious fear, this nature accepts that no good could ever come from resisting or challenging someone in a higher position. Unfortunately the majority of humankind falls into this category, even if at an unconscious level. Most of us are willing participators in the global chess game in which the individual spirit is easily manipulated by those at the head of the hierarchy.

REACTIVE NATURE – POMPOUS

Rather than bowing its head to authority, the reactive side of this Shadow enters the fray by becoming obsessed with climbing up in the hierarchy and therefore has to push others down in their urge to reach a high position. These are pompous people who then assume positions of power over others and make sure that they remain timid in order to maintain control. Both sides of the 45th Shadow are, therefore, caught in the game of hierarchy — one as a victim and the other as the victimiser. The only real way to escape the entire game of hierarchy is to take a stand against the system itself, but not one that is rooted in reaction or anger. To break out of either reactive or repressive natures requires that one take a great risk and break the pattern of one's interaction within family, business, government and society. One has to stand alone as a living model of a new kind of system in which individual spirit is respected above all else.

THE 45TH GIFT – SYNERGY

FROM HIERARCHY TO HETERARCHY

Beyond hierarchy lies a far more expansive concept. It is known as *heterarchy*. Heterarchy is a level of self-organisation that already exists to an extent within many hierarchies. Whereas hierarchy demands a flow of information that is essentially vertical, heterarchy distributes information horizontally. In a hierarchy the flow of authority and information is always dependent on the permission or approval of someone above. In a heterarchy, however, responsibility and decision-making is distributed equally throughout the system. The flow of information processed through a heterarchical system is more effective than in a hierarchy. It has often been compared to the way in which neuron connections operate within the human brain. Heterarchy is based upon Synergy, the 45th Gift. Unlike hierarchy it recognises the need for both individual and group empowerment, so it operates at a far higher frequency than the hierarchical model.

In the business world, more emphasis is being placed on the heterarchical model, even though it is still utilised within the existing hierarchical one. An example might be a whole department of a business that operates with very little interference from those above and shows itself extremely capable and responsible. The 45th Gift of Synergy requires a leap of faith out of fear-based thinking and control by a single person. It is in many ways akin to the ideal at the core of democracy — that decision-making should and must be collective. More than this, in the business world synergy and heterarchy want to take us much further — into networks where businesses that previously competed now dovetail and cross-pollinate each other, thus moving more energy throughout the whole system. This leap out of the old fear-based territorial thinking also means that in a heterarchical model business, clients and customers are shared between the various businesses that make up the network. However, what this also means is that those who hold positions of power and authority have to risk giving their authority away.

It is this fractal energy of goodwill that has the real power, not profitability itself.

For someone in a position of true power to give complete authority to another is to perform an act of pure alchemy. This is what the 45th Gift is all about. When such sacrifice and empowerment is made to the right group, the rewards can be phenomenal. Even though initially a business may appear to lose money by interfacing in this way with other businesses, the amount of synergy and goodwill that moves through the human network increases exponentially. Rather than being guided from above, this new kind of business is allowed to discover its own self-organising intelligence. The 45th Gift is interested in the long term, so it is this fractal energy of goodwill that has the real power, not profitability itself. The goodwill generated reaches outwards in ever-increasing spirals that impact on the world through synchronicity and word of mouth. The result is that more and more people become interested in the business or its products. The bottom line of the 45th Gift is that in the long term it is more efficient to move together in synergy rather than waste energy in competition.

The 45th Gift will have its main influence on the modern financial world since its roots are in the overall control and distribution of food and resources. Once the coming change in human awareness takes root in us, the whole nature of modern finance can change to serve the collective rather than the individual. It is through the 45th Gift that world poverty and hunger will finally come to an end. The 45th Gene Key and the 16th Gene Key (whose Gift is Versatility) form the Ring of Prosperity, a high frequency means of generating vast abundance across gene pools. What these two Gene Keys point to is that the more we diversify our resources and unite them, the more successful and efficient we will become.

In the current hierarchical approach to business, networking still mostly manifests through one business buying another business and integrating it into the existing hierarchy, thus maintaining control. Once again it is only through the selfless and unconditional giving away of responsibility and authority that true power begins to travel through a network or organisation. You have to gather together the people who you trust and then you have to give them their freedom to pursue their gifts without interference.

The 45th Gift is also deeply concerned with education. In order for world poverty to come to an end, people must be self-empowered rather than rescued. This means that they must be educated in how to become self-sufficient at all levels. They have to use the technology that enables them to provide for themselves. In terms of global energy, the same truth also applies; we have to learn to harness the 16th Gift of Versatility in order to draw energy from a wide array of resources rather than drawing all the water from a single well — for example, fossil fuels. Synergy entails sharing world resources rather than the trading of those resources. At a subtle level, most trade is based upon fear. As the energy of goodwill and sovereignty is released through the 45th Gift, the whole purpose of money will gradually begin to change.

The key influence of the 45th Gift will manifest through the family. In the West we can already see the breakdown of the traditional family unit. In fact it is not family values that will decline but family politics. The idea of the family business will be pivotal in this respect but in a new and different way. In our current world model, family and business are directly at odds. Business today tends to divide the nuclear family because one or both partners have to leave the children at too early an age, either in care or at school. In the future, business will consider one of its greatest responsibilities to be the custodianship of healthy families. As seen and instigated through the 45th Gift, future business will involve not so much a family business but a family *of* businesses networked together. In this way, different clans and fractals of human beings will be united around a heterarchical business model. This could also provide a networked educational and support base for parents and children to draw from, rather than having to send children out to institutions that both alienate and homogenise them. Children can thus cross-pollinate with other families through play and, as they grow, through mentoring in a wide array of creative business possibilities. In this way, education, finance and family support all begin to come together in a single unified network.

These are just some of the startling insights and breakthroughs that are possible through the rising frequency of the 45th Gift of Synergy. Synergy involves an entirely new way of thinking that is so alien to our current culture that at present it is challenging for the mass consciousness to even conceive of.

THE 45TH SIDDHI – COMMUNION

THE END OF MONEY

Since we began in the 45th Shadow with the prime issue of food, we will end by viewing the same issue at its highest frequency. The traditional notion of communion is rooted in the Christian rite of sharing bread and wine in memory of Christ's last supper. There is a great mystery concerning the true meaning of communion. Ultimately Christ's actions during the last supper are deeply symbolic of the great revelation contained within the 45th Siddhi. On one level, the bread represents money and the wine represents the unification of all human beings through DNA — our blood. The drinking of the wine symbolically activates the ecstatic levels of human awareness and the transcendence of your individuality. These two themes — blood and money — are the core of the symbolism enacted in the Holy Communion. At the siddhic level of frequency, humans will recognise their oneness on the material plane through the transcendence of money. Human beings are the true currency, and money is our prime means of interaction as it directly reflects our frequency. If you act from a frequency rooted in fear, you will always try to maintain control of life through money. However, money also offers each of us the opportunity to test universal law. If you trust in the energy of goodwill above all else, then you will alter the way in which money moves in and out of your life.

When Christ spoke the words: "This is my body", he was speaking from a collective level. Humanity is a collective body that is united through its bloodlines in a common ancestry. From the level of the 45th Siddhi, we are each other's food. Money is the key to the enactment of this revelation. We have already seen how, at the Gift level of frequency, the redistribution of money through the world economy must alter to serve a higher purpose. This manifests through the new heterarchical model based on group synergy. However, at the siddhic level, communion entails the end of money altogether because it represents humanity's original and completed consciousness. We have seen that most trade is based on a subtle fear and is rooted in a version of conditional giving. Unconditional giving negates the need for money altogether.

In the 45th Siddhi, all the various strands of human DNA finally come together. It is the cosmic gathering point for higher consciousness. It is also the place where the individual, the family and the collective merge. The way in which this merging occurs is through the ultimate model of human and divine organisation — the synarchy. In the human genetic matrix there are many wormholes that interconnect the various Gene Keys in different ways — there are magnetic bridges and polarities as well as chemical codon links. Here we can see a sequential link between the Gene Keys as they are laid out in the original Chinese I Ching. There are profound archetypal links between Gene Keys that lie next to other Gene Keys. In this example we can see a strong connection between the 44th and 45th Gene Keys. These archetypal links were originally told in story form as one learned the meaning of the different hexagrams in the I Ching. In this case, we can see how Synarchy, the 44th Siddhi, forms a bond with Communion, the 45th Siddhi.

At the siddhic level, communion entails the end of money altogether because it represents humanity's original and completed consciousness.

We can further see how the 46th Siddhi of Ecstasy, next in the sequence, recalls the exalted state engendered by the symbol of the sacred wine.

In the journey of consciousness through this 45th Gene Key, we have seen hierarchy give way to heterarchy, and now we see their fusion and transcendence manifested through synarchy. Synarchy, as outlined and described in the 44th Siddhi, is the underlying Divine formation of human consciousness when free from identification. The problem with heterarchy as a model on its own is that, without a common point of focus, it may create synergy between all aspects within a certain system but it won't see beyond its own system. Heterarchy and hierarchy when fused together are transcended in synarchy. In a beautiful twist, hierarchy is actually the unfulfilled organising principle that leads to synarchy. In other words, when humanity has attained the mystical state of communion, it discovers that it is organised into a vast fractal pattern of human geometries arranged in myriad dovetailing human wheels and branches. This is *The Synarchy,* and one must understand that it is indeed a challenging concept to describe in words.

Synarchy is the overall pattern of humanity operating as one consciousness. It is the mechanics of the mystical state of Communion. Just as the communion of Christ is represented by a group of twelve with a mystical thirteenth at its centre, so the Synarchy of all human beings is built out of many such fractal families. Each of these twelve beings then has *their* own inner circle and so the fractal goes on. Each being occupying a pivotal or hub position in the Synarchy is thus an agent of communion, since their actions, words, deeds and visions flow directly out into humanity, in essence building the Synarchy. Christ himself is the Divine example of a *hub fractal* — a conduit of goodwill who sacrifices himself for the sake of the greater community. As each of us follows this example, we taste the greater body of humanity and enact the mystery of Sacred Communion in our lives.

The leaders in a synarchy are not leaders in the normal sense because they are not made up of individuals but small groupings of humans operating out of a single awareness. In many ways, synarchy can be likened to the interlinked network of neurons and synapses in the human brain. This is precisely what is happening to humanity right now as the frequencies moving through the 45th Gene Key come more and more into play. We are discovering higher and higher levels of self-organisation. The great cosmic joke, however, is that we are *discovering* these patterns rather than consciously creating them. In other words, synarchy already exists as a blueprint hidden within hierarchy, the 45th Shadow. As fear leaves the human genome, we will discover that we are already arranged into a synarchy. It is a gradual process of discovery however, limited by our time-processing brains. For those at the siddhic level of consciousness, time no longer exists. Through the eyes of the Christ consciousness, the kingdom of heaven is all around us at all times.

In the extensive transmission of the 22nd Gene Key, there exists in outline form the bones of a future mystery school designed specifically for the Synarchy. At the core of this lies a secret teaching known as *The Nine Portals of Planetary Initiation.* These initiations follow the passage of human consciousness as it evolves through and beyond its current genetic form. Within this system, the initiation representing the absolute zenith of human consciousness is the Sixth Initiation of Communion. In the words of the 22nd Siddhi:

The Initiation of Communion also shares its name with the 45th Siddhi, which describes the great mystery of the taking of the sacred sacrament. Communion involves the direct imbibing of Divine consciousness at the altar. In entering into this field of frequency, you are transcending all sense of being separate from others. This is symbolised by the blood of Christ and the final breaking up of the karmic residues held in your DNA. For the Grace of Christ to enter you, you must be willing to make the ultimate sacrifice — to give up your lower bodies and their desires, feelings, memories, dreams and knowledge and to be taken over by a greater being who has all along been waiting within you. To enter through this great Initiation is to die into the second aspect of the sacred Trinity — the Christ.

In the far distant future, at least from the viewpoint of our current awareness, the 45th Siddhi will flower throughout humanity. Not only will this bring an end to money, but it will slowly bring an end to food. As our collective frequency mutates our biological bodies, so the body of humanity will evolve to live on the frequencies of light. This is outlined in the 6th and 47th Siddhis and is programmed into the future of humanity through the codon ring known as the Ring of Alchemy. The mutation of our biology actually heralds the end of humanity as we know it, as we become synthesised into an even greater communion within the cosmos as a whole.

☷ **46th** Gene Key

A Science of Luck

SIDDHI
ECSTASY
GIFT
DELIGHT
SHADOW
SERIOUSNESS

Programming Partner: 25th Gene Key

Codon Ring: The Ring of Matter
       (18, 46, 48, 57)

Physiology: Blood

Amino Acid: Alanine

THE 46ᵀᴴ SHADOW – SERIOUSNESS

THE RAINMAKER

Sometimes the best way to describe something is through a story. In fact this isn't so much a story as a description of an ancient archetype. It centres on a rainmaker, a sorcerer whose unique gift is to influence local weather patterns and cause rainfall through magical means. In ancient times (and still in many places today) when a region was experiencing an extended drought, the locals would send for the rainmaker.

In our story the rainmaker is a little old man, and when he arrives in the village he is offered anything he needs. After all, the future lives of the villagers and their families depend on his success. Without rain the crops will not grow and there will be nothing to eat. However, the rainmaker says that all he needs is a hut to sleep in and to be left alone for a few days. Knowing well that his every move is being watched by the curious villagers, the old man begins setting up his paraphernalia, whatever it may be — perhaps some strange looking device, or a series of offerings to the appropriate gods. Some rainmakers just disappear into their hut and wait, appearing to do absolutely nothing.

After a few days, if the rainmaker is genuine, it will begin to rain. The villagers heap praise upon him and his magical powers. His reputation grows because wherever he goes, he appears to make it rain. However, despite his far-reaching fame our rainmaker has a great secret known only to himself. He knows that in truth he has no special powers over the weather. His secret is that he has discovered his true purpose in life — he is a rainmaker, and wherever he goes it happens to rain. He doesn't make it rain; he is simply attuned to places where it is about to rain. This is why he doesn't have to do anything other than show up wherever he feels like going.

This simple story encapsulates all that is wonderful about the 46ᵗʰ Gene Key and also contains the quintessence of the true meaning of this whole work. Hidden inside your DNA lies your higher life purpose, and when you find that higher purpose, everything is laid out for you by the spirit of Divine Grace.

*Living
without
knowledge
or memory
of the love
of the higher
planes means
that you can
only take life
too seriously.*

The Shadow archetype of the 46th Gene Key is Seriousness. Seriousness is the most wide-spread of all diseases on our planet and is a primary cause of much ill fortune. When you live your life from this Shadow, you carry a black cloud above your head wherever you go. It always seems to rain when you don't want it to because you are out of synchronisation with the whole. You create obstacles for yourself when you become too focused either on the future or on the past. Seriousness is about worrying or expecting or wishing life to be other than it is right now. Seriousness takes you away from life and love and into issues of control and separation.

The 46th Gene Key governs your relationship to your physical body. It is part of the codon group known as the Ring of Matter, which programs the developmental process of incarnation from the point of conception to the age of 21. Specifically, the 46th Gene Key relates to your first seven year cycle in which all your Shadow patterns are hotwired into your physical structure — your posture, your breathing patterns and your relationship to the physical world through touch. It takes a child seven years to fully incarnate on the physical plane. Certain genes are switched on and other genes are switched off during this period. Thus all the future patterns of your physical health are laid down in your early years.

Regardless of your physical circumstances during this period, your future life is fashioned out of the frequency field of those around you and how they deal with life. The greatest gift anyone can give a child is a loving, tactile and virtuous upbringing. This puts a huge responsibility on parents because it is the parents, custodians or teachers who create the living aura into which the child incarnates. When the child learns through their physical aura that life is safe and loving, their body relaxes into its natural internal harmony. What many people don't realise is that it is the body that must feel safe, not the mind. Almost all human health problems can be traced back to this primary period of imprinting. If you have an ongoing health issue then at some point in your first seven years your body stopped feeling safe and somewhere inside your DNA a gene or set of genes was either switched on or off. It is this early physiological wiring which feeds the 46th Shadow of Seriousness. When parents are unable to trust in themselves, they become too serious about life, and they transmit this frequency into the aura of the child.

Just as the 46th Siddhi allows you to enter into the field of ecstasy, so the 46th Shadow demands that you live in the field of agony. At low frequencies, you live as though you were submerged in matter, unable to access the joys and beauties of the spirit. Living without knowledge or memory of the love of the higher planes means that you can *only* take life too seriously. There are even those who take the spiritual path too seriously, and no matter how impressive their achievements you can see from their faces that they lack the true radiance that comes from living a lighter, more carefree life.

As we have seen, the 46th Shadow programs the way your body feels through the way your parents felt about their bodies. Unless your parents were firmly anchored in the spirit of self-love, they could not help but transfer their seriousness onto you as a child. Parenting at a low frequency always involves conditioning through control, whether conscious or unconscious. We want our children to be happy, but we do not know how to be happy ourselves. In fact, the freedom and spontaneity of our children is a constant reminder of the depth of our own conditioning and unhappiness. This is one reason why modern parents find it so hard to be parents; they have forgotten how to be happy deep within their bodies. In truth, they have forgotten how to play.

The formula for a wonderful life is so simple — tread lightly and don't worry so much. Life will take you where it wants you to go. But we humans tend to live in our minds rather than in each present moment, and our minds always live within time. We are genetically coded to be over-serious because of the pairing of the 46th Shadow with the 25th Shadow of Constriction whose repressive expression is Ignorance. Ignorance is not bliss, as the saying goes. We only display ignorance when we are too serious about our lives. We do not trust life but try to take conscious control of the events that happen to us. It is our ignorance of how easy life really is that prevents us from enjoying the fruits of the 46th Gift, which as we will learn, are manifold.

REPRESSIVE NATURE – FRIGID

All repressive natures are in one sense or another frozen. Frigidity here is used not so much in the sexual sense, but in a much broader context to describe the freezing of one's sensuality. These people hide from life out of fear of their own bodies. If you do not like your body your vital energy turns sour. Such people lose touch with the juiciness of life. You can see this reflected in their lifestyle, clothes and particularly in their faces, which manifest their fear through a permanently pinched expression. When such people begin to enjoy the beauty of being in their body rather than worrying about what they look like, they release the inner warmth buried deep within and their whole being begins to thaw.

REACTIVE NATURE – FRIVOLOUS

Frivolity is an over-reaction to seriousness. These people pretend to really enjoy life, and from the outside they appear to take nothing seriously. However, if you scratch the surface of such a person, you will soon find that they are in fact highly emotionally reactive and harbour a huge amount of anger. This anger will explode sooner or later because these people have an investment in being seen as light-hearted, laid back and jovial. When the façade is broken, usually when someone is honest with them, they reveal how deadly seriously they actually take life. As a pattern, frivolousness runs away from the truth, and such people usually reflect this in their relationships, which are generally short-lived and numerous.

THE 46TH GIFT – DELIGHT

"THE RING OF NO MATTER"

To escape the low frequency of seriousness, it takes a single, simple quality — acceptance. Acceptance, the 25th Gift, is the programming partner of the 46th Gift of Delight. These two Gifts grow out of each other. In order to accept something about yourself, you first have to come out of your ignorance and own it. Acceptance equals ownership, which leads to Delight.

Delight is the sense of freedom that emerges from an appreciation of the richness of being alive. This Gift is about *feeling* alive within the realm of matter.

It is funny that our English word for matter is tied into the expression: *it matters* or *it doesn't matter*, because this is exactly what the Gift of Delight is based on — an inherent understanding that nothing really matters except life and love. Therefore, at a higher frequency, the Codon Ring of Matter might also be known as the *Ring of No Matter*! At higher frequencies, your attitude becomes much lighter as you go through life. No matter what your circumstances, you can accept and recognise them as being a part of grace. In other words, when you are being shown something in your life — once you have seen it and let it go — you have more energy available for being delighted. Understanding this Gene Key is pivotal to your understanding the core of the 64 Gene Keys themselves. You must approach this transmission lightly, allowing it to wheedle its magical way into the labyrinth of your DNA. If you take it less than seriously and remain open-hearted and open-minded, you may experience the delight for yourself as it spreads further and deeper into your life.

*Luck is
what
happens
when
you stop
interfering
with life.*

Each of the 64 Gifts represents a different kind of genius, and people with the 46th Gift in their Hologenetic Profile move through the world incredibly smoothly. They seem to have a genius for being in the right place at the right time, and others often see them as lucky. However, very few people really understand what luck is. As with the rainmaker in our earlier story, luck is what happens when you stop interfering with life. Luck is nature's way of telling you that you are in harmony with the whole. The 46th Gift is also the gift of serendipity, of allowing good fortune to occur to you, and yet it can only happen when your attitude becomes lighter and less grasping.

When you meet a person whose frequency has risen to the level of their Gift, there is always something striking about them. In the case of the 46th Gift, they have an unusual ability to let go of the past and surf the present moment. These are the kind of people who never have any worries in life, and this creates a tangible magnetic draw to them. They are mysteriously attractive on a deeply physical level, as though life itself is somehow more amplified within them. Above all, they are soft and sensuous. The energy of delight is rooted in a deep sensuousness that comes from being very comfortable in their bodies. This being comfortable has nothing whatsoever to do with how they look — they could be fat, thin, ugly or beautiful. It comes from loving life more than anything else in the world.

The 46th Gene Key also contains within it the germ of material success and failure. This much-sought-after secret is founded upon the principle of synchronicity — the law that binds together all free-moving objects in space and time. By *free-moving* what is meant here is self-accepting. When you move through life freely with an attitude of delight and openness, whatever happens is correct. Whether you can see why it is so in the moment is not important. Success is only clearly understood in hindsight. If your attitude remains open and accepting, you will see that the universe is at work in you. Say, for example, you train for half your life to win a gold medal at the Olympics, only to fall ill the day before you are due to compete. You fall in love with the nurse at the hospital, get married and live happily ever after.

The point is that if you hadn't missed your gold medal, you might not have seen the real purpose of your life. At the Gift level, you will rethink all standard definitions of success and failure.

Synchronicity is an energy field available to absolutely anyone, whether they have the 46th Gene Key in their profile or not. The only prerequisite for tapping into synchronicity is this delight. In other words, you must remain open to surprises, let go of where you want life to take you and trust in a force beyond your control. Delight invokes grace. This is about enjoying the ride. This doesn't mean that you should have no goals in life or that you have to end up being some kind of aimless drifter. It means you hold your goals so lightly that they can be dropped if necessary at any point. To have your dreams come true in life, you have only to remember this one thing — stop being so serious!

THE 46TH SIDDHI – ECSTASY

THE ORGASMIC INNER WORLD

The 46th Siddhi is one of those Siddhis that really is challenging to put into words. Ecstasy is not a word that most people consider when they think of themselves and their own lives. And yet it is this block at the mental level that keeps you at a frequency far from ecstasy. That is the trick with these words for the Siddhis. If you can open your heart to receive the frequencies they represent, the frequencies begin to seek you out. Siddhic states are all a matter of magnetism. You have to expand yourself far enough to become one pole of the energy field — to be the negative pole or the attractor. This is exactly what we humans are — receiving dishes for universal frequencies. If you can expand your consciousness far enough, then the highest frequencies actually descend into you.

Ecstasy is the highest frequency of the 46th Gene Key and it occurs only through the heart. It is in fact the true nature of humanity. At this level of consciousness, which is really beyond all levels, the ecstatic realisation of your true nature is so strong that it silences your mind the moment it is no longer needed. You may think it is impossible for a *normal* person to live in a state of ecstasy and function in the world. Not true at all. Ecstasy *is* your state whenever your mind comes to complete rest. In other words, as soon as your mind is required in the world, the ecstatic state fades into the background and your activity moves to the foreground. Then, as soon as your activity is complete, the ecstatic state returns. In this state, your mind is as silent as your life path dictates.

People who manifest this Siddhi cannot go unnoticed in the world. There are Siddhis that can be hidden from most people's awareness, but the 46th Siddhi is not among them. The Jacuzzi of love that swirls around these people is so tangible that it can even be felt by people of the densest frequency. If you are highly developed in your own frequency, you can physically feel such a person's aura from miles away. People who manifested this ecstatic Siddhi in the past have left its imprint in the place where they lived, even though they may have passed away hundreds or thousands of years ago. It is not that this 46th Siddhi is any more powerful than any other Siddhi, because in essence they are all the same, but that this manifestation of divine consciousness lies so close to the physical realm and emerges so strongly through the physical body.

The purpose of the Ring of Matter is to be penetrated completely by the spirit. When this occurs, ecstasy is the organic result. Children who are brought into this world surrounded by the energy of delight are children who need never leave the state of ecstasy. Ecstasy also comes in waves, in orgasms that travel through the *noosphere* or universal quantum field in which we live. The more people who experience such orgasmic energy coursing through their higher bodies, the more our planet will be transformed. One day, very soon, entire communities around the world will experience group awakenings in this orgasmic way. The orgasmic energy will actually pass like a wave through the fractal lines of humanity.

There is actually no work more important on this earth than play. When we become truly playful, we reshape the nature of our collective reality. This is the great truth of the 46th Siddhi and is its vast importance to today's modern, over-serious world. Only ecstasy can silence the mind, only ecstasy can solve the problems of the world, only ecstasy can lead to world peace and universal love.

Those who have allowed themselves to be overcome by the 46th Siddhi are so overflowing with their own ecstasy that they rarely have been able to find words for their experience.

Only ecstasy can silence the mind; only ecstasy can solve the problems of the world; only ecstasy can lead to world peace and universal love.

Those who *have* found words have nearly always spoken in poetry, which is the language of love. Such men and women — the ecstatics — have not always found it necessary to remove themselves from the world, but have found their love within the world of ordinary life. The 46th Siddhi revels in the experience of the marketplace, the family and the ordinariness of the world. These people do not really care what happens to them because the heart makes none of the distinctions that the mind makes. Like a swallow darting about the skies, the heart swoops down upon life, playing and delighting with every waking moment and experimenting with every conceivable experience it can find. The heart doesn't care a tick about success or failure, past or future, life or death. It knows only that it is alive and beating right now, and that realisation floods the being with the sweet, liquid wine of ecstasy.

Your ability to attune to the ecstatic nature of your body depends upon your ability to let go of your mind and open your heart. It depends upon how grateful you feel towards existence for giving you this body and the experience of being alive. If you have felt this ecstatic feeling even once in your life, then you can recreate it. And even if you haven't felt it, you can open yourself to it. It lies within you right now, at the exact moment of your reading this. It will always remain with you, waiting quietly within the very ventricles of your physical heart. All you have to do is invite it back into your life.

SIDDHI
TRANSFIGURATION

GIFT
TRANSMUTATION

SHADOW
OPPRESSION

47th Gene Key

Transmuting the Past

Programming Partner: 22nd Gene Key

Codon Ring: The Ring of Alchemy
(6, 40, 47, 64)

Physiology: Neocortex

Amino Acid: Glycine

THE 47TH SHADOW – OPPRESSION

THE MAGIC MIRROR

The 47th Shadow provides human beings with one of the greatest keys to the mystery of suffering. Within this 47th Gene Key lurks a vast reservoir of darkness in the form of inheritable ancestral anguish. The 47th Shadow is the storehouse of human karma. What we term as *karma* really refers to memory, but not *memory* as we commonly understand the term. The memory of the 47th Shadow is genetic memory carried in our blood. However, this is not readable memory that can be deciphered through logical sequencing. In genetics, the 47th Shadow refers to what is called *non coding DNA*. Also known more commonly as *junk DNA*, these chemical signatures are evolutionary artefacts passed down through our collective ancestral bloodlines. Although no clear function has yet been ascribed to this DNA, it accounts for a stunning 98% of our genome. This ancestral storehouse does indeed serve a huge purpose as yet unseen by science — it is the very turbine of human evolution. The hidden reason and function of this DNA is revealed in its entirety through this 47th Gene Key.

The reason that science cannot decode the 47th Shadow is because history is not linear but random, and random patterns cannot be solved through logical methods. Random patterns can only be resolved or read by fractal geometry, which reads chaotic patterns using holistic laws. The fact is that every human being carries the entire evolutionary memory of humanity within their body. This also means that we are unconsciously driven by this memory. When a person finds a way to access their deep unconscious, they are really accessing the collective unconscious. If they are not mentally and spiritually prepared for this event they will very likely become deeply disturbed or delusional. The only language that can comprehend this world is the language of archetypes, such as the language used by Jungian psychology, the arcane symbolism of the alchemists or the ancient totemic language of the shaman.

*The very
thing that
terrifies you
the most is
your route
to a higher
evolution.*

The language of archetypes — which is what the 64 Gene Keys are based upon — allows us humans to look into the hidden side of our nature — to step into the perilous shamanic underworld of the collective unconscious without becoming personally identified with what we see there.

When the shaman goes into the underworld of the unconscious, he (or she) uses the language of his own totems to navigate this realm. In doing so he can intuitively locate the specific shadow aspects of a person's unique genetic coding and bring them to light. Whether he interprets these aspects as demons or animals or anything else is irrelevant. What matters is that they represent the specific fear patterns of a certain individual. The true psychoanalyst works in the same way. Instead of attempting to find reasons for our psychoses and fears in our childhood, he or she understands that they are really archetypal aspects of one single, universal primal fear. We each come into the world with certain Prime archetypes that represent our fears and, in addition to this, our life story presents other archetypes for us to address as we follow our destiny in the world. The 47th Shadow is an access portal through which these deep fears flow into our waking consciousness. This is why the ancient Chinese named this hexagram *Oppression*, because such fears become major burdens in our life if we do not face them directly.

Very few people willingly confront their Shadow archetypes in life. They do not want to stare down the barrel of the 47th Shadow because the deeper they look, the deeper the rabbit-hole seems to go. Thus they simply live in denial or reaction to these unconscious fear patterns, not realising that their outer lives are a direct reflection of this repression. To really look deeply into your fears is to undergo a major transmutation, one which eventually becomes cosmic and even mythic at the siddhic level, as we shall see. But transmutations are not a comfortable business for people at the Shadow level of consciousness because to go through such an alchemical process you have to let go of all definitions of who and what you are. What we humans don't realise is that the grail we seek is hiding beneath our fear. The very thing that terrifies you the most is your route to a higher evolution. This is why all mainstream religions are founded upon the duality of good and evil. In keeping evil separate from good, we are actually denying the very part of our nature that will allow us to directly experience the Divine. Evil is nothing but our fear manifested in the world.

The programming partner of the 47th Shadow is the 22nd Shadow of Dishonour. Your denial of your own Shadow is not only a dishonouring of yourself, but of life itself. If you cannot honour yourself, then you certainly cannot honour others. It means that you listen to someone through your own agenda, so you only hear what you want to hear. You carefully edit out all that junk DNA that you don't want to look at in yourself. This whole process between these two Shadows creates a biofeedback loop that conceals more and more repressed memories. We are literally reinforcing our self-limiting patterns. The ancients referred to this as the accumulation of karma. This has led to the belief that one can offset this negative karma by doing good deeds to accumulate positive karma. However, such practices never lead to true freedom because your unconscious Shadows remain unaccepted and repressed. Religions such as Catholicism found a way of relieving the oppression through the process of confession.

Still, confession simply lays the responsibility of your Shadows onto an outer manifestation rather than assuming responsibility for your own deep inner process. Thus the very urge of life to transmute itself is efficiently suppressed.

The bottom line with the 47th Shadow is that unless you face your own unique fear archetypes, you cannot open the portal to inner Transmutation that is the 47th Gift. If you do not look your adversity square in the face and realise that it is the very manifestation of your own archetype, you miss the whole point and purpose of your life. Most of the world is still waiting to begin this process. Many who have begun it have become trapped in some therapeutic system that takes away their responsibility, thus further stalling the process. Facing your Shadows is a lonely business and it happens through the process of your natural destiny. It is the thing right in front of your nose that you try to avoid the most. It will not yield to being fixed. It will not be resolved through your mind and it will keep coming back at you again and again. It cannot be sidestepped or given over to another. Your suffering is a magic mirror — it's yours to own, to appreciate and to accept. Only when you finally stop trying to avoid it, can it show you its magic.

REPRESSIVE NATURE – HOPELESS

For those with a repressive bent, the 47th Shadow provokes a kind of mental collapse. This is the inner oppression in which the mind is utterly overwhelmed with one's life. Because the pattern of oppression shows no signs of being broken, these people simply give up on life at a certain point. They bow to the oppression and accept it as normal. In other words, they give up all hope that their life can get better. Such people live lives of quiet compromise, locked as they are in their closed circuit loop and sealed in by their inability or unwillingness to face their fears.

REACTIVE NATURE – DOGMATIC

Those with a reactive temperament externalise their oppression by projecting it onto others. These people use their minds to control their environment, becoming very dogmatic in their viewpoint. Because of the undercurrent of fear in such people, they try hard to solidify their mental patterns in order to give them the illusion of security. Whether the dogma be a scientific view or a religious one, it effectively freezes them and brings change to a standstill. This kind of nature finds it very difficult to relate with others unless those others buy into their dogmatic viewpoint. Likewise, any kind of freethinking or independence of mind or spirit in another is incredibly threatening to such people.

THE 47TH GIFT – TRANSMUTATION

THE ROYAL ART

It is through the 47th Gift of Transmutation that the true purpose of your hidden genetic programming — the junk DNA — is finally revealed. Beneath what the left-brain scientists misread as chaos hides one of the great secrets of creation. What geneticists know is that all life evolves through the process known as mutation — a random organic process in which mistakes are made when long strands of genes are copied. Until particle physics came into being, the word *transmutation* was associated entirely with the ancient science of alchemy. Transmutation involves the complete changing of one element into another, which involves a radioactive process at an atomic level. One might therefore say that mutation is a process of gradual change that may or may not result in transmutation.

Human beings are in fact a rolling wave of consciousness with no fixed identity. We are programmed to continually hit the shores of our limitations and dissolve into something else.

When a human being faces his or her own Shadows, they open the door to the process of Transmutation. Hidden in the vast programming of our non-coding DNA are the scripts for the entire history of life on earth. These scripts are jumbled and twisted around and around each other and are indecipherable to current forms of pattern recognition. However, these scripts create a huge swelling inside the individual — a pressure to continue life's universal storyline through your own life. This is why every human being has an innate sense of their own destiny. Whether you live out that destiny depends on your willingness to surrender to the pressure of this vast unconscious force coming from within. This is the force of Transmutation and it wants you to surpass yourself. If you let go into your fear of dying, that greatest fear, you will discover the Gift of Transmutation. Human beings are in fact a rolling wave of consciousness with no fixed identity. We are programmed to continually hit the shores of our limitations and dissolve into something else. For the majority of human beings, safe in their illusion of separateness, this is the greatest terror of all.

As one of the vital ingredients in a chemical genetic family known as the Ring of Alchemy, this 47th Gift of Transmutation is dangerous. The ancient alchemists discovered the archetypes of transmutation and named them, often using the changing of colours as symbolic of the stages of the process. Although many of these people mistakenly thought of alchemy as a physical practise whereby a base metal might be transmuted into gold, a few understood precisely what alchemy really was. Alchemy — the Royal Art — is the natural destiny of human beings who live their life totally, embracing everything and holding nothing back. It is the art of living dangerously. This does not mean that you necessarily take outer risks — the danger is to the illusion that there is anything fixed about you. The true human is indefinable because he or she is constantly surpassing all definitions. Alchemy is life. Transmutation is what drives human beings to keep going beyond their wildest dreams. To be in the process of transmutation is to be truly alive.

The process of transmutation has actually been described very well by the alchemists. It consists of an endless series of small and often indecipherable mutations leading to a finite number of transmutations. These transmutations are huge turning points in your life. To transmute is to make a quantum leap into a whole new dimension. If you follow the script hidden in your DNA, then you will experience these dimensional shifts in your inner and outer life. Many cultures' mythologies and mystical systems have described this process of spiritual evolution because it is universal. The only prerequisite for the process to keep moving is a continual surrender to your fears. The moment one fear has been embraced, another reveals itself and your life moves you to confront the new fear. We each have to take the lid off our Pandora's box and in doing so, we will discover the layers of oppression that lie within us. As these layers are dissolved one by one, you discard one by one the illusions that you think are protecting you from dissolving yourself.

Through this ongoing process of Transmutation, you gradually sift through the junk DNA, and in doing so, you begin to see your life on a much wider level of reality. The only way to read these codes is to live them. As you do, they unwind and reveal their true purpose — to return you to the very source of consciousness from which you came. Those with the 47th Gift in their Hologenetic Profile are people who are deeply aware of this alchemical process in human beings. The only way to transcend suffering is to move more deeply into it, embracing every feeling and event that comes to you. This is the way of deep immersion in the currents of life. It is the way of surrender.

THE 47TH SIDDHI — TRANSFIGURATION

THE TRUE MEANING OF THE CRUCIFIXION

The 47th Siddhi represents the culmination of the 47th Gene Key in human form. Transmutation never really ends. It simply outgrows its casing, and this is what occurs through the 47th Siddhi of Transfiguration. This word *transfiguration* is almost entirely associated with Christ's ascension and resurrection. It refers to his shining countenance when he reappears after his crucifixion. The life of Jesus Christ is in fact the perfect mythic enactment of all the stages of alchemical transmutation ending in this state of transfiguration. If we can escape the endless dogma and opinion that surrounds the figure of Jesus and view his life at this symbolic level, we can see the great secret of all human life. Christ's life symbolically represents the life of every human when they hold back nothing and embrace everything.

There have been descriptions of similar transfigurations throughout the world. Notably, the Tibetans have many records of people attaining the *Rainbow Body*. The ancient Taoists of China also have records dating back many centuries of masters who attained the state of transfiguration. This 47th Siddhi reveals the ultimate purpose of the non-coding DNA within the body, which actually does code for a process within the physical body after all. The force of transmutation strips you emotionally and mentally to such an extent that it eventually reaches down into the physical matter of your body.

*The whole of
our planet —
humanity, all
the creatures
and the earth
herself — will
eventually be
transfigured.*

The power of the myth then takes over and the very cells of your body begin to transmute into the pure light frequencies from which they are made. The elements that make up your body were made in stars, and you turn back into a star in your own mini supernova. This is the final state described by the alchemists — the *unio mystica* or sacred marriage in which all your constituent elements dissolve back into one another, and out of your base matter symbolic gold is formed.

Like all the siddhic states, transfiguration is not a common phenomenon. You can see how powerful an influence it has over the world through its manifestation at the end of Jesus' life. Other cultures where it has happened have been mainly insular so, popularly, it has been considered legend. One might ask why it hasn't occurred somewhere more recently in the age where it could be widely reported or even filmed. The answer to that question lies within the script inside our bodies. What is certain is that it will occur again and probably on a far larger scale than has ever been seen before. Humanity stands upon the threshold of one of its great transmutations. As a species, we are moving through the symbolic time of our crucifixion, and crucifixion precedes transfiguration. Our crucifixion means everything that is old must leave the world as a new light emerges. What must be sacrificed in the coming centuries is our modern world with most of its trappings. As with all transmutations, it is a time of great uncertainty, great fear and great excitement.

On a cosmic level, transfiguration doesn't end. The whole of our planet — humanity, all the creatures and the earth herself — will eventually be transfigured. Transfiguration on a social level must include the entire planetary organism because despite our current perspective, we are really the mind and eyes of the earth rather than being separate from it. There will come a time when all the elements that make up our planet vibrate at such a high frequency that the world as we know it will disappear into a field of vibrating light.

Before this unprecedented and fantastical event, which is in the far future as we measure time, the phenomenon of individual transfiguration will become more common. This is due to the activation of the 22nd Siddhi of Grace, the programming partner to this 47th Siddhi. These two Siddhis operate as one, which means that for an individual to attain the state of Transfiguration, he or she has to be touched by Divine Grace. No one can say when, where or to whom Grace will occur, but it is an aspect of the Divine Feminine principle, which is a part of the current cycle of incarnations. Divine Grace can only touch those who work through the 47th Siddhi because these individuals have to take on the collective karma of the whole of humanity. This is the final great transmutation enacted through the crucifixion. The crucifixion represents the individual's descent into hell, the underworld of the collective unconscious and the complete submersion in the ancestral pain and suffering of all beings. It is to face the collective fear of humanity that lies within the heart of every human's DNA. One who reaches this level of self-sacrifice calls upon himself or herself the Holy Spirit or Divine Grace, which precedes and enables resurrection and final transfiguration.

≡≡ 48th Gene Key

The Wonder of Uncertainty

SIDDHI
WISDOM

GIFT
RESOURCEFULNESS

SHADOW
INADEQUACY

Programming Partner: 21st Gene Key

Codon Ring: The Ring of Matter
      (18, 46, 48, 57)

Physiology: Lymphatic System (Spleen)

Amino Acid: Alanine

THE 48ᵀᴴ SHADOW – INADEQUACY

EQ AND IQ

There is no darker place represented anywhere within human DNA than the 48th Shadow. This Shadow gives rise to one of the deepest of human fears — the fear that we are inherently inadequate. Human beings as a rule have no idea of their true capabilities. We can look around and see individual examples of great men and women who have displayed remarkable gifts and who have sometimes achieved the miraculous. However, humanity as it is today has not waked from its dark dreams. We stand at the cusp of one of the greatest turning points in our evolutionary history, and each of us will have to look deeply into this primal fear if we are to make the great leap that now lies before us.

Only during times of great collective crisis is it possible to see the potential of the 48th Gene Key. It seems the medium of crisis is required to bind human beings together. We see this uniting power during times of war, which often activate the higher frequencies of this Gene Key and allow groups of people to operate as a single entity, overcoming great odds and performing incredible feats rarely seen in peacetime. This phenomenon says much about the nature of the 48th Gene Key, which at its source is a power rooted in communion and service. Because humanity is now facing its greatest ever crisis —namely the destruction of our own environment — the potential of this 48th Gene Key is urging us to dig deep into our souls for collective and practical resolutions. In the years ahead, we are going to have to understand this 48th Gene Key and the hold that its shadow frequency has over us.

Because of its role within the codon group known as the Ring of Matter, the 48th Gene Key is one of four Gene Keys that govern our developmental cycles as children. The 48th Gene Key imprints us throughout our second seven-year cycle from the age of seven to fourteen. This second cycle relates to our emotional development and explains exactly where our feelings of inadequacy come from. As we incarnate into our emotional or *astral* body, the prevalent emotional patterns of our parents and the world at large are imperceptibly imprinted within our aura.

The 48th Shadow seeps into our genes to undermine us through a deep sense of emotional inadequacy. As we go through the tender time of puberty, society's conditioning sends us very confusing and contradictory messages about how to handle our emotional and sexual nature. Most people have little idea how profound and delicate this developmental stage is for young people, and for the most part they are left to deal with it on their own. The result is that few emerge unscathed.

If we don't know how to handle emotional states with equanimity, integrity and clarity, we never fully enter adulthood, but remain at some level children.

Until quite recently, emotions were generally seen as something that undermined intelligence, which was understood as being equated only with rational thinking. Thankfully, emotional intelligence — also known as *EQ* — is now increasingly recognised by more and more people. The fact is that your EQ is a perfect counterbalance to your IQ, and together they make up a well-rounded intelligent individual. Most people have never learned to take full responsibility for their emotions. They get caught in the drama of projecting their emotional states onto others. The 48th Shadow is responsible for creating these *emotionally illiterate* generations. If we don't know how to handle emotional states with equanimity, integrity and clarity, we never fully enter adulthood, but remain at some level children.

The frequency that the 48th Shadow releases into the cells of your body emerges as a profound uncertainty about the future and your ability to handle it. When coupled with its programming partner, the 21st Shadow of Control, these two Shadows program human beings to try to stay in control of every area of our lives. We create a false reality of details, timelines and systems, all designed to make us feel more secure. The irony is that nothing external can take away our core fear of inadequacy. An even darker side to this Shadow concerns the manipulation of this fear as a means of controlling others. The feeling of emptiness experienced at the Shadow frequency drives human beings to try to fill this inner void through the acquisition of knowledge. But knowledge cannot take away the fear. Knowledge has both a dark side and a light side. Whereas the light side transforms knowledge into wisdom (its higher counterpart), the dark side becomes addicted to knowledge as a means of distraction and false security.

Human beings have become adept at selling the dream of security and the fear within the masses buys into it. All systems of knowledge based on logic promise security and the more complex the system the more people tend to believe it. This often happens with modern science. The problem is not science itself, which is a wonderful tool as long as it is used in the quest for the spirit of truth. The problem is often the *scientists* whose findings and theories are used to bolster their own personal agendas and emotional inadequacies. Such people proffer an illusory security screen that covers reality while purporting to explain the universe we live in.

The 48th Gene Key is quite simply beyond human understanding. It is a portal to the infinite, and there is nothing like infinity to raise the spectre of inadequacy in the human mind! The original Chinese name for this archetype is *The Well*. When you look down the well, you have no idea how deep it is or what lies at the bottom. The 48th Shadow is like a bottomless black hole. It is the primal fear of the feminine and was enacted in archetypal form during the great witch-hunts that took place across Europe during the Middle Ages.

Out of the 48th Shadow humanity's deep paranoia is born — these are the *men in black*, the aliens, gods or governments who seem to be manipulating our lives. It is the fear that someone else might be in possession of knowledge that can be used to control you. Of course, the true source of the fear is within, but that does not stop human beings from projecting it outside ourselves onto all manner of people and phenomena.

Modern science is only one arena that promises us security from our deepest fears. Systematised religion, economics and education all try to create the feeling of collective security for the people on our planet. So long as you keep people involved in some sort of system, the deep fear moves into the background. Another major manifestation of this 48th Shadow is the urge to create more and more wealth. This urge too is rooted in our fear of inadequacy. The Ring of Matter programs us to try to escape the fear through the material realm, rather than turning inward to the source of the discomfort itself. However, no amount of material wealth can make us feel secure, since the fear itself is rooted in the physical structure of our DNA.

Ultimately, we humans are a part of the mystery of the universe, and there are places where the mind is not destined to travel. Every human being must eventually turn inward and look into our deepest fear — the fear of the void. When we gather our courage and make that leap, we will discover something wonderful. The void is not empty and cold but warm, loving and bursting with light and wonder.

REPRESSIVE NATURE – BLAND

There are two types of people within the collective Shadow frequency — those who duck their heads and give in to their fears (repressive), and those who project their fears externally onto others (reactive). The repressive nature of the 48th Shadow forms the collective blandness of the mass consciousness. These are the sheep whose fears are buried beneath the surface by whatever the system tells them. The majority of people on our planet fall into this category. This consciousness is too afraid to look into the fear within its body, so it settles into the fixed patterns given to it by society. Of course, life arranges events in all people's lives that force them to face their fears. However, unless such events precipitate a major awakening, the repressive nature tends to bury its head even deeper in the sand once such periods pass.

REACTIVE NATURE – UNSCRUPULOUS

The other side of the 48th Shadow is seen in those who feed the general sense of inadequacy in others. These are people whose rage does not allow them to admit that they are also afraid and so they become victims of the system in a different way. These people use their knowledge to manipulate the fears of others while hiding behind the system. They keep their own fear at bay by externalising it, and they increase the general level of fear by their unscrupulous actions. Such behaviour greatly reinforces the Shadow frequency. It is because of these people that humanity feels a general paranoia. We know that there are people in positions of power throughout the system who do not care about others at all, but we cannot always see and identify them. This gives us the feeling that all of society is impersonal and beyond our control.

THE 48TH GIFT – RESOURCEFULNESS

THE LIGHT AT THE BOTTOM OF THE WELL

For all its darkness and fear, the 48th Gene Key holds great hope for humanity, and this hope mostly lies in the hands of parents. We have already seen that the 48th Gene Key governs our emotional developmental cycle in childhood. All human emotional issues are rooted in this phase, so it is extremely important that children between the ages of seven and fourteen are given a stable emotional environment in which to develop. Because the emotional or *astral* body is a subtle layer of the human aura, it learns primarily through the astral bodies of its parents. If a child's parents have not raised the frequency of their DNA beyond the Shadow consciousness, their dysfunctional emotional patterns imprint the astral body of the child. Almost all teenagers are wounded by their parents' subtle bodies well before they become teenagers. In psychology we tend to think of conditioning as behavioural, and it is, but it takes place at a far subtler level than most psychology acknowledges. When you emerge at the age of fourteen into your powerful teenage cycle, you either emerge whole, imbued with natural wisdom, emotional intelligence and internal stability, or you emerge with a profound sense of inadequacy that characterises your teenage years.

There is, however, light at the bottom of the well. As parents heal their own emotional issues and raise the frequency of their DNA, they pass on healthy emotional patterns to their children, who then pass these on to *their* children. In the last fifty years this pattern has brought more and more healthy adults into the world. There is no role of greater importance or service to humanity than that of being a parent. It is not only the fastest track to your own healing, but is also the fastest track to the healing of the whole world. Every healthy adult is a powerful resource for the healing of our planet, because such people are not afraid of their true feelings and fears. The astral body of our whole planet is in the process of being healed by the increasing waves of these people operating at the Gift frequency.

The secret of moving beyond the reach of the 48th Shadow can be found in a single word — trust. As you learn to trust life on a broader scale, the well begins to reveal some of its secrets. Life invites you to begin by trusting the Shadow frequency itself which means that you have to enter into your fears. Since the 48th Shadow is a very physical fear vibration deep within the body, this can be very uncomfortable. This fear does not come and go but is always there, fixed and focused inside you. By entering into this field of fear, you actually lessen your mental anxiety even though you cannot remove the fear. It is a process in which you slowly learn to stop being afraid of fear itself. Instead of hiding from the darkness inside you, it is as though you finally lower a bucket down the well and then pull it up to see what it contains. What you receive is a wonderful surprise. Out of the well come all manner of solutions to all manner of challenges in the world around you. You will be amazed that so much light can emerge from such a dark place. This is the essence of the Gift of Resourcefulness.

As you learn to trust that an answer will always emerge at the right moment in your life, your sense of fear and anxiety gradually begin to fade. The wonderful thing about the Gift of Resourcefulness is that it is self-fulfilling. Each time the bucket comes up out of the well, it contains exactly what is needed at that particular moment. In this way, trust leads to deeper trust, reinforcing inner security through repetition. Gradually your fear of inadequacy is proven to be an illusion. It doesn't matter where your projected fear hides — it could be your fear of being alone, your fear of not having enough money or your fear of running out of time — deep within you lies a self-sustaining and unlimited resource that can always provide a solution to your imagined fear. These solutions come to you when you arrive at the moment of fear open, un-reactive, vulnerable and unknowing. It is through your unknowing that the pearl will be delivered to you, so you must neither repress nor react to your fear; you must simply sit with it.

This state of surrender into not knowing the answers is hugely powerful in its honesty. It is about trusting in the great force field that lies beyond your normal sentience. The force field is always there but your body usually reacts according to its conditioned programming coming through the collective fears of the Shadow frequency. As you learn to trust your not knowing, life resolves itself effortlessly and beautifully and a natural process of deconditioning takes place inside you. Generally speaking this process takes a minimum of seven years, since that is how long it takes for the body to learn or unlearn cellular patterns. Your real resources lie in your gifts and talents — the inherent gifts that are wired into your DNA. The icing on the cake is that your resourcefulness unleashes a great tide of inner creativity that provides extremely elegant solutions to all manner of questions and issues in your life. The force field of which we have spoken is nothing less than the pattern of life as a whole. As you begin to trust life and your place within the greater weave of this pattern, your natural gifts spontaneously emerge. You begin to realise that you are capable of far more than you ever dreamed possible. When you move in rhythm with life it reveals its own hidden timing which is always perfect. It may not always conform to your mental dreams of how your life might or might not work out, but it does leave you feeling deeply fulfilled.

Every human being is designed to prosper because life itself is prosperous. Prosperity is an entirely different phenomenon from wealth. To prosper, all a human being needs is a little more than enough, whereas to be wealthy means to have a lot more than enough. Wealth by this definition is formed out of the need for security, which has already been shown to be an illusion and is rooted in the fear of the Shadow frequency. Whatever your destiny, life will arrange for you to have a little more than you need. For some that means very little and for others it means a very large amount because different humans live out different human myths. One can tell when a person is fulfilled in life because they find exactly the support they need when they need it, rather than stockpiling in order to try to create false security.

Another vital aspect of the Gift of Resourcefulness is its integrative power. All of your inner talents and resources are ultimately designed with a built-in programming agenda — to offer service to the whole. In this sense all true talent is holistic.

It is through your unknowing that the pearl will be delivered to you, so you must neither repress nor react to your fear; you must simply sit with it.

Your inner well exists only to serve others. It is they who come and throw the bucket down the well and draw out *your* resources. Metaphorically all human beings are designed to slake each other's thirst, which means that you have to interact honestly and selflessly in order to prosper. In serving the whole you serve yourself in the most efficient way possible. This is why resourcefulness has the power to unite even the greatest enemies. When human beings pool their resources they become truly powerful. We saw this at the Shadow frequency in the case of human beings uniting during common crises or wars. As the frequency of the mass consciousness gradually rises, we will begin to create new ways of doing business in the world. In the future we will see the growth of a service-based human culture rather than a greed-based culture and as this culture emerges, humanity will begin to operate as a single entity in perfect harmony with a higher universal rhythm.

One final insight that emerges from this 48th Gift concerns the way in which human beings currently understand energy. We have created a modern world based on the dynamics of explosion. The internal combustion engine is the invention responsible for the greatest growth curve in human history. We now face an energy crisis as we pass the threshold of *Peak Oil* and our oil reserves and fossil fuels begin to dwindle. As the 48th Gift awakens throughout humanity, another breakthrough will become possible — the harnessing of energy through the dynamics of implosion. We have seen that the archetypal nature of this 48th Gene Key is feminine and inward, rather than masculine and outward. There is a whole new field of understanding waiting to dawn in physics, and as we come to look differently at the world we will find ways of unlocking energy without the need for ignition. It is most likely that such a breakthrough will come through a new understanding of the force of gravity. When we learn to see through the eyes of the 48th Gift, we will unlock the secret of free energy — the infinite resource at the core of creation. This is the breakthrough that will truly herald a new global age, as it will resolve all our energy needs simply and quickly, and will make every home or community clean and self-sufficient.

THE 48TH SIDDHI – WISDOM

THE BEYONDNESS OF BEING AND NOT BEING

The 48th Siddhi of Wisdom is one of the truly great Siddhis. Wisdom has been revered and sought by all cultures since the beginning of time. There are many definitions of wisdom, most of them based on the notion of some kind of inner knowledge that allows you to see beyond the confines of what is considered *normal* awareness. Collectively, the 64 Gene Keys represent an encyclopaedia of the 64 archetypes or codes within the human continuum. At the siddhic level, however, this differentiated continuum ceases to exist. Each Siddhi uses its own paradoxical language to address that which lies beyond words. In this context, the 48th Siddhi is the great archetype of the beyond itself. It is something all human beings aspire to, and at the same time it is something that terrifies us. The 48th Siddhi opens up the void inside you. In a human being, it is the primal state of not knowing.

It is a most delightful and infuriating paradox that wisdom arrives through not knowing rather than through knowing. Knowing (i.e., knowledge) is something we can always gather from life and experience. To know takes great effort and exertion, whereas not knowing is already there inside you. Everything about wisdom is about being unsafe and *unsecure*. There is a great difference between being insecure and being unsecure. To be insecure is to be kidnapped by fear and taken on an illusory journey that promises final redemption but can never deliver it. That journey is the human journey, and it is ultimately discovered as meaningless. There are simply no solutions because there is no problem in the first place. To be *unsecure* on the other hand is to have embraced your urge to escape fear in every way possible. It is to realise that your body itself is not afraid to die. It is absolutely natural for your body to die. Neither is the mind really afraid to die, since the mind is just an aspect of the functioning of your body. What then is it inside us human beings that is afraid to die?

The answer to this question is the source of true wisdom. It is the ultimate solution to all our human questions and problems. Only life can answer this question. No words can point to the grand illusion of our existence. When the body dies, its constituent elements will return to the great web of life and continue to be endlessly recycled throughout the universe. What then remains of you? The answer to all such questions is another single question, and no matter what the question, the return question is always the same: *which* us? *which* you? *which* me? There is no me, you or us. To be unsecure is to be more secure than is possibly imaginable. To be unsecure is to surrender all your inner questions back into the infinite. There is a wonderful story about the great Indian sage Bodhidharma arriving before the Chinese Emperor with a shoe on his head. When the emperor asked him the meaning of the shoe, he replied that he wanted the emperor to know right away what kind of a man he was dealing with. This story is symbolic of the unsecure nature of true wisdom. It will not and cannot be comprehended with the mind.

From time immemorial, humans have intuitively grasped that wisdom is feminine in essence. The great Goddesses of many cultures are representations of this understanding. But wisdom itself is beyond opposites. The nature of the feminine simply points the way towards this embodiment. The images that have been used to describe wisdom are thus feminine in nature — water, wells, springs, valleys and darkness are all examples. Water is one of the greatest symbols for the nature of wisdom because its nature is paradoxical — it is empty yet full, weak yet strong, resistant yet yielding. Ultimately, it shapes itself to the vessel it is held in, and when the vessel has gone, it dissipates yet lives on. One who is truly wise is like water in all these ways — you are wise because you do not know you are wise, you are powerful because you do not care about power, you are fearless because you do not really exist.

The question of all seekers of wisdom is how to attain it. This is the question at the heart of all the mysteries and of all the great religions and sciences. Modern science continues to try and find the single unified theory that will resolve all questions about the universe. Humanity does not yet realise that the answer to this question will not resolve the universe — it will *dissolve* the universe! The answer lies within the question itself, just as it lies within the individual. The only true manifestation of wisdom is complete ordinariness.

You are wise because you do not know you are wise, you are powerful because you do not care about power, you are fearless because you do not really exist.

When this Siddhi dawns within an individual (a paradoxical sentence in itself) that individual ceases to exist but becomes an aspect of the whole, functioning spontaneously and innocently. The irony is that this is exactly what the individual was before wisdom occurred. In other words, wisdom changes nothing inside us. It is this realisation that brings about wisdom!

There is nothing wiser in all of humanity than the physical body. When a human being discovers this secret they begin to tap the source of universal wisdom. Wisdom is rooted in utter trust of your body. Through the eyes of wisdom, everything within the sphere of human experience can be reduced to simple physical sensation — even thinking. The body must be allowed to feel the way it feels, it must be allowed to think the way it thinks, it must be allowed to act the way it acts. Nothing the body does can ever be wrong. It is the false concept that there is a way to be harmonious and a way to be inharmonious that is the root of all human dilemmas. There is nothing but harmony, and there is nothing but wisdom. You must come to realise, through whatever experiences the body follows, that your acts, thoughts and movements emerge from the totality rather than having an independent source. Not only is there no individual choice in our lives, there is no *chooser*, so the whole concept of freedom or determinism goes out the window.

We began this 48th Gift by seeing how inadequate human beings can feel when faced with the many challenges we each have in life. We are designed to feel this inadequacy because it is the beginning of our journey. There is wisdom even within this inadequacy because it is felt by the body. Your mental anxiety is also an aspect of your bodily wisdom, as are your desires, fantasies, anger, contempt and lust. Everything begins with the body and ends with the body. If every feeling is allowed and lived out fully and in trust, then this deep inner vibration of fear within you must eventually fizzle out. The fear is really the fear of fear itself, and when it is looked squarely in the face then all other fears are simply reduced to bodily sensations that continually arise and disappear again. At this stage of deep wisdom, you can no longer discriminate between physical sensations. The feeling of intense bliss for example is no different from the feeling of lust or even physical pain. The body simply follows its own wisdom and this wisdom dissolves that which identifies with the sensations. As we said earlier, what the 48th Siddhi leaves behind is an absolutely ordinary human being.

☰ 49th Gene Key

Changing the World from the Inside

SIDDHI
REBIRTH

GIFT
REVOLUTION

SHADOW
REACTION

Programming Partner: 4th Gene Key

Codon Ring: The Ring of the Whirlwind
(49, 55)

Physiology: Solar Plexus

Amino Acid: Histidine

THE 49ᵀᴴ SHADOW – REACTION

REAPING THE WHIRLWIND

Here in the 49th Shadow we find the sleeping genetic trigger for the process that begins our collective ascent to higher consciousness. In the greater percentage of human beings this trigger is dormant, ensuring that the frequency passing through our genetics remains consistent and stable — in other words — asleep! In order for a human being to awaken to a higher reality, this trigger or switch lying in the 49th Shadow must be activated. When this occurs, the first stages of evolution begin inside you as an acute change in your emotional patterns. As we look further into the 49th Gene Key we will examine this early awakening process in more depth but before we can journey to that point, we must first understand what it is that keeps us from evolving and how widespread this phenomenon is.

The 49th Shadow of Reaction is one of the most powerful of all the 64 Shadows in terms of how it governs human behaviour. Unless there is some glimmer of awareness in this Shadow, your raw emotions will absolutely rule your life and the decisions that shape your life. For many generational cycles, this part of your DNA has been slowly and steadily evolving. In its rawest form it manifests as the ability to emotionally cut yourself off from others although, as we learn at the higher frequencies, this notion that you can be emotionally separate from another person is one of the great human illusions. Originally its more ancient purpose was to enable us to kill — both the creatures around us and each other. Until relatively recently, we needed to kill animals in order to survive. Human survival is generally based on efficiency and that used to mean that the quickest way to feed yourself, especially if you were an early nomadic culture, was to kill whatever animals were in your vicinity. However, as we settled into communities and developed a more agrarian lifestyle, we found other means of feeding ourselves that provided more security in the long term. Nowadays it is perfectly possible for human beings to survive indefinitely on a purely vegetarian diet. With our settling into communities and tribes based around a more agricultural lifestyle, the 49th Gene Key continued to evolve. In many ways it has made us more sensitive to our environment and to each other.

However, this Gene Key has only evolved so far. The more tribal your community, the greater is the fear for your safety. The tribal gene pool by design is self-sustaining and is threatened by other gene pools. A darker side of the 49th Shadow is its propensity to kill human beings who appear to be threatening. The more tribal your mentality is, the easier it is to emotionally detach from others. Therefore what we see out of this Shadow is the human tendency to view the outsider as inhuman in order that we can kill them. Today some of the world has developed beyond the tribal gene pool mentality, but the greater proportion has not. The war that is currently raging across the world stage — between the newly emerging global consciousness and the old traditional tribal ways — is largely due to a collective mutation taking place within this 49th Gene Key. There are those who are sensitive to life and those who are not, and the gap between the two genetic groupings is ever widening. It is important to see clearly that there are insensitivities on both sides of the fence — the global consciousness can be highly insensitive to the tribal and vice versa.

The social, political and economic issues tied to this 49th Shadow are extremely complex. One reason is that this Shadow is also responsible for our most ancient spiritual beliefs and customs. It is out of our ability to kill that our most basic need for spirituality has arisen. The very principles, totems and taboos of our tribal societies have evolved through the way we justify the killing of others. All these issues arise from our reaction to the other. Reaction is the key. One tribe reacts to another out of its emotional tribal identity and the result is a war. Reaction is an ancient tribal reflex that still dominates the world consciousness. Even those at the head of powerful western governments are still usually governed by such emotional reflexes. At its source, all reaction comes from a one-sided subjective belief founded upon the assumption of good and evil. As long as you see your own people as good and others as evil, you remain a prisoner of the 49th Shadow.

The 49th Shadow and its spectrum of frequencies are genetically related directly to the all-important 55th Shadow. These two Gene Keys share the same genetic codon group, known as the Ring of The Whirlwind, which ties them intricately together at a chemical level. To understand the powerful mutative process that is currently occurring through this aspect of human DNA one must also understand both of these Gene Keys. The mutation through the 55th Gene Key is manifesting through the individual, whereas the same energy passing through the 49th Shadow is giving rise to a socio-political and economic revolution. In this respect, it is much easier for us to see the results of the mutation through this 49th Shadow than it is through the 55th Shadow. The former can be seen through the changes and indeed crises in our communities and the world headlines, whereas the latter is the quieter inner revolution that must be experienced within your own individual life.

You can see the mutating nature of the 49th Shadow most clearly through your relationships. All relationships connect us to the tribal consciousness of the planet. If you want to understand what is occurring to this aspect of world consciousness, look no further than your most intimate relationships. The 49th Shadow is the Shadow of Reaction, and reaction cannot occur without a relationship. It is a knee-jerk response to an outer stimulus.

As long as you see your own people as good and others as evil, you remain a prisoner of the 49th Shadow.

The bedrock of the 49th Shadow is the married couple. It is irrelevant whether or not you are married legally. What is important is that there exists an intimate emotional and sexual coupling. The power of the 49th Shadow can be seen most clearly at an individual level through the constant reaction patterns that take place between the sexes, even in same sex couples. These patterns are built into the matrix of the relationship; otherwise there would be no sexual fire to ignite the relationship in the first place. The fuel of Reaction is the fear of rejection. This fear governs all emotional and sexual patterns at an unconscious level. The more sensitive you are, the nearer the fear is to the surface, which can be both a blessing and a curse. If you have some awareness, you may be able to see your own reaction patterns at work every time you have a disagreement with your partner. With a lot of awareness, you may even prevent yourself from reacting at all, whilst still feeling the emotional charge racing through your body.

At a genetic level, unity is seen through our connection to our wider family, community or even our God. One of our greatest fears is of being severed from this feeling of unity. It is mirrored in an unconscious memory of being separated from our mother at birth, the deepest of all rejection. All of your unconscious reaction patterns are fed by this fear, and you can see how this leads to the 4th Shadow of Intolerance, the programming partner of this 49th Shadow. Intolerance leads to reaction and vice versa, because we cannot handle the amplitude of our emotional fear when it is triggered by another. This 49th Shadow is a vestige of an age in which communities lived separately from each other and in fear of their survival. Naturally many parts of the world still operate like this today. However, what we are seeing now is the emerging of an early form of global awareness that will eventually become more and more sophisticated and integrative. Behind the diversity of our tribal groupings, we are all in fact one single world tribe, genetically sub-related to a single mitochondrial Eve. As our awareness penetrates down to this level, we will see a remarkable phenomenon — the merging of the collective view with the tribal view.

On an individual level, the current mutation passing through the 49th Shadow is changing the face of human relationships. It has always been said that our relationships are a mirror, but it was not always understood to mean that they are a mirror of the world. As the new awareness centre of the solar plexus begins to open its early petals, we will begin to stop the ancient reaction patterns in their tracks. We are learning not to react to our fear of rejection, and this fear is losing its grip on us. The new awareness that is coming will give us a physical and emotional knowing that we are interconnected through our auric field and that there is no possibility of rejection or abandonment. Our sensitivity will become incredibly refined. This will change everything — in our societies and our individual lives, but most of all, in our relationship to each other. Once the trigger of awakening within the 49th Shadow has been activated, which is now beginning, the process will precipitate a chain reaction — a genetic whirlwind that will shake the bedrock of human civilisation to its roots.

The fuel of Reaction is the fear of rejection.

REPRESSIVE NATURE – INERT

The repressive side of the Shadow of Reaction is absolutely no reaction. In many people, because of early childhood conditioning or some shocking event, proper emotional functioning is curtailed and results in a shutting-down or dampening of one's emotional chemistry. This manifests as a kind of emotional inertia or dryness in which the deep fear of rejection is so buried that it appears to be absent. These are people whose emotional lives may appear the most stable of all, but they are lacking in *juice*; their sexuality has dried up and there is no fight left in them. All emotional zest has been compromised for the sake of a false harmony. Many relationships follow this safe pattern in which the partners hardly communicate with each other — at least not in any depth and without ever exposing vulnerability. Such relationships exist only at the surface of life and conceal a huge, rarely admitted disappointment.

REACTIVE NATURE – REJECTING

All reactive natures are essentially founded upon this fear of rejection. This aspect of the 49th Shadow rejects before it feels rejection. These people push others away at the first sniff of someone coming too close to them. As their relationships become more intimate, the fear becomes greater. These people usually terminate their relationships before they can be hurt. Therefore, they often end up alone, preferring not to run the risk of being hurt. Of course in spending their lives alone, whether they want to or not, such people never feel a true fulfilment in life because by design they are meant to be in some form of committed relationship. The only stable relationship that such people could have is a relationship in which the two people rarely see each other. They may even live together, but do so without ever really having to communicate.

THE 49TH GIFT – REVOLUTION

THE SILENT REVOLUTION

As the higher frequencies penetrate the 49th Shadow, we will move through a time of unprecedented change and upheaval. This is the Gift of Revolution. The 49th Shadow will be the first aspect of the human Shadow to mutate right down to the genetic level. The implications of this are huge because as fewer and fewer humans are victims of their emotional reactions, violence will rapidly decrease in the world. One also has to consider the environmental impact of this mutation occurring across all cultures. Men and women of peace are infiltrating all races more than ever before. The revolution is worldwide — it will continue to impact us as individuals through our relationships, which will in turn impact our families, communities, our nationality and our entire identity as a species.

Revolution occurs wherever stagnant energy is brought into awareness. It is a direct result of your realising that you are ill. Given the potential violence and intolerance within the 49th Shadow, you can be sure that it will not let you go without a fight. There is a genetic revolution underway right now, and one of the first things it will do is throw out the old genetic material.

One way to interpret this is that everything that is rotten in the 49th Shadow will be brought up to the surface. When we view humanity from the level of our genetics, it allows us to look at things objectively, without our conditioning and prejudices. It means that we can look at what is happening in the world and understand why things are occurring. This understanding (coming from the programming partner of the 4th Gift) is what enables us to overcome our tendency towards reaction. However, there is a very powerful force latent within the 49th Gene Key that must continue to find an outlet. The same energy that leads to killing at a low frequency retains its destructive capacity at a high frequency. At a high frequency its purpose becomes the destruction of all things of a lower frequency. This is the ideal archetype of revolution.

At a social and political level, revolutions bring a dream with them and that dream is always to implement some form of radical change to government. Unfortunately this invariably involves the destruction of all that came before, and usually all that was good about the old system is also destroyed. The purpose of the 49th Gift is to bring this explosive new energy and awareness into the world — not so much at an individual level (which is the role of the 55th Gift) but at a collective and cultural level. Along with the 49th Gift comes an enormous longing to reform the ineffective ways in which society operates. People strongly influenced by this Gift are frequently involved in some form of inter-racial social reform. The 49th Gift does not produce the kind of revolution that history has so often seen — it gives birth to revolutionaries, not to reactionaries. This urge to improve the world is anchored in a deep understanding *of* the world.

People with the 49th Gene Key in their Hologenetic Profile have a deep understanding of the limitations imposed upon civilisation by our emotional inability to see beyond our tribal creeds. Their role is to assist in pulling down these old perspectives based upon fear and territory. However, while this Gift seeks to throw out the old ways, it also understands that certain aspects of the past must be kept intact and nurtured to fulfilment. The 49th Gift is a power that is already beginning to flood the awareness of many people across the globe. You can easily tell the difference between the reactionaries and the true revolutionaries. The reactionaries base their reforms on their anger and fear, whilst the agents of the 49th Gift are not victim to their old emotional prejudices. They do not provoke further reaction but seek to resolve conflicts and at the same time implement radical changes and ideas based upon a grand vision of the future. The nature of the Gift frequency is beyond fear and is anchored in a deep goodwill towards all creatures.

The nature of the Gift frequency is beyond fear and is anchored in a deep goodwill towards all creatures.

Above all, the 49th Gift understands how the current system works, which means it also understands how long it will take to change. The vision carried through the 49th Gift is an imprint emerging from the collective unconscious, and this deeply mystical truth is what unites all those working at the Gift level. As more and more people either raise their frequency naturally or are born into future genetic vehicles in which the 49th Shadow has been neutralised from birth, the world culture will begin to see the shape of these reforms spreading throughout our societies.

However, this part of our world DNA is currently perhaps the most volatile battlefield for the global mutation that is occurring. There is so much fear stored in our DNA that it will all have to come out in the end. In our time scale, it may be some time before world culture begins to find a sense of calm but when seen from a genetic evolutionary level, this change will occur in the blink of an eye.

THE 49ᵀᴴ SIDDHI — REBIRTH

THE FORKING OF THE SPECIES

The 49ᵗʰ Siddhi represents a huge leap in consciousness because it engenders a state of mystical divorce. The shift in perspective from the 49ᵗʰ Shadow to the 49ᵗʰ Gift is vast, but the dimensional shift from the 49ᵗʰ Gift to the 49ᵗʰ Siddhi is like moving into hyperspace. The very same energy configuration within our DNA that allows human beings to kill other life forms actually carries the impetus that will give birth to our total freedom. This is the deep chemical connection between the 55ᵗʰ Siddhi and the 49ᵗʰ Siddhi — together they create the rebirth of freedom, or the freedom of rebirth. These two Gene Keys are bonded together in the codon group known as the Ring of the Whirlwind. It is the 49ᵗʰ Siddhi that will rebuild our world after the 49ᵗʰ Gift has begun to disassemble it. To understand how this works one has to see the limitation of the Gift state of consciousness. At the Gift level, the 49ᵗʰ Gift of Revolution is still prone to going around in circles. Once we have escaped the vibration of fear, our world will indeed be different. It will improve so greatly and so rapidly that it will make time itself dizzy.

But a further secret lurks within your DNA, and its seed is here within you right now. The secret lies in the expression *mystical divorce*. Revolutions by the nature of their name keep coming around. After the coming shift in consciousness, humanity will settle into a new pattern and a totally new cycle. The genetic filtering process that is just beginning will continue for many generations. For a new mutation to take hold of humanity, an immensely complex process has to take place. In genetics, certain genes appear to have an unmistakeable effect on our behaviour or *phenotype* whereas others do not. Such genes are known as *penetrant genes*. The *penetrance* of the mutation in the codon relating to the 49ᵗʰ and 55ᵗʰ Siddhis will be profound, and human behavioural patterns will shift dramatically. However, the means by which a mutation spreads throughout the entire gene pool of a single species is limited by many factors, among them the presence of recessive genes — genes which effectively slow down the spread of mutation.

This suggests that it is highly unlikely such a mutation will overtake the whole human species. The most likely occurrence is that it will split our current species. To draw an analogy, it might be helpful to imagine the world 40,000 years ago populated by two very different branches of hominid — Neanderthal man and Cro-Magnon man. Cro-Magnon man forms the earliest known branch of Homo sapiens whereas Neanderthal man is from a far older branch of the species living as far back as 350,000 years ago. For unknown reasons, the older species, Neanderthal, became extinct.

The 49th Siddhi conceals an archetype that appears to be a part of all human evolution — that of Rebirth. In other words, every once in a while along the evolutionary chain a new species is born out of the old species. In paleoanthropology, this is known as the *Eve Theory* or single origin hypothesis. However, despite its origins, the new species — like the mythical phoenix — has nothing in common with its parent. It rises up out of the genetic material of the old and takes a whole new direction. This is the core of the 49th Siddhi — and it is the meaning of the term *mystical divorce*. Revolutions keep going round at a certain level of frequency, but evolution is a spiral requiring sudden leaps. At such times, revolution gives way to rebirth.

We have seen that the 49th Gift is deeply concerned with the socio-political infrastructure of our civilisation. The 49th Siddhi brings some further insights to this. The first insight is that the world in its current form cannot be fixed, no matter how profound or far-reaching the revolution may be. The very bedrock of our modern society is founded upon a species that has always made decisions rooted in fear. In this respect the whole civilisation is rotten from its core. The only way for a new future to be created is to begin from scratch. The 49th Siddhi is harsh in this respect, but its sights are set upon a far distant goal, and that goal can only be accomplished with a new beginning — a rebirth. As this Siddhi dawns, a new civilisation will be built as the old one continues to crumble. Two types of human beings will coexist and both will live from a totally different awareness. The old genetic fractal of humanity will still live from fear, so they will doubtless fear the changes that they see all around them. One can see the early form of this pattern occurring in the world even now.

If all this is really the case, you may wonder, what are we to do as individuals? If such a rebirth is really on the cusp of occurring and if the future is already genetically predetermined by a collective evolutionary urge then does it matter what we each do at all? If even the most revolutionary of impulses cannot fix the current world, then what is the purpose of following our Gifts? In many ways, this is one of the biggest questions posed by this book.

The answer is as simple as it is profound. This evolutionary leap into a whole new way of functioning *depends* upon us following our Gifts. If we cannot create the waves of the new revolution at all levels of society, then the rebirth that takes place at the zenith of consciousness, in the realm of the Siddhis, cannot take place. The rebirth is the organic flowering of the revolution. Just because we cannot fix the world as it is, does not mean that we cannot make the world a better place. Our vision of the perfect future is precisely what creates the necessary frequency shift that will trigger the genetic forking of our species. It *will* happen because it *must* happen, but we must still create that happening. This is the paradox.

Just as there is an evolutionary force pushing upwards within all matter, so there is an involutionary force working its way down from the realms of spirit towards the material plane. We can see this involutionary force clearly in the programming partner of the 49th Siddhi— the 4th Siddhi of Forgiveness. Forgiveness is an energy frequency working its way down into form, clearing and releasing everything in its path. This genetic cleansing makes the rebirth of a new species possible.

You cannot attain the siddhic state of consciousness without being totally reborn.

The entire realm of the Siddhis is the realm of Rebirth. You cannot attain the siddhic state of consciousness without being totally reborn. Every human being who has ever arrived at true enlightenment has experienced such a rebirth. The siddhic state requires both a mystical and genetic divorce from that which came before. This is why those who have attained such states are genetic anomalies during this current stage of evolution. They are the rare flowerings of consciousness that occur as the early echoes of our future work their way down into human form. They force the physical DNA to mutate prematurely in order that the human form can accommodate the future awareness. Such is their power. Because you are reading these words, you are potentially one of these early flowerings. Is that not a beautiful thing to ponder?

☷ 50th Gene Key

Cosmic Order

SIDDHI
HARMONY

GIFT
EQUILIBRIUM

SHADOW
CORRUPTION

Programming Partner: 3rd Gene Key

Codon Ring: The Ring of Illumunati
(44, 50)

Physiology: Immune System

Amino Acid: Glutamic Acid

THE 50TH SHADOW – CORRUPTION

LOST IN TRANSCRIPTION

For millennia, the wisdom of all human cultures has suggested an incredible possibility — that the micro-cosm exactly reflects the macrocosm and vice versa. As the famous mystical axiom states: "As above, so below." If this is true, which to a holistic mind would seem logical, then deep within the human body we should be able to find the answers to existence itself. Indeed, the Gene Keys themselves are founded upon just such a suggestion — that within these sixty-four genetic building blocks of DNA can be found an archetypal code that explains who we are and where we are heading, both as individuals and as a species. Ensuing from this suggestion, it would appear that certain archetypes within the human genome also have powerful links to other aspects of genetics as it governs the inner process through which life is created and maintained. The 50th Shadow stands out in particular as an archetype that mirrors the process in genetics known as transcription.

The 50th Shadow is the Shadow of Corruption. More often than not, we think of political and social corruption, where those in positions of authority misuse and abuse their power for the purpose of personal gain. In examining this 50th Shadow however, we need to consider the word corruption in another context — that of data corruption. Data corruption is an expression that denotes the incorrect translation of computer data during transmission or retrieval. Even though this 50th Shadow has to do with the social values of human beings, by thinking of corruption in this more impersonal way we can understand the word at a far deeper and more objective level.

In life there are natural laws that govern human beings. An example is our tendency to operate in hierarchies, behaviour which is born out of our mammalian ancestry. As well as these natural laws, there are man-made laws that attempt to maintain a certain level of order within our societies. It is because we still operate out of a hierarchical consciousness that we need such social laws to govern our hierarchies.

*When the
frequency
running
through a
whole society
decreases,
chaos and
corruption
rule.*

However, social hierarchies create a climate of deep division among human beings, since they give birth to comparison, which in turn leads to greed, desire, jealousy and inevitably social corruption. The human tendency to operate in hierarchies derives from our need to compete with each other, which comes from an older part of our brain that favours individual survival, or at best tribal survival. The Shadow frequency is created out of these archaic aspects of human awareness that are based on fear. However, other possibilities exist as the human brain thankfully continues to evolve. At the level of the Gift frequency, this 50th Gene Key sees an entirely new reality for human social interaction, one which is currently in the very earliest stage of manifesting in the world around us.

At the Shadow frequency, data that is processed by the human brain is translated through the medium of fear. As this occurs the data is corrupted, leading to a slanted manifestation in the world. In genetics, this 50th Shadow has much to do with the way in which RNA copies DNA. RNA is a chemical substance similar to DNA whose role is to transcribe aspects of your genetic code in order that new proteins can be formed. In other words, RNA is a messenger that reads and copies the instructions for life. In the process known as transcription, messages can be misinterpreted before they are translated. This is exactly what occurs within human society — fear causes misinterpretation leading to a reaction from one party, which triggers the same fear and a counter-reaction in another party. The result is Chaos, the 3rd Shadow and programming partner of this 50th Shadow.

The ancient Chinese name for this 50th hexagram of the I Ching is generally interpreted as *Cosmic Order* and its symbol is a cauldron. The sages of old evidently understood the archetypal role of the 50th Shadow at its highest frequency. When the frequency running through a whole society decreases, chaos and corruption rule. There are many examples of isolated societies and tribal groups that lived at a high frequency and enjoyed a peaceful existence. However, problems occur when different races, families and customs converge and try to live in the same general territory. It is then that the old fearful brain thinking once again springs into action. This fear misinterprets and corrupts data. Corruption can therefore be understood as the symptom of a collectively created reality. Corruption requires hierarchy in order to exist, and hierarchy is a low frequency attempt to maintain order within a community. By and large, the moment you create a law, one might say that you also create a rebel.

As we have seen from the metaphor of DNA and RNA, this whole issue of corruption is born of an erroneous translation of natural law. The 50th Gift at its higher frequency holds the blueprints of all human harmony, but through a process of flawed transcription rooted in the old part of our brain, this equilibrium cannot occur. This process can be understood even more deeply through the 44th Shadow of Interference which, together with the 50th, creates a chemical bridge in our DNA forming the Codon group known as the Ring of the Illuminati. The 44th Shadow sets up an interference pattern that directly leads to social misinterpretation. One can see how deeply involved this 50th Shadow is with the complex world we see around us today. Because of the speed at which technology has evolved, different human races with different laws and beliefs are thrown together into the collective cauldron.

Despite the speed of our technological revolution, the old hierarchical consciousness still prevails across continents and cultures, giving rise to a kind of international hierarchy ruled by the wealthiest countries. At the Shadow frequency this elite hierarchy is often perceived as a secret world government bent on controlling the world — a kind of dark conspiratorial illuminati.

The world we see through the news headlines is actually a world that is struggling to move beyond the 50th Shadow. Corruption is rife throughout the socio-political and economic systems across our planet. The wealthiest countries are trying to maintain global equilibrium through hierarchical control, whilst at the same time being riddled with their own corruption. Almost everywhere one looks, the fear-based thinking of the 50th Shadow is at work. However, as we humans will eventually learn, you can never tackle corruption by targeting individuals, individual groups or even individual countries. To restore equilibrium to this planet, we must look deeper than corruption itself. Corruption is merely the by-product of a deeply flawed worldview, and until one of the core problems — hierarchy itself — is tackled, corruption will and must continue to flourish.

REPRESSIVE NATURE – OVERLOADED

The 50th Shadow of Corruption creates two types of human beings — those who are victims of the hierarchy, either through reaction or submission, and those who take advantage of it for personal gain. Those who submit to hierarchy are those who fear it and, in such people, this fear creates repression. These people are overwhelmed with the weight of the world and with the responsibility of protecting those closest to them. They are the victims of the hierarchy itself, and form the great majority of those caught up in the system. Unable or unwilling to escape the system, they usually live lives in which their dreams are compromised due to their perceived social responsibilities. This low frequency creates a stalemate because it is in their dreams that their creativity resides. Only when the deep fear is faced can their creativity force an outlet for them to transcend the system.

REACTIVE NATURE – IRRESPONSIBLE

Those who take advantage of the hierarchy are as much victims of the system as those of whom they take advantage. In such people, the fear emerges as unconscious anger, which can be expressed either up the hierarchy at the leaders or down to those below. These people are irresponsible in the sense that they feel they are not accountable for the results of their actions. They are the empire creators, the industrialists, the competitive businessmen and women who are obsessed only with gain and status. They are also the rebels, criminals and corrupt officials at all levels of society. Collectively these people are the expression of the ugliness inherent in all hierarchy, even though they live out such lives for the most part unconsciously.

THE 50TH GIFT – EQUILIBRIUM

THE NATURE OF SELF-ORGANISING INTELLIGENCE

To understand the 50th Gift is to see a great hope for humanity arising in the years to come. Within the 50th Shadow lies a hint of this hope in the form of a new social model that must lie beyond hierarchy. If the data passing through the human brain is corrupted via the 50th Shadow then one must ask the question: "What does it look like when it is uncorrupted?" The answer is that within every human being is an inherent code for creating harmony, both on an individual level and on a global level. Here in the 50th Gift is found the inner fulcrum upon which balances the quality of our lives and our future possibility for world peace. As the Gift of Equilibrium, this 50th Gift also holds the key to deep inner peace.

No single Gene Key can be truly understood without an understanding of all the other Gene Keys, which is why it is so important to come to know the essence of each one. A single molecule of DNA contains every one of the 64 Gifts in archetypal chemical form, so every one of them also has a resonance within your being at some level. It is interesting to note the relationship between this 50th Gift of Equilibrium and the 59th Gift of Intimacy. In a certain sense one could say that the 50th Gift holds the blueprint of equilibrium whilst the 59th implements it. You can also follow this connection into the Shadow consciousness to find the deep connection between the 59th Shadow of Dishonesty and the 50th Shadow of Corruption. Intimacy is the key to manifesting equilibrium in the world. Intimacy in this context refers to honesty in interaction with others. The vital role of honesty is to create a clean group aura in which all hidden agendas are laid on the table. Without this, no true equilibrium can ever be reached.

Carl Jung said: "Nothing has a stronger influence psychologically on their environment and especially on their children than the unlived life of the parent." This is highly relevant to the 50th Gift, which has much to do with the way in which values are passed on from one generation to another. Repressed secrets or emotions lying between the parents will emerge through the life of the child, creating an atmosphere of unrest, unless both parents take responsibility for the manifestation themselves. Obviously, most parents do the opposite and assume it is the child that needs fixing, helping or disciplining rather than themselves. This universal law applies particularly to children in the first seven years of their lives. Beyond this phase behaviour is fully imprinted and the child may indeed need some form of deep understanding to shed the emotional load that has been unconsciously placed upon his or her shoulders. Through attending to the mirror of their children's behaviours, parents with children under the age of seven years have a golden opportunity to clean the aura of their relationship and bring deep-seated equilibrium into their family.

This same law can be applied to larger social groupings, businesses and entire communities. At high frequencies the 50th Gene Key creates powerful ripples and electromagnetic currents that have the remarkable effect of bringing groups into equilibrium. Wherever a 50th Gift is found in a person, a point of equilibrium exists within the fabric of the social group, be it a family, a business or even across a whole race.

These people can create harmony in a group without even thinking about it.

Those people who carry the 50th Gift as a prime Gene Key have a great responsibility in the world. If the cauldron of the 50th Gift represents the social values of a particular community or family, then the role of the 50th Gift is to balance the ingredients in the pot, in order to create a state of equilibrium. These people are like cooks who know exactly how to read the needs and requirements of any group and adjust the ingredients accordingly. Although this gift sounds a complex one requiring a host of interpersonal skills, the truth is that these people can create harmony in a group without even thinking about it. Their true power is rooted in the high frequency that moves through their aura.

We have seen that the 50th Gift holds a kind of genetic blueprint for social equilibrium. Because of this it is a deeply complex archetype. Maintaining social equilibrium involves fulfilling the needs of both the individual and the group as a whole. In this capacity, the vision that the 50th Gift holds is based upon the notion of heterarchy, a social model beyond hierarchy, and discussed in more depth in the 45th Gene Key. In a nutshell, heterarchy is based upon the principle that if an individual is not in balance then the group has no chance of finding balance. Heterarchy thus places individual equilibrium before all else. It does this through encouraging a process of individual creative empowerment where each individual is given an extraordinary gift — trust. By trusting the individual to take responsibility for his/her own contribution to the group, a very powerful energy field builds within that group.

Within the heterarchical model, the 50th Gene Key's programming partner and codon group also come into play in an important way. The 3rd Gift of Innovation engenders the spirit of playfulness and creative freedom that allows the heterarchy to empower itself. Only when individuals are trusted with their own freedom does a community truly begin to radiate health. The 3rd Gene Key actually governs individual cell mutation within the human body. On a social level this translates as the empowerment and harmonisation of a whole group, one cell at a time. Group equilibrium is also enhanced through the 44th Gene Key and its Gift of Teamwork, which together with the 50th Gift, forms the basis of the heterarchical model. True equilibrium and teamwork is always a self-organising phenomenon that can never be successfully implemented through external control. Herein lies the great secret to humanity's future collective harmony — it can only be achieved through individual freedom and trust in the self-organising intelligence that emerges from that freedom.

THE 50TH SIDDHI – HARMONY

THE GATHERING OF THE ILLUMINATI

The 50th Siddhi is the natural culmination of the process that occurs within the 50th Gift. Creating equilibrium is an ongoing process of balancing the opposites, but eventually the slide from one side to the other becomes smaller and smaller until perfect balance is achieved. This is true harmony — a vibrational code that brings all elements within any system into perfect and permanent resonance. Equilibrium can be lost, but harmony is constant and infinite.

Equilibrium can be lost, but harmony is constant and infinite.

Harmony is the true nature of the universe and of everything in it. It is simply a pre-existing field that must be entered into rather than created. At the Gift level, consciousness can experience individual equilibrium as part of a wider social equilibrium. Feeling a real sense of belonging in a group can induce a deep state of inner balance at a cellular level. However, at the siddhic level, this balance is taken much further.

The 50ᵗʰ Siddhi produces experiences that harmonise human awareness with celestial or universal consciousness. To experience true harmony in a human body, you first must have dissolved all sense of separateness. Only then can you melt into the so-called *harmony of the spheres*. It is only here in the 50ᵗʰ Siddhi that one really understands the higher meaning of the symbol of the cauldron. In many cultures, the cauldron has been a symbol of the form or vessel that receives higher consciousness into itself. At the siddhic level all imbalances within the subtle bodies must be destroyed in order that the cauldron of the physical body can become a vacuum. When this occurs within the DNA of a human being, the body becomes a deep resonant chamber for the highest universal frequencies. Such frequencies actually have the effect of dissolving the body itself so that the true cauldron is experienced as the universe. In some cultures the cauldron has also been seen as a great drum and the movements of the planetary and stellar bodies are the rhythms played upon this drum.

It is also through the 50ᵗʰ Siddhi that we see the true underlying nature of human society — that we humans are fundamentally elements of music. From the level of the 50ᵗʰ Siddhi, everything is experienced as music. Individual lives are notes threaded along a universal score that gradually becomes sweeter and sweeter to the ear as evolution reveals deeper and deeper harmonies. This higher social harmony will progressively reveal itself as human beings move beyond the domain of Shadow consciousness and realise the harmonic blueprint that underlies all human civilisation. As this is realised, and manifested as the Divine Synarchy spoken of in the 44ᵗʰ Siddhi, this codon awakens throughout the human genome and we human beings will become aware of the harmonic geometries of the higher evolutions. The higher subtle bodies of humankind are designed to self-organise into potent and coherent formations of awakening beings. This is the Ring of the Illuminati at work.

The higher subtle bodies of humankind are designed to self-organise into potent and coherent formations of awakening beings.

Humanity has always sensed the presence of the Illuminati or *chosen ones*. We have woven them into our legends, dreams and even our conspiracy theories. However, the Illuminati are a commonly misunderstood ideal. The term does indeed refer to a higher evolution of celestial beings or ascended masters, but those beings are not separate from us. They are our higher nature personified. As your higher bodies transmute your lower instincts and emotions, so you come into the frequency field of their higher harmony. This field draws you into itself and gradually transforms your cellular DNA into its highest mode of functioning. We now live in the time of the gathering of the Illuminati — a time of unprecedented planetary change. As the awakening becomes progressively more widespread, the awakened higher bodies of individuals will come into a heightened communal harmony that quickens the frequency of the whole of humanity. Eventually the Illuminati will assimilate all human beings into its ranks and humankind will finally experience itself as a single unified cosmic being.

When the 50th Siddhi begins to break through in human beings, an all-powerful current of harmony begins to shower the people with whom they are allied. Deep and hidden issues are brought to light in order that social equilibrium can occur. As the Siddhi occurs in one human being, it triggers the Gift in all those around them because the power of the celestial frequency drags consciousness out of its lower manifestation at the Shadow level. Through this 50th Gene Key humanity is destined to arrive at cosmic harmony, although such a concept will require a sustained period of equilibrium before the scales finally activate the higher harmony. We know this higher harmony as oneness with God.

The 50th Siddhi holds another secret, which happens to be the basis of all transformation and also has a particular relevance for cellular mutations in DNA across all species. This 50th Gene Key represents a tipping point in all transformational processes. Its presence always signifies a quantum leap from one state to another. People with this Gene Key in their Hologenetic Profile can therefore catalyse considerable shifts in the consciousness of those around them. A single person manifesting this Siddhi marks a turning point in the whole of human evolution. Because of its genetic pairing with the 3rd Siddhi, each time the 50th Siddhi appears in the world, it precedes a cellular mutation in humanity. One way of predicting such a shift is to look for an increase in the appearance of social heterarchies — groups of self-empowered people working creatively together without the limitation of a hierarchy. These are the early signs that the 50th Siddhi may already be appearing in the world. Remember that the original Chinese name for this 50th hexagram of the I Ching was *Cosmic Order*, and this is precisely what the 50th Siddhi brings to humanity.

All human individuals should take heart from the messages contained within this 50th Siddhi. Along with the 6th Siddhi of Peace, it is one of the true master keys governing and protecting human destiny. Beneath the veneer of your outer life, harmony prevails, even though you may not be able to see or feel it. The blueprint of your awakening is held securely within the vaults of this 50th Siddhi. The deep laws of the cosmos — epitomised by this 50th Siddhi — govern everything that we humans do in the world, from our loftiest sacrifices to our most despicable acts. Whoever you are, your acts gradually move you toward Divine harmony over the course of your life. For most of us there will always be moments when we touch upon the echo of this great harmony deep inside us. For some, the harmony will become a habit, born of a love of freedom and a vision of cooperation. If there is one thing you can say about this 50th Siddhi it is that as it grows in power, it will orchestrate groups and communities of human beings into a beautiful symphony in which we become the instruments, and consciousness itself is the music that plays through us.

☷ 51st Gene Key

Initiative to Initiation

SIDDHI
AWAKENING

GIFT
INITIATIVE

SHADOW
AGITATION

Programming Partner: 57th Gene Key
Codon Ring: The Ring of Humanity
(10, 17, 21, 25, 38, 51)

Physiology: Gall Bladder
Amino Acid: Arginine

THE 51ˢᵀ SHADOW – AGITATION

THE PORTAL OF FEARS

The 51ˢᵗ Gene Key and its spectrum of frequencies contain some startling secrets regarding human behaviour as well as leading you towards the process or experience referred to as awakening. One of the most recognisable human traits to emerge through our genetics is our innate human competitiveness. Until you reach the highest level of frequency in the 51ˢᵗ Siddhi, human beings are driven to compete with each other. Depending on how you channel this energy, it can either lead to unity or it can lead to division. Together with its programming partner the 57ᵗʰ Shadow of Unease, this 51ˢᵗ Shadow of Agitation creates enormous disturbance and insecurity within human beings. Until you can elevate your consciousness into and beyond the higher frequencies of the Gift level, you will always feel this sense of agitation inside you at some level.

The reason that the 51ˢᵗ Shadow creates such a disturbance in the energy field of human beings is quite simply because life is beyond our control. Unforeseen events will occur in your life from time to time and these events can radically change your destiny. At the Shadow frequency the perceived randomness of life creates a deep insecurity in human beings because of this underlying fear that something bad could happen to you at any moment. This insecurity is compounded through direct evidence as we witness it happening to the people around us. For example, in London during the Blitz in the Second World War, bombs would drop at random throughout the night. Most streets in central London were hit at some time or other and entire families were instantly killed whilst neighbouring properties often remained relatively unharmed. In life, the question of *which house gets hit* plagues all human beings. Because we see it happening to others, this generates the unconscious fear that it might happen to us next.

Shocks are a part of all human life, but at the Shadow level where fear reigns, the possibility and fear of shock continually unnerves us. The hallmark of all Shadow frequencies is a profound lack of trust in life itself. And trusting in life is neither an intellectual nor even an emotional issue. It is in fact purely physical.

*The hallmark
of all Shadow
frequencies
is a profound
lack of trust
in life itself.*

Trust is something that is either felt within the cells of your body or not. Without trust, human beings stay in a state of agitation — we tend to be jumpy, nervous and stressed. We either shy away from life out of fear or we rush at it out of rage or panic. Your frequency determines how you view shocks, as well as how you handle them physically and emotionally when they do come. At a higher frequency, shocks are like wormholes to a new and potentially higher dimension. Shock directly challenges the very bedrock of your reality and your attachment to that reality. In this sense, the true role of any shock is to chip away at your sense of separateness from life and release you from the false security of the Shadow consciousness.

The 51st Shadow devotes all its energy into trying to stave off the inevitable. It lives in denial of the ultimate shock — the fact of physical death, and in denying death it actually throttles life. Only one who has fully embraced the certainty of death is truly alive. Without the perspective of death life loses its true sense of value, which is precisely what desensitises human beings. Given that the 51st Gene Key is also responsible for the human competitive spirit, we see humans fight only for themselves and their own benefit without any sense of higher purpose. The competitiveness of the 51st Shadow is about needing to be first, not in order to better yourself, but to feel superior to others. The human competitive spirit can be a very ugly thing at the Shadow frequency because it can so easily brush aside others in its relentless push towards the top. It literally views all others as a means to rise higher, and in that sense it uses other people for the sole purpose of its own advancement. In this sense, superiority is a feeble attempt to allay the inevitable. All those who have risen to the top of the hierarchy throughout history are eventually humbled by death.

The 51st Gene Key as a genetic archetype is rather special. It is unique in that it represents a portal. Where this portal leads depends entirely upon the frequency you program into your genetics. There is a terrific amount of energy within the agitated state of the 51st Shadow. Without a sense of inner purpose or direction this can be a dangerous Gene Key. Agitation never leaves you alone but continually prompts you to do something in order to get some kind of reaction. It will do anything to release its restless, electrical energy. This can lead to all kinds of reckless and foolish acts, done simply for the sake of releasing the agitation. The 51st Shadow can drive people to do things that others would never dream of. This leads to a strange phenomenon that can occur to people strongly influenced by this Shadow — they can actually experience a state of fearlessness. However, this fearlessness is not a true fearlessness based on trust. It occurs when the state of agitation becomes so strong that it wants to extinguish itself, and in doing so it drowns out the fear within the body.

The 51st Shadow impacts the collective through highly unusual and often dangerous individuals. Anyone strongly influenced by the 51st Shadow is either in danger of falling into a deep state of depression or of becoming a liability in the world. Many of these people may have little or no respect for life, so they either lose all hope or they act out their sense of futility in some extreme way. These people are the agents of shock in the world, deepening the mass consciousness fear that nowhere and nothing is safe. The fear that at any moment something terrible may happen is further compounded by the mass media through its constant coverage of just such events.

It is remarkable to see how just a few individuals living out the extreme of this Shadow can exert such a huge influence over the rest of the world. As we have said, the 51st Shadow is a portal, and at the Shadow frequency this portal leads you to perceive the world only through the lens of your fear.

REPRESSIVE NATURE – COWARDLY

When refracted through an introverted nature, the 51st Shadow manifests as a kind of cowardice. These are people who steadily lose hope concerning life and, instead of lashing out like the reactive nature, slowly shut down and turn inward. These can be highly depressive people with little or no enthusiasm for anything in life. The cowardice here lies in the repression of a very powerful nature. These people simply allow fear to rule their life, even though they have the power within them to pull themselves out. The real fear is of facing the fear itself. Ironically, by facing the fear it is shown to be an illusion. However, such natures rarely find the individual courage to break the illusion, preferring instead to wallow in a repetitive cycle of self-pity. To break out of this cycle of depression these people have to wake up to the fact that only they can rescue themselves.

REACTIVE NATURE – HOSTILE

The extroverted version of this Shadow manifests as hostility. This hostility is the result of the dangerous pairing of a fearless rage and a deep sense of futility. These natures bring outer shock into the world, obviously in varying degrees. At an emotional level these people have no real sense of respect for human beings. They tend to be drawn to any sphere of society where competition is worshipped, from business to sports. They are generally loners with deep-seated emotional boundaries. They often throw themselves into highly dangerous or risky situations where they may even face death. With the reactive nature, the profound sense of agitation that underlies this 51st Gene Key projects itself onto others. Thus these people also provoke hostility in others, but without any real agenda. They simply do it because they cannot help themselves and do not care. The only way for such a pattern to be broken is for these people to channel their agitation into some creative project that will ultimately lead to a feeling of fulfillment.

THE 51ST GIFT – INITIATIVE

STIRRING GOOD FORTUNE

Whereas the portal through the 51st Shadow leads to hell, the 51st Gift leads to a place of great personal empowerment and genius. The 51st Gift is engaged whenever the human competitive spirit is put into the service of creativity. Every time you have the courage to follow your own independent creative juices, you have stepped through the 51st portal of Initiative. This 51st Gift is a key place within the human genetic matrix because it contains the activation code for individual empowerment. To follow your initiative is to step off the beaten track and to follow the dictates of your own inner being. There is no safety net when you follow your own

It is not possible for a human being to awaken without first stepping fully into their creative independence.

destiny in this way — it is a giant leap onto a path that no one else has ever travelled. The mass consciousness is both in awe and afraid of those who follow this path in life. The collective way is the way of security, but the way of the individual is mysterious and fraught with uncertainty. As we shall see, it is also the only path to true awakening. It is not possible for a human being to awaken without first stepping fully into their creative independence.

Those who do step through this portal into the 51st Gift are deliberately turning their backs on all who have come before them. They are closing the book on all previous teachings and wisdom and are going out to discover what *their* own individual truth is, rather than anyone else's. This is a mythic path — a path often seen as a journey into the underworld in which the traveller will confront many challenges and tests. It is also a path that eventually arrives home — it is a path of heart leading into the heart. Just as the 38th and 39th Gifts are archetypes of the warrior, so too is the 51st Gift. However, this is a different kind of warrior. This has nothing to do with fighting collective fears and it has little to do with Honour. This is the battle with your own fears, and unlike these other two Gene Keys, this battle does not involve a struggle — it involves a leap. The leap into the 51st Gift is the leap into the *higher self*. It is the shock of awakening from one level of being into another.

The 51st Gift is the highest expression of the human competitive spirit. At this level you no longer compete with others but with yourself. Even in the field of competition — whether that be political, financial or recreational — you use others as mirrors to reflect your own level of excellence rather than pushing them down so that you might rise higher. In sports, for example, the 51st Gift will find its own innate genius and strength by realising its *difference* from others. This 51st Gift is about harnessing your difference from others rather than trying to beat someone else at their own game. When you are truly being yourself, you unlock the magical force of genius that belongs to you and no other. To take your own initiative is to disregard everything you have ever learned or heard. There is no other way than this towards true genius. This is a path *through* fear, and the fear cannot be sidestepped. The specific context in which the fear arises is the perfect and deepest fear for that person. Whatever your deepest fear is, you will meet it here in the 51st Gift, and you will transcend it.

The 51st Gene Key is a key component in the genetic family known as the Ring of Humanity. Creative initiative is the path of every human spirit. Every one of us must at some point in life leave the crowd and head off into the uncharted wilderness of our heart. This is the true path and destiny of humanity. The vitality and courage in initiative stirs powerful responses from the quantum field. Thus the more you believe in yourself and act from this belief, the more life supports you. To overcome your own sense of unworthiness and step beyond competition is to embrace the magic of higher dimensions. Those who take the initiative engage the forces of good fortune and synchronicity. In this sense the 51st Gene Key will always reward your perseverance when you trust in your own heart. It will shock you with good fortune!

People with the Gift of Initiative are the first to do things in the world. Even though they may have followed others in their past, when it comes to the leap, they will always follow their own path. In this way, individuals throughout history stand upon each other's shoulders and keep the human spirit evolving. Just as the 51st Shadow creates agents of low frequency shock in the world, so the 51st Gift creates agents of positive shock in the world. These people come to awaken the collective consciousness out of its fear-based modality. Wherever individuals are lost, the people of the 51st Gift will challenge them to tread their own path. It is important to understand that people of the 51st Gift are not leaders — they are initiators. They come to catalyse new processes among humanity, or they live lives that are so unique and courageous that others are inspired to do likewise.

The 51st Gift has a particularly important role to play in the world of commerce by providing the competitive drive within organisations. In large groups or organisations the 64 Gene Keys each take on different genetic roles. The 51st Gift operates through the entire culture of the business rather than at the level of individual achievement. The more individuals there are with this genetic activation, the more competitive the spirit of the company will be. Most people are confined to living out of its Shadow frequency, which is more likely to make the company over-competitive and create a sense of collective agitation felt by all the employees and especially the management teams. However, as we can learn from the 64 Gene Keys, the secret is not really about quantity but quality. Thus a single person manifesting the 51st Gift can turn around an entire organisation through the collective morphogenetic field of that company. The presence of such a person in a pivotal position can propel the organisation into a totally new level of functioning in which individuals are empowered rather than kept down and controlled.

THE 51ST SIDDHI — AWAKENING

THE THUNDER OF YOUR GIVING

The phenomenon of awakening has captivated human beings for millennia. We are entranced that certain people seem to go through experiences that permanently alter them and put them in touch with a reality that we cannot see but can still sense. Of course there are many interpretations of the word *awakening* within the halls of spirituality. It may seem to many that there are grades and levels of awakening. Many of the new interpretations of awakening view consciousness as a ladder that can be ascended rung by rung. Even this work on the 64 Gene Keys presents the evolution of consciousness as a rising of energetic frequency through your genetics, resulting in changes in the operating system of your body and awareness. There is also a huge body of provisos at the spiritual level coming from the mouths of both the genuine teachers and the false prophets. To add yet another layer to the story, humanity is currently at one of the most potent crossroads within our genetic and spiritual evolution. At such times, the voices come thick and fast — the voices of doom and the voices of hope — each declaring their truth.

The 51st Siddhi sweeps all these requirements and systems aside. It makes everything simple and clear. For the 51st Siddhi there are only two states of consciousness — awake and asleep. Of course from within the siddhic state this division into awake and asleep is simply a matter of rhetoric, but it also holds true on the material plane. Something occurs to a person before they enter a siddhic state — something unpredictable and without cause — something momentous. No one can put into words what happens to someone as they enter a siddhic state. There are things that must always remain a mystery, even to science. There may be sciences of awakening before awakening, but there are no sciences after awakening. Therein lies the paradox. The 51st Siddhi is beyond spiritual jargon and systems. It has a language of its own and that language is simple and shocking. It says that until you are awake, you are asleep. There are no levels within awakening. To the 51st Siddhi, all ordering of consciousness is absurd — it is simply done by people who are asleep grading the patterns *of* their sleep!

What then is awakening and how do you tell when someone is awake and when someone is not? These are the great questions of the sleeping world. If you were to ask someone who is awake these same questions they would probably say that it doesn't matter, and that until you are awake you will never know why it doesn't matter; such things only matter to those who are not awake. The fact is that awakening does make you different — not in terms of your deepest consciousness, but in terms of your physical genetic vehicle and its functioning. By the same token you could argue that awakening might one day be genetically manipulated by science. In theory, this might even be possible. However, as long as human beings are asleep they will not bother throwing their valuable resources into such spiritual aspirations. The lower frequencies are based on self-interest and awakening does nothing in that direction. Added to this, awakening is such a delicate evolutionary process. It is not the result of individual effort but is a spontaneous evolutionary leap made possible by wider evolutionary forces operating throughout the entire universe. In short, awakening is a mystery that prevents its own nature from being solved or replicated.

From within our deep sleep, we are desperate to understand this phenomenon of awakening. However, in the language of the 51st Siddhi, it is a mystery that will not yield to pressure. We may wonder if there are signs that show a person is close to awakening, and we have certainly tried to create paths towards awakening. There are no signs. Awakening has happened to nice people and to nasty people! Nor are there paths, despite what even awakened beings have said or laid down for us. To be awake is a grand dilemma. No matter what you say it will be misinterpreted, so in the end you simply say whatever you say and trust in the vibration underneath it. Awakening is also easy to fake. Anyone with a powerful aura can claim awakening or even believe that they *are* awake. Mystical experience can also be mistaken for awakening. There are so many types of mystical encounters and visionary states that occur to human beings all the time, but awakening is something entirely different from all these. Many of the greatest mystical visionaries and even the great revelations or systems that come through them still do not touch the true field of awakening.

True awakening simply sees through everything and has nothing whatsoever to do with behaviour or experience. Neither does awakening come and go. When it comes, it stays forever. As you read each of the 64 Siddhis described within this book, you begin to have a good notion of how diverse the behaviour patterns of awakening really are. The 51st Siddhi is the ground beneath every one of these Siddhis, and no one can attain these states without true awakening. In words, awakening is the permanent dissolving of the separate self and occurs through a physical mutation within the body. No amount of meditation or spiritual practise can bring about this mutation, even though they may or may not precede awakening. It is not possible to influence this mutation because it has nothing to do with your experience or behaviour. This is the grandest spiritual illusion of them all.

Human beings do not readily take no for an answer. We like to feel that we can *do* something. However, this is the one area where you cannot do anything other than be yourself. Awakening occurs to someone who is so much themselves that it simply has no choice *but* to occur. The irony here is that you cannot work at being yourself. It is simply an innate gift. You can see how the Shadow of Agitation bleeds through this Gene Key at the highest level. Even when you are awake, you can still see the spiritual agitation that your presence causes others! The fact is that the siddhic state is a leap out of all levels, and it is a leap into the void. Nothing can prepare you for that leap, and nothing definitive precedes the leap. Neither is this leap a leap that you *take*. It is more as though the leap takes you.

The 51st Siddhi contains many other secrets; among them are the secrets of initiation. The universe we inhabit is constantly initiating all forms. The world of form is constantly impacting into itself — atoms collide with atoms, asteroids collide with planets, humans collide with each other in relationships. It is a game of penetration in which everything is trying to penetrate everything else. Every penetration is in a sense a shock and results in a mutation or transformation of some kind. Death itself is the transformation of one form back into its constituent forms. Before awakening, forms are still forms, so we are always at risk of some kind of impact or shock. Such shocks are really initiations in which the boundaries of your world are shaken. After awakening, all forms are experienced as interpenetrating each other, so there are no longer any shocks or impact. There is no localisation of awareness in the body in order for a shock to take place. Initiation always demands an environment, and awakening brings an end to the very concept of environment.

As we have seen, there is a great deal of confusion concerning the concept of awakening. It is a permanent mutation at a physical level, resulting in a fundamental change in the perceptive apparatus in the body. The ancient Chinese represented this 51st hexagram through the symbol of thunder, and it really does take us by surprise. Awakening is always a surprise. It comes once and only once, and after it has come, it stays forever. Once you are awake, you cannot go back to sleep. The thunder within this Gene Key is the thunder of life itself — it is the thunder of your giving. The awakened one becomes a vessel that knows only how to give of itself, there being no friction preventing life from flowing through. This giving is love, but it may or may not manifest as affectionate love. In truth, it is cosmic universal love.

There is one final thing to say about the 51st Siddhi that is of great import at this time in our evolution. Thus far we have only considered the 51st Siddhi at an individual level, but it also operates at a far wider level. As an aspect of the Ring of Humanity, this Gene Key ensures that one day, all of humanity will awaken.

Awakening is always a surprise. It comes once and only once, and after it has come, it stays forever.

Indeed, in the times to come this Siddhi will begin to awaken entire gene pools. There will come a time in which relationships will awaken, then families, then entire communities until finally the thunder will be heard throughout the human genome. At this stage, humanity itself will awaken as an individual cosmic being and will see its own true nature. As with individual awakening, this is not a gradual phenomenon, even though it may appear so. It is a sudden unexpected shock in awareness that is preceded by a mutative chain reaction through your cells and DNA. When humanity awakens, it will move beyond time. That single moment when humanity opens its true eyes for the first time will be the greatest single event in the history of awareness itself.

▦ 52nd Gene Key

The Stillpoint

SIDDHI
STILLNESS

GIFT
RESTRAINT

SHADOW
STRESS

Programming Partner: 58th Gene Key
Codon Ring: The Ring of Seeking
   (15, 39, 52, 53, 54, 58)

Physiology: Perineum
Amino Acid: Serine

THE 52ND SHADOW – STRESS

THE PHENOTYPE OF FEAR

The 52nd Shadow is responsible for one of the great phenomena of the modern world —stress. Stress operates on many levels although its primary impact takes place on the physical plane, inside your body. This 52nd Shadow and its programming partner the 58th Shadow of Dissatisfaction represent the deepest genetic binary pattern that undermines human health. This is especially true at a collective level. For that reason, we need to understand something very important about stress. It is a collective pressure and not a personal pressure. It is an energy field generated by every human being alive, rather than being rooted in any specific personal issue. This means that stress is also deeply connected to our environment and to those around us.

Our actual environment is formed by the boundaries of the human aura. The average human aura will extend to fill half of a normal sized room, which means that generally speaking, when you are in the same room as someone else, you share each other's auras. The more people you have in the same area, the greater the collective auric body grows. In densely populated areas, such as towns and cities, it is virtually impossible to escape the collective aura of others even in the sanctuary of your own home. In more rural parts of the world the pressure of the local aura is far less and the degree of stress is therefore far less. However, because of the huge population currently existing on our planet, it is no longer possible to entirely escape the vast collective energy field of humanity. Billions of auras interlocking together form a great sheath that covers our world. This sheath or *skin* is what prevents human beings from seeing and experiencing the true unity of all consciousness. It is the construct known in the East as the *maya* — meaning the great illusion.

Human DNA is a fascinating substance. The chemical codes contained within DNA, although locked in from birth, are highly sensitive to vibrating energy fields. Geneticists refer to the genetic makeup of an individual as the *genotype*, whereas they refer to the manifestations of that makeup as the *phenotype*. So, to an extent, the frequency of energy passing through the genotype determines the manifestations of the phenotype. This means that your environment greatly affects how you feel and behave, and most importantly, it influences who you think you are. At the subtlest level, your environment is created by the subatomic world of vibration. The frequency you are tuned to actually determines this environment, regardless of your physical geography. In the case of the 52nd Shadow, the collective or extended phenotype of humanity is shaped by fear, which creates an environment that puts the physical organism under great stress.

The only way to escape this collective programming field is to rise above it at an energetic level, which is no easy task. You must somehow raise the frequency of energy moving through your genotype. When this occurs, the phenotype — the way your nature is experienced and expressed — will also change. Most human beings will rise above this energy field for brief periods of time and then fall back into it again. It is extremely rare for someone to rise permanently above the field of stress. The secret is to change your inner environment — your feelings and your thoughts — thus changing what you see and hear. If all you hear is noise and all you see is chaos, then that will determine your experience. However, if you anchor yourself into a higher frequency, the same life can be experienced as though you were living in an entirely different world.

Stress is a state of physical pressure brought about through the lowering of your frequency. One of the key manifestations of stress is the inability to escape mental anxiety. What happens is that you give the mind the authority to make decisions that you hope will end the stress. However, this nearly always ends in disaster because the mind's activity is actually the manifestation of the stress itself. This classic biofeedback loop reinforces the very state you are trying to escape. This 52nd Shadow is deeply connected with the function of your adrenal glands and the classic *flight or fight* response in human beings. (The activity of your endocrine system is directly related to the frequency passing through your genotype, and as we shall see when we get to the 52nd Siddhi, the human glandular system is not just limited to manufacturing hormones such as adrenaline that are rooted in survival and fear.)

One of the keys to understanding the 52nd Shadow is to be found in the name of the 52nd Gift — the Gift of Restraint. The low frequency expression of the 52nd Shadow is rooted in the inability to keep yourself from reacting to fear. Sometimes the fear driving this Gene Key can be so deeply embedded in the unconscious that you have no concept of it, at all. Nonetheless, this fear pervades our environment and your reaction to it produces one of two responses — you either collapse or you run. Given its chemical connection to the adrenal system, this Shadow is deeply implicated in physical activity or lack of activity, the two human responses to stress. As far as the 52nd Shadow is concerned, there are two types of people — those who cannot sit still and those who are stuck. In our modern age, we tend to think of stress as being a chaotic, frenetic energy dynamic, but it also has this more repressed aspect.

As far as the 52nd Shadow is concerned, there are two types of people — those who cannot sit still and those who are stuck.

Natural life rhythms follow seasonal patterns — life knows when to work and it knows when to rest. However, the collective power of the 52nd Shadow keeps human beings from trusting in this flow of life and the Shadow frequencies of this 52nd Gene Key enhance one extreme or the other.

We can clearly see this manifesting in the world today. The West tends towards the extreme of restless activity, expansion and improvement with little or no collective sense of purpose and the East has traditionally moved in the other direction away from activity and improvement and more towards the religious and spiritual realms. Even though today in a reversal of roles the East is becoming more like the West and vice versa, the underlying phenotype of humanity remains based on fear and the two polarities of this Shadow continue to manifest. Until human beings experience their unity with all of creation they will create and be driven by this underlying stress — the physical manifestation of our deepest and most hidden fear.

REPRESSIVE NATURE – STUCK

The repressive nature tends to physically, emotionally and mentally collapse under the pressure of the stress created by the 52nd Shadow. These people never really get off the ground. The Gift of Restraint remains locked in this extreme shadow manifestation and leads to a deep feeling of being stuck. These feelings can so pervade the body that one enters deep states of depression and apathy. As these people give up on themselves, the adrenal system can actually atrophy leaving them physically impeded in some way. Once one has sunk into these states, it can be very difficult to escape. One of the only means of lifting oneself free of such a state is to find a way to be of service to others. There is a deep genetic need in the 52nd Gift to help others and when this urge is engaged, gradually one can find one's life force returning.

REACTIVE NATURE – RESTLESS

These are people who are unable to sit still. Reactive natures by their nature try to escape their fear through activity. However, these people mask their fear through projecting their frustration and anger onto others. The restless aspect of stress is based upon over-stimulation of the adrenal glands, producing more energy than one needs. However, the consistent secretion of adrenaline into the body actually causes great damage over time. Thus these people often burn themselves out in life. The secret for such people is to see through their own mental dynamic and discover the deep fear that drives them. By confronting and embracing this fear, they will gradually erode its hold over them and learn to activate their Gift of Restraint.

THE 52ND GIFT – RESTRAINT

ECOLOGICAL TORQUE

People may tend to devalue the 52nd Gift because of the way it sounds. Restraint does not after all sound particularly exciting or dynamic. Ironically, however, this is one of the most remarkable Gifts in the whole human genome. This Gift, like all the Gifts, is about striking a balance between two extremes of energy — in this case, the energy of activity and passivity. In fact, no Gene Key is more fundamental to human life than this 52nd Gene Key. It determines the source signature of all activity in the world.

If you begin something from a place of fear, that seed of fear will infect every aspect of the expansion of that activity. Even though the smallest human actions can sometimes lead to the creation of vast empires, if the seed of fear is there at the beginning, then like a virus it will pervade the structure and it will eventually bring that structure to the ground.

The 52ⁿᵈ Gift contains a secret of ecology. All true endeavours require a profound understanding of restraint, a phenomenon natural to all living systems. If a system is to last, it must learn to feed itself whilst it grows. If it is to proliferate, it must learn to diversify and take in an ever-expanding source of nutrients. However, for a system to be really successful it must have one vital factor right at its core from the very beginning — it must serve more than just itself. The 52ⁿᵈ Gift of Restraint requires a great deal of patience and an understanding that everything in nature moves at its own speed. Particularly in the beginning, things appear to move slowly. The moment we try to rush an idea, we disturb the ground in which it is sown. You can see from this how easily humans fall victim to stress.

The more selfless your intention, the more power it will have.

The 52ⁿᵈ Gift, you may divine, is a Gift that concerns human organisations. At its deepest level, it contains the seed that will one day bring all human beings into a perfect organisational unity. For now, this Gift concerns our own trust in life. To begin anything, you must first have a clear intention. The more selfless your intention, the more power it will have. If you begin with the right intention, then everything will follow, but you must resist the temptation to interfere with the process out of fear. The intention is the seed, and the seed contains all the necessary ingredients and properties that will be needed in the journey ahead. The seed even contains the specific fragrances that will attract the right allies at the right times. It is also true to say that the greater the power, the longer it takes to germinate. The seed of a yew tree and the seed of a sunflower are similar in size. However, whilst the sunflower will grow to its full size within the space of a few months, the yew tree has depth and complexity and will begin at a different pace and follow its own timing. It may take ten years just to reach the height of the sunflower, but it may live as long as five thousand years. So it is with all human ideas and actions.

Every human being alive contains this intention — it is the seed of your individual destiny. To get your intention clear, you must ask yourself the question: "How can I be of the greatest service to humanity?" Then you must live the answer. Because we humans cannot see the details of the journey ahead, we have to trust in the direction that our intention takes, even if it does not make sense to us at the time. This is the power of Restraint — to allow your life to unfold without urgent demand.

We also saw from analysis of the 52ⁿᵈ Shadow that another aspect of this Gene Key is the belief or feeling that we are stuck. Only if you lose touch with your innermost core can you be truly stuck. As long as you remain in touch with the seed of your true intention, you may experience an outer pause, but you cannot be stuck. Deep within the seed, or indeed the plant, new shoots are being born. It is usually at this stage when humans become restless and try to force a direction, damaging the delicate buds that will form the next phase in the evolution of your intention. Ironically, it is during the moments of outer pause, when all energy seems to have come to a halt, that the greatest growth occurs.

The stored up energy and potential within a single human being is infinite in its proportions. However, this potential must be managed economically and allowed to grow and expand in an organic fashion. In this respect, the Gift of Restraint is about non-interference. When you apply this to your own life, you have to accept yourself as part of a far greater flow than you can see, and as such you have to accept that there are times when you *will* feel deeply restrained. The acceptance of this restraint and its manifestation as patience is a truly powerful gift. There is an apt parallel here with the developing life of a young child. Children contain the seed of their own future, and if allowed to develop at their own pace and in their own way without too much interference, they will eventually flourish. Every parent has to find the delicate balance between laying down healthy boundaries and trusting in the life force as it moves through the child.

It is through restraint that human power can be harnessed in a creative way. The 52nd Gene Key is a member of the codon family known as the Ring of Seeking. You will learn as you journey through each of the six Gene Keys of this codon that they are all concerned with pressure. It is this internal pressure that drives evolution. There is a vast amount of life force literally wanting to burst out of all human beings from within this aspect of our DNA. If you scan across the names of the Gifts in this Codon Ring, you will get an idea of the power stored inside you — Magnetism (15), Dynamism (39), Restraint (52), Expansion (53), Aspiration (54) and Vitality (58). The 52nd Gift of Restraint stands out among these as the only one keeping all this pressure in check. It is of huge importance in regulating your life and in maintaining a degree of internal rhythm and structure. It is this Gene Key that actually generates the torque that allows all systems to gyrate and evolve.

THE 52ND SIDDHI – STILLNESS

THE STILLING OF THE WAVE

An interesting phenomenon occurs at the highest level of frequency that may help us understand why the siddhic state transcends frequency altogether. If you consider what frequency really is — the oscillation of energy waves at different speeds and intervals — you will find that there is a paradox when you take frequency to either of its extremes. Energy waves oscillating at lower and lower frequencies would eventually stop altogether and you would experience nothingness. At the other end of the spectrum, energy waves vibrating at higher and higher frequencies eventually become so close together that they merge to create another kind of nothingness. This nothingness represents the siddhic state. Obviously there are many words used to describe this state: Bliss or Universal Love or — in the case of the 52nd Siddhi — Stillness.

The Siddhi of Stillness can greatly help us to understand this concept of transcending frequency. Paradoxically, both ends of the spectrum lead to the same state. At both extremes of the spectrum, we experience stillness. Many spiritual systems or great teachers have referred to the ultimate enlightened state as *nothingness* or void.

The Buddha was particularly fond of this terminology. In fact this 52nd Siddhi has a real taste of the Buddha about it. The ancient Chinese named the 52nd Hexagram of the I Ching *Keeping Still Mountain* and one is reminded here of the image of Buddha sitting in absolute stillness beneath his bodhi tree, waiting for all phenomena to dissolve and the true reality of enlightenment to shine through.

When a being attains realisation through the 52nd Siddhi, some intriguing things occur. Since all frequency and energy patterns are experienced as stopping, you find yourself sitting at the heart of all creation. All phenomena are experienced as wheeling around you as you become the still-point of existence itself. The incredible sheath of fear and stress created by the world aura can no longer touch you since you occupy a space outside of all vibration. This is why mystics use terms such as the *spaceless space* to describe this state. Along with the stillness comes the experience of the 58th Siddhi of Bliss, the programming partner of the 52nd Siddhi.

These two great Siddhis, the 52nd and 58th, mirror one of the great universal concepts of geometry and physics — the torus. The torus is a multidimensional geometric figure that lies at the heart of all space-time. The torus demonstrates the universal laws of energy dynamics based on torque and spiralling forces. At one end of the torus is the black hole, representing the yin pole that sucks, contracts and enfolds all energy and matter into itself. At the other end of the torus is the white hole, representing the yang pole that releases, creates and expands all energy and matter outwards to create space-time itself. The torus is a magnificent figure that unites centrifugal and centripetal forces, bringing implosive and explosive dynamics into the same system. Within human DNA, the torus at the heart of all being is directly experienced through the state of enlightenment, which unites stillness (the black hole) with bliss (the white hole).

This experience of utter stillness and bliss is derived from a higher functioning of your genetics. The human endocrine system is essentially an alchemical factory in which chemicals and hormones are combined and created. As we saw with the 52nd Shadow, the glandular functions of the body secrete the hormone adrenaline whenever this Shadow is activated through fear. People living within the illusory field of the low frequencies are actually addicted to this hormone and the fear that creates it. However, at the siddhic level of frequency (or lack of frequency) the body actually creates hormones and neurotransmitters that are extremely rarefied. Every Siddhi has its unique glandular secretion and, in the case of the 52nd Siddhi, a hormone is secreted which induces bliss and stillness throughout the body.

For millennia humans have sought these magical secretions and have mythologised them as elixirs and panaceas that can be created chemically. In recent years, certain substances and drugs have been created that imitate or activate these hormonal secretions, an example of which is the drug ecstasy. What we need to understand is that the process of mutation leading to such hormonal secretions is a highly delicate organic process that is catalysed by a far more subtle process occurring at an energetic level. As this process has passed through many organic stages without interference, the body creates a physical counterpart to the subtle process that has taken place.

In the case of the 52nd Siddhi, the experience of time standing still is reflected through a neurotransmitter that effectively stops thought when awareness is not required for communication.

If you have read the 55th Gene Key, you have read about a collective phenomenon that will soon be coming into the world known as the *stilling of the wave*. This phenomenon will come about as a result of a physical genetic mutation that is predicted to sweep through humanity beginning around the year 2027. This mutation will trigger a sequential process involving the flowering of the 64 Gene Keys at a collective level and the expanding emergence of the Gifts and Siddhis in the world. The expression *stilling of the wave* refers to the stilling of the chaotic emotional energy field that currently dominates our planet. One of the Siddhis we will see coming into the world more and more is the 52nd Siddhi. Remember that this 52nd Gene Key is a seed archetype and it concerns the unfolding of intention into individual destiny through service. As all Siddhis are collective in nature, the 52nd Siddhi also contains the seed of a great new beginning, and the original intention and dream of mankind — to realise our position as the still point at the heart of creation itself.

The 52nd Siddhi contains within itself the power to focus humanity into a single unified pattern. It emanates a collective energy field that will eventually pacify the human emotional system. The 52nd Shadow of Stress is closely related to time. Much stress is caused because people experience time as moving so fast that they panic and try to catch up with it. The aura of a single being manifesting this 52nd Siddhi will literally stop the thoughts of everyone in their vicinity. It is a Siddhi that increases the wavelength between the particles of the subatomic world, and in doing so it slows down perceived time. Every Siddhi floods the energy field of humanity with its essence, so the presence of even a small number of beings at this level of consciousness will change the way in which all human emotions operate. This kind of influx of Stillness into our planetary aura will make it possible for many millions of people to follow their right destiny into far wider service of the whole.

The aura of a single being manifesting this 52nd Siddhi will literally stop the thoughts of everyone in their vicinity.

53rd Gene Key

53rd Gene Key

Evolving beyond Evolution

SIDDHI
SUPERABUNDANCE

GIFT
EXPANSION

SHADOW
IMMATURITY

Programming Partner: 54th Gene Key

Codon Ring: The Ring of Seeking
(15, 39, 52, 53, 54, 58)

Physiology: Urogenital Diaphragm

Amino Acid: Serine

THE 53ʳᴰ SHADOW – IMMATURITY

THE FALSE CULT OF THE INDIVIDUAL

There is an old folk saying that most of us have heard many times before: "Begin as you mean to go on" — and although a cliché it is a piece of wisdom very appropriate to the 53ʳᵈ Shadow, which concerns the energy inherent in all beginnings. Before you begin anything new in your life you need to ask yourself: "What is the true essence of this beginning?" The majority of people don't realise that most beginnings contain subtle traces of fear. If these grains of fear are present at the very source of your endeavour, you will have unwittingly planted the seeds of its eventual demise. At the shadow frequency, fear is an inner yoke inextricably bound to human intention, and intention is like the arrow that you load into the bow of your actions. No matter how much work you put into something, if the arrow is bent it will never quite hit the mark you were intending. In the context of this Gene Key, what we call *immaturity* is really nothing more than the human tendency to keep loading bent arrows into your bow. Because of this, no matter how well meaning, your actions at a low frequency can only lead to further disharmony.

Since the 53ʳᵈ Shadow is so closely linked to its programming partner the 54ᵗʰ Shadow of Greed, the most common expression of immaturity often involves power and money. Thus in the world of commerce we find fear to be at the root of almost all businesses, even those that claim to be service-based. The world has not yet really seen what can happen when a business puts service before survival, even though there are a few early examples surfacing in the world today. In human business at the Shadow frequency, growth is praised above all else, even though too much growth is unsustainable and damaging to the environment.

However, the 53ʳᵈ Shadow does not only concern business. It is a genetic reflex at the root of our whole civilisation. Because of this fear reflex within us, we cannot comprehend the great laws of nature, the prime law being that of abundance. When nature is left to her own devices, she flourishes while at the same time never losing touch with the overall picture. In nature, if one species becomes too prolific, then a counteracting force responds to the imbalance and restores equilibrium.

Man, too, is a part of nature and is subject to these same laws, although we behave as if we are outside of them.

As one of six internal pressures forming the Codon Ring of Seeking, the 53rd Gene Key is responsible for creating a great deal of stress at the Shadow frequency. The stress it causes in our modern world is directly reflected in the desire to become materially rich. Great individual wealth is unsustainable unless you have a higher purpose that requires it. Keep in mind that there is a vast difference between prosperity and wealth. Wealth is a stockpiling of money based upon fear and greed, whereas prosperity is a flow that expands and contracts with universal rhythms. Prosperity adjusts itself automatically to the needs of your higher purpose. Wealth is in no way equivalent to fulfilment. In fact, it most commonly leads to the opposite. The essence of the Ring of Seeking is to lead you out of your immaturity by showing you the true nature of your desire, greed and fear. Thus we learn in time that the fulfilment we seek is within us rather than outside.

The Shadow of Immaturity is rooted in the human tendency to see ourselves as separate from nature. The human mind has enormous difficulty seeing itself as a collective organism that is deeply embedded in nature and the earth. If one of us commits a selfish act or an act rooted in fear, it reinforces that act throughout the totality, which in turn strengthens its vibration in the world. This is what immaturity is — an aspect of the whole that does not yet realise that it *is* the whole. However, human beings have always sensed the inherent balancing force woven into creation. It is reflected for instance in the Buddhist and Hindu doctrine of karma — the law that every cause results in an effect that directly influences our own future. A common oversight here is that it is not only we as individuals who affect our future, but we as a collective.

In order for human beings to evolve, we have to pass through the phase of our development in which we learn that we are a single unified organism by witnessing the damage we inflict on that organism. We are like a child who pushes the mother until the mother disciplines the child. We must drop our primal fear of death above all else. Even our spiritual yearning to mentally project ourselves beyond this lifetime in the form of a reincarnating soul or a separate spiritual being is subtly rooted in our fear of dissolving back into pure consciousness. It is the fractal pattern of evolution that moves onwards — not our attachment to it. Death cleanly severs our attachment to our individuality, and yet for millennia humanity has been too afraid to

We have created a great cult based on the individual, even though the individual is itself an illusion.

really see this. We do indeed sense the continuity of life, but we also insist on projecting our individuality into it. We have created a great cult based on the individual, even though the individual is itself an illusion.

Why is it that we human beings cannot accept our mortality? The answer is simple — life would appear too frightening. Life has no morals. Life has no concept of individual justice. At the level of the absolute there is no *individual* soul that survives death, even though the existence of a reincarnating higher causal body has relative truth to it within the framework of the maya (see the 22nd Gene Key). Any mystical experience we have that falls short of an immersion in pure consciousness is a subtle projection of our individual need to go on existing. The fact is that the 53rd Shadow of Immaturity builds all these illusions into our minds. Life is simple and pure and doesn't require our projections. There is only the continuity of consciousness following its bloodlines, its fractal lines and its collective evolutionary mythology. These truths are often shocking to the mind and its complex system of beliefs and projections based on fear.

It is through these two Shadows of Greed and Immaturity that humanity must finally awaken to its nature as a single organism. Like a child, we must eventually grow out of our self-obsession and reach maturity. The original Chinese name of this 53rd hexagram in the I Ching is *Development* and so it is. Just like a young child, humanity is an organism that is too self-obsessed to be aware of the consequences of its actions. Although we are still immature, it is written into our DNA that we will gradually discover our actions do carry consequences, even though those consequences may only be realised by future generations. As such, individuals who are judged as evil need to be understood as aspects of the totality displaying its immaturity. Our desire to punish any individual may seem natural at the Shadow frequency, but it represents a level of self-delusion when seen from a collective level. Instead of punishing itself, our organism has to learn to understand itself better.

What then does the 53rd Shadow mean for you as an individual? It means that what the Buddha said is absolutely true — everything is constantly passing away and beginning again. You display your immaturity when you try to impose an external view or doctrine on life. Everything you do from the Shadow frequency is rooted in the fear of non-existence. Like a weed burrowing deep into your psyche this fear prevents consciousness from fully penetrating your form. The grace of the 53rd Shadow is in showing you that you are immature until you begin to understand how deeply your actions, thoughts and words are rooted in this fear. As you come to understand this vast and sweeping insight, your heart begins once again to open and trust in life just as it is — with no opinion, no judgement, no attachment and above all, no fear.

REPRESSIVE NATURE – SOLEMN

When the incredible vital energy to begin new things is repressed, it creates a very solemn human being. These people usually become fixed into a single activity their whole lives. There is a deep sense of sadness in such people, as though they might implode at any moment. It takes an enormous pressure to be serious in life, and it betrays a vast unconscious reservoir of fear. These people also find it near impossible to embrace new things in life. They try to hang on to control by keeping everything exactly as it is. Such people do not cope at all well with change and it tends to drive them even deeper inside themselves where they wall themselves off from the world. These people often end their lives with a great sorrow surrounding them.

REACTIVE NATURE – FICKLE

The reactive side of this 53rd Shadow never sits still long enough to allow a person to evolve. Instead they move from one thing to the next without following through with anything. These people are always beginning new things but without any sense of commitment in developing them further. The only reason such people begin something new is to escape their greatest fear

— of being trapped in a cycle in which they might have to face themselves. Ironically, they do remain trapped in a repetitive cycle of beginnings that lead nowhere. The lives of such people may look exciting at times but they lack any real depth or fulfilment. The reactive nature is about the unconscious expression of one's fear as anger. Because these people are not honest in who they are, they trigger angry reactions wherever they go, further justifying their urge to be fickle.

THE 53ʳᴰ GIFT – EXPANSION

SIMPLICITY THEORY

From reading and contemplating the 53ʳᵈ Shadow perhaps you can sense how spiritual this Gene Key actually is. It really does represent the driving force of evolution as consciousness burrows deeper and deeper into the material life. Life itself knows only expansion. Even when it chooses to contract, it does so only in order to expand further in a new or different direction. From the point of view of the frequency of the 53ʳᵈ Gift, all that exists is this perpetual evolutionary impulse to expand. Consequently, humanity is eventually destined to expand beyond itself. Of course, it is possible that humanity could destroy itself, but even if it did so, it would happen in order for life to expand even further in a new direction. Everything that comes before sources what is to come — that is the law of this 53ʳᵈ Gift and is the basis of the law of cause and effect.

True expansion always involves evolution. To the person who has embraced the high frequency of the 53ʳᵈ Gift, expansion is all you know. Your endeavours regularly outgrow their own forms, just as life regularly outgrows the forms that it inhabits. In the business world, there are organisations that simply expand and there are organisations that expand *and* evolve. Over-expansion in business is a sign of the Shadow consciousness at work. When you expand too much in a single direction, universal principles will cause the opposite effect of contraction. This is why empires and monopolies ultimately crumble.

True expansion also embraces the notion of *fractal growth*. Fractal growth in business only occurs when the people within the business — who represent the consciousness of the organisation — also evolve. True growth expands beyond its comfort zone — it is continually transcending its last level. When this is allowed to happen, the business grows in many directions simultaneously rather than in a single direction.

Evolution follows what modern science terms *Complexity Theory* — the theory that as living systems evolve, they become more and more complex. It is indeed true that the more elements you integrate into a system, the more complex it appears to become. However, just because it seems complex to the mind does not mean that it becomes less efficient. In fact, evolution demands that systems become *more* efficient and efficiency is based on simplicity rather than complexity. Synthesis only appears to be complex when you remain stuck at a low level of frequency and try to understand it intellectually.

The Gift of Expansion demands that individuals transcend their personal opinions, views and attempts to understand what is going on. It takes enormous trust to allow your life to truly expand because it appears to the mind to become more complex, whereas in fact it is moving into greater and greater synthesis. At certain points within your expansion, your awareness itself takes leaps that allow you to grasp the synthesis. Until these leaps come, you simply have to hold on and trust in the process.

On the other hand, the Shadow frequency, driven by greed, wants expansion without having to expand itself, which is why true expansion is actually relatively rare. Expansion is a process of transcending and including. The wonder of expansion is that each new level of integration is built upon the levels that came before, which are thus included in the synthesis itself. This can be applied to any aspect of life, from computer science to business to spirituality. Expansion happens when consciousness is permitted to penetrate form in greater depth. It is about penetration — the more you expand, the looser the form becomes so that you begin to glimpse the consciousness hovering behind it. Of course, the ultimate expansion is the expansion of human awareness itself, which underlies all other forms of expansion.

In a human being, the expansion of frequency can only occur in one way — through the heart. This 53rd Gift can be beautifully explained through the Indian concept of *bhakti* or devotional energy. This bhakti at the heart of evolution is continually outgrowing its own forms. The Gift frequency gives rise to the beginning of the transcendence of the mind. As you allow your awareness to expand, it opens your heart. From the Gift level of awareness all you see is the evolutionary impulse being born, living and appearing to die. There is no difference between the consciousness that inhabits a tree and the consciousness that exists in a human being. The only difference is the awareness operating system within each. The tree experiences life through its sap, its roots and its leaves, whereas we humans experience life through our bodies and our minds. Seen in this way, all life is following the same path towards a higher evolution. Death is not a contraction, as we might perceive it to be, but simply another expansion inwards.

All you might therefore say about the 53rd Shadow of Immaturity is that it isn't yet aware of its environment. Like a child who is not yet self-aware, it doesn't know how it affects the world. Self-awareness *is* maturity, and only when humanity is self-aware as a unified organism will it have grown up. The secret to this 53rd Gift is thus to give way to the *bhakti* — the ever-expanding-transcending-and-including energy of life itself. It requires that you allow yourself to be swept aside by life, letting go of all definitions of who you are and where you think you are going. Once you make this shift into your heart, you will find your life to be much simpler. You activate the frequencies of real prosperity by aligning yourself with the universal evolutionary impulse, and by surrendering in this way you are continually replenished and expanded. This is nature's Simplicity Theory, except that unlike Complexity Theory, it is no theory at all but a self-proving universal law.

In a human being, the expansion of frequency can only occur in one way — through the heart.

THE 53RD SIDDHI – SUPERABUNDANCE

THE END OF EVOLUTION

The 53rd Siddhi has given rise to one of the greatest misunderstandings in the history of spirituality — that of rebirth and karma. It is widely held among eastern mystical traditions that the soul is born, dies and then is reborn in another body over and over again. Through accumulated good deeds, the individual soul eventually attains transcendence of karma and becomes liberated or enlightened, at which point it cannot take another incarnation but returns to its limitless source. This is the basic dogma of reincarnation or the transmigration of the soul. In the Buddha's words:

"As there is no self, there is no transmigration of a self;
but there are deeds and the continued effect of deeds.
There is rebirth of karma; there is reincarnation.
This rebirth, this reincarnation, this reappearance of the conformations
is continuous and depends on the law of cause and effect."

From the Buddha's words, it is quite clear that he has been greatly misrepresented down the centuries. He states clearly here that there is no individual self or soul to be reborn, but that what reincarnates is the karma of your actions. This means there is no such thing as individual karma. Your actions go into the collective unconscious where they stir up mirror counter-effects. All deeds of a selfish nature reinforce the collective Shadow frequency and all deeds that bring about synthesis strengthen the collective higher frequencies. This process of reincarnation is explored in some depth in the 24th Siddhi of Silence. The great truth that comes through the 53rd Siddhi is that life consists of endless beginnings but no endings. This is the true meaning of Superabundance. Life goes on creating new forms whose actions determine the nature and destiny of future forms. There is no continuity between the forms themselves other than the genetic mechanism creating them. What continues, and what is superabundant, is consciousness itself as it endlessly penetrates the collective and writes the story of evolution.

In this context, the Siddhi of Superabundance is perhaps not quite as glamorous as it sounds. Material abundance is far more likely to occur at the Gift frequency, as at that level there is still a basic interest in your personal destiny and the destiny of others. The energy of *bhakti* — of giving to others — creates a huge upsurge in the collective energy field, which stirs all sorts of beneficent energies to return to you. In this respect you could say that the 53rd Gift conceals the ultimate secret of material abundance. However, at the siddhic level your identification with your body, destiny and self totally dissolves, leaving you in a mystical state of pure emptiness and availability. In fact, superabundance in many ways is closer to the concept of emptiness. It is a place where there is no further expansion because there is no evolution. If you are identified with the form of the world, you are identified with change, because all life is programmed to evolve.

To one who has awakened through the eyes of the 53rd Siddhi, humankind is both a being and a becoming. The form endlessly evolves and expands, but the consciousness never changes. And here we humans have made a basic error — we have identified with the instrument of our awareness, which expands according to the laws of evolution. We are indeed entering a new evolutionary phase right now in which human awareness seems ready to make a great quantum leap in terms of its expansion. But even though awareness may expand, *consciousness* does not and cannot expand since it is already everywhere and everything and even *everywhen*.

This is the crucial truth to understand. Beneath the form, consciousness never changes or evolves or expands or contracts. It simply is.

Superabundance refers to the concept beyond abundance — a space in which life is witnessed impassively by life. At the Gift level one rides the currents of evolution, which is always exciting and thrilling because you are at the cutting edge of awareness. At the siddhic level however, all excitement and all responsibility are gone. There is no longer an edge since all is experienced as a game in time and space. There is no personal comment anymore; there is no interest in destiny or evolution or expansion since all such concepts are seen for what they are — places of identification which conceal the truth. It is here in the 53rd Siddhi that we find the allegorical meaning of rebirth. You cannot take another birth because you realise that you have never been born! The mind is no longer following the arc of evolution but finally rests in its true nature as superabundant being.

People born into the 53rd Siddhi kill beginnings, or they kill endings. Whichever way you look at it, they bring an end to the paradox of human cycles. Superabundance is a space beyond frequency but we can only describe it as a very high frequency at which there is nothing left to do in the world. The 53rd Siddhi represents the state beyond and behind awareness known as pure consciousness. Therefore it can be said that consciousness in action manifests as expansion and evolution whereas consciousness at rest is the true underlying nature of everything. These people of the 53rd Siddhi were once undoubtedly seekers themselves since this Gene Key involves the gradual expansion of human awareness. However, at a certain point you have to let go of your seeking as the awareness inside expands outside the form. Up to this point, expansion has been gradual, although it often makes small leaps and forays into higher states. At this final stage, the stage known as realisation or enlightenment, the great and final leap occurs — it is the shocking leap into pure consciousness and the end of evolution itself.

Beneath the form, consciousness never changes or evolves or expands or contracts. It simply is.

▦ 54th Gene Key

The Serpent Path

SIDDHI
ASCENSION

GIFT
ASPIRATION

SHADOW
GREED

Programming Partner: 53rd Gene Key

Codon Ring: The Ring of Seeking
(15, 39, 52, 53, 54, 58)

Physiology: Tailbone

Amino Acid: Serine

THE 54TH SHADOW – GREED

FOR LOVE AND MONEY

The 54th Shadow is one of the great pressures that drive humanity. It is the drive to want more and at its shadow frequency this Gene Key becomes blind greed. It's important to remember at this point that none of these shadow frequencies are really negative. There is nothing wrong or bad with greed. It is simply an aspect of human nature and as such it has an evolutionary purpose. The purpose of greed is to pressure human tribal groupings and individuals to be materially successful. If you look at any modern first world country, you can see how far greed has advanced our civilisation. The primal energy behind the 54th Shadow was essential for the survival of early human tribal cultures, and indeed we can still see it operating in the developing world, where survival is often based directly upon where you are within the material hierarchy. Obviously you can also see how greed has a tendency to focus on a single individual, community or race to the detriment of all others. This means that at a certain evolutionary point, greed will have served its purpose and will need to be transcended. This is precisely what happens at the Gift level of frequency — when greed becomes aspiration. Greed in this sense refers to the desire to accumulate more material, whereas aspiration refers to the desire to attain something of a more spiritual nature.

If greed is pursued to its furthest limits without transcendence, it becomes self-destructive. We see this happening in the modern world today. When greed reaches its zenith it either becomes destructive, both to individuals and to the planet, or it gives individuals a new perspective. When people have acquired genuine wealth and stability, they often turn to more spiritual arenas for sustenance. This is how these two genetic programming partners — the 54th Gene Key and the 53rd Gene Key — operate together. If a society doesn't mature it becomes top-heavy, like the huge multinational business organisations we see in the world today. When organisations balloon like this, they inevitably take on a life of their own, draining resources from the planet.

Natural law demonstrates that an organisation dedicated simply to material accumulation will eventually crumble under its own weight, but, unfortunately, not before it has caused enormous destruction.

Running in tandem with the 54th Shadow, the 53rd Shadow of Immaturity ensures that any organisation, group or individual will remain stuck at a self-serving frequency. The 53rd Shadow blocks energy from making a quantum leap from an old cycle into a new cycle at a higher level of frequency. It is called immaturity because it simply never learns. Both these Shadows are deeply involved with money and the creation of wealth. The 54th Shadow has the added genetic imperative of operating through a hierarchical pattern, which means that it associates wealth with being at the top of the hierarchy. As such, the 54th Shadow courts recognition from those higher up in the hierarchy. In today's society this becomes the need for external symbols of status — a shiny new car or a huge house or the latest of anything and everything. One of the hallmarks of this Shadow is that it isn't just about *being* successful; it's also about *looking* successful.

Greed is an energy that will compromise its own integrity in a flash to get what it wants, and this is its downfall.

The essence of the 54th Shadow's success at material accumulation lies in its ability to create relationships that further its own material resources. In the modern world, this is about business. In business, success often arrives through the development of fruitful relationships — these can be within the business itself, with other agencies such as the media or with clients. Word of mouth still remains one of the most powerful tools for ensuring the successful transformation of a business, and a great problem generated by the 54th Shadow concerns the frequency of its transmission. Greed or desperation is an energy field that can be sensed by others, and as such it engenders distrust and closes down opportunities that might otherwise have been fruitful. There is a fine line between greed and Aspiration, the 54th Gift. Greed is really aspiration without trust. Because all frequencies attract similar frequencies, greed cannot really trust in its own allies since they are sure to be equally self-serving.

The 54th Shadow does not know how to attract the attention it needs to rise up within the hierarchy, even though it thinks it knows. Greed is an energy that will compromise its own integrity in a flash to get what it wants, and this is its downfall. In order to get recognition, greed must mature enough to become ambition, which does not have the same sense of desperation that goes with greed. Ambition may also be self-serving, but it has evolved enough to realise the pitfalls of pure greed. Greed in itself can lead to material success, if it pushes and pushes. However, such success cannot evolve into the higher aspect of aspiration and so stays in a low frequency loop based on material accumulation that will eventually backfire. We all know the phrase: "Money can't buy you happiness", and this epitomises the biofeedback loop created by excessive greed. As with all the Gene Keys, true happiness lies in continual transcendence.

Genuine ambition has a built-in genetic urge to keep transcending. This is why the affluence of the western world is beginning to naturally transform into a more spiritual aspiration. However, greed itself is based on fear, which fuels the need for ownership and the accumulation of material possessions.

Because of the fear at its roots, greed cannot afford to acknowledge anyone outside its immediate circle of support. It acts in direct competition with other groups or organisations and if successful, draws resources away from others, even when it has more than enough.

The 54th Gene Key is a member of the codon group known as the Ring of Seeking, which is *the* great pressure codon inside human DNA. Each of the six Gene Keys within this group drives an aspect of our human evolutionary movement. It is interesting to note how variations in frequency reorient the flow of energy through our DNA. For example, at lower frequencies all this genetic pressure is externalised through the 54th Shadow as the urge to seek and accrue material wealth. However, at a higher frequency the very same dynamic is sublimated and internalised to become the urge for meaning and purpose. There is nothing inherently wrong with the urge to seek material wealth, but it is a path that inevitably leads to disappointment, aloneness and misery. For many who attain material success, it acts as the trigger to seek something higher, but for many others it remains an addiction whose promise of fulfilment in the future prevents them from enjoying the beauty that surrounds them in the present moment.

REPRESSIVE NATURE – UN-AMBITIOUS

The repressive side of this Shadow is quite simply about the repression of ambition. These people may begin things with ambition and ardour but then often become disillusioned and give up. The wonderful drive behind this Gene Key thus becomes stymied and leads to passivity. This is a nature that is haunted by the fear of never being able to attain its goals, so it decides that it is better not to even begin the journey. The result is usually deep depression. The other side of this is reflected in people who deny their material needs and seek their spiritual nature to the detriment of their physical nature. This is a profoundly physical Gene Key and has to begin its journey of transformation from the foundation of the physical realm and the physical body.

REACTIVE NATURE – GREEDY

The reactive side of this Gene Key manifests as an obsession with material accumulation. These are the people whose greed becomes an expression of their rage. Their greed continues feeding itself, being an insatiable urge that never escapes its own low level of frequency. It can possess these people so completely that they become blinded by their need to own more and make more money. Such people lack one vital thing in life — the feeling that their lives are of benefit to others. Their greed closes down their hearts making them incapable of true relationships. Thus they often become materially successful but can only see other people in terms of possession and ownership. Inevitably such natures try to control others through their power and their anger, and as such never lead truly fulfilled lives.

THE 54TH GIFT – ASPIRATION

MATERIAL AND SPIRITUAL LIQUIDITY

When the ancient Chinese named the 64 archetypes that we now know as the Gene Keys, they did so through the medium of their cultural milieu. In the case of the 54th Gene Key, they named it *The Marrying Maiden*. The precise wording of the I Ching concerning this hexagram refers to a concubine marrying a man who already has another wife. The resulting text concerns the human need to understand how to align harmoniously within the existing family hierarchy. At the highest level, this image epitomises the 54th Siddhi. You have to gain the trust of higher forces through continual aspiration. Only after sustained effort (which is natural to the 54th Gene Key) does the spontaneous eruption of higher consciousness and ascension naturally occur.

When the 54th Shadow transforms into a higher frequency, the 54th Gift of Aspiration is born. Aspiration here refers to the energy to aspire to something beyond the material realm. Aspiration contains within it the seed of all higher consciousness. At the Gift level, aspiration has to do with working with others for the benefit of others. This Gift concerns the way in which energy is invested. At the Shadow level, any accumulated energy went back into the drive to accumulate more, with no other real purpose. At this higher level of frequency, accumulated energy is recycled and used to support people lower down in the hierarchy. In this way, a truly healthy model is created. The roots support the branches, twigs and flowers and the fruit fertilises the roots. The 54th Gift is aware that all systems in nature are interconnected and therefore cutting off energy from one area will ultimately deplete your own resources.

The 54th Gift aspires to a higher vision of prosperity. It still operates within its own community or organisation, but it knows that the secret of real growth and expansion (the 53rd Gift) lies in mutually cooperative models between different groups or organisations. Aspiration allows you a much broader vision of how to create prosperity for everyone. Here we see ambition moving beyond the personal and into the communal. These people want their entire community to thrive in order that they can draw more and more people up the hierarchy away from the self-destructive patterns of the lower frequencies. The by-product of this cross-fertilisation with other communities is exponential growth for everyone. Today the 54th Gift is more in evidence as more and more high-level businessmen and women begin to think in these new holistic business terms. However, this new phenomenon is just being born into the world. People are only now realising that businesses with a holistic perspective can be even more successful than greed-driven businesses.

This shift in perspective from self-serving to community-serving is giving birth to a whole new paradigm in business itself. For the first time since the industrial revolution, people are asking what the real purpose of business is. Instead of being an end in itself, business is being seen as a means to create a better and more sustainable world. Through networking with other businesses at a similar higher frequency, enormous change is possible in our world.

As these so called *cultural creatives* begin to work together and dovetail their energies and resources, they open up entirely new vistas that could potentially transform our entire planet. When enough people with aspiration overcome their individual fear and competitiveness, they will create a deep and lasting equilibrium in the global economy. The real urge behind aspiration is to seek out this higher harmony, and in the material world this means moving money from places where there is too much to places where there is not enough.

The Gift of Aspiration has some other fascinating qualities connected to the transformation of energy into more rarefied frequencies. At a genetic level, the 54th Gift has to do with the way in which memory is transferred and stored through the liquid of our cells. Memory itself is deeply affected by frequency. The 54th Shadow remembers only the ancient genetic fear of extinction if it fails to rise up in the hierarchy. This fear is physically transferred through subtle pheromones released by everyone under the spell of the Shadow frequency. The more successful someone is at the Shadow level, the more people in the world they frighten. The moment you operate out of fear or greed, your odour changes and you lose the trust in the relationship. The 54th Gift also transfers its higher frequency through the medium of physical presence, but in doing so it immediately puts others at ease. You can smell a supportive energy just as easily as you can smell fear. Thus the 54th Gift has its own biofeedback loop that creates trust wherever it goes.

Because the 54th Gift attunes to a higher communal vision, it only attracts people who resonate with this same subtle frequency. Therefore, it is essential that people with this Gift meet people face to face at the beginning of their relationships and that they do their business in the flesh. The 54th Gift has a deep instinctive understanding of money and energy through this genetic resonance to the liquid nature of memory. Wherever energy or money is kept in a liquid state and allowed to flow between people and organisations, prosperity is engendered. When it is frozen or amassed for too long it prevents further expansion. The same laws that apply within the human body also apply to society and economics.

As the energy of the Gift of Aspiration reaches higher frequencies it becomes more spiritual. Thus these people also understand energy flow and transformation as taught by various yogic systems and disciplines. Ancient models such as the Chinese system of Feng Shui are instinctive to the 54th Gift and do not even need to be studied. The flow of prosperity is directly proportional to the flow of energy at all levels within your life, beginning with your own body, which is the foundation of all journeys to higher consciousness. The 54th Gift is the base plate for all systems of alchemical transformation from lower states of frequency into higher modes of consciousness, and as such it is an extremely powerful and influential gift.

The flow of prosperity is directly proportional to the flow of energy at all levels within your life, beginning with your own body.

THE 54ᵀᴴ SIDDHI — ASCENSION

PHYSICAL ALCHEMY

The 54ᵗʰ Siddhi is a relatively well-documented Siddhi compared to many of the other 64. It is one of the Siddhis that is actually triggered through the continual pressure of aspiration. This is the Siddhi of seekers. Through research into the 64 Gene Keys and the associated science of hologenetics, it has been found that a disproportionate number of mystics and enlightened sages have the 54ᵗʰ Siddhi as one of their Prime Gifts. Perhaps the best known of these is Paramahansa Yogananda, one of the great mystics of the 20ᵗʰ century, who had this Siddhi as his Life's Work. The 54ᵗʰ Siddhi concerns the notion of Ascension — the continual alchemical transformation of matter into spiritual essence. Here the base energy of ambition is experienced at its highest frequency and becomes the continual pressure to keep ascending up the hierarchy. At these levels however, we are no longer talking about social or material hierarchy, but the spiritual evolutionary ladder which consciousness itself ascends on the return to its own source.

One of the common allusions to the 54ᵗʰ Siddhi concerns the Hindu concept of *kundalini* — the so-called *serpent power* lying coiled at the base of the spine. The goal of most transformational yogic systems is the awakening of this primal energy, which is said to rise up and activate each of the higher centres or *chakras* as it ascends through the human body. Many people have discovered to their detriment how dangerous premature or forced awakening of kundalini is to the human nervous system. Some people have indeed suffered severe psychological disorders because of the terrific currents of energy that such sustained yogic practice can unleash. Most yoga systems have accommodated to this danger over the centuries by offering step-by-step guidance to such meditations with a great deal of physical preparation and cleansing early on. With this 54ᵗʰ Siddhi, as we have seen at all levels, there is a need for deep grounding within the material realm.

What is so interesting about this Siddhi is not its manifestations, which have been very well documented, but the way it has spiritually conditioned so many people. There exists enormous confusion in the world concerning spiritual paths, especially now that the West has become such a melting pot for all the great mystical cultures and traditions. There is a particular path for every single person, and any other path than your own, especially at higher levels, can cause delusion and can even be dangerous. Every single person who has attained a siddhic state of realisation speaks through their specific Siddhis. However, if the sage is speaking through a Siddhi that is not a part of your own genetic makeup, you can easily become confused. Anyone who has attained a state of realisation knows this dilemma. It is impossible not to realise it. Thus the great sages attempt to convey the energy behind their state rather than its trappings.

The 54ᵗʰ Siddhi of Ascension is a very particular path. It concerns the transformation of the gross currents within the body into more and more refined currents. It is the essence behind many of the great tantric and alchemical yogic systems. However, in order to work with these systems, you must have the necessary aspiration. This is not something that should be forced.

Aspiration is a predetermined energy that carries the seed of the final flowering within it.

Aspiration is a predetermined energy that carries the seed of the final flowering within it. Any other energy will not carry this same seed. This is not to say that people cannot have heightened experiences of consciousness through such systems. They can. But only if you have the seed of this Siddhi already within you at a genetic level can you attain the final permanent flowering through these kinds of techniques.

True ascension is activated automatically through the lower frequency of aspiration. At a certain point a spontaneous physical transformation begins to wrack the body. Techniques at this stage are over and done with. The process of ascension takes over and you are absolutely helpless to stop or interfere with it. All fluid memory must be erased — personal, cultural, and genetic — it all has to be burned from the body. The kundalini energy has often been likened to a fire. Actually it is more like liquid fire, as the water molecules within the physical body are vaporised to create a *steaming* process in which spiritual essence is refined. After a certain period of time — sometimes spent in intense physical agony — stillness reigns over the body and you experience the purity of consciousness without thought. Even at this stage the steaming process continues as the physical body becomes more and more translucent.

Ascension is an incredibly physical sequence of events. It has nothing whatsoever to do with modern systems involving visualisation or cerebral meditation. It usually follows many years of seeking by someone whose very path is to seek and aspire to higher planes. So many people have been conditioned to seek higher consciousness through these kinds of routes, but they truly do not belong there. Neither does the seeking itself activate the ascension process, since it is merely the manifestation of a totally mysterious and intensely felt urge to know more. For those who do not belong on this path, it is a truly dangerous one. For those who do, it is effortless and spontaneous. One aspires, and sooner or later, one ascends — it really is as simple as that.

▤ 55th Gene Key

The Dragonfly's Dream

SIDDHI
FREEDOM

GIFT
FREEDOM

SHADOW
VICTIMISATION

Programming Partner: 59th Gene Key
Codon Ring: The Ring of the Whirlwind
(49, 55)

Physiology: Solar Plexus (Dorsal Ganglia)
Amino Acid: Histidine

PART 1 – THE GREAT CHANGE

INTRODUCTION TO THE 55ᵗʰ GENE KEY

Did you know that you have a unique genetic sequence that dictates the unfolding of your true nature? The purpose of this book is, as the title states, to unlock this higher purpose hidden in your DNA. The greater self — the cosmic part of each of us that transcends our mortal body — has lain secretly hidden inside humanity for aeons. Because it is inside your body, right under your nose as it were, the mass of humanity has never thought to look in such an obvious place for a lasting sense of peace and fulfilment. Up until now, the inner journey has been for the select few — those bold adventurers and courageous pioneers of the inner planes. Because of this our true divinity has seemed far away from the common man, whose more immediate concern has been to survive and manage life in the external world.

All of this is about to change. As you travel through the many layers of frequency within this book, you will begin to have an idea of how rich, how beautiful and how diverse is the journey to awakening. In the myths of all cultures, keys have been left that speak to us about the coming time of the Great Change. Humanity is feeling this change now because it is taking place now. Within a relatively short evolutionary period, the world we are living in will be transformed into a world that most of us would consider pure fantasy. You are alive at a deeply romantic moment — it is the moment in which the prince kisses Sleeping Beauty and, suddenly, she fully awakens. And as she awakens the world is transformed. This Great Change is the central theme woven into every single sentence in this book. If you are reading these words, your inner guidance has seen fit to remind you of this extraordinary prescient event, either to confirm or trigger your own personal process of awakening.

Therefore, allow yourself to confirm your own journey — which has brought you to this exact point in time — right now as you read these words. We human beings are following different vectors through time and space and each of our paths must at some time converge at this single point deep inside the body.

There is a place inside your DNA whose sole purpose is to trigger this awakening. The 55th Gene Key describes this place, but more than that, it allows you to contemplate and quicken the actual process of awakening. The 55th Gene Key and its sister transmission, the 22nd Gene Key, embody the most potent and profound message in this book. The 55th describes the evolutionary force moving from matter to spirit, and the 22nd describes the involutionary force moving from spirit to matter. Together these two Gene Keys capture the quintessence of the Great Change.

As you read the words and ideas behind this Gene Key, allow them to burrow into the deepest recesses inside your being. There are dormant memory codes inside you that are specifically designed to be activated and awakened by this transmission. As you allow it to penetrate, you might like to note the feelings, thoughts and impulses that stir inside you. Even if you feel resistance to this Gene Key, that too should be allowed and respected. Awakening is a process with its own mysterious timing and sequence. Therefore I invite you to breathe deeply, perhaps allowing yourself a deep sigh from time to time, and above all please enjoy your flight into this wonderful world where romance becomes reality...

Welcome to the heart of the Gene Keys transmission!

THE 55TH SHADOW – VICTIMISATION

INDRA'S NET

The 55th Gene Key and the mythical journey from its Shadow to its Siddhi truly form the heart of this entire work on the 64 Gene Keys. There is no more poignant nor contemporary note sounded within the whole genetic matrix than this odyssey through the seething underworld of victim mentality and out into the pure clear air of freedom. Of all the Gene Keys, this is the one we humans most long for and this is the Gift we will soon be given at a collective level. The entire reason behind the timing of this work on the 64 Gene Keys lies here within the 55th Shadow. It has been the theme of humanity since the development of the human neo-cortex gave us our capacity for self-reflective consciousness. It is the theme of the victim.

On an energetic level, you truly do always reap exactly as you sow.

The 55th Shadow of Victimisation and its programming partner, the 59th Shadow of Dishonesty, program all human beings at a cellular level to a single effect — they ensure that every individual becomes his or her own worst enemy. There is a universal law, colloquially known as: "What goes around, comes around", and it is this law that the 55th Shadow fails to see. The essence of this universal law is expressed through the famous biblical axiom: "As you sow, so shall you reap." These timeless clichés are generally interpreted as applying only to the surface of life rather than to the deeper energetics. It often appears that many who succeed in life in fact do so to the detriment of those around them. Likewise, the most innocent and openhearted people can be beset by terrible trials that appear to have no explanation. Thus it can seem on the surface that "as you sow, so shall you reap" has little or no substance and it is generally relegated to the realms of folk wisdom.

Because of this tendency only to see the surface of things, the mass consciousness of humanity misses one of the deepest secrets of life — on an energetic level, you truly do always reap exactly as you sow. It simply takes longer for this to become apparent in the material dimension. The highest expression of the 59th Siddhi is the quality of Transparency and as Transparency testifies, in the end you cannot hide from yourself. Thus the key to the 55th Shadow and its transcendence depends upon one factor — your attitude. It is not about what happens to you. It is about how you handle it.

LOSING ATTITUDE

There are really only two dimensions to the concept of attitude — you either behave as though you are a victim of circumstance, or you take full responsibility for your situation. Although this sounds simple, there are several layers of complexity to it. Since we are examining the 55th Shadow, we will look first at what happens when you play the role of the victim of circumstance. The 55th Shadow is located deep in the dorsal ganglia of the solar plexus, and it is about emotions. At low levels of frequency, human beings look outside themselves when they experience either an emotional high or an emotional low. We need to attach a reason to our emotional states. At the high end of the emotional spectrum, we believe that true joy is an effect rather than a cause. Because of this deep-seated belief, we spend most of our lives chasing whatever we think causes the effect of joy — it may be a perfect relationship, lots of money, fame, the perfect place to live, even our God. At the low end of the emotional spectrum, the game we play is blame. We blame anything from the food we have just eaten to our partners to the government for the reason that we feel bad.

This human tendency to look for outer causes for our moods is the greatest addiction on our planet. It is rooted in an essential core belief that we are victims of our material reality. This core belief sets up a low frequency pattern that is reinforced over and over again. In other words, with this inner attitude at the fore, we become caught in a web of our own making. What traps us is our longing. When we are down, we long to be high and when we are high, we long to hold onto the feeling. Thus the very feelings we seek create a perpetual hunger for fulfilment that can never be attained. The addiction is the search for fulfilment, not fulfilment itself. Hence the old chestnut about finding heaven — if you ever found it, you would hate it because the thing you loved was the *hope* of fulfilment rather than the state itself.

Here we find the secret of frequency itself — it is rooted in your unconscious attitude to life. Because your true attitude remains unconscious, there is no technique, per se, to raise your frequency. All that is required to reach your *genetic escape velocity* — the frequency that pulls you out of the Shadow state and into the Gift — is understanding. Understanding must dawn within you at the level of pure being — the understanding that you have become the unwitting victim of your own unconscious belief patterns. When this understanding dawns, you immediately begin to transcend the Shadow state. The great spiritual teacher Gurdjieff stated this so simply and beautifully when he said: "In order to escape from prison, one must first understand that one is *in* prison."

*High levels of
frequency do
not necessarily
lead you to
"spiritual"
experiences.*

FAKING FREEDOM

As was mentioned earlier, there are many dimensions to the 55th Shadow of Victimisation. The web that holds us at a low frequency has many subtle twists and turns within it. As the saying goes, one of the devil's greatest tricks is to get people to look for God. The trickiest aspects of the 55th Shadow concern spirituality, and this has particular significance during the current historical epoch. Spirituality itself can become the focal point of the victim consciousness because it can so easily give you the idea that you can do something to free yourself from your Shadows and your emotional suffering. This notion has led to the greatest illusion of them all — the illusion of another spiritual reality somehow outside our own sphere of experience. If we look at this notion with clarity, we see the very same pattern of the longing to find fulfilment. If you create an unattainable reality, then you can spend your whole life longing for that reality without ever having to directly experience it.

For many religious or spiritual people this can be a hard truth to stomach. True enlightenment is not what we really want at all. It isn't exciting in the least, but is utterly ordinary. In spite of this, most spirituality is built upon the pursuit of the extra-ordinary. High levels of frequency do not necessarily lead you to *spiritual* experiences. In fact, high levels of frequency tear down the very illusion that there is such a thing as *spiritual* experience. In the contemporary new age culture, spiritual materialism is rife — that is to say, people now have a brand new drug called the pursuit of truth.

It is important to understand that there is nothing wrong with any of these things. If you are drawn to seek something higher, then something is pushing through you, leading you somewhere. If you follow it through to its natural conclusion, it will reveal your true path in the end. For some people, seeking is a direct path to transcendence, but for others it simply serves as a distraction that draws them further away from their true nature.

The 55th Shadow prevents the spiritual seeker from following their urge through to its natural conclusion. It does this by identifying either with the form of the teaching, or with the teacher, or with the path itself. Therefore we see three basic categories of people on a spiritual path — those who are trapped by the structure of a particular teaching, those who are trapped by the magnetic power of a particular teacher and those who are trapped by their own constant compulsion to be a spiritual tourist. All three of these spiritual traps are authentic stages of any path that will eventually lead to true freedom, but all three also masquerade as freedom itself. These are some of the subtlest levels of the Shadow of Victimisation. As we shall see when we examine the Gift and the Siddhi of Freedom, true freedom has nothing to do with how we spend our time on the material plane. True freedom is not an effect. It is a kind of ever-expanding spaciousness that arises spontaneously inside you as you come to understand how deeply victimised you really are by your own core beliefs.

THE DYING OF DRAMA

One of the greatest allures of mankind is the ideal of romantic love. In this context we are not only referring to the longing of one human heart for another, but to the ideal of romance in a wider context. This is the idea of *life* as a romance. The very basis of romantic love is that it can never be truly fulfilled, but must continually flow from cataclysmic dips to ecstatic swells, creating a wonderfully rich symphony of human emotion and drama. All human performance art, from the loftiest play to the most mundane TV soap opera, stands as a metaphor within the metaphor of life itself. Because of the 55th Shadow, we are all victims of life's drama. We are caught in an incredibly intricate net woven from the strands of our pain on one side and our pleasure on the other. We actually love and hate the net simultaneously, but above all we are addicted to it, as we are addicted to all high drama.

Within your dream world and beneath the net, your greatest longing is given expression. Under the net you can live out your dream — you can soar, dance, weep, suffer and above all you can love. And yet, your love within the net is a deeply limited love, a love that never escapes the confines of its own illusions. Under the net, you fall in love or you fall out of love — either way, you remain a victim of your projections, expectations and inevitably your disappointments. The net cast by the 55th Shadow choreographs the ebb and flow of your breath patterns as you rise and fall on life's melodies. Sometimes, you sink into melancholy and your whole life force seems to come to a halt — your breathing itself becomes thinner. At other times, you fly on a sudden mood swing and your heart beats faster and your breathing fills your chest to bursting. This is what we believe freedom to be. In between the extremes, melodies give way to cadences; shifts in tempo give way to phrases, notes, trills, pauses and every conceivable kind of feeling. We live our lives submerged within these waves, and there is no end to our emotional processing.

As regards the 55th Shadow, the coming mutation will trigger the end of our addiction to the drama of life and the beginning of our discovery of what true freedom really means. At its core, the 55th Shadow conceals the longing of consciousness to return to its own unity, although this is most commonly expressed as the romantic longing for the perfect soul mate. There is a wonderful Hindu myth known as *Indra's Net* in which the cosmos is seen as an infinite lattice with a jewel at every junction in the net. Within each individual jewel every other jewel is perfectly reflected. The 55th Shadow casts a veil over these jewels, thereby keeping humanity trapped within the net and unable to experience the unity of all things. With the coming shift, our awareness will finally be able to slip through the strands of the emotional net that has kept us in a state of victimisation for so long. In doing so, we will glimpse for the first time the truth of our collective unity shimmering in every one of Indra's endless jewels.

THE SOLAR PLEXUS – THE SECOND BRAIN

The emotional matrix of the 55th Shadow lies within the domain of the solar plexus region in the human body. The massive complex of nerve ganglia in this area has often been referred to as the

Our emotions have more power over us than our minds, and the world around us continually testifies to this fact.

second brain. It operates independently from the cranial brain through its continual governance of vascular and visceral functions within the body. The sheer voltage of our emotional states, particularly at their extremes, far outweighs the subtle cognitive processes of reason that we hold in such high regard and which emanate from the cranial brain. Relatively little is known about the exact nature of the neuro-circuitry within the solar plexus or about its mechanics and true capabilities. What we do know is that despite our best efforts, our emotions have more power over us than our minds, and the world around us continually testifies to this fact.

THE REAWAKENING OF THE SOLAR PLEXUS – THE RIDDLE OF THE AEONS

Much of the 55th Gene Key transmission concerns the future of our species as it moves through the Great Change, and as such it is a profoundly prophetic piece of writing. However, one of the greatest pieces of the puzzle is actually found in our very distant past. Generations of mythologists, folklorists, archaeologists, mystics and historians have spoken of another race of human beings that existed before our modern recorded history. Indeed, all our great human myths and fairy tales are encoded with information about a lost golden age that was extinguished or lost in some form of vast cataclysm, deluge or flood. To the psychologist, these myths have always been seen as metaphorical and archetypal psychic yearnings to return to the security of the womb. But what if they are actually folk memories held within our ancestral DNA? The 55th Gene Key has much to say about this.

Although many ancient cultures have devised methods of mapping vast spans and cycles of time, one pattern holds true at the simplest and most mythical level — the trinary pattern. Every great work of art or allegorical story has at its core this archetypal trinary time-flow, which divides all human narrative, internal and external, into threes. Woven into the very structure of the human psyche is a profound resonance to these fundamental patterns. There is always some kind of initial fall from grace, followed by a journey of discovery and trial, culminating in a final redemptive victory. When we apply this mapping to the whole of human evolution we see our human story depicted as three great Aeons, vast spans of time marked by four major leaps in consciousness.

THE THREE AEONS AND THEIR EVOLUTIONARY STAGES

The Three Aeons and their four turning points describe the evolutionary arc of our entire planetary consciousness. Essentially, this threefold pattern describes three distinct evolutionary stages culminating in a fourth transcendent stage (a tetrahedral geometry of consciousness).

THE THEORY OF INVOLUTION AND THE SEVEN ROOT RACES

According to the Gene Keys, life is an interplay between two primary forces — the current of evolution and the current of involution. The way that we are trained to think in the West focuses mainly on the objective, external world rather than the inner, subjective reality. For this reason we tend to lay greater emphasis on the evolutionary current, which has become the basis of the modern scientific approach. However, many mystical and esoteric traditions from around the world have also considered life from the other aspect — which sees life as an *involutionary* process in which consciousness is gradually incarnating deeper and deeper into the form, shaping our evolution as it does so. This view (known as *Emanationism*) holds that at every stage of our planetary and personal evolution there is an unfolding hidden purpose, which reveals itself in successive steps. As Divine life *involutes*, so human and earth life evolves, and as we strive upwards in our consciousness towards the higher frequencies, so we are able to draw them down into our lives on the material plane.

Within the greater pattern of the Three Aeons lies another pattern based on seven substages, known as the seven Root Races. In this Theory of Involution each Root Race represents a major stage in our planetary development. In esoteric tradition, the Root Races are often seen as literal races of human beings who preceded our modern human. Within the Gene Keys Synthesis, the Root Races are viewed as the unveiling of the subtle layers of the living spirit of Gaia, our earth. In other words, from the involutionary perspective all aspects of the form of our planet and indeed our universe are seen as being imbued with levels of consciousness, from the densest mineral to the subtlest gas. The earliest Root Races thus represent the subtlest forms of Divine consciousness as it steps down its frequency in order to enter into physical manifestation (the First Aeon). At a certain point in the evolutionary story of our planet, consciousness enters so deeply into the material realm that it forgets itself entirely (the Second Aeon). Then comes the *Remembering* and the mythic return to paradise as consciousness transforms the material realm, integrating all dimensions back into itself thus completing its epic evolutionary arc (the Third Aeon).

THE SEVEN ROOT RACES AND THEIR CORRESPONDING PLANES OF REALITY

The seven Root Races also relate directly to the seven planes of reality and the seven subtle bodies of the aura, thus giving us both a narrative and a timeline for the *involutionary* stages of consciousness. (For more information on the Seven Sacred Bodies and their corresponding planes, you can read the 22nd Gene Key.)

The First Root Race – The Polarian – Monadic Plane

The Second Root Race – The Hyperborean – Atmic Plane

The Third Root Race – The Lemurian – Buddhic Plane

The Fourth Root Race – The Atlantean – Causal Plane

The Fifth Root Race – The Aryan – Mental Plane

The Sixth Root Race – The Trivian – Astral Plane

The Seventh Root Race – The Pangaian – Physical Plane

THE FIRST AEON – PREPARING THE GARDEN OF GAIA

The first two Root Races, known as the Polarian and Hyperborean, represent the crystallisation of the form of the earth itself — in other words, they span the time period in which our planet actually formed. Corresponding to the Monadic Body (the ultimate source of

Divinity), the Polarian Root Race represents Divine Idea or Will before it descends into form and experiences separation. The Second Root Race — the Hyperborean — relates to the Atmic Body, which is the body of light. This refers to the coalescing of our planet out of the minerals and elements from the sun. This phase includes the formation of the earth's atmosphere and the gradual refining of its elements and gases until our planet was able to support life. It was during these first phases of our evolution that the constituent elements of all planetary life were endowed with subtle forms of consciousness. In some traditions these are known as the angels of the elements and the devas of the mineral realm.

THE SECOND AEON – THE FLOWERING AND THE FALL

The Third Root Race, known as the Lemurian, represents the birth of animate life as it emerged from the waters and populated the earth. This was the Eden phase in which Divine Essence manifested its superabundance as all the kingdoms of nature. The Lemurian consciousness of our planet was and still remains a single unified being existing on the Buddhic Plane, known as the plane of ecstasy. This is also the plane of many of the Devic realms — the manifestations of consciousness inherent in all living forms. It was during the Lemurian phase that the first human beings were conceived and born.

The Fourth Root Race — the Atlantean — represents humanity before the fall. This race of human beings, sometimes known as the *Adamic race* is little more than a dim memory to modern human beings. We modern humans were separated from our true source by a series of cataclysms, which have become mythically known as *The Fall*. The truth of this *Fall* has been handed down by our indigenous cultures through their stories and creation myths, which have in turn found their way into our modern cultures and beliefs. The Atlantean Root Race and its culture and environment were completely obliterated and evolution was literally *rebooted* and begun again in a new and different direction. However, Atlantean consciousness existed and still exists on the causal plane, which is the plane of the archetypes — the quantum language that lies beyond the logical mind. The original Atlantean consciousness, unlike the modern human, was centred in the solar plexus and did not experience itself as separate from the source of all life, but was the very heart and mind of Gaia herself.

THE THIRD AEON – THE FIFTH RACE AND THE KALI YUGA

In every great tale there must be a fall. In Indian Vedic tradition, the evolutionary stages are known as the *Yugas*, and the darkest stage of all is known as the Kali Yuga, after Kali, the Dark Goddess of Time and Change. Our current Root Race — the Aryan — is now coming towards the end of the Kali Yuga, the epoch after the fall. The Aryan consciousness exists on the mental plane, and our main instrument of awareness is our continually evolving brain. Ironically our greatest gift, our ability to reason, gives us the illusion that we are separate from each other and from our environment. However, the Third Aeon is about the long walk home. Ever since we fell from grace, we humans have been seeking the way home. We seek through science, we seek through religion and, above all, we seek through love.

Our current Root Race — the Aryan — is now coming towards the end of the Kali Yuga, the epoch after the fall.

THE GREAT CHANGE AND THE SIXTH RACE

The time of the Great Change is now upon us. As the Fifth Root Race prepares to give way to the Sixth Root Race — the Trivian — time itself appears to be speeding up. The Sixth Race has long been prophesied by mystics and sages. Corresponding to the involution of Divine essence into the Astral Plane — the realm of emotion and desire — the Sixth Race will bring transformation to our entire planet. As Divine Consciousness continues to descend deeper and deeper into form, it reveals its true nature. The coming epoch will see the sublimation of human sexuality and desire into unconditional love. The Sixth Race will be triggered through the 55th Gene Key and its mutation within the solar plexus centre, the seat of human emotion. The Trivian Race heralds the reawakening of this centre and will allow human beings to experience once again the universal quantum field connecting all beings. This reawakening is not a retrograde movement into a past golden age but a new integration of the three lower planes — the physical, astral and mental into their higher counterparts — the causal, buddhic and atmic.

The Seventh and final Root Race — the Pangaian — really lies beyond words. It represents the integration of all the kingdoms of Gaia into one single vibrating presence. It is where spirit and matter become one, and where the Divine monadic essence shines through the physical plane, allowing it to ascend. It is the kingdom of heaven coming to earth.

THE ENDING OF THE TRILOGY AND THE RETURN TO EDEN

Humanity and our whole planetary consciousness now stands at its greatest threshold to date — the final phase of the Trilogy and the eventual resolution of the Riddle of the Aeons. This is such a rare event in the unfolding of consciousness that it brings about vast shifts at every level of life. What is coming seems so fantastic that our minds cannot stretch far enough to encompass such a reality. As our fairy tales tell us, the third time is the charm and always brings redemption. Indeed, all our great myths, films, romances and dramas culminate with some form of synthesis. Without this synthesis our hearts feel incomplete. It is always at the very end, when we have all but given up hope of redemption that liberation comes. It comes in a great tidal rush, playing out as a trinary pattern of testing and release that is so familiar to us that we ever find ourselves yearning for that happy ending. We yearn for it so deeply because it is stamped within the genetic structure of all life forms in our galaxy. And because it is in our DNA, our final destiny must and will be to witness the rebirth of Eden and live peacefully in the garden *for ever and ever after.*

REPRESSIVE NATURE – COMPLAINING

The 55th Shadow of Victimisation has two main forms of expression. The repressive nature manifests as complaining. Complaining is an unconscious mindset in which one makes oneself the central victim within one's own drama. Every time one complains, outwardly or internally, one effectively dis-empowers oneself. The repressive nature tends to complain inwardly, taking a pessimistic view of life, whereas the reactive nature tends towards finding a specific external target to blame. When one gets caught in the frequency of complaining, one is caught in the

It is always at the very end, when we have all but given up hope of redemption that liberation comes.

net of the drama of life — the *maya*. The energy of the complaint itself serves to strengthen the illusion that life is so very hard. Apart from reinforcing itself in this way, complaining also causes sustained general wear on our physical organism. Freedom occurs when we see through our deepest unconscious patterns to the heart of this energetic.

REACTIVE NATURE – BLAMING

The other common form taken by the 55th Shadow is blame. The reactive nature externalises its complaints by specifically blaming something or someone else. When we blame another, we fire an arrow that removes self-responsibility for our situation. In this sense, we *invest* aspects of ourselves within other beings, giving away our true power and presence. All blame is an expression of anger projected outwardly, but in this sense it is not pure. Pure anger is a release of the primal energy of fear that may be triggered by an external source yet does not target that source. The moment one blames another, one is again the victim of one's own drama. It is impossible to blame another for one's fate and simultaneously realise that one is simply an actor in a play. The very act of seeing through the seriousness of life releases the energy of blame. From the reactive side, true freedom occurs when the arrows of blame are caught in mid-flight before they reach their target.

PART 2 – THE MUTATIVE PROCESS

THE 55TH GIFT AND SIDDHI – FREEDOM

THE SPIRIT OF THE FUTURE EPOCH

Contemplation of the 55th Gift inevitably leads one to ponder the future of humanity and of our planet. Over the following pages, we will look both at what is occurring to humanity now and what will occur at the time of The Great Change that lies ahead of us. In reading the codes contained in the 64 Gene Keys, of less importance are the details of the unfolding of this Great Change. Any consideration of the details can at best be based on conjecture and opinion. However, through a profound resonance with the core frequency behind this work on the 64 Gene Keys, it is possible to capture the spirit of the coming epoch. As you may begin to divine from the following pages, the ripples of this mutation will affect all corners of life on our planet.

The other factor to bear in mind concerning the coming change is the speed at which it will overtake us. In evolutionary terms, it will happen overnight, although in practical temporal terms, it will happen gradually and almost imperceptibly. We are talking about a genetic mutation that will slowly colonise our species. In other words, the old human will literally be bred out of humanity. This means that very soon there will be children born among us who carry the full mutation and who will spread it through the gene pool. These children will be different from us. They will not emotionally engage with us on a victim level but will hold a high frequency that will, over time, transform the families into which they are born. We will consider their role more fully towards the end of this Gene Key.

THE RELEASE CODE OF HIGHER CONSCIOUSNESS

Down the centuries much has been written, spoken and taught concerning the nature of higher consciousness. We are now entering an era in which more and more people will have direct access to an experience of true higher consciousness. Eventually, what occurs during this 21st century will spread to the entire collective and catalyse an age that even now we can only dream of. Up until now, the process of awakening has been understood and explained (except in a very few cases) at an individual level only. Teachers, sages and gurus have communicated their truths in a style suited to individuals. The emphasis has almost always been: how can *I* awaken?

Two main elements within this question are now rapidly passing their sell-by date. The first is the question of *how*. As we shall see, the 55th Gene Key will bring an end to the question of *how*. Secondly, the individual element of *I* will be gradually phased out by the coming changes to humanity. We are entering the age of *we*. Only when we have fully absorbed the truth that humanity is one collective *we* will the final irony be delivered to us, as we once again become a deeply mystical *collective I*.

A vast and dynamic change in the form of a genetic cellular mutation is coming to human beings. It is being triggered by this 55th Gene Key and its related amino acid histidine. At a chemical level, your own body is in the process of preparing for this mutation as you read these words. The process is underway at a collective level too, and not a single human being is immune to it. At its highest level, the 55th Gene Key is the release code for higher consciousness. There are many profound implications to this process and there is a specific sequence to its unfolding. As we examine this Gene Key in further depth, we shall try to look ahead to the kinds of changes we are facing and how they may affect us both as individuals and as society in general.

THE PIVOTING OF AWARENESS

Within the Spectrum of Consciousness — the linguistic matrix underpinning the sixty-four Gene Keys — the 55th Gift is unique. When you look down the columns of words that represent the frequencies of each Gift and Siddhi, you will see that the word for the 55th Gift — Freedom — is the same as the word for the Siddhi. This is the only place in the Spectrum where this occurs, and it has very great import. The 55th Gift is a pivot around which human awareness is launching a new faculty — the ability to travel through physical space. This development will change everything about the world as we know it. Once awareness has been set free in this way, what we have previously seen as higher consciousness will become our ordinary state. This is why the word for the Siddhi and the Gift are the same. The Spectrum of Consciousness itself will be torn open from this point and, one by one, each of the Gifts will be freed from its Shadow and will merge with its highest potential in the Siddhi. As the Shadow energy rises up to the Gift, so the Siddhic energy descends to the Gift. It is as though the genetic wheel of fortune is going to ratchet into a whole new gear the moment it reaches a specific cog in its path. This cog is the 55th Gift. From that moment on a new force will enter the world, with new laws and new ramifications for us all.

*The form is
the skin, the
fruit is the
awareness
and the seed is
consciousness.*

Up until now in human history, awareness has been confined to the individual human form. We experience awareness as movement, as feeling and as thought. Unless someone reaches a heightened or siddhic state, they cannot access awareness outside of their body. However, throughout history, expanded awareness has spontaneously flowered in certain human beings, giving us a taste of our future. In a siddhic state of consciousness, awareness is the connective tissue between organisms — it is the interface between consciousness itself and the world of form. The form is the skin, the fruit is the awareness and the seed is consciousness. In the simplest terms, awareness is the key that opens the door between what we call God and man.

HEAVENLY HYDRAULICS

To understand the true nature of the coming awakening, one can find a beautiful parallel in the life cycle of the dragonfly. Dragonflies spend much of their early lives underwater. As underwater insects they are known as nymphs, and unlike most pond insects they do not have to come to the surface for air. For most of their lives, nymphs live totally underwater where they are highly successful predators, feeding on anything from leaf litter to small fish. During this stage of its life, the nymph goes through a series of *moults* — stages of maturity in which it sheds its skin but still remains as a nymph. These stages of the insect's life can last several years, and during this time it has no idea whatsoever of what kind of future lies ahead. It is undergoing a series of *hidden* mutations. Then, one day, all of a sudden, some sleeping gene deep within is triggered and the nymph does something totally out of character — it finds the stalk of a nearby plant and climbs out of the water. For the first time in its life it tastes air and direct sunlight.

Once the nymph has left the safety of its underwater environment, the sunlight begins to work on it, catalysing what will be its final *moult*. It is at this stage that the true magic occurs, as the advanced creature hidden within the nymph cracks the outer larval skin. Over a period of several hours, four crumpled wings appear and the distinctive slender thorax begins to uncurl. What is of great metaphorical relevance at this stage concerns the element of water. As the emerging dragonfly rises out of the element of water and prepares to be reborn into a new life in the element of air, the water that still lies within its body becomes the key to the transformation process. Through a process of hydraulics, the water within the nymph's body is pumped into the emerging wings and thorax, causing them to unfurl and spread out for the first time. In other words, the dragonfly assumes its aerodynamic shape by means of the water from its old life. This water is what drives its mutation from nymph to dragonfly. As soon as all the water is expended and the dragonfly is fully extended, it takes to the air and begins its new life.

The life cycle of the dragonfly is a perfect metaphor for the awakening of the 55th Gift and its Siddhi. The raw energy of your emotions becomes the vehicle for the unfurling of your future awareness, and once that awareness is born, your life exists forever on a higher plane. This metaphor also shows us that as a species we must dive deeply into the emotional field where we will go through a series of mutations of which we are generally unaware. Whilst we exist in the world of emotions, we will have little inkling of the life that lies ahead of us. When the 55th Shadow finally mutates fully, the collective awakening will truly begin in earnest.

The Early Stages of the Awakening Sequence

We will examine the specific timing and sequence of this awakening at the end of the section. At present, in relation to the dragonfly metaphor, we are in the stages of climbing up the stem out of the water and into the sunlight. The world genetic stage is in the full sway of this drama, and because of this it can be a deeply confusing time. You may already be having glimpses and premonitions of what is to come as your body and psyche becomes the battlefield for this mutative process. Particularly if you carry the 55th Gift as one of your prime Gifts, you may be highly susceptible to wild fluctuations in your normal rhythms, energy patterns and emotions. This is a deep process of integration that will last a considerable time, but that will gradually become more stable.

The early stages of the awakening (up until 2012) are set to be the most volatile phase of the process. It is during this phase that our emotional systems are literally being broken down. Two Gene Keys are strongly connected with the 55th Gift — its programming partner, the 59th Gift of Intimacy and the 39th Gift of Dynamism. They are also highly active in this process of awakening. The 39th Gift and its Siddhi of Liberation will challenge every emotional facet of our nature. You can see here the direct relationship between these two higher states of Liberation and Freedom. The 39th Siddhi actually provokes the final state of Freedom. Liberation is a dynamic process whereas Freedom is a release. Equally powerful is the 59th Siddhi of Transparency, which awakens simultaneously with the 55th Siddhi. You can see the hidden agenda beneath the awakening process in this Siddhi — we humans are being forced to become transparent, like the dragonfly's wings. The 59th Gift of Intimacy is the first step along this road. We have to allow life to crack our hearts open through our relationships.

We already know that the 55th Gene Key is about romance, which is why the awakening of the 55th Gift is about relationships. After this awakening has occurred, one will no longer exist as an individual. Awareness will operate collectively. The early shattering of the sense of separateness will take place through your intimate relationships. From now on, the more you try to hide from others, the more suffering you will bring upon yourself. Every hidden agenda must be aired and destroyed. The mind's obsessive grip on separateness must also be destroyed. This is the end of the age of selfishness. There will be many who resist this mutation, as there must be. They are not a part of what is to come and this must be respected. Through these people old energy will leave the world. There is no choice in this — it is a matter of the collective selection of appropriate genetic material for the future human.

After this awakening has occurred, one will no longer exist as an individual. Awareness will operate collectively.

The Vaporisation of the Victim Consciousness

As we have seen, the 55th Shadow is rooted in the notion of being a victim, and in particular of being a victim of emotions, whether your own or someone else's. After the awakening of the 55th Gift, the concept of emotions *belonging* to someone will be absurd. Emotions operate in a wave frequency, and at a collective level there is only one wave which connects us all together. That some people may generate this wave and others may receive it is simply mechanics.

Like the symbol of the dragonfly, our new awareness will lift us above the dark waters of the victim consciousness, but this will not be a simple transcendence. We will not become less human in this process. In fact, the process can only be triggered by such a deep submersion into our human wounding that it becomes the very catalyst of our transcendence.

The process of awakening is a process that has long been known. It is most accurately described through the esoteric science of alchemy. In traditional Taoist alchemy, there is a secret formula known as the *Kan and Li*. Kan means water and Li means fire. In this alchemical formula, the solar plexus is seen as a cauldron, and emotional energy is the water within the cauldron. The fire beneath the cauldron is awareness (also known as *chi*), and this awareness is said to *cook* the emotional energy (also known as *jing*). The result is a process of steaming in which a third transcendent force arises through the process. The Chinese call this third force *shen*, which means spirit. Western alchemy uses similar archetypes but in a different cultural way. In the west we tend to see the two forces as a man and woman within us — animus and anima. These two copulate in a mystical union giving birth to a magical child, often seen as mercury.

In the language of the 64 Gene Keys, the Shadow state is the raw material of eventual transcendence. Without diving deeply into our Shadows and freeing the awareness from its roots, we will never experience the elation of the vaporisation of the victim consciousness within us. Only on those blissful vapours can we rise above the emotional depths and surf the collective wave.

THE RISING OF HUMANITY'S GIFTS

There are two main phases to the awakening triggered by the 55th Gift. The first phase is represented by this rising up of mass consciousness out of the Shadows of victim consciousness. As this occurs, we will see the world as we currently know it gradually changing shape. Up until now, only a small percentage of human beings have managed to escape the Shadow state and deliver their Gifts to the world. Only a very, very few have attained the siddhic level of consciousness. This is all exactly as it should be. Each frequency band depends upon that below and above it. In other words, the more people there are transcending their Shadow state, the more chance there is for someone at a higher level to make the leap to the siddhic level. It may take 100,000 people living at the Gift level to provide the momentum for one being to make the leap to the siddhic level. By the same token, one person at the siddhic level provides the collective frequency for thousands to escape the lower frequency of their Shadows and begin to live out their Gifts.

When a person is freed from the Shadow state, they become a creative conduit for life itself. They also begin to fulfil their true destiny within the whole. The final destiny of the whole is represented by the 50th Siddhi and the 6th Siddhi — Harmony and Peace. This means that as a person begins to do what they love to do in life, they begin to co-create these conditions on the physical plane. As a process, this may take hundreds or even thousands of years to reach its final phase.

When it does, like the dragonfly, our entire planet will mutate to its next phase of evolution in another reality, represented by the 28th Siddhi of Immortality.

This word *Freedom* is truly a dimensionless word. As we begin this process of transcending our Shadows, miracles can occur in our lives. Freedom is the spirit of the 55th Gift — it is the spirit of humanity. As your awareness stretches, the spirit of freedom breaks down the barriers in your life. The *fractal lines* open all around you and energy that was lying choked in a certain dimension suddenly precipitates unforeseen beneficial circumstances in your life. Every aspect of your life is interconnected, so a breakthrough at the source of your being will ripple out into all areas, some that you perhaps didn't even remember existed.

THE THREEFOLD AWAKENING SEQUENCE

In Part 1 of this Gene Key we looked at the threefold pattern inherent to all universal rhythms and we charted this through the evolutionary phases of the Three Aeons and our current sub-phase known as the 333, which holds the master genetic sequence for closing the Aeons. In the last 20 years or so the world has indeed undergone vast changes in its inner structure. As we move through this incredible portal we can identify three distinct dates or markers that set the trajectory of the process of planetary awakening and fusion. These markers are points of shift written into the vibrational score of our evolution. They are 1987 — the Harmonic Convergence, 2012 — the Melodic Resonance, and 2027 — the Rhythmic Symphony. These three phases of Harmony, Melody and Rhythm form the imprinting field for the complete restructuring of all vibrational life on our planet.

As we begin this process of transcending our Shadows, miracles can occur in our lives.

1987 – HARMONIC CONVERGENCE

Much has been said of the Harmonic Convergence. It represented a crossing over point in consciousness in which an unprecedented event occurred. Triggered by a supernova within a neighbouring galaxy, 1987 bore witness to the beginning of the Age of Synthesis. A series of unprecedented celestial alignments made possible a shift in human brain chemistry, allowing us to finally perceive the unifying Truth behind all the great teachings of the ages. It is important to understand that these marker points are not events, but ongoing developmental processes. The Harmonic Convergence is still taking place today on many levels as huge previously separate spheres of human endeavour come together. We are now seeing the beginning of the synthesis of all the human sciences and arts, of the left brain and the right brain, of the male and the female, of the East and the West. The harmony is, as the great sage Heraclitus said, a hidden harmony, but it is now becoming increasingly more apparent.

2012 – MELODIC RESONANCE

As the most talked-about date in recent times, there is not much left unsaid about 2012. Let us then put it in the context of this threefold awakening sequence. In metaphorical language, 1987 was the gestation period, 2012 is the birth and 2027 is the coming to fruition of the new order.

The real meaning of melody is rooted in the understanding of romance. Melody is the aspect of music that catches the emotional breath and causes human beings to dream. 2012 marks the alignment of humanity as a single organism through the breath and the reawakened awareness of the solar plexus centre. Whatever deep dreams and longings you hold inside you will be seeded and locked in by this date, as we come into resonance with the heart of humanity through the reawakening of the Atlantean/Edenic awareness.

2012 also marks a dividing line in human evolution. If by this date you are not in resonance with the dream that is moving into form, your DNA will be locked out of the story. This is in fact a perfectly natural occurrence. Much of current human DNA must be phased out in order for a new form to be constructed. Therefore, over several generations we will see a great number of old patterns leaving our world. For some considerable time this means that we may have the appearance of two separate realities existing simultaneously — those who are still living within the old systems and those who are building the new ones.

2027 – RHYTHMIC SYMPHONY

Many different mystics and ancient calendar systems have long predicted the great turning point of human evolution coming to a climax in this current age. The Human Design System, one of the great systems that laid the foundation for this work on the 64 Gene Keys, uses the 64 codes of the ancient I Ching as a kind of genetic clock to measure the timing of potential future mutations in human DNA. As a genetic clock, it predicts that a huge genetic mutation will sweep through the solar plexus system of humanity beginning in the year 2027. As such, the year 2027 is a difficult year to put into words. The coming consciousness shift will be an implosion of extremely high frequency siddhic consciousness. Certainly after this date nothing will ever be the same again.

Between 2012 and 2027, a core awakening fractal of humanity will lay the foundations of a new world that will reshape this planet from the inside out.

Beginning in the year 2027, our planet will begin a gradual process of falling silent into a state of awe. Between 2012 and 2027, a core awakening fractal of humanity will lay the foundations of a new world that will, over many generations, reshape this planet from the inside out. Old systems will crumble as the new order arises unscathed within their midst. This time will mark the phase of recreating Eden, which never left this planet but remained as an energetic blueprint. Harmony and melody will be synthesised here into a divine universal rhythm. For the first time humanity will hear and be the virtuoso soloist in the great symphony of the spheres. At a certain point in our future beyond 2027, we will finally discover the wonder of simply being. There really will be nothing left to do on earth but enjoy the garden — something our species has not yet managed to do.

THE MARRIAGE OF THE SACRED COUPLE

As we touched upon previously, the coming shift will precipitate the end of the age of the individual *I* and the beginning of the age of the collective *We*. There are multiple stages in this process, and the first stage concerns a major shift in the frequency of relationships on our planet. The awakening will bring many new phenomena into the world, one of which we have always dreamed of but never yet attained — the ideal of the sacred marriage. The contemporary institution of marriage is an attempt to capture this ideal on the physical plane. However, marriages and relationships until now, even the clearest and purest, have not been able to fully embody the principle of marriage at its highest potential — the actual sharing of the same aura.

In order for the ideal of the sacred couple to exist on the physical plane there must first be a melding of awareness. This is the *unio mystica* or *coniunctio* spoken of by the alchemists. Enlightenment or realisation has always been a state that flowers in individuals and, historically, the world has never seen an enlightened couple in its truest sense. We may have seen symbolic examples, and there are certainly couples who have experienced these states together for short periods of time. However, the first stage in the breaking down of barriers between these human forms will be the healing of the yin/yang split between man and woman. The ancestral pressure between the sexes is so vast that it has thus far prevented true melding.

When the first relationships experience dual enlightenment, we will know that the deepest wound of all has finally been healed — the wound symbolised by the division and fall of Adam and Eve. These sacred marriages will have an unbelievable energy field around them — in fact, they will be at the core of whole new communities. Such experiences will herald the end of sexuality as we know it today because the same genetic force that repels the sexes is the force responsible for mating. In other words, the human sexual force will gradually be sublimated into creativity and higher consciousness. Over time, this means that the population of our planet will go through a steady and consistent decline.

The ancient symbol for the 55th Gene Key is the cup of abundance or the sacred chalice. At the Shadow level of consciousness this cup is never full — one side of the relationship is always pulling and the other pushing, one needing and the other rejecting. This situation is caused because of the human tendency to blame, which creates a constant dynamic in the relationship whereby both partners drain each other.

In the human relationships that are coming, the cup is neither half-full nor half-empty. There is only one awareness in the relationship, so the cup is always overflowing. We will no longer fall in love, but we *will* rise in love. The great love that exists between the yin and yang will finally shatter the illusion of our separateness and release the endless font of energy from the nucleus of creation itself. Ultimately, it will be through the extended families and communities of these sacred marriages that the new awareness will spread.

The human being is nothing more than a symphony of interwoven rhythms, tempos and sounds.

THE MUSIC OF CHANGE

Many scientists have found similarities between the structure of DNA and music. Parts of DNA and protein sequences are often repeated with very minor changes. This imperfect repetition has often been likened to the compositional structure of music, particularly classical music and music from the East. The idea that the human body itself is musical is not so far-fetched. We are a delicate framework of rhythms and melodies — our brainwaves, blood circulation, heartbeat, endocrine cycles and the very fluid of our cells all breathe according to a very consistent rhythm. At an even deeper subatomic level our molecules and their atomic structures also vibrate at very high frequencies and are designed around universal geometries. Seen in this way, the human being is nothing more than a symphony of interwoven rhythms, tempos and sounds.

The 55th Gift is deeply connected to sound and the way in which our bodies and emotions respond to sound. The ageless connection between the human emotional spectrum and music is rooted here in this Gene Key. Perhaps one of the most poignant analogies between the structure of DNA and music concerns the triplet. DNA is structured in triplets made up of combinations of base pairs. The triplet is the key structural foundation of the entire genetic helix. In music, the triplet represents something quite extraordinary — it represents the pure longing of life itself. The musical triplet is always trying to resolve to another note, and in this sense it has the effect of leaving the human heart hanging in the air. This longing is exactly what is expressed through the 55th Gift — it is the longing to create more. Unlike duality, trinity is not a straight line — it doesn't rest but repeats, always free and always fresh.

As the great change comes to human beings, the ancient fear within us will subside and we will hear a new kind of music. We will vibrate at a higher frequency that will chemically lift us free from the old genetic fears. We will become one with the music of life and experience the entire spectrum of emotions from the light to the dark, without fear and without shame. This is a new kind of music for humans — there are no paths to follow and no need for systems or structures to keep us safe. Those old ways are leaving the world. The new human being will no longer try to escape the pure longing of life within him or her. We will no longer fear true freedom because we will operate from an awareness that is beyond the mind and its concerns about the future. Ultimate freedom has nothing to do with your life circumstances — it is the freedom of allowing the self to dissolve into the waves of the ocean. It is the freedom that is born through one's absolute trust in life.

POETIC GENETICS

The highest expression of human language is poetry. True poetry captures the hidden essence of what cannot be said in words. The secret is in the rhythm, the cadence and the tonal frequency. To be a poet, imagination has to free itself from the structure of language. Likewise, the true nature of humanity cannot be caught or homogenised into a logical framework. Our true nature is wildness, and it is this wildness that scares people. The moment you think you have pinned life down, it mutates. We humans are in a deep process of transcending the mental game of trying to understand life. The ancient Indian sages called the world we inhabit *maya*

— an illusion. Our problem has always been that we try to understand this maya through an instrument (the mind) bound by laws that prevent true understanding. You cannot use an instrument within the maya to understand the maya.

This new awareness in humans will bring an end to so many things. One such ending we will witness will be the end of the question of *how*. As a species, we will no longer be obsessed with intellectual understanding. This will also mark the end of the spiritual seeker. We will no longer fix our awareness on our structures and systems. We will no longer hunger at *any* level. Like the poet or the musician, we will enter into the mystery itself. Humanity is actually at the very early stages of transcending our genetics. As our consciousness begins to rise up on the pure awareness of our emotional system, we will finally see through the veil that has long held us captive. Once we are free from our minds in this way, we can truly create great poetry of our lives. We are entering into an era of great beauty — it will be a transcendent era in which creativity will rule and life itself will be experienced as art.

THE POSSIBLE AND LIKELY EFFECTS OF FUTURE GENETIC MUTATION

As we transit into this coming phase, particularly in the years following 2027, there are many things about the world that will change. Because of the nature of mutation, there will be sudden quantum leaps that are followed by long periods of integration. All change on a social level takes time and some of these phases may last hundreds of years.

PHYSICAL CHANGES

The secret of the 55th Gift at a physiological level lies in a single element — salt. Salt has long been known for its purifying properties and its ability to leach toxins from the body. Every single cell within your body contains salt, and its balance within the body is a major key to health. Everything connected to this 55th Gift is rooted in its literal and metaphorical relationship to water. As we learned through the 32nd Gene Key, water holds memory. When your emotions become really intense, you release memory through the salt in your tears and/or sweat. What is beginning to happen to humanity now, and what will become more and more intense, is a process whereby ancient memories are being chemically released from our bodies. Heightened emotional awareness will gradually draw the toxic genetic memories out of the human form. At a physical level, this will occur through your sweat, your tears and your urine.

In the same way that seawater evaporates to leave behind its salt, so human beings are going through a process of evaporation and distillation. At a chemical level we are beginning to change. A new network of neuro-circuitry in the solar plexus is superseding the reptilian fear-based neuro-circuitry of the old brain. As the 59th Siddhi testifies, human beings will gradually become more transparent as the body no longer produces the old chemicals created by fear. With the closing down of certain chemical processes associated with the hindbrain, the body's needs will radically change. Without the toxins created by fear, the body will need far less salt, and it will become much less dense.

Humanity is actually at the very early stages of transcending our genetics.

DIET

As the human body's need for salt decreases our digestive system will begin to mutate. This is after all a solar plexus mutation. As our digestive system mutates to accommodate a higher frequency passing throughout our DNA, it is likely that our diet will also change. The body will not only stop craving salty foods, but will actually reject them outright. It is also probable that gradually humans will stop eating meat, and we will certainly not be able to tolerate the high salt intake provided by modern processed food. As our children inherit the mutation through their DNA, they may well be born with a physiological allergy to salty food and/or meat. All these changes are the *result* of the mutation and will come in their own time. During the current transition period human beings actually require even more salt than usual in order for the toxicity of the past to be collectively purified. This is the hidden reason behind the current worldwide revolution in processed food. Nature knows exactly what she is doing, and we should take heart from this.

Digestion is rooted in the mineral realm — in the way in which the body uses and dissolves trace elements from food and water. In the future, in a totally new way, we will become highly efficient at drawing and combining the elements from food. The mechanical means for this will be through our moods. In other words, our bodies will tell us exactly what we need to eat and when, through the medium of our mood. One of the most likely effects of the mutation will be that we will simply not feel hungry as often as we do now, with the result that we eat far less. Added to this, our bodies will begin to find other means of absorbing higher frequency food through air and sunlight. Eventually, far down the road when the final pieces of the cosmic chess game are in place, the 6th Siddhi will flower within the collective, making our skin totally translucent, thereby allowing us to live purely on light.

EMOTIONS AND DECISIONS – THE STILLING OF THE WAVE

Some of the most radical changes to humanity will concern the emotional system itself. Presently, human beings are victims of the whims of their emotions. Their decisions are out of harmony with their true nature, creating a collective energy field of chaos. As the mutation takes hold, what we now call emotion will have a completely different role. It will no longer be experienced as emotion at all. It will be a means of communication. The people in whom this mutation manifests will not be caught in the emotional drama of life. They will still feel every single nuance of the emotional environment deep within their bodies, but their awareness will ride on top of these waves instead of being lost *within* them. The result is that they will feel extremely calm, and one of the ways in which they will be recognised is through the peacefulness in their eyes.

Every person carrying the mutation will effectively still the wave in his or her environment. As more and more people are born into this awareness, their collective presence will slowly tune the rest of humanity to a different dimension — a dimension of endless clarity and stillness. This will also deeply affect the way in which human beings make decisions. Decisions will no longer be subject to the shifting patterns of emotional chemistry. Decisions will emerge instantly and with great clarity as the collective chemistry across the planet becomes calmer. Such decisions will no longer *belong* to individuals but will emerge directly from the harmonious nexus of the collective itself.

The process of the stilling of the wave will ultimately lead to an era of world peace. As a metaphor, this process is akin to an orchestra warming up before a concert — all one hears is a cacophony of different tones sounded on different instruments at random. This is the current state of humanity. When the mutation arrives, the conductor taps his stick on his podium until every instrument is silent. Only when silence has been attained can we hear the hidden harmony that is the true nature of humanity.

THE ENVIRONMENT

Many people today are very concerned about the environment of our planet and the great damage that is occurring due to the huge pressures of globalisation. Before we look at the good news that lies ahead of us, it is important to understand why humanity currently appears to be causing such damage to itself. To understand such things one has to look at the bigger picture. The planet is our greater body, and just as the physical body of human beings is going through a genetic mutation, so is all life. All of life is a delicate network of interwoven threads. It is not possible for one species to undergo a major mutation without affecting all other species.

Our current generation is the sacrificial generation. Our collective body is purging humanity of its ancient toxins. From the dietary level, we have seen that the west in particular has such a high salt intake that a huge swath of the population is obese. Fat is the fuel of mutation, and this mutation is leaching out the collective Shadows of humanity. Stress is another symptom of increased activity through the solar plexus. Mutation puts the physical vehicle under great stress. At every level of society the ancient human wound is being expressed — through business, government and the environment itself. This is the true meaning of the myth of the flood. The flood is coming, and it will separate the Gift consciousness from the victim consciousness.

Global warming and pollution are classic expressions of the human wound seen on a broader level. These kinds of phenomena represent the last raging of the victim consciousness, with the earth herself playing the role of victim. There is enormous collective fear surrounding the issue of what we are doing to our environment, but the irony is that if humanity were not in the grip of this world mutation we *would* inevitably destroy ourselves. The rising of the 55th Gift as a physical genetic mutation is literally creating a new species. As our spirit settles and our awareness allows us to experience unity with each other, we will also experience unity with all creatures. The new awareness will particularly connect us to animals in a direct way, since their awareness already functions collectively. Although they carry different genetic equipment, their true nature is that of the 55th Gift — Freedom. Not only will we stop eating animals, but for the first time we will experience ourselves as one with them. In everything we do, freedom will play the central role.

A major key to the future will be the huge decrease in human population that occurs due to the shift in frequency of the emotional/sexual apparatus. The busy world we see today will

Only when silence has been attained can we hear the hidden harmony that is the true nature of humanity.

fall silent — huge tracts of the earth will be reclaimed by wilderness. The sense of space and freedom that is the essence of our planet will return. As we have seen, the nature of freedom is wildness. We will not have to do anything to heal the planet. There simply will not be enough of us to do harm, so nature will find her own wild balance. The animals will be free to roam, the plants and forests will be free to spread and blossom, and man will be free to enjoy simply being alive. The very force that has driven man to where he is now — the force of fear — will be gone.

As we said earlier, it is always difficult to foresee the details of what life will look like in our future. What we *can* see is the spirit of the age. It is more than likely that man will continue to harness the incredible technology he has thus far created, and with the rising of our inherent gifts, we are bound to improve on it exponentially. The future will not be a retrogressive period of going back to our primal roots. It will be more a co-creative stewardship of nature. In essence, human beings have always been gardeners, and that is really our role on this planet — to complete the beauty of nature by adding our own spirit to it.

The real work of transmutation at a planetary level will be done by the oceans. All the toxins created by man will find their way into the water cycle and over time be purified by the salt within the world's oceans. Once again, one can see the elemental power that lies behind this 55th Gift as well as the mystical meaning behind the coming age of Aquarius — the age of the water-bearer.

FUTURE TECHNOLOGY AND THE NEW SCIENCE OF SYNTHESIS

When contemplating humanity's potential future technologies and their use and effect on our world, we should bear in mind that the coming mutation will directly affect the way we think. Since our primary awareness is shifting to the solar plexus area, all future insights and breakthroughs in science will come from this awareness rather than from our logical mind. This will entirely change scientific approach. Instead of beginning with doubt and then working to resolve that doubt through scientific method, we will begin with certainty and use logic to confirm and deepen that certainty. This will give birth to a new era of science and technology, and the future science will be a science of synthesis. Science will work hand in hand with art, music, mythology, and psychology and, of particular importance, it will be rooted in the physical structure and understanding of the body.

It is a universal law that our external findings mirror our inner development.

The central axis of all future logical systems involved in this grand new synthesis is sacred geometry. Geometry is the central organising model that allows the human mind to correlate all patterns within the holographic universe. For example, it is now being shown in advanced physics that the geometry of the 64 is not only present within the tetrahedral structure of DNA but also underpins space-time itself, as well as being the foundation of music. With the help of advanced computer technology, we can now generate highly complex models of our universe using the laws of fractal geometry. Using this geometry it will become possible to unify all the sciences and arts into a single integrated cohesive whole. Such a synthesis can only be made possible through mass collaboration across a wide variety of fields of expertise.

With our awareness opening up within the solar plexus, the new physics will set off in a completely new direction. Our most bountiful natural resource is the sun, and it will likely become the true source of our energy. One of the great western sages of the last century, Mikael Aivanhov, spoke of future humanity becoming a *solar civilisation*. In the hologram of the universe, the reawakening of our own inner sun in the solar plexus will have its reflection in our technology. It is a universal law that our external findings mirror our inner development. This statement has even greater implications for our future. As we transcend the very structure of our own DNA, we will break free from the gravitational pull of the lower frequencies. In science this will be reflected in the new technologies such as plasma physics, which will shortly enable us to transcend physical gravity and bend time and space.

The new science will begin to take humanity into a future that at this moment may well seem like science fiction. Once we have technologies that harness the powers of gravity, we can travel outside our solar system and begin to explore our galaxy and universe. This will mark the phase in which earth finally becomes a player in a far wider field of intelligence than we currently comprehend. All these breakthroughs are in fact much closer than most of us dare dream. More than likely the technological foundations will be laid for this completely new epoch of humanity's story in the first half of this very century.

GOVERNMENT, POVERTY AND MONEY

To grasp the future social structure of humanity you need to have a clear understanding of the nature of fractals. (You can learn more about this subject from a thorough contemplation of the 44th, 45th and 49th Siddhis. Each of these concerns the different levels of change that will revolutionise the way human beings interact on a collective level.) It is evident that humanity will eventually become linked together by a single pervading spirit, in much the same way that we are linked together today on the material plane through the Internet. The creation of the electronic World Wide Web is a precursor of what is to come at a genetic level. The nature of this human spirit is freedom, which means that freedom will become the only real human agenda.

As the human spirit becomes free, another critical Gene Key will flower collectively — the 50th Gift of Equilibrium. This Gift is one of the most important in terms of how human beings serve and support each other. Through the 50th Gift, humanity is slowly being brought into a state of cosmic harmony. At a social level, the presence of this Gift in different societies and racial groupings will bring about a new kind of order. This Gift will precipitate the gradual demise of corruption and crime. This means that appropriate assistance will move between the developed countries and the less developed countries and the problems of poverty will eventually be overcome.

The future of money can also be seen quite clearly through an understanding of certain Gifts and Siddhis — in particular the 45th Siddhi. Money is essentially a physical expression of victim consciousness. It represents human fear. Our relationship to money is therefore our relationship to fear. There is nothing that reveals a hidden agenda more quickly than the subject of money.

Our relationship to money is therefore our relationship to fear.

Nearly all money, given or received, carries a hidden *charge*. Only money that is given or received unconditionally has no charge attached. As money is handled in a cleaner way, it will become energetically laundered and manifest one of the great cosmic laws — that to give is to receive. The most successful businesses of the future will be based upon the 45th Gift of Synergy. These kinds of businesses will no longer be based on competition and fear, but will be transparent and highly efficient. Greed and fear are actually highly inefficient.

As the highest aspects of the 45th Siddhi come into play, money itself will eventually come to an end. When this finally occurs, it will be the greatest symbol of true freedom that our planet has ever manifested, and it will inspire a worldwide celebration the like of which has never been known. As we have seen, this 55th Gene Key is a part of the codon group called the Ring of the Whirlwind. Together with the 49th Gene Key, it will cause dramatic changes at all levels of our society. Interestingly, this codon ring codes for an amino acid called histidine, which is released during physical orgasm. The whirlwind moving through the human genome can indeed be likened to a collective orgasm — a spiralling force of consciousness that ripples through the body of humanity, taking us to higher and higher levels of unity and ecstasy.

DEATH, MEDICINE AND THE SIDDHIC SUPERNOVA

Freedom is the only true medicine of the future. There are many levels of freedom, but the ultimate freedom is freedom from believing you are separate from life. The coming awareness through the 55th Gene Key heralds the absolute end of the fear of death. In fact, the 55th Gene Key does more than end this fear — it proves that there is no such thing as death. This fear of death actually lies within the 28th Shadow, which has a deep connection to the 55th Gift. As the 55th Gene Key mutates, so will the 28th Gene Key, at least to the Gift level of Totality. The secret to optimal health lies here in the 28th Gift because it is about the free flow of life force through the physical body. As we humans transcend the old fear, life force will once again flow through our bodies unimpeded. The sheer power and vitality of this energy carries enormous healing potential, and it will literally eradicate all the diseases that riddle humanity.

The true nature of illness and disease is rooted in this core fear of death.

The true nature of illness and disease is rooted in this core fear of death. With the uprooting of this fear, we are truly entering into an age where medicine will be superfluous and will slowly become defunct. Naturally, as the old diseases are purified, some of them may mutate and even become more widespread for a period of time. This process will probably last several hundred years. True healing is connected to our ancestral DNA, and in order for a single person to be in perfect health, their entire genetic lineage must be burned clean. This cleansing takes place through the presence of the Siddhis, as they manifest in the world. Every time a Siddhi manifests through someone, it sends a shockwave of purity back down the entire genetic fractal line. People who bring the siddhic frequencies into the world also take into themselves the collective Shadows of their ancestral lines.

We stand on the cusp of a siddhic supernova. The number of those in whom the Siddhis are manifest will very shortly increase in the world as a great incarnation takes place on the physical plane. This incarnation represents the third aspect of the Holy Trinity — the Divine Feminine spirit. However, this incarnation will not be a single being — it will be a collective spirit using a specific constellation of beings, each occupying a *core fractal*. (For further detailed information on the role of core fractals you can read the 44th Siddhi.) The process of the incarnation of the Divine Feminine will last many generations but its final result will be the purification of all the fractal lines throughout humanity, resulting in the burning up of the collective *karma* stored in human DNA and the eventual eradication of all disease on the physical plane.

CHILDREN AND EDUCATION

The final area we will look at is in many ways the most important of all. It is our children who hold the future in their hands. Many of the young children coming into the world today are carrying the seed of the future mutation in their blood. It will be their children who give birth to the new awareness, beginning around the date 2027. One of the amazing things about children is the clear and innocent way in which they process the emotional wave. Our current generation of children really reflect the chemical changes that are to come, and their emotional nature in this sense is quite unique. These are children whose emotional swings or physical manifestations should not be seen at a personal level, but at a collective level. Of course, these children need to be given the usual boundaries that any parent gives a child, but with a greater sense of freedom as well. The mutation is coming to our emotional system and as it moves through our chemistry it will tend to throw up erratic behaviour patterns and unpredictable emotional phenomena. The real key is for parents not to assume that there is something wrong with their children but to give them even more love and to be extremely patient.

As for the children of future generations, some of them will carry the mutation and some will not. The mutation will appear all over the globe. It will not be difficult to spot those who carry the new awareness because they will not display the emotional symptoms we usually associate with young children growing up. Parents will find a new kind of peacefulness emerging within their families simply from having one of these children in their home. All these children will display unique gifts from an early age depending on their Hologenetic Profile and their Prime Gifts.

One of the great changes these children will bring will be to the education system. Because their main centre of awareness is located outside of their brain, they will appear exceedingly bright. Once the mind is transcended true genius emerges. Their method of learning will be closer to osmosis than repetition, and their memory will be extraordinary. One might have the impression from all of this that these children will be very vulnerable within our existing society. However, this is not true. Because of their gifts, they will not require special treatment or schooling. On the contrary, they must integrate into normal life. Wherever they go, they will draw the experiences to them that allow them to further enhance their gifts. Their strength derives from their very transparency. They will be moved by a force so powerful that it is beyond our current understanding. It would be impossible for such children to ever feel alone or victimised.

Once the mind is transcended true genius emerges.

The presence of these children throughout society will reveal the limitations of existing educational systems. One of the practices from our past that may very likely make a comeback will be the system of apprenticeship — where children with specific talents become apprenticed to specific teachers and learn about life through the world rather than from behind a desk. At every level, the 55th Gift of Freedom will make itself known. For a child, freedom is about play. It is through play that a child learns about his or her world. Thus it is likely that the children of the future will no longer be sent to schools at an early age, but will be given the amount of space they need to truly flourish.

As these children grow up, they will initiate the first great pulse of the new epoch into the world. It is therefore imperative that they are fully integrated into the existing social structures of the planet. Many of them may become teachers themselves, or doctors, lawyers, business people and other *normal* professionals. Because they can feel life at a holistic level, they will be able to introduce subtle reforms that create a ripple effect at all levels of society. Everything they touch will become more efficient. These will be people who hold absolutely no fear within their system, but can sense the fear in everyone else. This level of empathy will make them masters of relationships. Slowly and imperceptibly, these children and their children's children will transform our planet. As was mentioned earlier, higher consciousness is literally going to *breed* the victim consciousness out of humanity.

CONCLUSION

All of the above is an intuitive exploration into the archetypal codes contained in the 64 Gene Keys as seen through the lens of the 55th Gene Key. As such it contains the frequencies of our common future rather than the specifics. It is the frequency lying behind the entire work on the 64 Gene Keys that is of prime importance. There are those who resonate fully with these frequencies and those who do not. This book is written for those of you who do resonate with such higher frequencies. Everybody has within their being a barometer of Truth, and this manifests in different ways within different people. If you have felt the breath of truth within these words then you are one who is ready to dive deeply into your own shadows and own them fully. Freedom comes at a price, and that price is transparency. You must own every negative feeling and tendency within you and take full responsibility for it. You must draw back the subtle arrows of blame, and you must locate every hidden vestige of fear within your being and embrace it *without* fear.

Once you are transparent and truthful with yourself and others, this seed of the future awareness can take root in you. Even though it may not be present in your genetics as a physical mutation, you can still resonate strongly to the energy field behind it. Furthermore, if you are open-hearted and humble enough, this awareness will inevitably awaken within you, and it will use your life as a launching pad to soar into the skies of freedom and prepare the ground for the true world of high romance that is to come.

䷞ 56th Gene Key

Divine Indulgence

SIDDHI
INTOXICATION

GIFT
ENRICHMENT

SHADOW
DISTRACTION

Programming Partner: 60th Gene Key

Codon Ring: The Ring of Trials
(12, 33, 56)

Physiology: Thyroid/Parathyroid

Amino Acid: None (Terminator Codon)

THE 56TH SHADOW – DISTRACTION

THE WORLD MASK

As you delve more deeply into the 64 Gene Keys, you will begin to appreciate how these codes are interwoven across many different dimensions. As a hologenetic reflection of the cosmos in which we live, the 64 Gene Keys allow you to travel through the infinite reaches of your inner universe. The 21 Codon Rings — universal genetic groupings inside your body — are one of the greatest mysteries within the structure of your DNA. Chemically, the codons synthesise the sixty-four genetic triplets into groups that codify the twenty-one major amino acids. As archetypes however, the Codon Rings reflect a mysterious symmetry at play within the universe as a whole. Within this interwoven geometry, the Ring of Trials, which includes the 12th, 33rd and 56th Gene Keys, lays out the great dramatic script of evolution itself. These three Gene Keys do not code to any amino acids but instead relate to a set of specific instructions known as *stop* codons.

If you were to track through your own DNA you would come to these special places every now and again in the reams of coded information that make up the fabric of your being. These three Gene Keys and the 41st Gene Key (known as the *start codon*) share a vital genetic role inside you. And there is more to them than just instructions, despite what biologists may think or see. As the building block of living matter, DNA is designed to mutate in order to continue evolving. The genetic code itself changes shape and adapts its function over great spans of time and the Gene Keys also change their functioning. The 35th Gene Key is an example of this. It sits alone in the genome in a similar fashion to the stop and start codons, but it is very unique as you will discover when you explore its rather unusual nature. The fact is that the 35th Gene Key used to function as a stop codon in a far more primitive phase of our evolution. However, this aspect of your DNA has mutated over the course of evolution and now, at its highest frequency, it allows human beings to take a short-cut through their genetics, resulting in phenomena that we generally regard as miracles.

The Ring of Trials also lends itself to such possibilities, which means we do not know what its Gene Keys may be capable of once they have been awakened.

In the modern world, we appear almost desperate to distract ourselves from who we really are.

However, at the Shadow frequency you can see exactly what they do and how they affect you. As the first of the Three Great Trials, the 33rd Shadow sets the pattern of your forgetting and conceals your true universal nature. This trial means you must journey through time and space across many incarnations until you come back to a memory of your greater self. The second trial, laid down by the 56th Shadow, involves maintaining the illusion of your individuality by keeping you distracted through your five senses. As you remember more of yourself, you will come to realise the extent of your addiction to the external world, and over time you will turn your energies inward and break this powerful addiction. Finally, the 12th Shadow of Vanity comes at the end of your evolution. It is the last great trial and tests the depth of your surrender to the ultimate. In this trial you must give up everything you have attained in all your countless journeys, and in so doing you will attain final transcendence.

Now that we have a clearer background to this 56th Gene Key, we can enter more fully into the field of its transmission and learn how devastating its Shadow theme of distraction can be. Let us begin with an example of the power of this distraction. Over fifty percent of humanity has never used a telephone. Allow the implications of that statistic to sink into your consciousness for a few moments. If that many people in the world are still distracted by issues of survival, what are the other fifty percent of people who are more fortunate doing about it? The answer is — almost nothing. They are too distracted themselves by the minutiae of their own lives — by mortgages, telephones, restaurants, television, politics, computerisation and just about everything else you can imagine. Distraction greatly impedes evolution, but it also finally brings you into a fuller appreciation of your own misery. In the modern world, we appear almost desperate to distract ourselves from who we really are. At the same time we now have so many distractions that we are becoming increasingly aware of the extent of our addiction.

You will see below that the reactive nature of the 56th Shadow is over-stimulation, and this is the essence of distraction — as long as you are over-stimulated through your senses, you do not have to feel your own discomfort. Over-stimulation or under-stimulation keeps you numb. The Shadow of Distraction places a mask over the world, which prevents you from seeing life as it truly is. As we constantly leave our centre and travel out through the five senses, we become victims of our material lives. Unlike many of the 64 Shadows which keep you a victim of your own thinking, the 56th Shadow ensures that you will be a victim of someone else's thinking — in other words, environmental conditioning. Whether through your country's government, television or media, or simply the belief systems of your religions, cultures, teachers, parents or peers, the world constantly tells you how you should think. It is not surprising that we are easily distracted from our real dreams and ideals as we get washed into the belief systems of others.

Distraction works in one of two ways. The most common type of distraction is outer distraction. In other words — the outer world of the senses distracts you from your inner world of feelings and the reality of the higher realms. In this respect, we tend to blame the outside world and the people in it for what happens to us, instead of realising that our circumstances reflect our inner state. The well-known new age epithet "you create your own reality" is actually only half true. You do not create the actual events of your life — but you do influence their playing out through your attitude. When you blame people in the world around you, you set up a victim frequency pattern that reinforces itself over and over in your life. If, on the other hand, you are able to accept everything that happens to you regardless of whether you enjoy it or not, you set up a surrender frequency pattern that allows you to move through life with great fluidity and beauty, and your life reflects it.

The other less common form of distraction is inner distraction. Inner distraction is when you are so inwardly focused that you forget the outer world — you live in a fantasy world of your own making with no real anchor in the material world. In this sense you look through a lens whereby everything meets the criteria of your fantasy. You see what you want to see, but you do not see the truth. This is where we can see the power of the 56th Gene Key's programming partner, the 60th Shadow of Limitation. The 60th Gene Key is about the importance of structure and form, and for some people this can seem a distraction from their fantasy. The Gift of the 60th Gene Key is Realism, which means that you must accept the present moment as it is without any projected mental overlay. The moment you become distracted from what is actually happening, you severely limit the outcome of the events taking place in the present moment.

It is easy to see how the 60th Shadow of Limitation feeds the human need to be distracted. Whenever you feel that you are limited it means that your mind has trapped you. Instead of freeing yourself from this discomfort through facing and accepting it, your tendency will be to run away from the feeling as fast as you can — whether that means opening the fridge, turning on the TV or picking up the telephone. The world mask keeps us engaged, entertained and distracted by the dramas going on all around us, as well as within us. And, perhaps saddest of all, the world mask keeps us poor — for when we are distracted by something that simply keeps us in the same state of inert numbness, we are truly impoverished.

REPRESSIVE NATURE — SULLEN

The repressive mode of the 56th Shadow is sullenness. To be sullen means to be under stimulated. It is a collapsing of our spirit into a kind of numbness. This state is one often associated with teenagers, who can often enter these kinds of long-term sulks. Many adults who had difficult childhoods also find themselves victim to these *dead* spaces, which become entrenched as patterns in the endocrine and nervous systems. This is how the physical body over time manifests an emotional pattern through its chemistry. In adults with the 56th Shadow this repressive side can often manifest as the *lemming* syndrome in which people lose sight of their true aspirations and become enslaved to the tedium of mundane lives. You can see this Shadow reflected in the eyes of such people — they appear lacklustre and devoid of all joy.

REACTIVE NATURE — OVERSTIMULATED

The reactive side of the 56th Shadow is over-stimulation. This manifests as a constant need to maintain movement, at all levels of one's being. There is a particular need within this Shadow to satisfy the eyes and for anything that stimulates the eyes — from reading to watching television to fantasising to travelling. These people can lead completely internal lives. On the outside they may seem perfectly normal, but inside they harbour all kinds of fantasies. On the other hand, these people can also lead lives that are focused only on the outer world, in denial of the inner world altogether. The 56th Shadow is reflected in all behavioural patterns that allow us to avoid feeling the reality of who we are and how we currently feel. We keep moving, changing relationships, or trying new experiences. We simply do not know how to stop.

THE 56TH GIFT – ENRICHMENT

TURNING WITHIN

As with all the Gifts of the 64 Gene Keys, Enrichment pulls you out of the shadows and into the light of the higher frequencies. Enrichment is what life is all about. Distraction is fine, but only as long as it enriches us. Those who display the 56th Gift have learned the difference between what feeds the human spirit and what saps it. This means that they are no longer victims of distraction, but have learned the art of self-discipline. If you are reading this and think you are not a victim of distraction, there is a very simple litmus test. Is there anything you cannot say no to in life? If there is, no matter what it may be, you remain at some level its victim. Remember, the 56th Gift is not about abstention, it is about having the ability to apportion life as it enters through your five senses.

The Gift of Enrichment is not simply about having willpower. Enrichment is different from enjoyment or entertainment. You know that you would really enjoy that chocolate cake for example, but you may also decide that on that particular occasion, it will not enrich your spirit. Another day, the same cake *may* enrich your spirit. The point is that you are not a victim of your senses. The 56th Gift is about treading the fine balance between vice and virtue — between wildness and responsibility. Those with this Gift are neither addicted to abstention nor to overindulgence. They simply know above all how to get the most out of life.

This 56th Gift likes to sample the delights of the garden and baulks at nothing. It may take you into some unsavoury places and relationships. However, the 56th Gift has an alchemical flavour to it. The alchemy is about knowing how to use what we call evil as a means to transcend. Evil, as the 56th Gift knows, is simply an energy configuration that exists at a low frequency. The same energy at a higher frequency has enormous potential to be of service to the whole. Therefore, the 56th Gift will take in all manner of low frequency waves because it knows how to transmute their energy into joy and purpose. If you have this 56th Gene Key in a prominent position in your Hologenetic Profile, you have the rare gift of being able to show others how their problems are really wonderful opportunities. There is great lightness and humour in this Gene Key.

True enjoyment is rooted inside your being rather than in the external.

The 56th Gift knows a great truth — true enjoyment is rooted inside your being rather than in the external. As you begin to embody this truth, your awareness naturally turns inward. The same energy that would have become a distraction in the outer world turns inward towards your own source. As it does so it causes an inner transformation. Over a period of time, the 56th Gift actually trains you how to meditate.

You may not meditate formally, but you enter into a state of meditation where sensual desires are seen for what they are — illusory attempts at fulfilment. This does not mean that you become some kind of mendicant or ascetic, but it does break your addiction to seeking fulfilment in the external world. As this occurs, the life lived through your senses becomes highly refined. You even begin to develop your extra senses — the higher attributes of your subtle bodies beyond the physical, emotional and mental worlds.

One of the wonderful traits of those with this Gift is that they enrich others through what they have learned in life. The ability to discipline yourself when necessary leads to your being seen by others as a potential role model. If you have the strength to love yourself, others are automatically drawn to you. Essentially, the 56th Gift is about balance. If you have this Gift, then you can always balance fun with seriousness. You can party with the best of them, but the difference is you know exactly when to stop. This is the power of the stop codon inside each of us. It behaves as a seal that closes up the places where we leak energy. Indeed, in ancient China, the five senses were known esoterically as the *five thieves*. They were understood to be the places where our vital force or *chi* drains out from the body. As you learn how to *seal* your human tendency to forget yourself through your senses, you experience an inner flame growing inside you.

Enrichment means to suck the marrow out of life. The 56th Gift is a gift of feeling, of sensuality and aliveness. It means that you find wonder where others find monotony and you find beauty where others see ugliness. It is about appreciation and gratitude — the more grateful you are for every moment, the more the moment comes alive within you. Above all, the 56th Gift's greatest potential genius lies in communication. It is able to entertain and divert the attention of others. At a lower frequency range, this Gene Key may be found in advertising and political spin, and at a higher frequency, it may be found in comedians, entertainers or inspirational speakers. At even higher frequency bands, it is the great and ancient art of storytelling or myth-making — the sharing of personal experiences that have touched and opened one's own heart. At these higher levels, this 56th Gift of Enrichment is a Gift of love, for the more you enrich the lives of others, the more this Gift pours through your heart.

THE 56TH SIDDHI – INTOXICATION

THE DIVINE ENTERTAINMENT BUSINESS

The 56th Siddhi is actually highly amusing. This is the reverse art of Distraction. Those who display this Siddhi have disciplined themselves to be distracted only by the Divine, only by what is uplifting, only by the most luminous currents and emanations. This is the Siddhi of Intoxication. The root of this ancient word derives from the word *toxic* or *toxin*, which in turn comes from the Greek for the word arrow. The ultimate toxin is love, and this is where the myth of Cupid and his arrows of love comes from. People immersed in this Siddhi have taken the Gift of Enrichment to its ultimate zenith — they have allowed themselves to be pierced over and over again by love. The irony of this level of consciousness is that even though enrichment requires discipline, intoxication requires no discipline whatsoever to maintain!

*To be Divine
does not
mean giving
up sensual
delights.*

Siddhis only manifest after great breakthroughs in consciousness. In a sense they are rewards. The reward of the Siddhi of Intoxication is to find yourself in a state of permanent distraction — the distraction of pure love!

Mythically speaking, the 56th Siddhi has many archetypal parallels. All our human gods and deities have emerged out of the 64 Gene Keys. Out of this 56th Gene Key come all the great hedonistic deities — Dionysius, Bacchus and Pan being a few examples from the Greek pantheon. The 56th Siddhi knows matter in all its delights, trials and depths. However, it is not about material indulgence, but Divine indulgence! The 56th Siddhi wants humanity to experience the richness of life, so it often creates a synthesis of spirituality and material decadence. To the 56th Siddhi in its purity, decadence requires only a single sip of wine on the material plane to trigger the parallel intoxication on the higher planes. Those with the 56th Siddhi do not stop their enjoyment of material delights — they just revel in them in homeopathic doses! Because of this, the 56th Siddhi does not conform to any usual sense of holiness or spirituality. It will eventually teach humanity that to be Divine does not mean giving up sensual delights. It simply does not require them in any way whatsoever and therefore can truly enjoy all aspects of life.

The 56th Siddhi is a Siddhi that is extremely contagious. Just as the many distractions of the Shadow frequency are contagious — staring at the computer, watching television, taking drugs or drinking alcohol — so the highest level of the 56th Siddhi is equally contagious and very addictive. This is about being addicted to the highest frequency of love. Unlike the lower frequency manifestations of love, which are about chasing Cupid's infamous arrows in the external world, the 56th Siddhi finds the source of the toxin itself — your own superabundant heart. Intoxication is about being swallowed up by your own love. People manifesting such euphoric states have the gift of distracting others away from their self-destructive patterns and raising them to a higher frequency through love. Because this Siddhi is deeply rooted in the Shadow of Distraction, these people also have an astute understanding of the laws that govern human beings. Having passed through the Gift of Enrichment, they know about the human preponderance for excess. They know just how to talk to you and just how to infect you with their love and humour. Like drunks on a higher plane, these people simply revel in their own aura of love, and as such they can be irresistible. They have no agenda other than sharing their own good fortune with whomever and whatever crosses their path.

The 64 Siddhis are an encyclopedia of attributes that most people would describe as *holy* or *divine*. The 56th Siddhi is closest to what we see as madness or drunkenness. These are people whose flowering has so spectacularly erased their sense of continuity in life that it takes a huge energy just to keep them from bursting into laughter. They effervesce with life — they tease and tickle your spirit — they cannot be understood or encapsulated by any form of logic. These are the Divine drunks who occasionally come staggering into the world of form. Such people are a reminder from the Source that life is for love, for beauty and for fun. It is a game played by fools, and each of us must come to see our own folly and embrace it in laughter and acceptance.

The 56th Siddhi flows directly through the heart. This is the highest art of entertainment and laughter. It is one of the great poetic Siddhis. This 56th Siddhi has nothing to do with discipline in the sense that we understand the word.

Those rare beings that manifest this state are a law unto themselves. You cannot understand them with your mind, but if you sit with them and lap up their laughter, you yourself may become drunk on their exquisite frequencies. They are like wine that has matured to perfection and their only wish is to go on drinking from the love that purls endlessly from within their heart. The being intoxicated by this Siddhi has realised an astonishing fact about existence — there is no point to it. All their beliefs and searches have now come to an end. The state that remains is one of wonder and delight. For this person, life is nothing but entertainment since there is nothing more for them to learn or do or achieve. When Intoxication bursts through from the higher planes, all learning turns into wonder. You continue to absorb the delights of life, but your learning has stopped — simply because learning suggests evolution, and this Siddhi ends the game of evolution.

Since it follows the 55th Siddhi in the ancient coding sequence of the I Ching, the 56th Siddhi gives us some clues as to where humanity is going after we experience our collective awakening, as described in the 55th Gene Key. In essence, humanity moves into the entertainment business. Once we have achieved our attainment, there is no role for us other than the audience entertained by existence. True entertainment involves laughter, inspiration, wonder and ultimately intoxication. The programming partner to this Siddhi, the 60th Siddhi of Justice, will begin to awaken at around the same time as the 56th. As it does, the world will begin to correct itself and its imbalances. The codes we have taken for granted for so long will begin to break down — our economy, our institutions and systems of law and governance, the very fear of death that has built our modern world — all will begin to crack and decay. As the old system cracks, those at the higher frequencies (and they will grow in numbers every day) will release a great rush of love and intoxication through the collective solar plexus of humanity.

In the ancient Jewish sacred text known as the Talmud, a strange and mystical prophecy reads:

> *"And in the Time to Come,*
> *The Holy One will make a banquet for the Righteous*
> *from the flesh of the leviathan,*
> *and its skin will be used to cover the tent*
> *where the banquet will take place."*

This prophecy concerns the awakening of the Synarchy at the heart of humanity. The *Righteous* refers to the higher frequency of the heart. Those who do not turn away from the toxins of the Shadow consciousness but transform them internally will unlock the secrets of higher consciousness. This is referred to as *eating the flesh of the leviathan*. The Christ consciousness must take the lower frequencies into itself and transmute them back into light. The finale is described as a great feast or banquet, in which the skin of the leviathan is used as the tent under which the celebrations take place. This wonderful and mysterious metaphor refers to the cracking of the maya or illusion that prevents humanity from knowing its higher nature. This is the *skin* — our mental awareness which conceals the truth from us. Furthermore, we will use that *skin* as a means of celebration. This is exactly what the 56th Siddhi does. Because it is no longer taken in by mental constructs, it becomes intoxicated by the wonders of the mind and its creations.

If you know someone who has this 56th Siddhi within their Hologenetic Profile, practise seeing this highest of levels hidden deeply behind their behaviour. There is a love hiding
within this person as awesome as anything anyone can imagine. If you have this Siddhi in *your*
Hologenetic Profile, your life is supposed to teach you about this love through the world and
its suffering. You should never shy away from suffering, for suffering is the toxin that you are
here to use to become intoxicated. As you let life's arrows pierce you one after another, you will
eventually find yourself so defeated that you begin to laugh. You will stop trying to direct life
and you will surrender. The glorious moment this happens to you, your entire consciousness
will shift from the horizontal to the vertical. You will begin to see that at this level, every single
thing in life can be enriching. It is all a matter of attitude.

"As it draws you to itself
What pleasure your suffering becomes.
Its fires are like water
Do not tense your face.
To be present in the soul is its work,
And to break your vows.
By its complex art
These atoms are trembling in their hearts."

Rumi

䷸ 57th Gene Key

A Gentle Wind

SIDDHI
CLARITY

GIFT
INTUITION

SHADOW
UNEASE

Programming Partner: 51st Gene Key

Codon Ring: The Ring of Matter
(18, 46, 48, 57)

Physiology: Cranial Ganglia (Belly)

Amino Acid: Alanine

THE 57ᵀᴴ SHADOW – UNEASE

THE FEAR-BAND FREQUENCIES

From the point of view of the 57ᵗʰ Gene Key, absolutely everything in life is acoustic. Even light can be reduced down to a sonic signature. Even so, the sound spectrum that we human beings can access is actually very narrow. The most sensitive mammals can hear sounds well beyond our abilities; we know for example that dogs can hear high-pitched sounds and creatures such as whales and elephants can hear sound frequencies far below our own spectrum. Other creatures such as insects interpret sound through their entire bodies or legs as pure vibration, which is of course exactly what it is. This entire work on the 64 Gene Keys is a human attempt to paint a picture of the universe of different frequencies that we inhabit and that move through us and around us. At the highest level, as we shall see, we humans are quite simply made up of layers of flickering and alternating sound waves.

The 64 Shadows are all states of consciousness governed by fear. To understand more precisely what we mean when we use the word fear, it may help to reduce it to a certain range of frequencies. If fear-based states fall into a particular waveband then we can see how easy it might be to adjust our own frequency and raise ourselves above this fear-band. This does indeed sound easy. However, there is one thing we have to remember above all else. Humanity collectively vibrates within the fear-band frequencies. Therefore, as human beings, we are each under enormous pressure to resonate within these same frequencies. Every human being is like an acoustic tuning fork. If we are placed next to a powerful audio output source, before long we will automatically begin to vibrate at the same wavelength as that output. On planet earth, this process is ensured through our childhood conditioning. The standard human output source, which is based on fear, is known as the 57ᵗʰ Shadow of Unease.

The ancient symbol for this 57ᵗʰ hexagram in the I Ching is the wind. As a symbol, wind has many dimensions. It is also a symbol for the pervasiveness of spirit because it moves invisibly around the world, touching everyone. When seen from the Shadow consciousness, wind can be brutal and even terrifying, uprooting and destroying wherever it goes. When the wind is up, it often conveys a sense of unease. This 57th Shadow represents a very deep and ancient fear — the fear of what might be coming, of not knowing what is in the wind. Human beings are genetically programmed to fear the future — it is wired into our DNA through this 57th Shadow. In our early prehistory, human beings functioned almost entirely through their individual attunement to frequency. If their intuition picked up something dangerous in the wind, their instincts immediately caused their bodies to move accordingly, whether that meant running or hiding or grabbing a weapon.

Today, modern man has developed in a different direction. We are now far more polarised in our brains than our bodies and most people make decisions through reason rather than intuition. This development has changed the 57th Shadow of Unease. Unease no longer functions as an early warning system restricting fear only to the moment when it is needed for survival. Now unease is translated by our minds. It is continuous and manifests as anxiety. Furthermore, because of this, it is enhanced through the universal morphogenetic field that connects all human beings as one. The mind has become stronger than instinct, and seeks to end unease through the creation of external security. And so the rat race of modern culture is born. The more mind-centred humanity becomes the more security it tries to create for itself and in turn the more paranoid it becomes. Security and protection have become a global obsession, even though they are a complete illusion. Life is as uncertain as it ever was, and even for the wealthiest and most protected of human beings, the unease still remains.

Humanity today lives within an audio-visual feast of fear. The shadow frequency bears down upon our minds like a huge pressure that we cannot escape. Neither can we turn back the evolutionary clock. The brain is already developed and our mind has such a powerful vibration that there is no way to stop it. We are caught in a global web of fear — so much so that our fear has become collective and we fear for the future of our species. The great contemporary symbol of our fear is money. Apart from those who have accumulated great wealth, the majority of human beings project their fear of the future onto money or the lack thereof. Ironically, those who have a great deal of money have discovered that it doesn't take away the fear. Fear simply relocates to somewhere else in the psyche. All of this fear and anxiety keeps human beings stuck in their heads from where it is very hard to escape. No form of thinking can take away the fear because the fear is there precisely *because* of the thinking.

No form of thinking can take away the fear because the fear is there precisely because of the thinking.

Without our recently developed neocortex, our early ancestors lived very much moment-to-moment in a reality we can no longer easily imagine. Our minds simply do not allow us to have the experience of living in the absolute present, even though all life, including our body, does live in the present. The mental rat race within each individual's mind has shaped the world we see around us today. In this context, fear is a very creative force, but it prevents us from rising above a certain frequency band as a species. We have come just about as far as we can within these relatively low band frequencies. If we tarry much longer, we will indeed enter the very phase of self-destruction that terrifies us the most. One thing is sure, the mind cannot figure its way out of our current situation either individually or collectively.

However, the good news is that we humans are not afflicted with a disease but are simply passing through a certain developmental stage of our evolution.

The 57th Shadow, paired and strengthened by its programming partner the 51st Shadow of Agitation, simply does not allow you to feel at ease in our world. This is one of the great shadow pairings that create interference and disease in the human body. All physical disease is rooted in the frequencies of fear. As humanity evolves beyond its current stage, it will eventually move beyond fear, which will result in the end of all disease.

Interestingly enough, the 57th Shadow conditions your life most powerfully during gestation, when you are a developing foetus in your mother's womb. The vibration of fear actually passes into you at the point of conception. It is then further enhanced by the auric field of your parents, and particularly that of your mother. During these nine months, the essence of your primary developmental cycles from birth to age 21 is hard-wired into your DNA. You can understand this in greater depth through studying and contemplating the chemical family known as the Ring of Matter, which contains the 18th, 46th, 48th and 57th Gene Keys. As we will learn at the Gift level, the 57th Gene Key plays a vital role in the transformation of our species to a higher frequency beyond fear.

Through this Gene Key we can see that all fear is greatly exacerbated through the anxiety created by the human mind. The 51st and 57th Shadows keep us worrying continually about the future and we stay tuned into a narrow frequency band that keeps us in an endless mental loop. Fortunately, there is a way out of this loop. There is a way of moving into a new threshold where fear has far less hold over us. This is the direction in which humanity is now beginning to evolve. Those who listen carefully through their intuition might hear something incredibly new coming on the wind. And yes, one should always trust what one hears on the wind.

REPRESSIVE NATURE – HESITANT

Hesitancy occurs when an intuition is suppressed by the power of the mind. The body knows what is correct in every cell of its being, but the mind immediately imposes its doubt, anxiety or opinion, thereby rendering the true perception powerless. In this way, all true alignment to the power of the now is lost and clarity — which is pristine and visceral — is repressed in the body. Spontaneous clarity is a state that exists outside the mind and can only be known through the purity of being. The sheer aliveness of a clear and instantaneous knowing is the cornerstone of one's true inner radiance and health. Hesitancy or indecision is the hallmark of the Shadow frequency. These are people who tend to become trapped by their own worries, which make them unable to really feel the certainty of spontaneous clarity and commitment.

REACTIVE NATURE – IMPETUOUS

Impetuousness arises as a human reaction to unease or fear. Its sole purpose is to try to escape or bring an end to fear through making a quick decision. Such decisions are not made from the state of clarity described above but are themselves rooted in fear. Because of the nature of impetuous decisions, they can only lead to more misery. Not only do they fail to bring an end to the feeling of uneasiness but they also manifest additional turmoil in one's outer life. A reactively made decision can only move in the opposite direction of evolution, therefore, against the flow of nature. This is not to say that such decisions are necessarily wrong. Life needs to create turmoil as part of its own awakening process. The key to escaping the loop that such decisions inevitably create is to detect one's own fear and experience it fully without reacting first. This witnessing is precisely what dismantles the pattern.

THE 57TH GIFT – INTUITION

ENTERING THE SYNFIELD

Of all the 64 Gene Keys, few have such a profound connection to individual human health as this 57th Gene Key. As the foundational aspect of the Ring of Matter, the 57th Gene Key governs the cycle of gestation, which in turn lays the pattern for your development throughout childhood. It is during this primary cycle that all your genetic programming is laid down. Your genes build your body at the frequency of the energetic field in which your mother lives. Therefore, every mother plays a crucial role in the biological, emotional and mental structure of the child. The mother is actually a co-creator of the incarnating child, and every thought, feeling and impulse running through her being will direct the DNA within the foetus. This obviously puts a huge responsibility on pregnant mothers and has significant implications concerning the transformation of our species through deep honouring of the mother and of the importance of her role during pregnancy.

The developing foetus lives in a world of frequency. It literally swims in and imbibes the tones, colours, sounds, emotions, thoughts and intentions of its environment. Even so, the foetus interprets these frequencies through the mother's responses to them. Thus the mother's frequency directly dictates the destiny of the future human being. Each of the three trimesters of pregnancy relate to the three seven year cycles of the child's development, the first being the physical, the second the emotional and the third the mental. In other words, in those first nine months, the first 21 years of your life are completely mapped out. Of course, no one is entirely a victim of the frequency of their mother. At any stage in any of the developmental cycles issues rise to the surface in order that they may be embraced, purified and healed. As your own frequency rises, so you will gradually heal the many layers of your inner being. Even so, it is important to realise how great a head start a child can get from elevated frequencies in the mother during pregnancy.

The 57th Gift of Intuition is basically your body's system for interacting harmoniously with the outer world. It is the low-frequency prenatal programming that later interferes with the clear operation of your intuition. Although all illnesses were imprinted during the gestation cycle, they can be healed through directly raising the frequency of your DNA. This essentially resets or reboots your entire genetic operating system.

As you address the Shadow states that such a process naturally brings to light, so you will witness a deepening sensitivity inside you to everything and everyone in the world around you. This is what the 57th Gift of Intuition is about — it is the natural guidance system of all human beings.

If you review the evolution of human awareness up until our present point in time, you may discover a vital insight on which our future evolution will hinge. This insight concerns the role of the inner masculine and inner feminine principles. If we go back in our evolution to primitive man, we can see how deeply developed was our instinctive awareness — our gut feeling or hunch. Our individual survival depended on the innate animal instinct functioning through our bodies, i.e. through the five primary senses and the mythical sixth sense — that ability to perceive something coming before our physical senses actually detect it. All human beings alive today have inherited this inner sixth sense if we but knew how to trust in it. In a word, your most powerful inner compass is your intuition.

Having developed our intuition — the feminine aspect of our psyche — humanity then went on to develop its masculine side, the mind. Whereas intuition listens and receives, mind explores and conquers. This is why our current epoch is so fascinating. We human beings must now remember our past and reconnect to the power of our intuition. Having done this, we will have to learn to trust our intuition over and above our mental faculties. In this way we will create a naturally structured internal psyche that mirrors nature. Intuition is how nature talks to and through human beings — it is the auditory canal through which the whole coordinates and communicates with its many parts. If we humans can attune ourselves to this gentle, subtler internal voice, we will finally begin to feel physically at ease. Furthermore, when our internal hierarchy is naturally formulated in this way, the genius of the mind can finally come into play and follow the dictates of nature herself.

The human mind is a truly extraordinary instrument. It is also a very dangerous instrument without the proper internal guidance. We can all see how destructive the mind can be when it is allowed free range without the sense of being connected to the whole. As humanity learns once again to trust its deeper feminine side, as it is beginning to do, the mind will naturally fall into its own rhythm. This revolution is already underway in the individual. Intuition emerges from the whole so it naturally leads to synthesis, and intuition backed by the intellect is capable of extraordinary things. The fact is that the more you trust in your intuition, the more integrated your life becomes. Your relationships open up and become softer, the path of your destiny is made increasingly clear and events move more smoothly, as though the entire universe were supporting you. This is of course exactly what is occurring.

Every time you trust in your intuition or make a decision based on it, you raise the frequency of your whole aura.

The process of learning to again trust in your intuition is nothing less than the dismantling of the illusion that you are separate from life. It is a return to the heightened sensitivity you had in your mother's womb. The more you widen this pathway inside yourself, the easier life becomes. Your fear and anxiety will persist in the beginning, but after some time intuition will become more natural and powerful inside you, as though an invisible force were overriding your old conditioned programming.

Furthermore, every time you trust in your intuition or make a decision based on it, you raise the frequency of your whole aura. Your awareness operating system changes gear and your body hums with life. The deeper you go into this new awareness the more the fear inside is transcended. At higher levels still, you will begin to sense vibrations through your whole body. One of the great revelations you may have through the 57th Gift is that fear is not inside you. It is a field that you may live in, pass through or ascend beyond. Evolving your so-called sixth sense is the first stage in this process of vibratory ascension because it allows you to access the universal quantum field or collective unconscious.

Once your body becomes lighter and vibrates at higher frequencies, you enter an amazing world known by many names in different traditions. This is the world of the gods and goddesses, or what the theosophists call the causal plane. On this level the higher mind begins to function, although it has little in common with the mind as we experience it at lower frequencies. The nature of the higher mind is clairaudience — the ability to pick up vibrations through your aura and interpret them through your brain. It is from this causal plane or *synfield* that all great revelations and spiritual knowledge are downloaded into human beings. Obviously such revelations occur at various frequency bands within the Gift level itself, and the purity of the messages depends on the frequency of the aura that is receiving them. However, the higher you go up the spectrum of frequency, the more integrative and synthesised the transmissions become. In the end however, regardless of the potential heights that this Gift offers human beings, this 57th Gift of Intuition shows you one of the clearest and simplest paths to move beyond the shadows of your fears.

THE 57TH SIDDHI – CLARITY

THE ART OF SOFTNESS

At the higher reaches of the 57th Gene Key, you begin to access the ability to tune your clairaudience beyond the borders of time. This marvellous gift thus allows you to bend time and intuit the future, which in turn changes the way you live your life, allowing you to relax more deeply into your being than ever before. Even at this incredibly high level of consciousness however, the subtlest traces of Shadow frequencies may still remain. Your abilities allow you to sense what is coming, but you are still functioning within the realm of duality. In seeing the future you acknowledge its existence and, as long as it exists, you don't function fully within the present. Your awareness at these levels has stabilised for the most part in the present, but it still flickers in and out of the present moment.

When the ancient sages spoke of the siddhis or special powers as potential obstacles to the path of liberation, they may well have been referring to the upper reaches of the 57th Gene Key. Because your frequency becomes very refined at this level, your intuition penetrates everything, including human beings. The sheer sense of power one feels from being able to sense the future or read a person's aura in this way can become an addiction to one operating at this level.

The subtlest fear becomes the fear of losing all that power by going any further. Of course, when the veil lifts and reveals the 57th Siddhi, you *do* indeed lose these powers, but not in the sense that one would normally understand. What you lose is the sense of power itself. Since you lose your identification as a separate being, the whole concept of individual power comes to an end.

The 57th Siddhi is the Siddhi of Clarity. Since they are programming partners, the 51st and 57th Siddhis always dawn together. Thus we only see reality clearly when we awaken fully. The only way to expunge fear totally from your being is to eliminate that being altogether, which is what occurs at the siddhic sphere of consciousness. At the Shadow level, we explored the idea that each human being acts as a kind of acoustic tuning fork, picking up the frequencies of the output source beside which it is placed. Similarly, we can see how this works at the Gift level as your awareness expands to pick up a wider and more integrated vision of reality. At the siddhic level, which really brings an end to all levels, the output source and the receiver eliminate each other. You could also say that they come into a harmony so perfect that it is experienced as silence. This state is the eternal now. It captures and conveys the very truth of immortality. There is no fear because there is no tomorrow and therefore there is no death. This is clarity.

We learned that the original symbol for this archetype is the wind. In the I Ching, it is usually translated as the *gentle wind*. This essence of gentleness is one of the greatest secrets to life. Consciousness is the gentlest, most subtle phenomenon. This is why the ancient sages so frequently likened it to water or wind, elements of such subtlety and softness that they can penetrate everything. Clarity is about seeing this softness at the heart of all things. In the acoustic field of life, everything rises from softness and returns to the same softness. When you live your life in harmony with this softness, you come into accord with what the ancients called the Tao — the transcendence of the opposites. Furthermore, as you open yourself to this gentleness, this manifestation of clarity, it will reveal itself in your life continually — through the sound of the wind furrowing the treetops, or a puff of cloud drifting across an ocean sky. The same softness is to be found everywhere since it is the spirit of life itself. If you allow it, it will transport you immediately into the world of the eternal now.

Among humans, the art of gentleness is the greatest of the lost arts. We do not realise that the more softly we lay our hand upon something, the more it opens to us and the more deeply we can access it. Our minds tell us that the opposite must be the case. This 57th Siddhi, like all the Siddhis, contains the mystery of the feminine principle, even though it is beyond duality itself. Those who really understand the true meaning of healing know that it is about this essence of softness. This 57th Siddhi holds the secrets of miraculous healing through its specific ability to attune to the subtleties of physical DNA. Such gentleness opens the heart and leads to transcendence. Despite what our minds assume, gentleness knows no weakness. It simply operates according to its own laws and timing. Operating beyond weakness and strength, it permeates everything. Clarity is the realisation that everything is linked through gentleness.

When someone attains enlightenment through this 57th Siddhi, it manifests through them in an extraordinary and beautiful way — they become a tuning fork for the Divine Presence.

Clarity is the realisation that everything is linked through gentleness.

If you sit in their company, the incredible softness of their aura begins to heighten your frequency very rapidly. In the presence of such a one, many people can suddenly experience Siddhis — auditory manifestations of higher consciousness. To stay beside such a being for a prolonged period will eventually dissolve your sense of separateness altogether. However, one needs to beware of sitting before an audio output of this kind! You must always approach these people with the right attitude — that of infinite softness. The 57th Siddhi tells us precisely how to approach a great master or an awakened being. It tells us how to approach *all* aspects of life. If you adopt this same softness of spirit throughout your life, then whether you engage with a master or not, clarity will eventually reveal itself and you will realise the true nature of being.

58th Gene Key

From Stress to Bliss

SIDDHI
BLISS

GIFT
VITALITY

SHADOW
DISSATISFACTION

Programming Partner: 52nd Gene Key

Codon Ring: The Ring of Seeking
(15, 39, 52, 53, 54, 58)

Physiology: Perineum

Amino Acid: Serine

THE 58TH SHADOW – DISSATISFACTION

DIVINE DISSATISFACTION

In the original Chinese I Ching, the inspiration for these 64 Gene Keys, each symbol or hexagram is represented by the combining of eight different types of natural phenomena such as thunder, wind, earth, fire and so on. The 58th Gift is symbolised by the single repeated figure of a lake. A lake is a beautiful and simple symbol on which to meditate because it immediately captures the very essence of calmness. Lakes signify emotional calm as well as mental stillness. Moreover, when you study the 58th Gene Key's programming partner, the 52nd Gene Key, you will discover that its highest and most natural manifestation is stillness. These themes of joy and stillness are therefore intimately and genetically linked. In contrast, the 58th Shadow — Dissatisfaction — is very unspecific. It is a lack of fulfilment rooted in the programming partner of this 58th Shadow, the 52nd Shadow of Stress. It doesn't refer directly to a particular emotional state such as sadness, boredom or frustration, or to a mental state such as anxiety or worry. It simply implies a lack of joy and a deep sense of restlessness or unease.

Whenever the 52nd Shadow of Stress breaks the surface of the still inner lake inside you, your natural state of being is lost. The question contained within this 58th Shadow then is: how does this happen? How and why do human beings so easily lose touch with their natural state? The answer lies in a single concept — the future. As you peruse and contemplate these 64 Gene Keys you will see that there are entire Gene Keys— for example the 10th and the 20th — which are dedicated to the experience known as *living in the now*. This simple expression is the core principle of almost all great mystical and spiritual systems and paths. So how on earth do you live in the present moment? One way to bring greater awareness to this conundrum is to understand why, how and when you leave the present moment in your own life. In this respect, the 58th Shadow and its Gift can lend you considerable insight.

Most people drawn to reading this work will already know that the main culprit within human beings is the mind. If you are at all interested in spirituality or self-improvement you probably have heard this over and over again — the mind is the problem. Out of this truth countless systems have been born — meditations, practises, affirmations and modalities — all with the single purpose of helping you transcend the mind and find inner fulfilment. Although it is indeed true that the mind is the root of the problem, there is great danger in directly confronting the mind because it is such a slippery mechanism. The problem with the mind is that it hankers after self-improvement, the very core of the 58th Shadow. It wants more than anything to give you the feeling that you can do something to bring about the state of joy. Unfortunately, in this respect, anything you do — any technique, any system, any strategy — can only bring about continued dissatisfaction!

The 58th Shadow stirs up a great genetic pressure within human beings, the pressure to improve something or to be of service in some way. Dissatisfaction is an energy frequency aimed entirely at the future. When not agitated, this Gene Key manifests its natural state — that of vitality and joy. This is what is so hugely ironic about the 58th Shadow. It provokes you to seek happiness in the outside world only to bring you to the conclusion that you cannot produce the state of joy because it already exists inside you. The 58th Shadow creates the illusion of the future. The real joke is that in being driven outwards in our quest for fulfilment, we human beings actually improve the world and help it to gradually become more synthesised. In other words, it is your very dissatisfaction that is of the greatest service to the whole.

When we examine the 52nd Shadow of Stress in conjunction with the 58th Shadow of Dissatisfaction, we learn that these two codes programme human beings on a trans-genetic level. In other words, they reinforce each other through the collective morphogenetic field. The more dissatisfied human beings there are on our planet, the more powerful this Shadow frequency becomes. Ironically, this is why the population explosion has led to vast improvements in the quality of our lives. In our modern world of today, the search for personal satisfaction and fulfilment has become practically universal. It is extraordinary to consider that our search for fulfilment creates and compounds the very stress it seeks to end. This is precisely how evolution operates — it makes us miserable, and in our quest to end our misery we unwittingly evolve.

For the individual, the 58th Shadow does a great service. It stirs up the quiet waters of your inner lake so that you must try to find some kind of peace. Eventually you realise that the future dream you are trying to capture does not exist in the outer world, and so you are driven back into yourself. It is at this stage that most people begin their spiritual search. It is simply a gearshift from trying to improve the world to trying to improve yourself. Even so, the spiritual quest is yet another fruitless search because it too is based upon the illusion of the future. Nonetheless, the process cannot be sidestepped. Pressure is pressure, and this 58th Shadow will not let you rest until you have exhausted your inner search. The pressure keeps pulling on your mind to do something to relieve the pressure. Added to this, in your mystical search you discover the many systems, teachers and paths that promise an end to the pressure. Inevitably you will try one or more of these. Many of us never manage to transcend this process and remain addicted to the idea of a perfect and peaceful future for the rest of our lives.

The Shadow of Dissatisfaction is based on the false promise that there is something you can do to bring about happiness. Even doing nothing as a reaction to this knowledge can become a subtle doing on your part. What this Shadow therefore does in human beings is make us realise how hopeless our situation really is! Despite the way it sounds, this Shadow is not negative in the least. This Shadow exhausts you until you finally come to your knees. In this sense, dissatisfaction is truly divine and contains the secret of grace. Once you begin to grasp this truth deep inside your being, the future you have been holding onto slowly crumbles, revealing for the first time a marvellous Gift welling up inside you — the Gift of Vitality.

REPRESSIVE NATURE – NONE

The 58th Shadow is the only aspect of our genetic makeup that cannot be repressed. The reason for this is that it signifies vitality — the life force itself, which is outside our control. If it could be repressed by human beings, we would cause our own deaths and as a species would never have been able to evolve. Fortunately however, life is far stronger than human beings and will not be denied. We are destined to face our dissatisfaction and can only react to it.

REACTIVE NATURE – INTERFERING

Such is the strength of life force that we humans can only react to it, whether consciously or unconsciously. This reaction most often expresses itself through our interfering with the natural flow of nature. Such interference causes us further discomfort and dissatisfaction. The greatest difficulties arise when we occasionally taste the spontaneity of our own inherent joy. Because we have all tasted this experience, we crave it and we constantly try to recreate it. Only when we realise that dissatisfaction is simply a low frequency manifestation of our life force pressuring us to evolve, does the great miracle happen — we finally stop interfering with life and experience our true joyous nature.

THE 58TH GIFT – VITALITY

THE JOY OF SERVICE

The moment you accept your plight as a human being eternally driven to seek fulfilment in the future, everything in your life changes. This deep understanding sparks the next process when the life force within you shifts gear into a new phase. The process begins in earnest as you grasp the depth of your own dilemma at an intellectual level. The understanding then passes from the intellect to the intuition and finally deep into the very heart of your body — into your DNA. When acceptance reaches down inside you to this degree it produces what can only be described as an atomic rupturing. The energy that has been seeking an outlet into the world through your dissatisfaction is turned back upon itself, which forces it into the atomic structure of the body. The result is a kind of implosion of inner vital force which catalyses extraordinary micro processes within your physiology. In short, you begin to become yourself once again.

At the Gift frequency, the same energy that you experienced as dissatisfaction now begins to turn into joy.

At the Gift frequency, the same energy that you experienced as dissatisfaction now begins to turn into joy. This joyousness lies within every human being and is nothing short of life expressing itself without resistance. At this stage of your individual evolution, something strange is breaking through inside you — the evolutionary urge to become self-aware. *Life* wants to become aware of itself, freed from boundaries and restraints. Thus it begins a process of dis-identification inside you. As you realise at deeper and deeper levels that the future truly does not exist, you prevent your life force from projecting itself into that future. Your mind becomes quieter, as though more space were opening between your thoughts. Your relationship to the future also changes. You realise that it is not and never has been in your hands but is maintained by a collective force of which you are only a minute aspect. This deepening realisation loosens your concern for what happens to you, even though a part of you still remains fascinated.

All of these changes in your inner attitude are brought about because of the shift in direction of the atomic vital force that is keeping you alive. Increased vitality really means increased freedom. No matter how old or decrepit your body may be, you begin to feel rejuvenated by the joy that begins to well up within you. You find yourself happy for no real reason at all. Such a force needs to find an outlet in the world and its most natural path is in finding a way to be of service to others. However, unlike the 58th Shadow, the 58th Gift does not try to help others in order to allay its own dissatisfaction and thus interfering with natural organic processes. On the contrary, the Gift of Vitality knows empathically how to work with life processes in a non-interfering manner. Vitality always recognises vitality and is particularly adept at helping to free blocked vital energy. In fact, the mere presence of such people can catalyse increased energetic flow through a system. These principles can be applied to any field of endeavour — from healing the physical body to the structuring of a bridge or even to increasing the profit of a business.

The 58th Gene Key is part of the genetic family known as the Ring of Seeking, a complex codon that codes for the amino acid serine. Each of the Gene Keys within this group creates a different pressure inside you. Together these six pressures drive you to seek answers that can bring an end to the longing inside you. All seeking is therefore partially rooted in dissatisfaction, and all seeking will eventually lead you inwards. It is this turning in that unlocks the huge reservoirs of vitality that are available to you when you relax enough. As you see that all seeking is simply life looking for itself, you begin to stop seeking. As the pressure within you falls away, the vital energy becomes clearer and more radiant within you. Many people who are awakening in this way actually go through some kind of healing crisis as the energy matrix of the subtle bodies comes back online. This can be a time of intense physical transformation.

Above all, the 58th Gift is an unstoppable force that once set in motion cannot be turned back. Because the 58th Shadow is uniquely irrepressible, when the frequency of this Gene Key hits the Gift level things tend to happen very quickly in your life. Not only does your life take on a new shape, but a far deeper process overtakes you — the process of moving into the heart. This is an infectious phase in which you learn to move in deep harmony with your environment.

At a collective level this Gift is very powerful as it will one day unite humanity to work towards a higher goal. In some respects it is already doing this across the globe and the modern trend towards globalisation is an example of this. The future of human evolution truly lies in the ideal of service. One day, service will be at the root of all business, economics and government, as humanity realises that it really is better for all of us. The joyousness we seek as individuals is intimately tied to the joyousness of the collective, which is precisely why the 58th Gift is delighted to work so tirelessly in that direction.

THE 58TH SIDDHI – BLISS

BEYOND FOCUS

As your identification with the future progressively loosens at the Gift level, you are brought more and more deeply into the present moment. Your core dissatisfaction is being transformed into vitality, which is then drawn back into your body. As the frequency of your vitality attains a higher and higher pitch, at a certain point it peaks and catalyses a spontaneous shift in awareness. As previously described, this vital force burrows so deeply into your being that it triggers a process within your DNA known as enlightenment. It is as though before this event your awareness sticks to your DNA like a kind of glue, giving you the impression that life is localised only inside you. Then, at the tipping point, awareness finally lets go of its attachment to your DNA and you experience a kind of death.

The final vestiges of the Shadow frequency pursue us even into the Gift frequency where they cling to us like lichen to rocks. However, the focused power of the frequency just before enlightenment has the wondrous effect of loosening these final tendrils within you and *something* lets go at the deepest level inside. Suddenly the life force is no longer focused inside you and you no longer feel a centre inside you. Because of this release in focus, everything reverts to its pure and pristine state and your being becomes once again a pure still lake. These waters will not be stirred up again because they are the waters of pure consciousness, whose nature is entirely unfocused. When consciousness comes, it comes into you like a flood, wiping you clean in less than a microsecond. It is one of the greatest mysteries of existence.

When consciousness comes, it comes into you like a flood, wiping you clean in less than a microsecond.

Once a being has realised enlightenment through this 58th Siddhi, they become entirely unfocused. Their awareness stretches into infinity, their gaze becomes unfixed and dreamy and their heart is so full that it bursts with love for all creation. Even though the awareness in such a being has become universally unfocused, their body becomes a point of intense focus for the process known as bliss. Bliss occurs as a by-product within a person in whom these events occur. Physically, the experience of bliss is brought about by the spontaneous release of certain chemicals within the brain, which are now produced continuously due to activations in your DNA. These activations allow you to witness unbounded the continuously breaking waves of life welling up within you.

Experiences of such heightened states are still relatively rare events in the world that we know. On average perhaps one person in every generation across our whole planet has such a transformation. Because the siddhic state is not brought about by any particular practice or activity but is born of understanding and grace, such experiences often occur to people who have no particular religious or spiritual practise. One enduring example of the spontaneous eruption of the 58th Siddhi is the renowned sage Ramana Maharshi who lived in India in the 20th century. Ramana experienced just such a spontaneous death at the age of sixteen with no prior understanding of such phenomena. The Siddhi of Bliss cannot be hidden or contained in any way, but radiates out from the very pores of one's being. One can still capture something of its essence by looking at photographs of Ramana Maharshi's eyes.

There are many viewpoints expressed by 'experts' about what enlightenment really is and how it happens. Each siddhic state is essentially the exact same experience in which consciousness expresses itself without resistance through your genetic makeup. There is enormous confusion amongst seekers and even teachers about the many possible manifestations of enlightenment. The experience of bliss can even occur whilst still in a Shadow state, since there are drugs one can take that activate these chemicals for limited periods of time. Heightened spiritual states also often include periods of blissfulness that can last days or even months. None of these experiences can be compared with the actual enlightened state itself. Some say that enlightenment has nothing whatsoever to do with blissful experiences and that such states are even traps that prevent the real thing. However, through the 64 Siddhis one can get an idea of how varied the expression of enlightenment or realisation can be.

One of the great problems among seekers is identifying with the manifestations of enlightenment and assuming that this is what it means or looks like. The actual manifestations are really inconsequential. That one vehicle is coded to suffer waves of continuous bliss and another is coded to become a paragon of honour or virtue is not the point. The point is to realise that there is no point! No matter what one thinks, does or says, the experience of enlightenment cannot be attained through seeking, even though seeking may come before it. The final state is outside of our grasp. Again and again the seeker must come up against this impasse. You must look into your constant dissatisfaction until you begin to loosen up and find the humour in your situation. In this way, you will finally come to realise that the Siddhi is present within the Shadow, just as the flower is present at all times within the seed.

SIDDHI
TRANSPARENCY

GIFT
INTIMACY

SHADOW
DISHONESTY

䷘ 59th Gene Key

The Dragon in your Genome

Programming Partner: 55th Gene Key

Codon Ring: The Ring of Union
(4, 7, 29, 59)

Physiology: Sacral Plexus (Sexual Organs)

Amino Acid: Valine

THE 59ᵀᴴ SHADOW – DISHONESTY

DOING THE GENETIC LAUNDRY

In looking at the 59th Shadow, we see the essence of all the problems underpinning the social structures within our world. There is no more topical issue within the entire human genome. The 59th Gene Key, the 55th Gene Key and to an extent the 49th Gene Key are the places within our DNA where a complete transmutation is currently underway. (The timing of this shift and its longer-term implications are discussed more fully in the 55th Gene Key.) One of the biggest insights into the archetypes within our DNA comes from a close understanding of the pairing of Shadows, Gifts and Siddhis. We see that humans appear to be programmed in binaries, and that every Shadow has a partner. Because the 55th Gene Key is currently triggering a genetic shift in our species, the 59th Gene Key (its programming partner) is doing the same. Whereas the 55th Gene Key is triggering individual awakening across our planet, it is the 59th Gene Key that is truly responsible for triggering planetary mutation at a genetic level.

At its base level, the 59th Gene Key is about sex and reproduction. It represents the drive in human beings to multiply. As such it is an impersonal force that selects our potential mates for us. The reasons for being drawn to a particular person are complex, but essentially, we can be sure that our genes have *their* survival in mind! Most relationships are therefore not easy by design. There are also other reasons why human beings are drawn to each other that are rooted in a more spiritual domain — karmic reasons for example. However, from a universal point of view, whatever forces draw us together at a cosmic level must operate through our genetics and our biology, so all these different perspectives ultimately come together.

Behind the 59th Shadow hides a deep distrust of others and a core fear of relationships. This is the Shadow that makes the world a lonely place because despite the number of people there are around us most of the time, we rarely communicate with each other in real depth. As we explore the 59th Gift and Siddhi, we will see what that really means.

What is unusual about the 59th Shadow is that the fear does not lie in the individual. It is in the aura between us. When you are alone the fear is not there, but as soon as another person comes into the same room as you, the fear of the 59th Shadow is there as a subtle undercurrent. It is even more amazing to consider why this fear exists at all — because it is the foundation of sexual attraction. This may come as a shock to many people, but the truth is that the moment you stop being afraid of someone, you transcend your sexual attraction to them. The fear provides the necessary friction that makes attraction possible. This is why the 59th Shadow is the Shadow of Dishonesty. As long as we keep something of ourselves hidden from the other there is always something left for them to fear. It doesn't necessarily mean conscious dishonesty. We are genetically dishonest. The extraordinary revelation that comes from the contemplation of this 59th Shadow is that our genes actually want us to be dishonest. This means that life itself has been holding humanity back from realising its higher nature. One needs some time to digest this last sentence. Life needs human beings to be afraid of each other in order that it can fulfil its genetic potential in this current stage of our evolution.

In order to fully grasp the wonderful depth within this 59th Shadow, one needs look at human evolution in a wider perspective. The fear that keeps human beings from realising their common genetic ancestry has led to the phenomenon of isolated tribal gene pools all across our planet. If we were not afraid of each other, we would have interbred immediately wherever we migrated. There would be no social, political or geographical boundaries — no countries, borders or wars. But, and it's a big but, there would also be no genetic differentiation, no art, religion or cultural colour. In short, we would have become a single amorphous mass and probably, we would not have survived. Everything in evolution serves its purpose and has its time. One phase gives way to the next. The world we see around us today is a direct product of our genetic fear of each other — with all its beauty and its horror. The time has come for the next shift. With the mutation of the 55th Gene Key underway, the 59th Shadow is being deeply undermined. In many ways, our fear of each other may seem to be on the increase as it is brought to the surface for all to see.

Everything about the Dishonesty inherent in this 59th Shadow *must* now be brought to the surface. The 59th Gene Key cuts deeper than race, belief or creed. It goes deeper even than the individual blood ties of family. It binds humans together as a single interrelated genetic family. The world of today, with its differentiated tribes, societies, nations and boundaries is at the very beginning of being utterly transformed. This transformation will not come about through social or economic revolution. It will not come about through the love of great charismatic leaders. All of these things may and probably will be keys in the transformation of the world, but the foundation lies here in the 59th Shadow. Fear will gradually be bred out of human beings. Without our fear of each other, the world will change dramatically. The greatest change of all will be to our human sexuality. We are on track to become an androgynous species in which the separate sexual poles will come together within the individual human being.

As the 55th Shadow demonstrates, we are trapped by our sexuality and by our animal nature. We are trapped by our inability to emotionally handle the collective chemistry engendered by our fear. This fear rules our systems of government and education as well as our relationships and individual lives. The 59th Shadow is rooted in our inability to come clean with each other. It is the source of all hidden agendas. At the moment, what we need to understand is that our world is undergoing a genetic transmutation and we are guinea pigs in the global genetic laboratory. All genetic material is currently being sifted and sorted. Any behaviour traits that are rooted in the old ancestral fear will be discarded. They will simply disappear from the human genome. This means that there are now personality types leaving the world, never to return. The old ways are having their last fling in the world, and we can see this in the battle between the collective way of the future and the old tribal ways of the past. This does not mean that all elements of the past will be lost. It simply means that all elements of behaviour rooted in fear or hidden agendas will be fused into a higher functioning of our DNA. One has to be able to read between the world headlines. This is not a time of the good guys fighting the bad guys as the politicians would have us believe. This is not about gene pools anymore — it is about behavioural integrity.

The genetic cleansing that we are currently experiencing on a global level will very likely go on for a number of centuries. It has to work its way out of our DNA through our bloodlines. However, as the human population decreases due to our mutating sexuality and the lessening of global fear, the world will naturally become a quieter and more peaceful place. The so-called *Judgement Day* is really taking place within our genes. No individuals are punished or rewarded. There is simply an increase in genetic material with a holistic perspective coming into the world, and there is a gradual eradication of genetic material that is self-destructive and isolationist.

On reading all of this, you may well ask the question: "Well what do I do with this information? Is there any point in my trying to apply any of this if it is already occurring as an evolutionary quantum leap?" This leads us to an interesting paradox. Is our collective behaviour influencing our genetic paradigm shift, or is the genetic change influencing our behaviour? The evolutionary or spiritual activist would opt for the former view whilst the scientist might opt for the latter. As always, both sides of the paradox are true. Each gives rise to its opposite. There is a spiritual force involving down into form that is causing our genes to mutate, and there is a genetic force evolving upwards, causing our behaviour to become more spiritual. The answer to the question of what to do lies in contemplation of the 59th Shadow.

If you move against evolution at a time such as this, you will meet an opposing force whose power is incomprehensible.

Since the world is evolving naturally in a particular direction, your behaviour will either flow in that direction or oppose that direction. If you want to hitch a ride on the currents of evolution you would do well to align your agenda with its program. You must begin to look into your fears and come clean with your hidden agendas. One of the main reasons for the writing of this book is to encourage more of us to look into the eyes of our inner demons and coax out these ancient fears. It is time to take out the genetic laundry. Wherever you are dishonest — with yourself, in your relationships or your work — there you must look with eyes open and unflinching. You must begin to work with the higher frequencies of this 59th Shadow to unlock its Gift. One thing is sure, if you move against evolution at a time such as this, you will meet an opposing force whose power is incomprehensible. Such a power will ultimately destroy you as it is destroying all such behaviour rooted in selfishness and separation.

REPRESSIVE NATURE – EXCLUDED

The repressed nature of the 59th Shadow is about feeling excluded. Feeling excluded is a classic example of the *victim* state as it blames others rather than taking responsibility for its own feelings. These people unconsciously exclude themselves out of a deep-seated fear of losing control. When you open your doors to others, then you always lose control. Such people actually manage to draw negative attention through their self-exclusion. Furthermore, when you are excluded you can maintain your defence mechanisms intact, even though you feel bad. The feeling of being excluded can actually become an addictive state. It gives you the illusion that you are still in control of your own emotional environment. However, the moment you do choose to include yourself, the entire illusion is instantly shattered.

REACTIVE NATURE – INTRUSIVE

The other side of the 59th Shadow is based on anger. The fear of being excluded is not repressed here but is expressed reactively as anger, which then becomes intrusive. These people try to push themselves in the door, physically and emotionally invading another person's aura. When they meet rejection, they become outraged and exclude themselves. It is the same strategy as the repressive nature, in that it tries to maintain control of the emotional environment through blaming others. The difference here is that the intrusive nature experiences rejection actively, whereas the repressed nature experiences it passively. Intrusive people can be addicted to unhealthy emotional relationships, so long as they find a willing victim to play their game. Such people will try to dominate their relationships in order to avoid rejection, which means that they can never stay with anyone who is honest enough to challenge them.

THE 59TH GIFT – INTIMACY

SUBLIMATION AND THE SERPENT

The transformation of the 59th Shadow into the 59th Gift of Intimacy is the subject of many ancient traditions and prophecies. Since the 59th Shadow represents the unbridled fertile power of your animal sexuality, then by raising its frequency you unlock the transcendental power of sex. For millennia mankind has sought to resolve its many issues around sexuality. For our religions in particular it has proved to be an enduring problem. Everyone knows that when sexuality is repressed, it can become an all-consuming power ultimately leading to distortions in human behaviour that are inherently unhealthy. The pure genetic force of pressure in this 59th Gene Key causes great personal and social dilemmas for us humans. When the ancient Chinese I Ching sages named the 59th hexagram *Dispersion* they knew exactly what this force was capable of. Its only real interest at the Shadow frequency is to disperse itself as widely and frequently as possible.

The 59th Shadow has been playing havoc with human attempts at monogamy for a long time. Indeed, women and men react very differently to the fear inherent in this Gene Key.

The traditional male reaction is to try to escape being trapped by a single woman (this fear of being trapped is the polarity of the 55th Shadow). The female reaction is to try to hold onto the male aura because of its promise of protection for offspring. Women intuitively understand the male's urge to disperse his genes widely. At the Shadow frequency a woman reacts by trying to keep the male as close as possible, thereby creating the opposite reaction in the male — that of wanting to escape and be free. Thus goes the age-old battle of the sexes.

Hidden within the 59th Shadow however is a deep evolutionary urge to transcend the lower frequencies and give birth to a higher form. Perhaps the deepest archetypal symbol of sexuality is the figure of the serpent or dragon, and within this alchemical symbol hides the key to the 59th Gift. Serpent or dragon energy represents the evolutionary urge to transform oneself, reflected in the serpent's shedding of its own skin. This sexual power in humans has always held the promise of higher states of consciousness and we have devised many techniques and systems that attempt to utilise it. These meet with varying and limited success, because although the evolutionary urge to transform is there within us, the timing for this mechanism cannot be forced. Nature always flowers in certain individuals early, just as certain flowers in the garden bloom first. However, to try and force our sexual flowering before it is ready can be dangerous. It will come in its own time.

As the 55th Gene Key mutates the human solar plexus system, it will allow the higher functioning of your sexuality to emerge. This will be a permanent functioning inherent from birth. To understand how this process is occurring, we must look in more depth at what happens to sexual power as it hits the Gift frequency and manifests true Intimacy. Intimacy may sound like a soft and cuddly kind of state, but the reality of the 59th Gift of Intimacy is somewhat different. Intimacy presupposes honesty and acceptance of the Shadow state. This means that the fears between the sexes must be acknowledged, understood and allowed to exist. This *allowing* then acts like a valve that opens up the full power of the sexual force, a shattering power that shakes the very fabric of your being. This is why it frightens us so — because it is such a raw power. What we don't usually realise is that the power of our sexuality, if embraced fully, will open us to a higher state.

The sexual power of the 59th Gift is highly contradictory; it is both creative and destructive. At the level of the human aura, it destroys any interference patterns that appear to separate people. In other words, it causes your identity or *ego* to be dissolved. As the core of sexuality across our planet, the 59th Gift has a deep underlying non-linear pattern. When modern science first explored Chaos theory, it began to understand the 59th Gene Key. The very nature of sexual energy is wild, organic and untameable, but it also follows universal patterns even within its chaos. The sexual force is a spiralling force, which is why all living creatures have similar foundational spiral geometries. This force creates what are known as fractal geometries throughout nature — self-similar systems that are all interlinked yet at the same time utterly unique. These creative transformational patterns and geometries are also found within human relationships. The power of human intimacy is the power of two human auras interacting to create a third aura, whilst in the process the two original auras are sublimated. The more differentiated the original auras are, the more potential there is for transcendence.

As this heart opening occurs, true intimacy is born and two people meet within a single awareness.

One always needs two opposite poles and the coming mutation of the human solar plexus system will essentially create a cauldron whenever two such poles meet. Within that cauldron the auric fields of the two parties are chaotically interfused and through an alchemical process a new awareness dawns above the solar plexus centre. The actual culmination of this process does not occur in the solar plexus area but in the heart. In traditional esoteric systems, we tend to view the centres of force within the body as separate as opposed to being part of a single system. The heart centre is in fact the higher glandular functioning of the solar plexus. As this heart opening occurs, true intimacy is born and two people meet within a single awareness. It is important therefore to grasp the true nature of the process known as sublimation of sexuality. It is a shattering process in which chaos must be experienced before the higher emergence is perceived. Since this process is currently occurring throughout the entire human gene pool, what we see in the world today may be understood as a reflection of this chaotic breaking down of the world aura. Only when it has reached a critical zenith will we begin to see the birth of the third awareness. This will be reflected in the material plane through the birth of a new kind of human being whose single purpose is to house this new awareness.

THE 59TH SIDDHI – TRANSPARENCY

THE RETURN OF QUETZALCOATL

The final phase of the evolutionary impulse latent within the 59th Gene Key will probably not occur on our planet at a collective level for many thousands of years. Having said that, the nature of time itself will change so radically that the whole notion of things taking a long time will cease to have any real meaning for us. The mutation of the solar plexus essentially equates to the opening of our planetary heart centre, and as most of us know, time stands still when you're in love! Beyond this transformation out of the Shadow and into the Gift lies a further transformation in which genetic mutations will take place within the brain chemistry of the pineal and pituitary glands. This higher process will lead to the dawning of the 59th Siddhi, the Siddhi of Transparency.

Transparency as we understand the word today is not even close to what it means at the siddhic level. Today when we speak of being transparent, we mean being open and honest in our communication with others. The Siddhi of Transparency actually involves the dispersion of all aspects of the self back into the sea of creation. In mythical terms, the 59th and 55th Gene Keys represent the force of prime yang and prime yin respectively. The 59th Gift is the seed or semen and the 55th Gift is the egg. When the frequency of this Gene Key reaches the siddhic level, these kinds of symbols cease to have meaning since they no longer function through duality. The semen of the 59th Gene Key is spent and the egg is no longer needed. What you have left is a state beyond description and beyond evolution. Transparency is what remains when evolution becomes meaningless. At a genetic level, the 59th Siddhi cannot exist within the physical body because it is the driving force *behind* evolution.

The state of Transparency carries no agenda or purpose. It simply acts as the conduit for awareness. Thus, the only purpose of the 59th Gene Key is to break down all barriers that stand in the way of union. When this has happened, this aspect of our genetics will no longer be needed.

The 59th Gene Key is the master key in the genetic codon group known as the Ring of Union. These four Gene Keys — the 4th, 7th 29th and 59th — have long governed human relationship patterns (and their dysfunction) on our planet. Their ultimate role is to bring the human family into realisation of its state of higher union. The sequence leading to Transparency begins with the process of Forgiveness, triggered by the 4th Gene Key. Once you have forgiven yourself, using your relationships as a mirror, you will discover your true aura of Virtue (the 7th Gene Key). Once you have found your inner virtue you begin to pour yourself out into others, recognising the same Divine source in every person you meet. It is this Devotion (the 29th Gene Key) that is the true doorway to Transparency. Devotion empties you out, bringing you into a purified higher state of awareness where you no longer relate to people or objects as *other*. Transparency brings everything inside your own being, eliminating the very concept of inside and outside. It is the greatest leveller within the whole sphere of life.

In nature, the 59th Gift is represented by the golden mean — a universal geometric form underlying all creation. However, if one tries to create a spiral out of the golden mean to represent this Siddhi in its entirety, one immediately meets a well-known mathematical impasse because such a figure would have no beginning and no end. This is representative of the 59th Siddhi — it cannot appear in nature or be understood by mathematics. It simply does not appear to exist, hence its Transparency. You can perhaps see why Transparency taken to a collective level entails the disappearance of the form of our universe. When this does eventually occur, as it must, it is difficult to say what form humanity will take, if indeed it takes any form at all. If humanity were to take on some kind of future form, it would certainly no longer be an outgrowth of evolution as we understand the word. Neither would it have the need to reproduce itself since its being would be eternal.

In the times ahead — and even perhaps today — the 59th Siddhi is imminent. Beings will come into the world manifesting the full power of this Siddhi. Just as mass consciousness will begin manifesting the 59th Gift of Intimacy as the norm, so a few will begin to manifest this most rarefied of vibrations. These will be people who have gone beyond all concepts of agendas.

The ultimate agenda is evolution — the idea that there is still some scope for growth. Although the form of our world will continue to evolve, the consciousness out of which it emerges can never and has never evolved. Only awareness can evolve and it is evolving very quickly through the transformations and mutations coming through this Gift and Siddhi. However, when awareness has reached its ultimate expression it becomes a clear mirror for consciousness, and this is what Transparency is. It reflects life without any additional comment. It is a state we humans have always dreamed of attaining — the ability to look upon the world without judgement.

There is an ancient Mayan prophecy that speaks of the return of Quetzalcoatl — a being symbolised by the feathered serpent. The ancient calendars have long connected this event to coincide with the year 2012 at which time, according to the Maya, time will cease to exist. Such prophecies are perhaps best not taken literally, but seen as alchemical markers emerging from the collective unconscious. The journey of consciousness through the 59th Gene Key is symbolised aptly by the figure of Quetzalcoatl. The feathered serpent is the same as the symbol of the dragon — it represents the harmonisation of the lower nature (the serpent) with the higher nature (the bird). This is indeed the new epoch we are now entering.

Most scientists agree that our genes have a hidden agenda to survive and that, no matter what we do, they will find ways to mutate if their ongoing survival is threatened. Genes operate below our conscious awareness. You might even say that God is hidden within our genes, but until we raise our frequency, we cannot experience God. We therefore must learn how to be transparent — both with ourselves and in interaction with others. Our first step is to let our barriers down and look deeply into our fears. Even our deepest genetic fears — such as the fear of losing our loved ones — act as subtle barriers that keep us from realising that our genes are not exclusive but inclusive. Every man and woman is our genetic brother and sister and no single individual, family, tribe or nation is an island. We are all one genetic tribe moving through a huge and transformative period in history, the end result of which will be the realisation of our unity.

We are all one genetic tribe moving through a huge and transformative period in history, the end result of which will be the realisation of our unity.

60th Gene Key

The Cracking of the Vessel

SIDDHI
JUSTICE

GIFT
REALISM

SHADOW
LIMITATION

Programming Partner: 56th Gene Key
Codon Ring: The Ring of Gaia
(19, 60, 61)

Physiology: Colon
Amino Acid: Isoleucine

The 60ᵀᴴ Shadow – Limitation

Closed-Circuit Thinking

The 60ᵗʰ Shadow represents one of the most potent forces responsible for pulling humankind in the opposite direction of evolution. It is the power of devolution and limitation and is the counterweight to life itself. When one consistently yields to its Shadow frequency, this Gene Key catalyses death. The 60ᵗʰ Shadow will shut down your life-receiving faculties, essentially freezing life into ever tighter patterns over the course of time. In our modern world, the 60ᵗʰ Shadow is apparent everywhere — it appears wherever innovation and imagination are stifled by red tape, or wherever men or women seem to have forgotten what it means to be human. The very fact that the greater proportion of the world population still lives in poverty while a select few thrive is due in no small part to the 60ᵗʰ Shadow.

Up until now in our human evolution, all of our dreams of a harmonious world have been thwarted by this one single force — the Shadow of Limitation. It sometimes seems that no matter how good our intentions are, there is a force that appears bent on keeping them from manifesting. But this 60ᵗʰ Shadow is also a great collective release code for higher harmony. It will not yield to force but waits for one thing — time. All of life follows cycles within time. This was the great revelation behind the original I Ching. Its 64 symbols or *hexagrams* hold the sequential codes for the unwinding of nature's great plan. Unless you can escape its frequency, the 60ᵗʰ Shadow will prevent you from seeing that life has a built-in mechanism for righting itself. At the Shadow frequency this will cause you to distrust the flow of life. Sometimes the flow of evolution appears to stand still or get *stuck* and this manifests as the human inability to break out of set structures and patterns.

The 60ᵗʰ Shadow and its Gift are about structure. The law of all form is that no structure can last — each and every one of them comes programmed to eventually decline and die. This alone is food for much insight in your life. The 60ᵗʰ Shadow concerns the human over-reliance on structure and the consequent death of something very special indeed — magic. Magic refers to events that do not follow logical, sequential laws.

*The 60ᵗʰ
Shadow is
the greatest
critic of
magic.*

Magic is spontaneous, highly mutative, unpredictable and uncontrollable. Its greatest quality is that it is beyond meaning or understanding. The 60th Shadow is the greatest critic of magic. Its single purpose is to control the flow of life and prevent anything original from occurring. One of the places where we see a very clear representation of the 60th Shadow is in the realm of human laws. We humans create laws that govern our societies and then we put in place sophisticated systems of legislation to implement those laws. The original purpose of our laws was to protect the innocent and ensure that justice is delivered to those who transgress these laws. However, the structures themselves often become so bogged down in their own limitations that they cease to operate efficiently or even fairly.

In the world today, the single greatest limitation to creating a universal society based on peace is red tape. There are so many laws in place in so many societies that they prevent us humans from seeing or behaving outside our little boxes. The laws themselves may be social, moral, religious or economic. The programming partner to the 60th Shadow is the 56th Shadow of Distraction, and it is easy to see how these two paired Gene Keys conspire to blind human beings from actually doing what it is that we really want to do. We are too bogged down in our own structures, and we become so distracted by these structures that we actually forget our original intentions. The world economy is one of the greatest structures of limitation on our planet. Money is an enormous limitation to humanity. Its existence requires so many laws that bind us and control our behaviour. As long as money exists on our planet, we will continue to ensnare our species within the ultimate material limitation. Until money is finally eliminated, true material freedom can never be experienced.

All structures eventually dissolve and falter. This law can be seen throughout history — it can be seen in the inevitable decline of every great empire that has passed into form. It is important to understand that it is always the structure itself that causes the decline. Structures are an essential part of life. However, all structures are expendable. The body itself is an expendable structure, as is the earth. In order for consciousness to keep penetrating the world of form, structures will always be necessary until the form itself has been transmuted. This is the key that we humans have to remember. The 60th Shadow creates an over-attachment and over reliance on form instead of highlighting the indwelling spirit or idea. The highest aspect of this 60th Gift is the notion of universal Justice — the 60th Siddhi, and you can see precisely what happens when man tries to capture this beautiful concept in a material structure — how easily it becomes perverted and how frequently it gets bogged down in legal small print.

The other side of the Shadow of Limitation is about the past. Whenever and wherever people remain stuck because of an outdated mindset, the 60th Shadow is at work. This is a Shadow that abhors the new, the innovative and the original. It is the nemesis of change and youth. This is an archetype deep within human DNA that has always tried to control and suppress the youth of the world. Many of the rigid structures of our modern school systems ensure that young minds are structured, homogenised and packaged from a very early age. One of the great limitations for children is to be forced to sit indoors behind a desk at an age when all their bodies want to do is run and explore life.

This early imposition of structure on the developing mind, body and emotions of children has created a backlash of reactionary behaviour from children at all levels of society. You cannot imprison youth without serious repercussions coming back to haunt you.

One of the greatest areas you can see the true power of the 60th Shadow is in religion. The moral laws imposed on our society through our religious systems are some of the oldest on the planet. Cosmic moral laws exist without the need for systems or structures to implement them. The more we humans enforce laws, the more reaction there is, and consequently the more we have to police them. This is the classic feedback loop of the 60th Shadow. It is a self-perpetuating nightmare of control and reaction. The 60th Shadow will insist for example that every word of the Koran or the Bible is to be taken literally. Strict adherence to these kinds of ancient codes or laws is deeply limiting and restrictive to the human spirit. We do not need to be told what is right and what is wrong. We have the means within us to determine this for ourselves. Any code or system that cannot be questioned or adapted to suit the times is a devolutionary force and is thus destined to eventual decline.

Of course the greatest limitation of all is human thinking itself. This is where the 60th Shadow tries to strangle life at the individual level. Because of our cultural conditioning, your mind gets used to thinking within certain mental structures and you become comfortable within those structures. There is a direct chemical link between the 60th and the 61st Gene Keys through their associated codon ring, the Ring of Gaia. Inspiration (the 61st Gift) is what occurs when your mind momentarily breaks out of its habitual structures. Most people, for example, do not think magically; that is to say, they do not leave the door ajar within their brains so that Inspiration may creep in at any moment. The main reason for the death of magic is closed-circuit thinking, where the mind follows familiar patterns or systems of thought around and around without the ability to think beyond them. It is out of this closed-circuit thinking that all self-limiting forms are ultimately brought into the world.

REPRESSIVE NATURE – UNSTRUCTURED

When the 60th Gene Key is repressed, the result is a deep lack of structure in a person's life. At a cellular level this can actually cause physical trouble or disease because such structure holds together the physiological functioning of the organism. The lives of such people never seem to be harmonious as they are out of kilter with their true nature. These people need strong structures in life — for example, family, career and a sense of direction. Without these kinds of structures, they are adrift in the world and their potential seems to be wasted. They never seem to be involved in anything that lasts, but keep changing their circumstances. They are afraid of structure and commitment, and this fear prevents them from finding the correct allies or environment for their gifts to bloom.

REACTIVE NATURE – RIGID

When the 60th Shadow is expressed through an angry nature it becomes deeply controlling and rigid. These people cannot allow others to question anything they say or do. They behave as though they were beyond reproach, and if anyone crosses them, they lash out in anger. The reactive nature of this Shadow creates an over-reliance on form and structure to the detriment of the original spirit or idea within the structure. These people assume that any new idea or way of doing things is a threat to their security and they react accordingly. They do not have any real understanding of relationships since they cannot allow others to transgress their own behavioural codes or opinions. Unless they are able let go of their stranglehold on life, they will eventually shrivel up and decline.

THE 60TH GIFT – REALISM

THE COMMON SENSE OF MAGIC

People with the 60th Gift are very sought out in the world, and for one main reason — they understand the limitations of the world of form and therefore know the laws of manifesting *within* the world of form. This is what Realism refers to in the context of this 60th Gift — it is the ability to balance youthfulness and wisdom — idealism with structure. It is all very well having a wonderful idea that will change the world, but without the realism of the 60th Gift, this idea will probably never amount to anything. Another synonym for the 60th Gift might be the gift of common sense. Common sense is inherent to all human beings, but it is lost whenever a person or society becomes too bogged down in frameworks and structures. It is through common sense that we human beings evolve and cooperate with creation efficiently and harmoniously. The power and potential inherent in this 60th Gift should never be taken for granted.

People with the 60th Gift understand that for anything new to take root in the material world, it must follow certain laws. The main law of manifestation in the material world is based on structure. Innovation without structure simply will not last. The seed needs its hard shell to protect it and the river needs its banks to direct it. The 60th Gift is about creating strong banks to direct and funnel the energy for change into the world. To be realistic in the sense of this 60th Gift means to understand that there is a great deal of red tape in the world and this has to be tackled. The classic metaphor for the role of this Gift is that of grafting a vibrant young shoot onto a powerful old rootstock. In this metaphor the 60th Gift is the rootstock, which represents the ability to work with the existing laws and traditions of the world and introduce change *through* them rather than in spite of them.

People with the 60th Gift are masters of creating structures, whether mental, emotional or physical. These structures in themselves are holding stations for new ideas and energy, and as such are of less importance than the energy that comes through them. This is what the 60th Gift knows — it knows that the structure will eventually mutate or even die, but the inner spirit will move on and if necessary find a new structure to carry it. The 60th Gift is like the spaceship that takes man to the moon. Once man has reached the moon, the spaceship is no longer needed, but its creation allows other more advanced structures to evolve.

To take this simile further, the Gift of Realism is about maintaining the balance between the idea of reaching the moon, and the huge physical and economical requirements needed to actually make that idea a reality. But the real essence of this Gene Key is about not losing perspective from either the idea side or the structure side. We have already seen how an over-reliance on the structural side ensures that most ideas are either bogged down and never happen, or else end up far from their original vision.

As we have also seen, there is a great capacity for magic in this 60th Gene Key and some may find it ironic that it is called the Gift of Realism. Actually realism in its truest sense always involves magic. We know from modern quantum physics that all matter is made up of vibrating energy fields. Therefore all structures in the world of form are really an illusion. They are frameworks through which something unexpected may occur. The only thing needed for magic to occur is some form of a structure and an open mind! This may seem a surprisingly simple set of criteria but it is rarely found in the world. We usually see people burdened by the weight of the 60th Shadow — that is, people toting great impressive structures and systems but so identified with and attached to their structure that it throttles the original idea. We humans are not comfortable with uncertainty and we do not trust in magic, thus we cling to our systems, religions, laws and modes of thinking.

One of the subtlest of structures is language itself. People with the 60th Gift can therefore be masters of language, as long as they do not become trapped by it. Language is not the territory, but is the means by which change can be expressed. You can talk about change forever and become obsessed with the ideas and thoughts themselves. But in order for something really new to enter the world, language has to be used in a playful manner, as a means of expressing a frequency. It is the frequency that holds the energy of change. Language is simply the means of resonating the music. We all know that in popular magic, the magician distracts us and then performs his trick when we are looking the other way. In the same way, the 60th Gift can use language or indeed any form as a means to distract people while the real energy passes into them unnoticed.

This 60th Gift has much to do with music, being acoustic in nature. All new forms that enter the world have to pass through this Gene Key. Human chemistry is musical at the deepest level, and humans experience this through their individual moods and energy swings. The 60th Gift requires deep acceptance of the uncertainty and unpredictability of the rhythms of life. These people understand that natural periods occur when nothing seems to be moving. Such periods can come on suddenly and go just as suddenly. They give rise both to our mood swings and to our bursts of sudden manifestation. The 60th Gift knows that there is magic in the darkness before manifestation and knows not to interfere with these essential life processes. To be realistic means to accept the natural limitations of being in form without being a victim of them.

The 60th Gene Key is a part of a trinity of Gene Keys (along with the 19th and 61st) that code for the amino acid isoleucine. This is the Ring of Gaia, one of the most fascinating of all the 21 codon rings. The Ring of Gaia prevents or allows awareness to move between all the

The only thing needed for magic to occur is some form of a structure and an open mind!

different life forms on our planet. The 19th Gene Key with its Gift of Sensitivity has the potential to open up a higher genetic functioning in human beings that will allow us to directly experience what it is like to live inside another creature. At the height of its sensitivity, this codon allows all earth life, both sentient and inanimate, to experience its quantum unity. The 60th and 61st Gene Keys and their Gifts of Realism and Inspiration hold deep secrets concerning this planet and our role within its organic structure. The 60th Gene Key represents the pure material density of the form of Gaia herself, whereas the 61st Gene Key points to the magic deeply embedded within the heart of the earth. The original name for the 61st hexagram in the I Ching is *Inner Truth*, and this is the beauty of the Ring of Gaia — everything, from the tiniest nano-particle to the cosmos itself, has the same shining jewel of inner truth at its core.

It is the responsibility and great privilege of humanity to be the outermost peak of the awareness of Gaia. We are her eyes and ears. We are her very mind. Everything of true value is hidden inside form — inside your body, inside the ferrous core of the planet, inside the vibrating fabric of your inner being. It is all rooted inwardly. This is why humankind must discover inner space, why we must turn inward for inspiration. All the answers to all our problems and challenges are hidden in the creatures and structures of nature, and all of those structures and creatures are embedded in microcosmic form inside every molecule of our DNA. Common sense is not in opposition to magic. It is common sense to remain open-minded to everything in the universe, because everything has at its core the same wondrous inner light.

THE 60TH SIDDHI – JUSTICE

THE EARTHSHIP MERKABA

The 60th Siddhi is an extremely rare Siddhi. There are some things in life that are quite simply a mystery, and this 60th Siddhi is one of them. It is called Justice but the true meaning of this word in a siddhic context is utterly different from our usual interpretation of it. As we have seen already, the 60th Gene Key concerns the laws and limitations that govern the world of matter. At the Gift level, there is the intellectual understanding that these laws are not quite what they seem (reflected in the modern scientific view of quantum physics) and that there appears to be a force operating outside the laws governing physical reality. This force may be called God, grace, magic, fate or even chance, but the fact is that it lies beyond our human control.

When a human being attains a siddhic state, it is a rare event. Because we are all connected through the universal field of consciousness, this event has repercussions for all sentient life. It doesn't matter whether the person in this state ever even sees another human being — their realisation creates a great swell within the ocean of human consciousness and the waves from this high frequency surge affect all human beings. Depending on the flavour of the Siddhi, that is, its specific coding, the upsurge in consciousness affects our universe in different ways. Someone who attains the 25th Siddhi of Universal Love, for example, activates a huge opening in the collective heart-field that surrounds and connects all beings and creatures on our planet and beyond. Such an event will lead to all kinds of breakthroughs in the human world of form.

When a being attains self-realisation through the 60th Siddhi, something quite extraordinary happens (and to the Shadow consciousness, something very frightening). The fundamental laws that govern physical existence are loosened and, in some cases, broken altogether. At a collective genetic level, this kind of event is called a *frameshift mutation*. When it occurs, the reading frame within the whole of human DNA shifts so that the way in which the genetic code is translated is totally different. At a macrocosmic level, this has not happened very often in human history, and for this reason this 60th Siddhi still remains a mystery. True Justice realised at a Divine level is not a phenomenon human beings can easily understand. However, the coming of Justice is lodged within the collective psyche, being reflected through the 60th Gene Key and its chemical counterpart within your DNA. Thus almost all cultures have myths or religions promising a day in the future when the Gods will descend to earth and pronounce their judgement on all living things. This myth, embodied, for example, in the Christian notion of Judgement Day, is a jaded human interpretation of the power of the 60th Siddhi — it is what we understand as Divine Justice.

Because we humans think in terms of our morality, we generally see justice in terms of punishment and reward. More often than not, we see it in terms of retribution. Even our most sophisticated spiritual and scientific systems think of life in terms of cause and effect or karma. This is due to the limitation of being caught in a dualistic reality. However, the true concept of justice is not bound by laws at all. It could best be described as *acausal*. When something is acausal it means that it happens for no reason that we can see or understand. In an acausal reality, causes do not lead to effects and thus there can be no justice as we understand the word. If someone were to kill someone in an acausal world, they would never be punished. There would be no such thing as duality; therefore no separate person would exist to commit such a crime. How then are we to imagine such a reality? The answer is that we cannot.

The 60th Siddhi really does represent Judgement Day. It is the end of our illusions of right and wrong, good and bad, fair and unfair. The 60th Siddhi breaks all the laws of our reality, beginning with the law of time. Whereas we see time in a sequence, the 60th Siddhi experiences all time in the present. This means that there is no event horizon or sense of linearity, meaning that the person manifesting this Siddhi can therefore travel through *time*. The second law that the 60th Siddhi breaks is the law that holds form itself together — what we call gravity. In other words, these people are able to travel not only through time, but also through space. The laws of gravity that hold atoms together and maintain

The laws of gravity that hold atoms together and maintain order within our cosmos are completely smashed by this Siddhi.

order within our cosmos are completely smashed by this Siddhi. Our popular mythologies contain many stories of eyewitness accounts of people flying or beings from other worlds coming to earth. Some of these accounts may be taken from manifestations of the 60th Siddhi. Many of the great esoteric schools — among them the Tibetan, the Taoist and the Egyptian — have detailed accounts of masters who attained the *body of light* or the *rainbow body*. These are examples where the laws that govern our world of form have been broken, and this manifests through a mutation of the physical vehicle of the body.

In India there is a mythic Himalayan yogi known as *Babaji*. Many myths surround this man who is said to appear from time to time across the centuries, materialising in form, appearing, then disappearing. Other cultures have similar legends and stories that fire popular imagination but are seen as superstition by the general consciousness. There is enormous magic hidden within this 60th Siddhi. It is true magic because it is acausal. It cannot be learned, mastered or imitated. Every time a being attains enlightenment through the 60th Siddhi, magic comes pouring through this being and they become a phenomenon, a symbol of the breaking of the laws of form.

At the collective level, the 60th Siddhi will be one of the very last Siddhis to dawn. This is why we have always interpreted it as Judgment Day. When the 60th Siddhi begins to ignite in multiple beings, it will be the ending of the world as we know it. Everything will indeed be revealed in its true sense. The laws that hold together the earth will crack and our planet will begin to transmute. The 60th Siddhi has been known by certain esoteric schools as the *merkaba*, or chariot of light. The true collective merkaba is the earth herself, the spirit of Gaia. When the earth disappears, it will do so suddenly, and all the cells that have held the consciousness that we call Gaia will move to another reality. Only then will we really know what Divine Justice means, and it will have no meaning whatsoever on an individual level. We have absolutely no idea what this event will look like or feel like, but those with the 60th Siddhi hold the key to this ultimate magical occurrence. And we should remember that from the point of view of the 60th Siddhi, this has already occurred. Every human being knows this is our future, as it lies secretly encoded in the deep limitation of our material DNA. However, certain Divine laws conceal those higher aspects of our own nature from our view. We have to finally evolve to be able to see them, and the two great laws that we must evolve beyond are gravity and time.

These two laws are not absolute but completely relative to awareness. The binary helix of our DNA is microcosmically composed out of these two laws. DNA holds memory and it acts as the living, evolving, local structure for sentient life. Evolution demands time and localisation demands space. When awareness passes beyond these two laws, they are seen not as false but as lower frequency signatures of higher Divine laws. Beyond our human form other realities and dimensions exist. The cosmos is teeming and abundant with inter-dimensional life forms. In more advanced forms of life the mirror of DNA is no longer a binary structure but a trinary helix. The triple helix unites all beings beyond space and time and frees awareness to travel unimpeded throughout the cosmos. However, the triple helix cannot support a carbon-based life form. It requires a far subtler structure or vehicle — the underlying geometry of the universe itself.

▤ 61st Gene Key

The Holy of Holies

SIDDHI
SANCTITY

GIFT
INSPIRATION

SHADOW
PSYCHOSIS

Programming Partner: 62nd Gene Key

Codon Ring: The Ring of Gaia
(19, 60, 61)

Physiology: Pineal Gland

Amino Acid: Isoleucine

THE 61ˢᵀ SHADOW – PSYCHOSIS

THE PRESSURE OF WHY

A standard dictionary definition of the word *psychosis* might read: "Any kind of mental defection or derangement." Generally seen as a mental affliction rooted in chemistry, psychosis is associated with disorganised thinking and delusional beliefs. Perhaps its key hallmark is that the individual so affected is usually unaware of their affliction. There are degrees of psychosis from mild to acute. The psychoses that are recognised are the more acute forms in which human beings are unable to function in society without endangering either themselves or others. However, what the 61ˢᵗ Shadow will show is that psychosis, which entails a loss of contact with reality, is the background consciousness of almost all human beings, including those that purport to understand the nature of the malady itself. At the risk of shocking the psychological community or indeed the wider community, psychosis is in fact the ordinary state of the mass consciousness of humanity in the world today.

In the 61ˢᵗ Shadow the true nature of the mind is obscured. The true nature of the mind is described, as much as it can be described in words, in the 61ˢᵗ Siddhi of Sanctity where it is seen as emptiness, or perhaps more clearly as infinite space. However, the mind as we know it today is sick, and it is sick for one reason alone — because it seeks itself. It is through the narrow gate of the 61ˢᵗ Gene Key that a great question permeates the world of human beings, the question of *why*. This question *Why?* is in fact an aberration caused by the limitations of our current awareness. It is also this question that has powered the direction taken by humanity for thousands of years and in particular over the past few hundred years. There is a tremendous pressure within the human brain to find an answer to this question that continues to well up from the deepest reaches of our unconscious.

Both right and left hemispheres of the brain grapple with an answer to this question but both are ultimately destined to fail. The right hemisphere seeks to end the pressure through religion whilst the left hemisphere seeks to end the pressure through science.

*Psychosis is
in fact the
ordinary state
of the mass
consciousness
of humanity
in the world
today.*

The key to our failure to relieve the pressure is found in the programming partner of the 61st Shadow, the 62nd Shadow of Intellect. The pressure from the 61st Shadow is neurologically and chemically routed directly into the human intellect, which is what turns the pressure into a question in the first place. The word *why* is the first word of all words, and out of it has been born all the different languages of humanity. This pairing of the 61st Gene Key and the 62nd Gene Key is a very mystical pairing.

Until humanity's final evolutionary leap takes place — the quantum leap triggered through the 55th Gene Key — we will remain in a state of psychosis. Our minds will continually drive us consciously or unconsciously to seek an answer to the question, or try to find release from the pressure behind the question. The arc of human evolution in modern times has been fuelled by this question, so you can see what a lofty purpose it serves. And yet it produces a deep dilemma — it prevents us from actually experiencing reality. As with all psychotic states, we are unaware of our own psychosis. Even those sensitive people who do become aware and can see an inevitable human slide into self-destruction are caught within the dilemma. The dilemma is that you cannot fix the psychosis from within the psychosis. The ancient civilisations called this perceptive cloak thrown over the world the *maya* — the great illusion.

The issue of the true nature of reality is all about frequency, just as this entire work on the 64 Gene Keys is about frequency. Only when the frequency passing through your individual genetic coding rises to a sufficient level can you begin to perceive reality, or as William Blake famously put it:

"If the doors of perception were cleansed, everything would appear as it is, infinite."

This is aptly demonstrated through the shift in perception that occurs between the 61st and 62nd Gene Keys at the Gift levels of frequency. At the very highest levels beyond that, reality is indeed experienced in all its infinite splendour. However, at the frequency level of the mass consciousness of humanity today, the 61st Shadow prevails and humanity is simply a victim of the pressure of this Shadow. We will do anything and believe anything or anyone who promises to deliver us from the pressure. Thus because it promises us respite from the pressure within our minds, religion is one of the biggest businesses on our planet.

The problems with religion are not usually sourced by the founders. Those rare beings who attained the highest frequencies in the past spoke directly from the siddhic reality. But the 61st Shadow holds a single great flaw, the flaw of worship. The moment you worship another, you set yourself below them in a victim position. You guarantee that your frequency will remain at that low level and be severely impaired. You need to understand here the difference between worship and devotion. Devotion contains the seed of individual dissolution, whereas worship requires a basic duality between you and your God. The only way to release the pressure of the 61st Shadow is to step right into the heart of the pressure rather than constantly trying to make it go away. But the pressure from the 61st Shadow is truly terrifying for the human intellect, which has formed an intricate mental construct around itself and called it reality.

Stepping into the pure pressure from the 61st Shadow will shatter the intellect completely, quite possibly inducing a state of acute psychosis in which you undergo a spiritual and/or mental breakdown or breakthrough. In order for this delicate shattering process to occur safely, a person must pass through an organic preparatory period in which the hold of the human intellect and its false reality is progressively loosened. This is exactly the process referred to through the 61st Gift — the Gift of Inspiration.

This pressure of the 61st Shadow also feeds the scientific mind. Thus, the scientific mind is also set up to fail in delivering a release from the pressure because the question of *why* is an unanswerable and rhetorical question. Logic is a closed-circuit system of thinking (a concept you can understand better through the 63rd Gene Key). Because logic always leads to stalemate, it simply cannot provide an answer to the ultimate question. Science cannot even truly answer the question of *how* because *how* is a derivation of *why*. Even the most advanced scientific thinking in quantum physics can no longer find answers within the domain of logic, and therefore logic bends its own laws, inventing dimensions that cannot be seen or proven. Today we are witnessing the last desperate throws of the intellect as it tries to understand a question that cannot be understood intellectually. Ultimately, as we shall see, the question can only be answered when it has become one with your consciousness, at which point it paradoxically dissolves.

REPRESSIVE NATURE – DISENCHANTED

Disenchantment occurs when one turns away from the pressure of the 61st Shadow. The repression of the question of one's true origin leads directly to imitative, conformist behaviour in the world. Disenchantment is an inner giving up that usually has its roots in our childhood conditioning. It is hiding from our own question because it terrifies us. The inner question scares us because if we pursue it, we will have to break away from all that feels comfortable. We will have to set off on a personal quest and that will take us away from the status quo. Furthermore, the inner quest is a dangerous path and can only be travelled alone. When the 61st Shadow is refracted through a repressive nature, these people go the traditional route — through education, religion or science. However, deep down inside them, the question remains and disenchantment becomes their true inner state. No matter how hard they try or how much they gain and succeed in the world, inwardly they remain unfulfilled and restless.

REACTIVE NATURE – FANATICAL

The reactive human nature becomes obsessed by the question coming through the 61st Shadow. Rather than pass through the gauntlet of the 61st Gift, these people become frozen and fixed on a single answer to their inner question. Such people find a safe and comfortable place in the structure, leader, creed or direction to which they have become attached. What they do is place an intellectual answer over the top of the question and they build their reality around that answer. By rigidly holding on to their answer, they also have to deal with the psychotic

The dilemma is that you cannot fix the psychosis from within the psychosis.

dilemma of having found "the only true answer to the great question of existence." They become fanatical about their discovery and usually become some form of missionary in its further propagation in the world. Beneath the surface of every reactive nature is a well of insecurity that bubbles to the surface as anger. This anger further protects them from having to deal with the true question within them.

THE 61ST GIFT – INSPIRATION

GOD IS PRESSURE

*Creativity
is the
single most
important
Gift for
drawing
humanity out
of its mass
psychosis.*

Inspiration is what happens when you stop worshiping God and start becoming God. The 61st Gift of Inspiration may sound like a common enough attribute, but the reality of this Gift is far from comfortable. Inspiration refers to a process that is very different from our normal understanding of the word. The word inspiration derives from the ancient Indo-European word for breath and is connected to the Latin *spiritus*, also meaning breath. The process of inspiration is a gradual releasing of your inner breath through the fabric of your reality and out into the world. Despite its wonderful creative manifestations in the world, inspiration involves a powerful dismantling of the inner realities that have been built by your mind.

Inspiration begins when you place yourself directly in the firing line of the inner question that lies within each human being. The ancient Chinese named this 61st Hexagram *Inner Truth* and as a pictographic symbol it represents that which is hidden deep within us. The path of inspiration often begins as an outer quest as you seek answers from the outside world through systems, teachers or disciplines. In the beginning, inspiration comes to you sporadically as flashes, in which you fleetingly perceive the nature of reality in a distilled form. Sometimes inner truth may be revealed to you for longer periods. Such shattering and powerful experiences usually alter the path of one's destiny. What separates inspiration from other heightened experiences is that after inspiration you are permanently altered. Even a brief flash of true inspiration will change the way your awareness operates. Through inspiration you are being prepared for an even greater experience than you can yet conceive.

The path through the 61st Gift is necessarily a very creative path, since true inspiration destroys some aspect of your inner delusion, thus releasing a great surge of trapped energy into your body and life. Such energy naturally seeks an outlet through creativity. Creativity is the single most important Gift for drawing humanity out of its mass psychosis. It unlocks the latent forces of inspiration inside you, pulling you away from a state of victimisation. Inspiration and creativity have their challenges however. The main one for human beings is patience. Inspiration cannot be forced or predicted, but comes when it comes and stays for as long as it stays. In between these heightened inspirational states, you may well become dejected or depressed. However, at a certain level of frequency you will reach a plateau, and at this level the inspiration itself gives you enough energy to maintain a heightened state. Once again, the key is some form of creative process.

By its very nature the Gift of Inspiration is spiritual because it serves to loosen the hold of your mental constructs, opening and expanding your capacity for love. Alongside the 61st Gift, the 62nd Gift of Precision enables the wordless experience of inspiration to be expressed in a language that others can understand. The 62nd Gift of Precision is adept at expressing the mystery of life with great intelligence, beauty and economy. One of the hallmarks of the 61st Gift in action is the originality of its expression and the seemingly endless stream of its activity in the world. Deep within the person in the midst of such a process something extraordinary is occurring — their hold on reality as they know it is slipping. Many people are not able to let go during these stages and cling to the forms they have begun to create in the world. Those who can continually release the definitions of their reality do begin to enter a more rarefied realm. They enter a stream of consciousness in which inspiration begins to annihilate their lower mind. At this stage, it may appear as though a greater being is somehow taking over your life. You may not know it, but you are approaching a door — the door to the greatest secret known to humanity — the door to the Divine.

One of the greatest statements ever made concerning the nature of divinity is the mystical axiom: "God is pressure." This revelation powerfully describes the process whereby the inner truth hidden within human beings is unlocked. You are a pressure machine! Deep inside your body, the pressure of the mystery of your being beats in the heart of every single molecule of living DNA. Through its associated codon ring, the Ring of Gaia, this 61st Gene Key represents a mystery hiding inside every single unit of matter in the universe. This mystery is the mystery of Christ consciousness, that quintessence of inner light that holds all things together. It is the creative evolutionary process that gradually unveils this inner light. Through the Ring of Gaia we can see how our ecosystem — the living, breathing diversity of the blue-green planet and all its life forms — is destined to discover its own inner truth as one entity. This codon ring shows that evolution is a force that is breaking out from inside us, and that to unlock its secrets we have only to look inward.

THE 61ST SIDDHI – SANCTITY

ENTERING THE UNKNOWABLE

In arriving at the 61st Siddhi, we approach a great mystery — the mystery of life itself. The inspiration pouring through the higher reaches of the 61st Gift is a stream of sparks from a great transformation occurring within every human being who approaches the inner door. As we have seen, the manifestations of inspiration are unimportant to the one who is truly inspired. These manifestations are simply the myriad colours of the experience itself and as such are wonderful to those outsiders who are witnessing the process. Sooner or later however, inspiration comes to an end. The process exhausts itself as it throws out all the mythic identities contained within your DNA.

As you enter the field of the 61st Siddhi, a huge, pregnant silence descends and all your mental activity abruptly ceases.

As you enter the field of the 61st Siddhi, a huge, pregnant silence descends and all your mental activity abruptly ceases. Here inner truth resides. In western mythic traditions, this is a place and is called by many names — the Ark of the Covenant, the Grail Castle, the Holy of Holies, the Celestial Palace. In the East, it is generally described as a state — enlightenment, nirvana, samadhi or some such term.

The 61st Siddhi is neither a place nor a state. It is the inner experience of sanctity — of being one with the Divine. Within this Siddhi, true reality dawns. The pressure that creates humanity's mass psychosis is no longer routed directly into the human brain but is *refused* and rerouted through the solar plexus system (which houses a far more advanced system of awareness). Once the pressure of awareness is removed from the brain, the question of *why* finally ceases, and all other questions emerging from it such as *how* and *who* also die. However, the pressure must go somewhere and indeed it does. Through the medium of the solar plexus centre, it goes everywhere. It is through the constantly vibrating wave frequencies emerging from the solar plexus that awareness is carried beyond the body into every corner of the universe. You immediately find yourself both completely empty and yet endlessly full.

The experience of Sanctity is paradoxically an experience with no *experiencer* and is one of the greatest mysteries of evolution. Once it has been entered it cannot end, and must be differentiated from all other mystical experiences or higher visionary states. Such states occur at the higher reaches of the 61st *Gift*. A human being in whom the 61st Siddhi is being revealed is nothing but a human vehicle manifesting pure universal awareness. This state of Sanctity is the underlying nature of reality and until you come to rest within it, you are always asleep. For those of us outside of this realisation, such a person seems to be a god. They ripple with divinity and exude the strange unearthly essence that humans refer to as *holiness*. They emit a frequency of such availability and at the same time such power that we find them either irresistible or terrifying. These are the people who we, in our psychosis, deify and worship. We do so to our detriment because the whole point of such a presence among us is to demonstrate the fact of every human being's inner divinity.

For one of us to settle into the realisation of our inherent divinity all we need do is entertain another who has already realised this inner truth. The being within the 61st Siddhi no longer knows any difference between you and him or her. Because this state of consciousness has always existed and will always exist within every aspect of the universe, such a person cannot help you in any way. They simply serve as a mirror of your own inherent divinity. So to be in the presence of the 61st Siddhi in another is no guarantee of anything and in some respects can lead to more confusion. What really has to occur is the deep understanding inside you that you are asleep, and this understanding can be greatly facilitated by a mirror. However, to remain with that mirror frequently leads to worship of the mirror, which prevents your own realisation. This is why, for example, the Buddhists say that if you meet the Buddha along the road, you must kill him.

The 61st Siddhi is about living life without answers. It is about simply becoming the mystery of who you are. Only when all answers finally fall away does the inner truth dawn. If there is a Siddhi that negates all words, descriptions and answers it is this 61st Siddhi.

Life is a mystery. Enlightenment is a mystery. Inner truth is a mystery. No matter what you do to try to solve the mystery, you will never come closer. There is no attitude you can adopt in response to the mystery, since all such things are subtle attempts at solving it. You simply must come to realise that you know absolutely nothing, and this profound revelation can only occur in its own way and at its own time.

For humanity in its current phase of evolution, the 61st Siddhi can only cause infuriation. It is better that we forget all about it and bring our focus to the Gift level, where our mind at least has something to grasp and something to aim for. Once you have pulled yourself clear of your psychosis it is simply a matter of time before the Siddhi also arrives. The vehicles we currently inhabit are extremely limited in terms of their awareness operating system, which favours the development of the mind. Therefore, we must respect the phase of evolution through which we are passing and enjoy what lies before us. The 61st Siddhi is really a matter of luck at this stage. If the vehicle you are sitting in happens to mutate, then you are in luck!

The future, however, holds a different story. The 61st Siddhi will become far more common-place in the next few hundred years, because humanity is in the process of making a quantum leap. The children who will begin to arrive on this planet will carry this Siddhi of Sanctity in their bones. They will radiate it through every pore, and they will carry it out into the world with them. They will be the living embodiment of life's mystery — of life's potential in form. They will bring a balance back into the world that will see the mind cease its turbulent search for an ultimate answer. In the final analysis, we *are* the answer.

62nd Gene Key

The Language of Light

SIDDHI
IMPECCABILITY

GIFT
PRECISION

SHADOW
INTELLECT

Programming Partner: 61st Gene Key

Codon Ring: The Ring of No Return
(31, 62)

Physiology: Throat/Thyroid

Amino Acid: Tyrosine

THE 62ND SHADOW – INTELLECT

THE STUPIDITY OF BEING CLEVER

In the original sequence of the 64 Gene Keys as laid down in the I Ching, there is an interesting geometry connected with both its beginning and its ending. As you may have learned, the first and last pairs of Gene Keys — the 1st and 2nd together with the 63rd and 64th — act rather like cosmic bookends to the entire evolutionary and involutionary process of creation. The first pair can be seen as a prologue, and the final pair an epilogue to the great drama that lies encoded within their boundary. When seen in this way, the true sequence of the drama of evolution begins with the 3rd Gene Key of Innocence and ends with the 61st Gene Key of Sanctity, which is also known in the original I Ching as *Inner Truth*. The 61st Gene Key (the programming partner of the 62nd Gene Key) has the feeling of being the finale to a great orchestral symphony. However, the 62nd Gene key is something entirely different. Where the Gene Keys themselves represent the book of life, the 62nd Gene Key stands alone at the end, as the index or glossary of all that has come before.

The 62nd Gene Key contains layer upon layer of coded information about the meaning and purpose of the cosmos. Behind its inner door, the 62nd Gene Key reveals what the Gene Keys themselves are and what they are for. They are the living language of light that lies at the foundation of the universe. The 64-bit matrix is the core structural principle behind all art, science and natural phenomena. All human languages and vocabularies have emerged out of this primary alphabet of consciousness. At the deepest level the 62nd Gene Key teaches you the holographic language of creation. Once you have learned this inner language you will see it repeated in fractal form over and over again in everything your awareness touches. It is precise and infinitely complex, and yet it is elegantly simple to learn, formed as it is from only six possible permutations of each of the 64 Gene Keys.

Many people may be surprised to see the word *intellect* represented as one of the 64 Shadows. The language of the Gene Keys holds true at all levels of frequency, and unlike human languages it is not a language that can be learned and mastered via the intellect alone.

To master the language of creation we have to embody it fully at every level, not just the intellectual level. However, in the modern world intellect is generally regarded as a sought-after and admired human gift rather than as something that might actually hamper human evolution! Therefore, it is important to clarify some terminology. Intellect is most often confused with intelligence and understanding. In the context of the Gene Keys intellect refers to the thinking capacity of the human mind, which bases all its suppositions on its two main objectives — the acquisition of facts, and the skill of manipulating those facts via language.

Intellect is the skill of manipulating knowledge, but knowledge is quite different from understanding. Understanding in this sense does not simply refer to the activities of the mind but to the whole experiential being. You can be a dunce at the intellectual level, but can still understand many profound truths within the heart of your being. Likewise, intelligence has nothing to do with intellect. In fact, these two attributes are often (though not always) diametrically opposed to each other. Generally speaking, the more intellectual you are, the less intelligence you use. In the context of the Gene Key language, intelligence is something that occurs without the use of the mind, although intelligence may also use the mind as a means to transmit itself.

The modern world that we inhabit is truly upside down. Our very schooling is designed to make us more intellectual and less intelligent. We are already intelligent as children. Your intelligence is stunning, vibrant and natural. Your natural intelligence is to be found in the way you move your limbs, in the brightness of your eyes and in the freedom of your self-expression. It is the reservoir of your future genius. But that genius is very efficiently curtailed from the moment you enter school. The more information you put into your head, the more sedated you become. In most cases, as you progress through the modern school system, layer upon layer of homogenised information is force-fed to you. You are even required to compete to remember it all! It is a very efficient system in terms of how deeply it represses human intelligence, but the bottom line is that it makes us all the same. Since we have each learned the same information, and learned it in the same way, our brains become neurologically programmed to operate in similar ways.

Intelligence is of the heart, whereas intellect is of the mind.

Like any of the human Shadows, the 62nd Shadow of Intellect is not inherently bad. Intellect is a wonderful quality if it is used correctly in the service of innate intelligence. But when intellect is put in charge of the planet, as it currently is, then intelligence cannot be seen except in the few people who have raised this Gene Key to a higher frequency. Intelligence is of the heart, whereas intellect is of the mind. This is the core sentence. The 62nd Shadow is obsessed with facts, and facts are born of language and names. Without a name, you cannot create a fact. Names and language in their turn create the software of the human mind. Without this software the mind is silent. In a world where facts are regarded as treasure, the mind reigns as king.

Our world is divided between the 62nd Shadow of Intellect and its programming partner, the 61st Shadow of Psychosis. That these two themes are genetically linked should speak volumes about the way we perceive the human mind. The pairing of the 62nd Shadow and the 61st Shadow reflects a fundamental split within humanity itself.

Through the 62nd Shadow man tries to solve the mysteries of life through his intellect — in other words, through science. Meanwhile, through the 61st Shadow he tries to solve the same mysteries without his mind — through religion. These two poles, which at their extremes become scientific intellectualism and religious fanaticism, are genetic sub-programmes that keep humanity operating at a low frequency level. Individuals who carry the specific imprinting of these Shadows are caught in the crossfire of these two underlying human themes. Such people can spend their lives defending one viewpoint whilst suppressing the other hidden aspect of their nature. If the 62nd Shadow is dominant, the female nature will be suppressed, and if the 61st Shadow is dominant, the male nature will be suppressed. This suppression of your inner male or female polarity is the main cause of disease on our planet, both at an individual level and at the collective level.

The 62nd Shadow really represents abuse *through* language. It is not we who abuse language, but language that abuses us. Language *relies* upon our frequency. At low frequencies, it completely takes control of our reality. The human intellectual capacity to read, write and speak is both our greatest blessing and our greatest curse. Problems arise when we identify our lives with our thoughts, which as we have seen, have been pre-programmed throughout our upbringing. Until we are able to step outside of this mental framework, we are controlled by language instead of being free to control it. All facts are relative, as modern quantum mechanics beautifully demonstrates, and this puts the entire conceptual framework of language and intellect on the chopping block.

The fact that science is on the cusp of undermining the very foundation of intellect speaks volumes about the extraordinary times through which we are currently living. However, the intellect is not designed to give itself up. It will fight on, disproving any theory that takes it out of its own factual territory. Only when humanity raises its frequency into the heart will it be able to use intellect to resolve this great paradox. Put another way, the only way you can scientifically prove that God exists is to *become* God. Knowledge must then become knowing, intellect must surrender to intelligence and as a man of true understanding once said: "The last shall become first and the first shall become last." What this really means is that the heart shall lead the mind, rather than the other way around.

REPRESSIVE NATURE – OBSESSIVE

When the 62nd Shadow turns inward, it becomes obsession. Everything about this 62nd Gene Key is about focussing on the small things. When fear manifests through this Shadow it uses details as a way to avoid feeling its own suffering. These people become lost in the mundane world, living a life of endless details with little or no creative outlet for their life force. At the lowest end of the spectrum such people can become mentally ill, obsessing on the tiniest things as a way of coping with life. Such people are held prisoner by their minds rather than by the details themselves. For the majority of humanity however, this Shadow serves to repress the natural sparkle of the individual by keeping him or her locked into a monotonous mindset where details are both one's crutch and one's enemy.

REACTIVE NATURE – PEDANTIC

These people use their intellect to attack everything outside them in an attempt to defend their own deep insecurity. Such people are entirely in the grip of their intellect which questions everything endlessly, and in particular delights in finding obscure details and facts that disprove and disempower others. The intellect in these people is usually highly developed and in some cases they gain great recognition through it. However, such people are also utterly unable to switch off their minds. They focus on other people in order not to look into themselves. The source of this complex is a sense of rage, which the mind keeps bottled up until something or someone triggers its release, a fairly frequent occurrence.

THE 62ND GIFT – PRECISION

THE GREATEST STEP

The 62nd Gift, the Gift of Precision, is far beyond the realm of the intellect, which as we have seen is based on mere knowledge. As you start to awaken through the higher frequencies of this Gene Key, you either begin to question the world around you (the repressive nature) or you stop questioning everything (the reactive nature). Awakening is magical in its ability to bring about a natural balance within your being. In other words, if you are caught in the obsessive behaviour patterns of the *repressed* 62nd Shadow, your mental abilities suddenly begin to come alive again, as though someone had cleaned the windows of your perception. You begin to question your own obsessive behaviour, which ultimately brings about a complete transformation in your attitude and usually in your lifestyle as well.

In the case of the *reactive* 62nd Shadow, your awakening will take you through a natural humbling process in which you realise that you are causing your own misery by focussing in minute and irrelevant detail on anything and everything but your own pain. When you begin to look into your own nature and take full responsibility for your behaviour, you will go through a wonderful softening process as your natural feminine side comes once again to the fore.

In both cases, we are seeing the Gift of Precision being reborn. Precision happens when natural intelligence strikes a balance between the heart and the mind, but with one caveat — that the heart, the feminine principle, is given control over life. The masculine principle, the intellect, then moves into service of the feminine principle, which is about intuition, listening and receiving rather than thinking, expressing and transmitting.

As the Gift of Precision grows stronger, it may seem as though the world gradually begins to come alive again. Intelligence recognises intelligence, and with the mind out of the way the invisible essence that connects all beings is once again felt. For example, when you look at a tree through the 62nd Shadow, you register only the facts that you have learned about the tree. You register its name, type and any other words connected to it — branches, twigs, leaves, and so on — but you never see the actual tree. The tree is intelligent, and to really know the tree, you have to use your own intelligence.

This means that you don't simply look through your eyes and mind. You take the tree into your being — you feel its aliveness, its mysterious aura, you actually breathe it into yourself. Precision is what happens when intelligence is born. It's not simply about being precise and exact at an intellectual level, it's a completely new way of seeing life.

Precision is inspirational and original (the 61st Gift is Inspiration), and such inspiration does not depend on your vocabulary. When the Gift of Precision describes something, it arranges facts in ways that are inspirational and exciting rather than dull and dry. As this Gift arises in a person, they begin to communicate with such economy and exactness that almost everything they say is beautiful, poignant and seamless. Such people soon develop keen gifts as communicators, speakers, writers, artists, actors or scientists. This Gift is designed to find the limelight. When the heart leads the way and uses logic to describe what it sees and feels, others cannot help but listen.

The 62nd Gene Key is genetically coupled with the 31st Gene Key forming the codon known as the Ring of No Return. This mystical name describes the evolutionary process that takes place when higher consciousness reaches the throat centre in human beings. The throat centre is where the greatest human initiation occurs. Once the higher involutionary currents begin to use your voice to transmit their truths, you begin a process of detaching from your own identity. The 62nd Gene Key allows access to the universal language of light behind all forms. When the pitch of your frequency allows it, the words you speak and their emanations change and begin to serve a higher purpose. This purpose is twofold: firstly it transforms *you* because it allows you to really vibrate your heart and transmit your love through language, and secondly, it brings more light into the world and spreads the transmission of your awakening to others.

There is deep magic woven into language. It is why the origin of the word *spell* correlates with the idea of casting spells that have power over others. Every word has an inner spirit — a code of light that lends it an independent force in the cosmos. The moment a word or group of words is given voice, vibrations radiate out into the universe. There is no return. At the level of the 62nd Gift, your use of words becomes much more precise because you recognise this great truth. The 31st Gene Key describes this gradual process of becoming a clearer channel for Truth itself. In essence, light is breaking through you, through your words. As your language becomes purer, you expose your heart to others and to the world. You are taking the greatest step — the leap from worship to embodiment. It is here that you will have to face the fear of humiliation as you give your voice up to your heart. Words spoken from love are deeply healing. At the same time they can stir up all manner of projections from the Shadow frequency. This is why there is no return once you begin to speak your truth — you have broken away from the lower realms forevermore.

The 62nd Gift represents the true power behind the Gene Keys themselves. At first it gives you a private inner language that allows you to initiate a deep transformation of your Shadow consciousness. At a certain stage in your journey however, you will feel the call of this language within you and its urge to escape into the world through your voice.

*The more
rarefied your
frequency, the
fewer words
are available
to you, so your
language takes
on a beautiful
simplicity.*

It may well frighten you to speak to others using the higher frequencies and tones, but this is the beginning of your higher initiation. Once begun it will gather momentum inside you until you feel yourself literally taken over by your higher self. You will no longer use words that protect others from themselves, but will speak the truth as you feel it arising from deep within you, clearly and without projection.

The fact is that the more rarefied your frequency, the fewer words are available to you, so your language takes on a beautiful simplicity. The 64 words that describe the Siddhis will become embedded inside you as the core language of creation. Such words, when felt deeply and spoken from the heart, can carry great *shakti* or transformational power.

THE 62ND SIDDHI – IMPECCABILITY

COSMOMETRY – THE LANGUAGE OF PERFECTION

The siddhic state of the 62nd Gene Key is impossible to understand intellectually. As the 62nd Siddhi dawns, all sense of continuity in your normal life is broken. In the true siddhic state, all trace of your prior genetic condition is erased. Enlightenment is not only a spiritual and mystical event but also a chemical one, in that the very biochemistry of your body undergoes profound alterations. When you disappear into the siddhic state (and that is precisely how it happens), what is left behind is simply pure consciousness operating through a specific genetic vehicle. This is generally referred to as the death of the ego. Once consciousness ceases to identify itself as existing, the behaviour of the human vehicle is beyond the grasp of the intellect. Such behaviour is said to be *impeccable*.

The word *impeccable* is often misunderstood when used in reference to true masters. The most common mistake we make is to believe that one who manifests the siddhic state must now behave in some kind of *holy* way. There is a widespread assumption that so-called Divine states of consciousness come with a sanctified code of conduct! If you truly understand the nature of the 62nd Siddhi of Impeccability, then you will see how laughable this notion is. To be impeccable means to be *exempt from doing sin*, which sounds as though it entails a certain level of purity and virtue. However, something profound needs to be understood here: one who has disappeared into the siddhic state has truly disappeared. There truly is no one at home. Pure consciousness is at play within this being, so whatever they say or do is beyond reproach. If a man were to commit murder in a siddhic state (actually, an impossibility), he would be beyond reproach. Of course, society would punish him and humanity would judge his action as evil or wrong, but this does not detract from the fact that he is beyond reproach because there is no *he*. You would only be punishing an empty body! This is the meaning of Impeccability.

As we have seen through the 62nd Gift, communication at the more refined levels becomes very precise, especially through the medium of language. The 62nd Siddhi concerns the use of language at a level beyond our ken. Every single word has its frequency and vibration, and the order and syntax of your diction creates a certain aura.

At the siddhic level, words do not come via the mind any more. They are not thoughts that become words. They emerge directly from the void, naked and pure.

If you look at the sayings of the great masters, you may often find that they say contradictory things about the siddhic state. For example, one may say that enlightenment can only be found through intense seeking, whilst another may say that nothing you do will ever bring about this state. There appear to be many discrepancies between the words spoken by sages. This is because each still filters his or her experience through their own background, language and culture. However, when someone manifests the 62nd Siddhi, they create an exact linguistic science of enlightenment. It is often amusing to note how closely related are the Shadow and the Siddhi. Thus, such people will use the intellect to defeat itself through pure logic. They will also use language to demonstrate how futile language actually is, for it is language and thought that underpin the very illusion of the human state in the first place.

The 62nd Siddhi may appear to be highly intellectual, and indeed it will appeal in particular to men, who are more mentally polarised than women. These are the sages, like Socrates, whose logical argumentation cannot be defeated because consciousness itself is using the left side of the brain with impeccable precision. In such cases, language itself can be used as a means to bring about surrender within another. These people are adepts at using language in extraordinary ways in order to highlight the grand illusion in which the human mind operates. But even more than this, the 62nd Siddhi brings the understanding that every cell within the universe is a Divine word. This profound Truth allows such a being to become a master of Divine language. Because the Divine alphabet is so simple, a true master can respond to anything and anyone anywhere with exact precision.

The science underlying all creation is sacred geometry. The eternal truths contained in sacred geometry unite all human sciences, arts and approaches to understanding through either the left or right brain. There is a *Cosmometry* behind life that choreographs all actions. The closer you come to full awakening, the more harmoniously your life flows with this cosmometry. When you finally merge into the ocean of consciousness you achieve complete oneness with the Divine Cosmometry. Everything you think, say or do is no longer done by you but by the whole. Therefore it is done impeccably. The real difference between the Siddhi and the Shadow is the complete absence of fear — which is the absence of self-identification.

Divine Cosmometry is the language of perfection. It is the manifestation of impeccability in which your every movement and breath becomes an emanation of the pure light. Your whole being is simply made up of living, intelligent lines of force that move and flow without resistance. There is no longer anything out of place within your being. There is no longer any need, agenda or discomfort. All is exactly as it should be, in you and in everyone else. The embodiment of this truth leads to ultimate peace. You have crossed the great divide in the throat and can never again return to the illusion of separation. You have entered into the language of light itself to see the one word in all words — the fathomless, the ineffable, the wordless — the impeccable beauty of being.

Everything you think, say or do is no longer done by you but by the whole. Therefore it is done impeccably.

SIDDHI
TRUTH

GIFT
INQUIRY

SHADOW
DOUBT

63rd Gene Key

Reaching the Source

Programming Partner: 64th Gene Key
Codon Ring: The Ring of Divinity
(22, 36, 37, 63)

Physiology: Pineal Gland
Amino Acid: Proline

THE 63RD SHADOW – DOUBT

THE PARADOXICAL POWER OF DOUBT

In arriving at the 63rd Gene Key, you have in effect come to the end of the 64 Gene Keys. In the mystical story of evolution, the 63rd and 64th Gene Keys mark the epilogue — the cosmic text that gives us a certain sense of completion. This completion however is not about closure. The I Ching and the Gene Keys form an infinite Mobius strip that endlessly cycles around in ever more expansive arcs. The original I Ching as traditionally transcribed is a sequence beginning at the number 1 and ending with 64. However, there are as many sequences as there are numerological combinations. Your own evolutionary sequence is completely unique to you and it is highly unlikely that it will ever be repeated in anyone else. This is the paradoxical wonder of the Gene Keys — they are a digital encyclopaedia of consciousness that can only be accessed and activated in a totally analogue and random manner.

The traditional sequence is held together by two sets of polarities — the 1st and 2nd Gene Keys at the beginning and the 63rd and 64th Gene Keys at the end. These four Gene Keys have great importance both genetically and mathematically. The mathematical configurations latent within the I Ching have been discussed by many great logical minds. The celebrated 17th century philosopher and mathematician Gottfried von Leibniz found confirmation in the I Ching for his invention of binary mathematics, the foundation of virtually all modern computer architecture. That the essential code for the 64 Gene Keys lies at the root of both our DNA and all computer programming is an indication of how profound and real these 64 codes of consciousness really are. The polarity formed by the 63rd and 64th Gene Keys contains paradoxes that can be as wonderful as they are frustrating for the human mind.

The relationship between the 63rd and 64th Gifts is the relationship between logic and imagination, or put another way, between your left brain and your right brain. As we shall see, they are so interconnected that they cannot easily be separated.

Einstein made the famous statement: "Imagination is more important than knowledge." The fact is that the laws of logic depend upon a completely illogical premise — the notion of infinity. Infinity is a logical impossibility and yet the human brain is designed in such a way that it can conceptualise this paradox. At the human Shadow frequency, these two great genetic archetypes, the 63rd Shadow and the 64th Shadow give birth to two conditions known as doubt and confusion respectively. Doubt is born when you lose touch with the wonder of imagination, and confusion comes about when you rely too heavily on logic. If you turn these concepts around, you could just as easily say that doubt is the foundation of logic whereas confusion is the breeding ground for imagination.

The 63rd Shadow of Doubt represents a huge question mark literally wired into the human brain. Your logical left brain is designed to see things in terms of repeating patterns and this ability has been a key ingredient in human evolution. As the cognitive functioning in our brains developed, early human beings began to teach themselves more and more effective skills to ensure their survival. Today we have become so efficient in the use of our tools that we can even read the logical patterns wired into our very DNA. Doubt can therefore be a very positive force in driving evolution forward. However, the problem with doubt is that it does not go away! Doubt sits there deep within the neurocircuitry of every human brain where it causes a huge unconscious pressure. When seen purely for what it is, doubt is simply the word we humans give to this constant mental pressure. Although this pressure behaves differently from one human to another, the one consistency is that it drives each of us to try to bring an end to the pressure. We humans cannot easily accept this never-ending feeling of being unsettled somewhere deep inside. For most of us, the need for mental certainty results in our taking on a firm set of beliefs or values that provide a barrier against the doubt inside our minds.

Doubt is the beginning and end of the human journey. It leads to the Gift of Inquiry, which in turn leads to the discovery of the Siddhi of Truth, the final answer to all doubt. However, the journey through the levels of this 63rd Gift does not lead you in the direction you might expect. The final resolution of logic lies outside the domain of logic. In this sense Truth comes as an Illumination from within (the 64th Siddhi) rather than an answer from without. Our conundrum is simply a matter of evolutionary circuitry — our brains have evolved to try to find release from the inner pressure. But in a sense, we have been conned by nature because there is no answer to our doubt. At the same time our attempts to end our mental anxiety have forced the brain to evolve even further. This is how our anxiety serves evolution.

The other important aspect about doubt is that, although it is truly impersonal, we humans make it personal. If doubt serves genuine inquiry then it becomes creative, but if it becomes personal or is internalised, it becomes destructive. If you do not use your mind in a creative way, the pressure either becomes a projection onto someone or something outside or it collapses inward and becomes self-doubt. It is self-doubt that so undermines the individual from trusting in his or her own inherent creativity. Doubt belongs to the collective. Individuals need to understand that self-doubt is a low frequency conditioning field that pressures all human beings to evolve to a higher life of service to the totality. When you are in the throes of self-doubt, you must weather the storm. The very best thing to do with self-doubt is share it with others. In doing so, you can embrace it and allow it to move on through. The worst thing you can do during a period of self-doubt is to take forced or premature action.

If the individual succumbs to the pressure of doubt, the other direction he or she can take at the Shadow frequency is to repress it through dogma and opinion. Since the 63rd Shadow is the root of scientific thinking, it can be used to construct a false logical reality that gives the mind answers to the doubt. When the logical mind is prevented from doubting, it also stops evolving. Without doubt, the logical mind is inherently unhealthy because it short-circuits its own nature, which is to keep asking questions. As was said earlier, when logic moves to the extreme periphery of what can be known it meets paradox. We can see this in the spheres of mathematics, philosophy or quantum physics. The human logical mind is simply not designed to be certain of anything other than paradox! Anything less than paradox takes place within the Shadow frequency and must therefore be logically false.

The beauty of logic is that it is itself a great paradox — it seeks an end to doubt but can never be satisfied in any single answer. The only satisfactory answer is one that defies logic! At the Shadow frequency, all the human mind wants is certainty, and if it cannot find it, it will manufacture it. Out of this very human fear of uncertainty, all dogmas and rigid systems are born. Anyone or anything that purports to bring an end to your uncertainty through external means is selling falsity. Such people or systems are actually a potential danger to you. Doubt is not the enemy. Your fear of uncertainty needs to be recognised and embraced because the deeper you can embrace uncertainty, the closer you come to the transcendent.

REPRESSIVE NATURE – SELF-DOUBT

When doubt is repressed it plagues us as self-doubt. Self-doubt is based upon comparison and does us no harm as long as we simply observe it as a mental conditioning pattern. Self-doubt only becomes destructive when we identify with it and believe its propaganda. If left to its own devices, self-doubt must eventually destroy itself. If we trust in our doubt completely then we must even doubt our own self-doubt! This is the paradoxical power of doubt witnessed as an evolutionary force. However, for most people self-doubt is never allowed to complete its own process. We give in to our fear and fix it at a certain frequency. When this happens, self-doubt eats away at us as anxiety. This anxiety constantly troubles us and has sweeping effects on our health and general well being. It even pursues us into our dream life and will not allow us to sleep properly. Unless our awareness dis-identifies with these mental patterns, we enter a terrible world in which nothing we do or say is ever good enough.

REACTIVE NATURE – SUSPICION

When the 63rd Shadow is expressed in its reactive mode, it is externally projected into the world and becomes suspicion. This kind of doubt is also ultimately rooted in self-doubt but is transformed into anger and taken out on others. The most common form of suspicion falls upon those closest to us. Under this 63rd Shadow, we project our doubts onto our partners, spouses,

The human logical mind is simply not designed to be certain of anything other than paradox!

bosses and even our own children. Whenever we make our doubts personal we embark on a wave of destruction that endlessly feeds back into itself. Other people's responses to our suspicions can be vehement or defensive, which further fuels our doubts about them. In this way logic appears to prove that our suspicions about others are true. This is a classic biofeedback loop common to many relationships. It can be so easily avoided if individuals either own their own self-doubt or project it back out into society in a creative exploratory way.

THE 63RD GIFT – INQUIRY

THE LADDER OF SELF INQUIRY

When the energy of doubt is harnessed cleanly it becomes the Gift of Inquiry. Inquiry is about remaining open without finding a definitive answer to life. All pure Inquiry is of great service to humanity. It inevitably unlocks many of nature's secrets by examining life in minute detail. The real purpose of doubt is to serve the higher good in an impersonal way. If someone has a deep genetic tendency to doubt or ask questions, they can greatly improve the quality of their life and relationships by channelling that doubt away from their personal life and pressing it into service. The 63rd Gift is about progression through levels and stages. This is the way in which logical understanding works. The continual pressure of doubt fuels the spirit of inquiry, which leads to deeper and deeper levels of understanding. However, the more deeply you inquire into anything, the more complex it becomes. You have to take it apart piece by piece, whether it is a scientific theory or an internal combustion engine. What is of prime importance is what you learn through the process.

As you slowly ascend the ladder of frequencies, the nature of your inquiry changes. You may begin by inquiring into some aspect of the world around you. This may be something you really want to understand or it may be something about the world that you want to improve or challenge. There is one golden rule with the 63rd Gift of Inquiry — if you inquire deeply enough into anything, you will arrive at the revelation that you, the observer, are intricately bound with that which you observe. This discovery actually heralds the beginning of the end of the logical approach, since it signifies the end of your ability to assess life objectively. At higher levels of frequency, you begin to enter the holistic domain of synthesis in which all phenomena are increasingly seen to affect each other. This new quantum world thus forces your inquiry into a far more subjective territory — that of your inner self. At a certain stage, the Gift of Inquiry must always lead you back into your self.

The 63rd Gene Key is one of four Gene Keys (the 36, 37, 22 and 63) that make up a genetic grouping in your DNA known as the Ring of Divinity. Because of its connection to the all-important 22nd Gene Key, this codon ring has much to do with the transformation of human awareness through suffering. The 63rd Gene Key plays a central role in this theme because it is your Divine doubt that drives you to follow a spiritual path in the first place. The deep discomfort of not remembering your true nature becomes more and more evident the more profoundly you inquire into your own nature.

As you discover the doubt inside you, then you become aware of it in all other human beings as well. This brings you to a great truth — that all human beings are made equal in this suffering (the 37th Gift of Equality), that doubt makes us human (the 36th Gift of Humanity) and leads us to a certain Graciousness (the 22nd Gift) in respect of how we treat others.

When you begin earnestly and honestly to question your own nature and purpose, you enter one of the great spiritual paths of all time, known in the East as the path of yoga. Yoga is essentially a progressive path of refining your physical and spiritual essence to such a height that you arrive at liberation or spiritual realisation. It is the system of internalising your life force so that it reaches higher and higher frequencies. Many different systems of yoga have been developed and adapted by cultures around the world. Some emphasise the physical body, some are meditation or prayer-based and some are rooted in service. What is common to all systems of yoga is the progressive sequential process whereby you raise the frequency of your vibration from a low state to a higher state. It is the Gift of Inquiry that urges you to understand all the states along the way.

Even at these higher states of awareness, the primary doubt that has always been with you stubbornly remains — the doubt about your very own existence. At our core, we want to know who and what we really are. For most people at lower frequencies this question simply exists only in seed form. The higher your stage of awareness however, the more possessed you become by this question mark inside you. Your Inquiry thus leads you through progressive levels of awareness and realisation and you may experience strange internal phenomena along the way. All this time however, you are still operating within the sphere of the 63rd Gift, which gives the impression that you are moving somewhere or evolving from something towards something. Even at these high levels, your Inquiry still has its roots subtly in the logical framework of your thinking. But, and it's the greatest *but* in all creation, all logic must eventually, by its very own nature, cancel itself out. All yoga leads in a single direction — to its own nemesis — to tantra, the way of surrender.

The higher your stage of awareness however, the more possessed you become by this question mark inside you.

These two genetic programming partners, the 63rd and 64th Gene Keys, represent the paths of yoga and tantra respectively. Whereas tantra is a path based on acceptance of what is, yoga is the path of doubt because it harnesses the great human spirit of inquiry. In the yogic path, technique is central. It assumes from the beginning that there is a basic split in human beings, and the sole task of yoga is to unify this split. This is what the word yoga means — to unify or *yoke*. All the great systems that allow human beings to progressively reach towards higher states have been born out of yoga. Because the human mind is most comfortable with the concept of logical progression through time, these kinds of paths have always appealed to most serious seekers and inquirers after Truth, the mysterious and hugely paradoxical 63rd Siddhi.

THE 63ᴿᴰ SIDDHI – TRUTH

TRUTH IS EVERY STEP OF THE WAY

The Gift of Inquiry, symbolised by the path of yoga, has a deep shock in store for those who are drawn to it. In the original Chinese I Ching, the two hexagrams corresponding to the 63ʳᵈ and 64ᵗʰ Gene Keys have very revealing names. The 63ʳᵈ is called *After Completion* and the 64ᵗʰ is called *Before Completion*. You might think that these names were the wrong way around since the 63ʳᵈ obviously comes before the 64ᵗʰ in the original sequence. However, all you will find here at the end of the I Ching is paradox, poetry and puzzlement. We are entering the territory of Truth with a capital *T*. No matter how elevated a person's awareness becomes or how advanced their yogic discipline is, until they reach this siddhic realm they will never realise the ultimate paradox that is Truth. The only way to arrive at such a state is to reach a position of inner surrender. After all your efforts and exertions to find Truth, you must eventually become one with your own doubt.

At the Shadow frequency there is nothing but Truth, at the Gift frequency there is nothing but Truth. Truth is every step of the way.

This is the shock of the 63ʳᵈ Siddhi — that doubt *is* Truth. This is a deep let down for your questing spirit. It is the ultimate state of failure and dejection. If logic is allowed free reign, it will always cancel out logic and conquer itself. In this sense all yoga can only ever lead to tantra. This does not in any way negate the value of the spirit of inquiry that propels the seeker of Truth. All male paths must eventually lead to the ocean of the feminine. The more logic is followed, the more mysterious and poetic it seems to become — until it reaches the impasse that logic cannot move beyond. Logic and its human counterpart doubt are designed to self-destruct, taking *you* the observer with them. When doubt is trusted at such a profound level, the doubter disappears into the doubt, leaving Truth. However, there is no sequential process to this — Truth is there all along within the doubt. It is simply waiting to be realised, like a pearl within an oyster. It needn't be prised out, but simply recognised. If you follow a sequential and linear path that ends in this recognition or enlightenment, the path itself has no bearing on the enlightenment, even though it appears to.

Truth is realised in many different ways. The other side of this Siddhi is the 64ᵗʰ Siddhi of Illumination, which arrives at the same conclusion in a totally different way — through the poetics of randomness. If you enjoy the logical path of progression, then that is your way. In many respects it feels a safer route for human beings since it gives you the feeling that you can prepare for Truth. Few are those who are drawn to the yin path of the right brain where there are no maps or charts to follow other than the confusion of life itself. The yogic path to Truth is a beautifully clipped garden in comparison to the wild jungles of the 64ᵗʰ Siddhi. The 63ʳᵈ Shadow of Doubt is the path of eventual exhaustion. Like the Buddha, you must study, discipline yourself, climb and exert yourself to get to the top, only to realise that the answer was there all along, even before you began.

The Siddhi of Truth is everywhere. There is nothing that is not Truth. Whatever happens to you, whatever you feel in every moment — however delirious or disturbing — it is nothing but Truth. Truth is all there is.

At the Shadow frequency there is nothing but Truth, at the Gift frequency there is nothing but Truth. Truth is every step of the way. This book of the 64 Gene Keys is a yogic voyage into every conceivable archetype within human consciousness, drawing together past, present and future. It is an exploration driven by doubt, by the need to question every facet of existence. Fixed in every nook and cranny of our DNA, it is the very emblem of the human urge to understand our true nature. And yet, from the point of view of Truth, this book is a failure. It cannot be anything else. All attempts at understanding are doomed to fail. If you read through each and every one of the 64 Siddhis you will see the scope of this failure. Every Siddhi encapsulates the same paradox, tilted at a slightly different linguistic angle. There are no answers to doubt. There are only patterns and the illusion of the one looking at the patterns. At every stage and in every word there is Truth.

This realisation that everything is Truth is well represented by the I Ching name *After Completion*. The realisation of Truth involves a complete relaxation of your entire being. The quest is at an end because the quest has deleted the questioner. The doubter has entered into the question so fully that the question has swallowed his or her awareness. All paths to Truth are therefore a complete fabrication, all techniques only distractions and all systems and concepts are ultimately worthless. These are unhappy words for the seeker, who still remains identified with his or her search, but even that is Truth. There is no such thing as distorted Truth or hidden Truth. Truth is simply here now, in every moment, at every stage of your life. It is eternal, undying, pure, incorruptible and so beautifully simple.

Truth really is the most precious and beautiful of the Siddhis in its human expression because of its vast, vast relaxedness. It is like resting under a tree on a warm balmy evening after walking all day long. You gaze out at the sun setting into the sea, or the wind nestling into the leaves at your feet, or the stream purling softly over time-lapped rocks. With Truth, the inner journey — the attempt to follow the doubt right back up to its source high in the mountains and moors — comes to an end.

Whenever you give yourself rest from worrying, you are treading around the fringes of remembering Truth. Whenever you smile or breathe a deep sigh, there you are coming closer to Truth. Whenever your face loses its hard edges and becomes softer, there Truth is being remembered. Truth is your natural state — your simple state, without complexity or concern. It is a drifting space, a total sinking into your being. As you may see from the true ending of this book written in your genes, it has no need of meaning and no concern for or possibility of an answer. In a single sentence, Truth is the eternal moment of the universe.

64th Gene Key

The Aurora

SIDDHI
ILLUMINATION

GIFT
IMAGINATION

SHADOW
CONFUSION

Programming Partner: 63rd Gene Key

Codon Ring: The Ring of Alchemy
(6, 40, 47, 64)

Physiology: Pineal Gland

Amino Acid: Glycine

THE 64TH SHADOW – CONFUSION

THE CHAOS OF THE ELEMENTS

With the 64th Gene Key we come to one of the greatest mysteries of existence — the mystery of inner light. When this light is obscured within a human being it results in Confusion, the 64th Shadow. Confusion is the great human Shadow state. It sweeps across our world like a great blanket, smothering, disempowering and screening the mass consciousness from the true nature of reality. As the last of the 64 Gene Keys in their sequential form, the 64th Gene Key offers us some final warnings. This is after all the Shadow of Confusion. As we see repeatedly throughout the 64 Gene Keys, the Shadows are not inherently bad in the sense of being evil. They are in fact the raw material of the higher fields of consciousness — like the nugget of coal that may hide a diamond of great beauty. As the confusion reveals its underlying nature and begins to coalesce into an organised etheric substance, it becomes in turn the wonder of human imagination. Finally, when the imagination transcends itself at the highest frequencies, the inner light at the heart of all creation explodes inside your being as spiritual illumination. This is the journey of every human being.

Confusion in itself is a perfectly natural state. The ancient alchemists referred to this state as the *massa confusa*, the chaos of the elements, likening it to the primal swirling that preceded the birth of the universe. Confusion is a state with neither order nor structure; it is a state rippling with pure potentiality. Only when the human mind attempts to interpret it does it become bewildering. If you are able to look into this primal state of consciousness without engaging your mind in any way, you will see the true nature of being manifested as Illumination, the 64th Siddhi.

Each of the 64 Shadows is born out of the human mind's tendency to identify with whatever it sees. This tendency creates a biofeedback loop between the two polarities of each Shadow state — in this case, the loop is generated between the 64th Shadow of Confusion, and its programming partner, the 63rd Shadow of Doubt. Here is how it works:

At every moment, your thinking reflects the way in which the inner body is feeling. If your overall frequency is low, you generally feel a kind of background unease throughout your physical, emotional and mental bodies. This unease is generated by the global frequency in which we all live — in other words, every human being feels the suffering of the whole world through the quantum field that connects us all. The more you listen into your body, the more you will attune to this collective sense of unease that is rooted in fear. Most people develop patterns from an early age to escape feeling this vast desert of world pain and the mind is the first line of defence. As long as we are addicted to thinking, we can avoid fully feeling it.

This suffering within each human being is rooted entirely in the past. It came down into you through your ancestral DNA and was transferred to you as a child through the coping strategies of your parents and peers. Your basic urge to flee from this pain will keep you from ever facing what you really are, and this fact, lying deep within your cells, gives rise to another of the great human Shadows — Self-doubt. We human beings doubt ourselves because we are not really ourselves in the first place. Instead we inhabit the confusion, and the more our minds try to cope with this confusion, the more we feed our own self-doubt. It is a biofeedback loop. At the general low frequency of the planet, the mind cannot escape itself, but instead keeps feeding its own illusions. Those illusions then play out through the course of events that we call our lives. Thus our true potential is never fully lived, or as Thoreau so aptly put it, we "lead lives of quiet desperation."

The moment the mind stops thinking, the confusion ends, which shows you what a sham it is.

If you take just about any human being and you scratch the surface of their awareness, you will very soon uncover layers and layers of repressed pain. The mind lives in a state of permanent confusion, trying to make it all go away. But confusion is a dead-end street. It is a falsity fabricated by the mind. The moment the mind stops thinking, the confusion ends, which shows you what a sham it is. As you will shortly see at the siddhic frequency, the state that we refer to as confusion is actually the holiest state of awareness there is.

To return to our discussion of this Gene Key, it is all about light. All the 64 Gene Keys are really about light. This light, which is the inner nature of consciousness, is inherent within all forms. It lies obscured until evolution allows it to naturally reveal itself. This is the process of alchemy — the greatest science of all sciences. The 64th Gene Key is an integral aspect of the Ring of Alchemy, which includes three other great alchemical Gene Keys — the 6th Gene Key, which causes confusion in your relationships, the 47th Gene Key, which locks you into a futile battle with your own mind, and the 40th Gene Key, which leads to a profound feeling of isolation. When you are confused, you are alone. The light that moves through the 64th Gene Key wants more than anything to enthuse and fire your mind with inspiration rather than confuse you. But it depends on your attitude, which is the litmus test for your general level of frequency. Therefore, when you are feeling low, it is always best to transmute your thinking onto a higher plane. If you are unable to do this, at the very least you need to wait out the phase and allow it to evolve of its own accord. If you are patient and un-reactive, the light *will* shine through in the end.

REPRESSIVE NATURE – IMITATING

It takes an enormous amount of energy to repress our inherited human pain. To repress that amount of pain you actually need a great deal of support from others, ironic as that may seem. This support comes in the form of the status quo — from the millions of others who also hide from the truth of how the world feels and bury themselves in activity and thinking. Half of the world represses their confusion through imitation — they do what their parents did, or they do what their friends and teachers do. Imitation is the archenemy of imagination. It is a massive illusory safety net devised by the collective to keep from feeling the state of the world as it is. Repressive natures combat their fear through imitating everyone else. Some of them even manage to look original whilst they are doing it!

REACTIVE NATURE – CONFUSED

There are those whose nature simply cannot handle repressed feelings. Certain types of physiology are not designed to deal with the sustained voltage of pain held beneath the surface, at least not for any significant length of time. In such people, the core human pain held in their DNA is externally expressed through their outer lives. Such lives are always incredibly confused. These are the people who simply cannot stay in relationship without being abusive. Whilst the repressed nature is a victim of the status quo, the reactive nature is a victim of their anger towards the status quo. Their lives are lived in an unconscious attempt to take revenge on life itself. These people react against the status quo, often in an aggressive and unpredictable manner. Out of this basic dynamic between repressed and reactive, you can clearly see how both the abuser and the abused are born.

THE 64TH GIFT – IMAGINATION

THE ART OF LIFE

The depth of agony felt and expressed by the human race is both born and resolved in the genetic coupling of the 64th Gene Key and its programming partner, the 63rd Gene Key. Despite the reality of the pain we are born into as we enter a human body, there is good news. The solution is so simple. The suffering we feel — though real within the context of our physiology — is an illusion when it comes to our minds, and our minds hold the key to its release. As we see at the Shadow frequency, the mind automatically collapses into confusion when faced with the pain underlying life. If you were taken out of your everyday life and placed in solitary confinement for even the space of a single week, you would soon feel the profound pain that lies within your body. With nothing to do and nothing to distract your awareness, the pain quickly comes to the surface. Interestingly, this is one of the roles of meditation — to allow this pain to rise to the surface of your awareness so that it can be transmuted. All it takes is willingness to feel the pain, and the miracle begins.

Once you realise that there is no point in trying to make the pain go away, you may finally begin to see the loop you have been caught in. It is your mind that continually takes you away from the pain. As your inner spirit decides to face who you really are, for the first time ever the mind's tricks are witnessed. In that witnessing the mind's long reign over you gradually begins to crumble. The more you watch your mind trying to understand and/or avoid the confusion, the more a shift inside you begins to occur. Thinking takes a great deal of energy. Something has to feed all those neurons. Thus the moment you take away the mind's food — your belief that it can help you — all the energy that was going into feeding it is gradually liberated. In certain rare cases, this energy can be abruptly liberated causing the phenomenon known as sudden enlightenment.

All natural energy or life force has one inbuilt programme — to grow and evolve. This is the nature of life. Thus when you release the quantum energy that was latent inside you, it begins to rise. As it rises, the evolutionary force takes a hold of your mind and begins to paint with it. This is how true creativity is born and the human imagination is unleashed. Imagination is the expression of unimpeded life force coursing through your genetics. This is the 64th Gift. Imagination is born out of confusion, but only when you embrace confusion without trying to change it. Imagination allows alchemy to occur inside your being. In the beginning, you may paint, write, sing about or simply tell the story of your pain and how it feels to you. It doesn't matter what form the pain takes. What matters is that it can be expressed and accepted. In fact, at the Gift frequency pain can be more than simply expressed — it can become art.

If you allow your pain, or the world pain, to be expressed through an artistic process, you will see alchemy in progress. Your art will follow a natural sequence of archetypes. You will begin by descending symbolically into the underworld and give voice and shape to the demons and fear frequencies that live there. As you allow your frequency to rise further, your demons will gradually reveal their hidden nature. The light will emerge from inside them and the frogs will become princes. All true art is alchemical. Artistic process that does not release the inner light is simply the art of the Shadow consciousness finding its way into imagination. As long as you are courageous and honest, your expression will continue to evolve naturally to the higher frequencies. Eventually, we all end up painting the angels!

If you allow your pain, or the world pain, to be expressed through an artistic process, you will see alchemy in progress.

The other great power of imagination is that it spells the death of imitation. To imagine means to go where no one else has been before — to break free from mental, intellectual, and cultural constraints and let your mind soar upon the wings of your heart. Imagination is abstract, illogical and wild. It creates wormholes between worlds, moving too swiftly to analyse itself for meaning or reason. To imagine is to leap, to bound and to whoop with delight outside of all logic and pattern. It is the source of all art.

Those who exemplify the Gift of Imagination understand light and the properties of light. Light makes images possible, and images are the fuel of imagination as the etymology of the word suggests. To imagine is therefore to see and envision — shape, colour, form and movement. It bridges inner vision with the outer eye. It allows your life to become a work in progress — a true work of art.

THE 64TH SIDDHI – ILLUMINATION

ENLIGHTENED POETRY

The 64th Siddhi is the Siddhi of Illumination. This is Imagination without the *I*. When that *I* surrenders its need to exist, magic shines through. The evolutionary force obliterates the identification with form and Illumination floods your being. When you approach a siddhic state, you often experience occasional glimpses of that state before it fully detonates inside you. There are many different types of spiritual and religious experience. Different teachers speak of different states. Some advocate specific disciplines to attain these states, whilst others say they can only be attained when discipline has been entirely dropped. All words spoken from the siddhic state carry the same Truth. They simply come through one of these 64 genetic slants. Therefore, if you attach meaning to the words spoken by a particular master, you will very probably go adrift. You only have a 1 in 64 chance of meeting someone who speaks the exact language that matches your own potential primary Siddhis.

Perhaps the best advice for anyone who has contact with a person manifesting a siddhic state concerns the difference between hearing and listening. Only when you stop listening to such a person will you finally hear them. In this context, we are not talking about mental listening. The mind listens in order to end its own confusion. It is as though someone is talking to you but you cannot hear them because you yourself are talking at the same time. But the moment you let go of the meaning of the words, your true being can hear what is really communicated in the illusory space between you. Siddhis can only be transferred through silence. It is the silences between the words that carry the transmission. If a teacher is really gifted, the pattern of their speech will calm the listener's mind, inducing a kind of trance in which Truth can be transferred.

For centuries, Masters have used confusion as a means to stop the mind in its tracks and transfer the wordless truth. Perhaps one of the greatest examples is the Zen approach, which gives the mind an unsolvable paradox known as a *koan* on which to meditate. These koans so confuse the mind that eventually it stops to rest and regain some of its energy. In that rest a leap of consciousness known as a *satori* takes place. If you have the 64th Gene Key prominent within your Hologenetic Profile, you may experience these sudden leaps in consciousness. Sometimes confusion is direct, like the Zen approach, and sometimes it is indirect. The indirect transmission of Truth uses paradox or poetry to pacify the mind. If you know how to listen to poetry, then you are always close to Truth. Poetry is a metaphor for the wordless. It dances along the banks of silence, teasing you to enter.

The 64th Siddhi of Illumination and its programming partner the 63rd Siddhi of Truth represent the two wings of tantra and yoga — opposite paths towards the same ultimate reality. These are the higher frequencies of art and science respectively. Whereas yoga is a path of discipline aiming at progressive attainment of higher Truth, tantra is the path of surrender, which deals in sudden leaps in consciousness.

*To be
illuminated
means to
think God's
thoughts.*

Those who manifest the 64th Siddhi are those who teach spontaneously. They will use anything they feel like using as an illustration of what it means to be one with Truth. There is no logic or pattern to such people or their teachings. They may even use logic as a device and then contradict it entirely through their behaviour or words. The tantric path is the easiest path to misunderstand because it cannot be followed with the mind, but only with the heart. It takes a certain degree of madness in a person to follow this path, uncharted as it is. It is the path of the poetic soul — the lover of wildness, of spontaneity, of paradox — the lover of the moment.

When Illumination floods your being, it also takes possession of your mind. In this sense, to be illuminated means to think God's thoughts. Such thoughts are beyond our very concept of thought. At this level, the mental body gives way to the causal body (see the 22nd Gene Key), which allows indwelling awareness to become one with thought itself. Such thought makes no sense to the mind, but burns itself white-hot into your very DNA. At the Divine level, your causal body (sometimes called your *soul*) becomes your physical vehicle. This great mystery is the reason why the initiatic traditions say that after enlightenment you can no longer incarnate in the world of form. You can incarnate as a subtle higher being, but not as a carbon-based life form.

Once the frequency of inner light illuminates physical form, that form begins to ascend. It may or may not be able to ascend physically, depending upon its genetic constituents. Physical ascension is directly dependent upon Divine Will (the 40th Siddhi), and as such it is always predestined. Only certain evolutionary human vehicles are equipped for this purpose. This is the meaning of Transfiguration (the 47th Siddhi) — the causal body effectively draws the physical body up into itself. Only the higher codes of the Ring of Alchemy involve such rare occurrences. The 64th Siddhi also gives rise to the *aureole* — the halo surrounding the head of the enlightened. This image, so prevalent in the religious art of so many cultures, is a direct reflection of the Siddhi of Illumination as it has occurred throughout history.

The Siddhi of Illumination is the direct expression of Divine Mind. Divine thoughts manifest instantly in the world of form — such is their power. Thus one in the siddhic state of Illumination directly experiences divine creation. If there is any identification with these thoughts, the indwelling awareness will see itself as a God or Messiah endlessly creating the world through thought. However, the highest state of Illumination negates any form of identification. Divine thoughts cannot be understood. They are simply conduits of different frequencies, woven together into beautiful ideas, poems, words or images. They have nothing to do with the body or brain that generates them — they are solely for the inspiration of others. Such a Siddhi makes no sense, and yet it also makes the most perfect sense of all. To be illuminated by the 64th Siddhi is to remain empty, while being constantly flooded with the rainbow colours of the inner aurora — it is to be an easel for the imagination of the universe itself.

The wind on the water
sings with a face
of forgotten words

GLOSSARY OF PERSONAL EMPOWERMENT

The following Glossary is designed as a contemplative tool in its own right as well as a reference for understanding the Gene Keys terminology. Each word contains an empowerment, as does its description, so that as you contemplate these words and their deeper meanings as well as their relevance to your life, they may over time help you to raise your own frequency. The Glossary also operates in a holographic way so that one term appears within another. The more words you therefore contemplate, the deeper the wisdom moves within you and the more clearly you will see the overall vista. Eventually the words themselves and their meanings may become embedded in you as a field of higher remembrance that transcends the language itself.

Absorption — A state of consciousness in which your aura begins to feed off its own light, thus perpetuating a very stable high frequency throughout your being. As you enter the state of absorption your DNA begins to trigger your endocrine system to secrete certain rarefied hormones on a continual basis. These hormones are associated with higher-brain functioning and involve states of spiritual illumination and transcendence. At such a stage, it is no longer possible for you to be drawn back into the lower frequencies for more than brief periods of time. Arising naturally out of contemplation and leading to embodiment, absorption occurs when you first begin to inhabit the buddhic body, after the Fourth Initiation.

Activation Sequence — The Activation Sequence is the primary genetic sequence in your Hologenetic Profile. The Activation Sequence describes a series of three leaps in awareness that unfold in your life as you activate the higher purpose within your DNA.

These inner realisations are called your Challenge, your Breakthrough and your Core Stability. Calculated from the position of the sun at the time of your birth, your Activation Sequence pinpoints four specific Gene Keys (known as your Four Prime Gifts) that form the vibratory field of your genius. As its name suggests, your Activation Sequence is a trigger that can catalyse a period of intense transformation in your life.

Amino Acid — A chemical constituent used by your body to build proteins. There are 20 major amino acids. You have within you the power to influence the various combinations of amino acids in your body, thus building a body of radiant health and catalysing the chemical foundation for all higher states.

Archetype — A compressed idea containing many dimensions of meaning, imagery and feeling. Archetypes resonate with deep universal themes in the unconscious of all human beings, regardless of race, genetics or conditioning. According to the Gene Keys, there are exactly 64 universal archetypes. When you continuously contemplate and identify with a particular archetype, it has the capacity to transform your state of consciousness from a low-frequency to a high-frequency.

Astral Body — The second major subtle layer of your aura, corresponding to the astral plane. Of all the subtle bodies, the astral body is closest in vibration to the physical body and its etheric counterpart, which means that your emotional life has the most powerful and direct effect on your physical health and vitality. The astral body is gradually developed in your second seven-year cycle – from the age of 8 until 14 – during which time all your major emotional patterns are laid down. As you contemplate the shadow consciousness field and how it affects and governs you personally, you are reaching down into your astral body and re-imprinting your basic emotional patterns with a higher frequency. This will bring about a major transformation in all your relationships as you become less reactive and more emotionally mature.

Astral Plane — The second of the seven major planes of reality upon which all human beings function. The astral plane is a subtle electromagnetic field generated by all low-frequency human desire and emotion. On the astral plane of reality, your every feeling or desire has an independent existence and can be understood as an entity with its own vibratory frequency. Through the law of affinity, you draw into your aura the astral entities that match the frequency of your feelings and desires. As you purify your emotional nature, it gradually becomes impossible for astral entities to influence your life and feelings. At this point, you begin to function on the higher octave of the astral plane known as the buddhic plane.

Atmic Body — The sixth major subtle layer of your aura, corresponding to the atmic plane and the Christ consciousness. Your atmic body, which manifests through the 64 Siddhis, is so vast that it defies comprehension. To enter fully into this body, your identification with the lower bodies – physical, emotional (astral) and mental – must be severed completely. When this happens, your cycle of incarnations will come to an end. The atmic body creates an increasing pressure on your lower nature as its light gradually filters down to illuminate the lower three bodies, resulting in a magnificent phenomenon known as the dawning of the rainbow body. Over time, this leads to a complete restructuring of your life as the inner light of the atmic plane dawns inside you, culminating in your full embodiment of divinity.

Atmic Plane — The sixth of the seven major planes of reality upon which all human beings function. The atmic plane is the higher-frequency octave of the mental plane and is the plane of your true higher self. On the atmic plane of reality, the entire cosmos is experienced as a living mind whose primary impulse is love. When you cross the threshold to this plane (through the Sixth Initiation), then all your independent thinking immediately ceases to be replaced by pure light. To contact your greater being on the atmic plane, all you have to do is focus consistently and intently on this inner light.

Aura/Auric Field — The true multi-dimensional body of all human beings, the aura is a broad-spectrum electromagnetic field emanating from and grounded by your physical body. Your aura acts like a prism that refracts various different wavebands of light. It is the electromagnetic expression of your chemical state at any given moment. This also means that you can change your physiological chemistry by means of your aura. Through meditating or concentrating on your aura, you can refine its frequency and impact your physical, emotional and mental wellbeing for the better. As the overall frequency of your aura rises, its magnetic power also expands and you even begin to transform your environment and those around you. Your aura always relays the absolute truth of your inner state, registering your every thought, feeling, urge and hidden pattern as a subtle emanation.

By working with the 64 Gene Keys, you are effectively purifying the many dense layers of your aura, making you more radiant, more sensitive and more compassionate towards all beings.

Awareness — An aspect of consciousness unique to all life forms. In a human being, awareness can be divided into three main layers, although in reality, they are all a single awareness – physical awareness, emotional awareness and mental awareness. At low levels of frequency, human awareness is confined to the human body – physical awareness remains rooted in survival and fear, emotional awareness remains rooted in desire and drama and mental awareness remains rooted in comparison and judgment. As you raise the frequency throughout your being, your awareness becomes more refined and shifts from the local environment to the cosmic. Physical awareness becomes divine presence, emotional awareness becomes universal love and mental awareness becomes silence and wisdom.

Bardo — The intermediary stage between incarnations. At a certain point before your physical death, you enter into the bardo sequence, in which the subtler layers of your aura begin a process of disengagement from the lower three bodies (physical, astral, mental). This sequence leads to physical death but also continues after death. In the bardo stages after death, our subtle bodies move through a series of alchemical distillation in which the low-frequency material that we have collected throughout our incarnation is separated from the high-frequency essence. The continuity of the bardo cycles and their sequences interconnect our many lifetimes until we transcend human evolution at the Sixth Initiation.

Bhakti — The subtle fluidic emanations generated by your heart when it devotes itself to the highest good. You generate bhakti through your aura as your inner radiance continues to grow. Bhakti is the refined essence of your lower three bodies (physical, astral and mental). Through living a life of devotion and service, you flood your own aura and the world with bhakti. Bhakti always rises and is the counterpart to shakti, which always rains down. Whereas bhakti is the evolutionary quintessence of all that is good in human beings, shakti is the involutionary quintessence of the Divine. The more you open your heart and allow it to speak and live in your life, the more bhakti you generate and the more shakti (divine grace) rains down upon you.

Buddhic Body — The fifth major subtle layer of your aura, corresponding to the buddhic plane. Your buddhic body is only accessible to you once your heart has completely opened. All true higher mystical experiences or revelations come through your heart and are therefore rooted in the buddhic body. Once your awareness is fully anchored in the buddhic body, the causal body dissolves and reincarnation in the normal sense is no longer possible. The buddhic body also corresponds to the mystical state of absorption (the Fourth and Fifth Initiations), which is what occurs as your awareness becomes stabilised in the higher frequencies. It represents the third feminine realm of the Divine Trinity – that of divine activity or compassionate action.

Buddhic Plane — The fifth of the seven major planes of reality upon which all human beings function. The buddhic plane is the higher-frequency octave of the astral plane and is the plane of devotion and ecstasy. In many ancient traditions, this plane of reality is known as the realm of the gods and goddesses. When you identify with a divine entity, you are engaging directly with the buddhic plane. Here your own individual identity becomes merged into the collective body of humanity, and all the pain and anguish of the lower planes are transformed into an all-encompassing love. As you purify your DNA of its ancestral memories and shadow patterns, so you gradually come to know a new life on the buddhic plane.

Causal Body — The fourth major subtle layer of your aura, corresponding to the causal plane. The causal body is generally known as the soul, as it represents that aspect of your consciousness that incarnates again and again into the world of form. Your causal body stores the collected goodwill of all your lives as a memory signature written in light. After death, your lower three bodies (physical, astral and mental) disintegrate and only that which is refined and pure is drawn up and retained in your causal body. As your causal body develops more lucidity through the process of reincarnation, so the higher bodies can use it as a means of directing the lower three bodies to higher and higher frequencies. In this respect, your causal body is the great bridge between the lower and higher planes.

Causal Plane — The fourth of the major planes of reality upon which all human beings function. The causal plane is the higher-frequency octave of your physical body and represents the plane of pure archetypes, where thought and feeling become one. The causal plane is the realm of synthesis, and all human genius rises into this plane in order to see the holographic nature of the cosmos. Even though the causal plane is beyond language, its energy and nature can be communicated through language as frequency. The 64 Gene Keys themselves are a transmission from the causal plane. The causal plane also forms the bridge that links the human evolution to the higher evolutions beyond humanity and as such, it holds the mysteries of death. To raise your frequency beyond the causal plane is to transcend death.

Codon — A section of DNA made up of three base pairs, and which codes for a specific amino acid. There are a total of 64 codons in human DNA. The 64 Gene Keys allow you to directly communicate with the 64 codons in the DNA of every cell in your body through the law of resonance. Through contemplation, absorption and embodiment of the Gene Keys, you will raise the frequency of vibration in the 64 codons, thus unlocking the secrets of your higher nature.

Codon Ring — A chemical family within your body made up of one or more codons. There are a total of 21 Codon Rings, each one relating to a specific amino acid or stop codon. The Codon Rings are transgenetic chemical families that operate across entire gene pools, drawing certain people naturally together in pairs, groups and ultimately, forming whole societies. The Codon Rings are the biological machinery behind what the ancients called karma. The manner in which they interlock forms the geometric unified field underpinning humanity, and as human DNA mutates in order to carry the higher frequencies that are coming with the Great Change, the 21 Codon Rings will bring humanity into the biological realisation of its true nature and unity.

Concentration — One of the three primary paths leading to the higher states of absorption and embodiment. Concentration is the yang path, represented by the ancient science of yoga. It utilises focused effort and willpower to bring about a series of transformations that gradually raise the frequency of your awareness.

Consciousness — Consciousness is all that exists. As the source and creator of reality, it is indivisible, omniscient, omnipresent and omnipotent. Consciousness does not necessarily involve or require awareness. It is the ground of all being and non-being.

Contemplation — One of the three primary paths leading to the higher states of absorption and embodiment. Contemplation is the central path, represented by the Tao. It utilises elements of both concentration (effort) and meditation (no effort) to bring about a heightening of your frequency. Contemplation takes place on all three of the lower human planes; there is physical contemplation, emotional contemplation and mental contemplation. Over time, contemplation transforms the physical, astral and mental bodies into their higher-frequency counterparts – the causal, buddhic and atmic bodies. Prolonged contemplation on the 64 Gene Keys is one of the quickest and easiest ways to activate the higher frequencies lying latent within your DNA.

Corpus Christi — One of the journeys making up the Gene Keys Synthesis, the Corpus Christi is the complete science of the rainbow body – the true underlying nature of all human beings. The Corpus Christi is a synthesis of transmissions, teachings and techniques that underpins the 64 Gene Keys. Representing the higher Mystery School teachings of the Gene Keys, it includes the teaching of the Seven Seals, the Seven Sacred Bodies and the Nine Initiations. Deep immersion in the teachings of the Corpus Christi assists you in grounding and embodying the higher frequencies of light into your everyday life. These are the teachings and techniques that allow you to draw the transmission of the Gene Keys layer by layer into the subtle bodies that make up your aura. Literally meaning The Body of Christ, the Corpus Christi prepares you to work with higher evolutionary frequencies by progressively purifying the many dimensions of your inner being, beginning with your physical body.

Cosmometry — The visual science of the sacred geometries underlying creation. Through understanding the foundation of cosmometry, your mental body (your mind) can surrender into the certainty of the perfection of all phenomena. Through accepting the laws of cosmometry, your astral body (your feelings) can over time become purified and loving. Through embodying the principles of cosmometry, your physical body can become radiant and relaxed as it comes into vibrational harmony with the entire cosmos.

DNA — Deoxyribonucleic acid. The multi-dimensional programming software of human consciousness. Your DNA is a supersensitive substance found in every cell of your body.

Depending on your attitude at any given moment, your DNA is the architect (via your endocrine system) of your physical, emotional and mental reality. As you raise the frequency of energy passing through your DNA, it unveils higher programming functions that were lying latent inside you. These higher functions give rise to the natural expression of your innate genius. At the most refined levels of frequency, your DNA synthesises hormones that allow your physical awareness to actually transcend DNA itself, culminating in the embodiment of your inner divinity.

Embodiment — The natural culmination of the process of concentration, meditation or contemplation. After you attain the state of absorption (the Fourth and Fifth Initiations), you eventually make the great quantum leap into full embodiment (the Sixth Initiation). Embodiment relates to what many traditions know as enlightenment or realisation. It involves the complete embodiment of the higher three bodies onto their corresponding lower planes. The process of embodiment begins from the moment you are born into a human body, and it follows the trajectory of your evolution. The more evolved you are, the more embodied you become.

Epigenetics — The branch of genetics devoted to the study of how and why environmental signals can mutate gene expression within the body. The revelations emerging out of epigenetics are already changing the way mainstream science thinks about evolution. While the old paradigm thinking was that our genes come prepackaged and govern all aspects of our behaviour, epigenetics is showing how our interactions with the environment can create lasting changes in our DNA, some of which are imprinted in our offspring for generations. The greatest import of epigenetics is that it demonstrates how interrelated all life is, thus building a greater understanding of the universe as interdependent and holographic. It is through epigenetics that we can see the true power of human attitude and its electromagnetic effect on our DNA. Different frequencies of attitude activate different expressions of genes, even though the sequence of those genes may remain the same. The greatest insight offered by epigenetics is still dawning in the scientific world – the timeless truth that consciousness creates reality.

Etheric Body — Sometimes known as the etheric double, the etheric body is the counterpart of your physical body, extending out beyond it into the aura.

As the closest of the subtle bodies to the physical, the etheric body is quite well understood by many cultures, in particular in its relationship to our physical health. Systems such as acupuncture or energy medicine work directly on the etheric body, which consists of a vast network of subtle pathways, meridians or nadis, which create the fundamental grid of the aura. Illness and disease in the physical body first manifest through blockages in these inner energetic pathways. However, the root of all human illness does not lie in the etheric body but is to be found deeper in the astral or mental bodies. As you purify your astral and mental bodies, your etheric body is directly impacted and blockages to the flow of chi or prana are cleared. This releases a huge amount of healing energy and vitality into your physical body, which over time becomes lighter, healthier and more radiant. In the teachings known as the Corpus Christi, the etheric body is seen as an integral part of the physical body rather than a separate subtle body in its own right.

Evolution — An impetus or higher will innate in all material forms. The current of evolution is responsible for gradually raising the vibratory frequency of all matter, even matter without awareness. As forms evolve, they gradually assume awareness and in time transcend their sense of separation and return to their formless essence. There are many spheres of evolution, all interconnected with each other, and human evolution is but one. Evolution represents that force within matter that always strives upwards towards spirit, as opposed to involution, its counterforce, which is the essence of spirit descending or embedding itself within the form.

Fractal — The holographic manifestation of light as it enters the material world and illuminates its true nature. A fractal is an endlessly repeating natural pattern that is maintained throughout the universe regardless of scale. For instance, the microscopic patterns within the membranes of cells in the human body are similar to those observed across the landscape of the Earth when viewed from space. Similarly, the geometric laws that govern galactic nebulae are visually replicated when you slice a fruit in half. The more deeply you realise the fractal nature of reality, the more embodied and loving you become. Every act you make in life generates a fractal wave pattern that affects all creatures in the universe. Through a process of continual evolutionary biofeedback you can refine your life to such a point that you are in complete resonance with every fractal aspect of the universe.

Fractal Line — When our current universe was conceived at the moment of the Big Bang, the crystalline seed of our evolution shattered into countless fractal shards or fragments. These fractal aspects of the whole radiated out in precise geometric patterns known as fractal lines. All fractal lines can be traced back to one of three primary fractal lines, thus seeding the trinity within all aspects of the holographic universe. As you move into deeper harmony with your true nature, you come into alignment with all beings within your seed fractal line, which catalyses great synchronicity and grace in your life.

Frequency — A means of measuring the vibratory nature of radiant energy such as sound, light or even awareness. The central premise of the Gene Keys Synthesis is that you can alter the frequency of light passing through your DNA, thereby speeding up or slowing down the force of evolution itself. Through deep contemplation on the 64 Gene Keys and their teachings, you can raise the frequency of your DNA and thus change the vibratory frequency of your aura, coming into higher and higher states of harmony with the universal field.

Frequency Band — In the Gene Keys Synthesis, the vibratory rate of your aura is mirrored in three frequency bands known as the Shadow, the Gift and the Siddhi. Although there are in fact many layers or bandwidths of frequency, this threefold language makes the Gene Keys simple to understand, contemplate and ultimately, embody. The three frequency bands are laid out precisely through the Spectrum of Consciousness – the linguistic map of the 64 Gene Keys and their frequency bands.

Gene Key — One of 64 universal attributes of consciousness. Each Gene Key is a multi-dimensional portal into your inner being, whose sole purpose is to activate your higher purpose and ultimately, allow you to embrace your own divinity. One way in which your higher purpose is activated is through sustained contemplation of the Gene Keys and their frequency bands.

Gene Keys Synthesis — The Gene Keys transmission made manifest as a new body of world teachings. The Gene Keys Synthesis is a holographic synthesis of many of the great lineages and strands of human spiritual teachings and paths. The Gene Keys Synthesis is one of the most comprehensive and accessible holistic systems for exploring, understanding and integrating the vast changes currently taking place across the face of our planet.

Genetic Code — The master code of all known organic life. The genetic code contains the programming hardware for physical, emotional and mental life. Your genetic code comes with layers of programming that can be activated by means of your attitude – your thoughts, feelings, words and actions. Low-frequency signals (rooted in unconscious fear) activate the older hardware of your reptilian brain, whereas high-frequency signals such as creativity and love activate the genetic hardware that allows you to experience and eventually embody higher consciousness.

Genius — The innate intelligence of all human beings. True genius (as opposed to intellectual genius) is a spontaneous and unstudied creative uniqueness rooted in unconditional love. Genius is the natural manifestation of a human life when it is allowed to expand without force. Genius is a hallmark of the Gift frequency band where self-forgiveness leads to a progressive opening of your heart, resulting in an explosion of creative energy throughout your being. The higher the frequency of your DNA rises, the greater your urge will be to use your genius in service to the whole. As more and more people join their genius together, the world as we see it today will be transformed.

Genome — The complete genetic matrix of any living organism. Made up of DNA, your genome contains the entire set of hereditary instructions for building and maintaining your life. At a quantum level all genomes are holographically related and interconnected, which means that when one species or even one individual mutates its genome, then all other species and individuals are subtly affected. All genomes are designed as open systems that can mutate their own programming according to environmental signals. Your genome is the physical repository of your sanskaras – the blueprints of your karma in this life. As you live out this karma, so your genome mutates until one day you transcend your DNA entirely and attain the state of full embodiment at the Sixth Initiation.

Gift (Frequency) — The frequency band relating to human genius and open-heartedness. As your awareness delves more fully into the Shadow frequencies, it unlocks latent energy held within your DNA. This energy is released through your physical, astral and mental bodies as light. Physically, this can lead to changes in your body chemistry and increased vitality.

Emotionally, it can lead to uplifting feelings, joyousness and a pervading sense of optimism. Mentally, it can lead to insight and great creativity. The Gift frequency is a process of gradual revelation as your true higher nature (the Siddhi) is unveiled. There are many states and stages within the Gift frequency band and it represents the quantum field where the forces of involution and evolution come together. One of the hallmarks of the Gift frequency is the ability to take full responsibility for one's own karma – that is one's thoughts, feelings, words and actions. At this level of frequency, one no longer identifies as a victim of any perceived external stimulus.

Golden Path, The — The master genetic sequence for permanently raising your frequency from the Shadow to the Gift frequency. An integration of the Activation Sequence, the Venus Sequence and the Pearl Sequence, the Golden Path describes the natural unfolding of human awareness as it matures beyond the victim patterns of the Shadow frequency. All human beings, sooner or later, must walk the Golden Path, as it symbolises the passage of the individuated soul through the first four initiations. As your lower three bodies (the physical, the astral and the mental) are gradually purified and brought into harmonic resonance, you will experience the opening of your heart and the releasing of your creative genius into the world. The Golden Path lays the foundation for living a high-frequency life.

Great Change, The — A phase of evolution in which all systems within our universe will make a quantum leap into a higher dimension. The Great Change refers to a specific time period in which human awareness is moving from being self-centred to being collective. In order for this shift to take place, a worldwide genetic mutation is underway within the human species. This unprecedented event is taking root within humanity between the years 1987 and 2027, and its repercussions will continue to evolve and transform our species for many hundreds of years. The result of the Great Change will be the gradual dawning of a new kind of human, Homo Sanctus – the sacred human. Because all systems and species throughout the universe are holographically interconnected, the Great Change is not local to our solar system alone but is one part of a vast ripple passing throughout the immensity of spacetime.

Hexagram — A pictographic binary symbol that forms the basis of the I Ching. The 64 hexagrams of the I

Ching are directly analogous to the 64 Gene Keys. Each hexagram is made up of six lines, either broken (yin) or unbroken (yang). The Gene Keys offer a modern interpretation of the 64 hexagrams as they relate to our core genetic structure and to the underlying structure of the universe itself. Each Hexagram or Gene Key is a portal to an encyclopaedia of knowledge and insight about yourself and your place in the universe. Through sustained contemplation on the hexagrams, their structure and interdependence, you can raise the frequency of light moving through your DNA and experience life at a new level of awareness.

Hologenetic Profile — A universal geometric matrix whose central purpose is to show the relationship between an individual and the whole. Your Hologenetic Profile is a personalised map of the various genetic sequences that will unlock or awaken different aspects of your genius. Unifying astrological calculation with an archetypal understanding of genetics, your Hologenetic Profile is the original blueprint that tells you who you are, how you operate and, above all, why you are here. As you contemplate your own Profile and its many pathways, sequences and geometries, it will activate and awaken the resonant faculties within the living field of your aura. As the central pathworking tool within the Gene Keys Synthesis, your Hologenetic Profile allows and invites you to bring the power of contemplation progressively deeper into your life. Because of its hologenetic nature, wherever you place your awareness within your Profile, you will activate all other pathways simultaneously. This means that all the sequences and pathways within your Profile serve the same purpose – to raise the frequency of your whole aura and activate the higher purpose within your DNA.

Holographic — The underlying nature of the universe. The foundation of the holographic view is the realisation that everything is a mirror of everything else, and that everything is dynamically interrelated and moving through time and space in perfect concert. When you finally come to experience your true nature as embodiment, you stand in the centre of the hologram of existence, and every single cell within your body resonates with this fundamental holographic truth.

Homo Sanctus — Literally, the blessed human, Homo Sanctus is the new human emerging into the world. Catalysed by the Great Change, Homo Sanctus is the new genetic human vehicle.

Although this new human may have the same genome as the existing model, it resonates from birth at a higher frequency due to subtle mutations that are taking place primarily within the solar plexus system. These mutations allow the activation of higher coding sequences within our DNA, which essentially make the vehicle immune to the Shadow frequencies. Homo Sanctus represents what many mystics have termed the emergence of the Sixth Race – a universal human who directly experiences the holographic unity of all beings through his or her solar plexus and heart. Homo Sanctus will emerge in the world gradually as a worldwide genetic mutation over many generations. The dawn of this new human will, in time, bring an end to the current age in which human beings have forgotten their true universal nature.

I Ching — The original prima materia of the Gene Keys, the I Ching is a sacred Chinese text dating from around the 4th century BC. Many commentaries and versions of the I Ching exist, and it is perhaps best known as a popular oracle. The Gene Keys are a natural culmination of all previous incarnations of the I Ching. They ultimately point to the truth that all sacred texts have their source inside us. The same truths intuitively grasped by the ancient sages can now be proven by modern genetics – that the universe is built upon natural codes and these codes can be deciphered and unlocked. The original I Ching was held in the highest esteem as a sacred text with the capacity to mirror living wisdom in every moment. Likewise, the Gene Keys point us inwardly to seek the source of our suffering in our Shadows and guide us in transforming that suffering into creativity and freedom.

Initiation — As consciousness travels through its human journey, it follows a fixed evolutionary structure. This structure is known as the Nine Portals of Planetary Initiation (as outlined in the 22nd Gene Key). As we travel in and out of form in our incarnative journey, we move gradually along this ladder of consciousness. At certain points in our journey, we move through initiations – periods of intensity in which we undergo huge transmutations. True initiation is not something that can be predicted or ritualised. It is a natural part of life itself as it evolves. Neither does it require any learning or religious or spiritual affiliation. True initiation is a rare process that one always undergoes alone. The great initiations are upsurges in the frequency of our subtle bodies, and when we pass through them, they are often dramatic and challenging to integrate into our lives. Not only do individuals pass through the initiations, but humanity itself also must pass along this same ladder of consciousness.

Involution — The means by which grace – that divine essence that lies beyond all understanding – incarnates progressively into form. Involution is the counterforce to evolution – that current which gives us the impression we are progressing and evolving of our own accord. From the point of view of involution, all things are predestined and there can be no individual free will since all events are simply playing themselves out according to a higher unravelling. As evolution evokes aspiration towards something higher, so involution invokes inspiration as something higher that already lies inside us waiting to be discovered.

Karma — Karma refers to the specific slice of suffering you have undertaken to transform during your lifetime. The agents of karma are your sanskaras – the specific manifestations of karma as it is played out in your life. From the point of view of the Gene Keys Synthesis, karma is understood both personally and impersonally. At the personal level, karma helps you to understand and accept the specific conditions of your life as the fruits of your past karma. Karma is therefore the fuel of awakening, as its transmutation allows us to ascend to ever higher frequencies. Karma is determined according to the level of frequency made manifest in our subtle bodies, and by the stage we have reached as we move through the Nine Initiations (described in the 22nd Gene Key).

On the impersonal level, karma can be understood in a truly cosmic way. All acts are performed by the whole for the purpose of the whole. Impersonal karma is karma that we have taken on in addition to our personal karma. As we transmute our personal karma, our hearts widen enough to begin transforming the collective karma of humanity. This occurs to a greater degree after the 4th Initiation of Matrimony. All human DNA carries this ancestral karma, and in order for full awakening to occur, it must be transmuted by compassion.

Logoic Body — The mysterious eighth body of the Corpus Christi. The logoic body represents the body that always lies beyond the concept of beyond. In the mystical teachings of the Corpus Christi, the ultimate state of consciousness is represented by the seventh body, known

as the monadic. The eighth, logoic body is the paradoxical expression of the void itself. After all the currents of evolution and involution have played out their cosmic dramas, then, once again, the cosmos as we know it will cease to exist. The logoic body represents the eternal cosmic pause known to the ancient Vedic sages as the Night of Brahma.

Maya — An illusory veil formed by the human mind that prevents consciousness from realising its eternal nature. All forms exist within layers and layers of sheaths. Our human perception is also subject to the limitations of our frequency bandwidth. As we attune our awareness to higher frequencies, it passes through the layers and sheaths of the Maya.

Meditation — One of the three primary paths leading to the higher states of absorption and embodiment. Meditation is the yin path, represented by the ancient science of tantra. The true essence of the meditative path is to simply watch, witness and allow. Through meditation, one gradually comes to the realisation that one's true nature dwells in a choiceless awareness. This great revelation may come as a gentle unfolding that raises the frequency of your awareness over time or as a sudden implosion that allows you a permanent experience of your divine self – or it may come as both.

Mental Body — The mental body exists at a higher frequency than your emotions and is constructed out of your thinking life. The mental body is greatly influenced by the collective mental body of humanity itself, which tends to pull our thinking down into the unfulfilled desires of the astral body. As your thinking revolves around higher impulses, the mental body gradually disentangles itself from the astral body and takes on greater power. The mental body can also be used by the lower consciousness to repress the natural impulses of the astral body, which can also lead to problems in health at all levels. A low-frequency and limiting mental paradigm sets up low-frequency emotional patterns within the astral body, whereas a high-frequency mental paradigm creates emotional clarity and freedom.

Mental Plane — The third of the seven major planes of reality upon which all human beings function. The mental plane is the frequency plane created and dominated by the energy of thought. On the mental plane, all thoughts and ideas have an independent life that human beings either attract or repel. The mental plane itself is made up of different strata of mental energies that resonate at different frequencies. A low-frequency mental paradigm, for example, is created by thought patterns that are self-limiting, creating division and separation. Such a paradigm activates neural pathways that are based upon fear and survival. A high-frequency mental paradigm is characterised by a mental openness that encourages insight and breakthrough from the higher causal plane. This kind of thinking is unifying, positive and sees where things are interconnected rather than where they are threatened. As you enter a higher mode of thinking based on how to be of greater service to others and to the whole, all manner of insights and gifts begin to dawn in your mind. Eventually, your awareness transcends the frequencies of thought altogether, and you rise above the mental plane and experience true clarity.

Monadic Body — The seventh major subtle layer of your aura, corresponding to the monadic plane. The monadic body cannot be said to be a body, rather it represents the point of origin of the current of involution or divine will. When human consciousness has attained its highest potential in the sixth, atmic body, then it is said that the rainbow body dawns. This refers to the monadic body, which has often been symbolised as a flowering or a rainbow, since it is an expression of unity and completion. For the monadic to be expressed through human awareness, the lower three bodies – physical, astral and mental – must be absorbed into their high-frequency counterparts – the causal, buddhic and atmic. When this happens, all layers and levels dissolve, all powers and Siddhis are transcended and surrendered, and you become once again a truly ordinary human being.

Monadic Plane — The plane of the first aspect of the Holy Trinity – that of divine will. The monadic plane is the event horizon where the twin currents of evolution and involution spontaneously dawn. Represented symbolically by the hub of the wheel of life and law, the monadic plane is the central organising principle around which all life is built. Infinitesimal in nature, the monadic plane exists down to the tiniest imaginable particle and imbues it with consciousness. In the mystical metaphor of the Corpus Christi, the monadic plane is the intersection of the lower and higher trinities and as such, it is not really to be understood as a plane in and of itself. It is the transcendence of all levels of separateness and the culmination of human evolution.

Mutation — An unpredictable event that breaks the continuity in any linear sequence, at any level within the universe. In genetic terms, mutations are mistakes made during cell replication. Mutation is the mother of difference, since it creates endless forks in the evolutionary impulse, leading to new and unseen processes. In our everyday lives, mutations also occur all the time. They occur whenever there is a break in the established patterns or rhythms of your life. It is our fear of mutation that fuels the shadow frequency field. For example, when you find yourself moving through a period of mutation, you will feel a profound uncertainty about yourself and your life. If you repress or react to this feeling out of fear, you will disturb the processes of good fortune that always accompany mutation. As you learn to surrender to the natural mutative processes in your life, you will unlock the powerful creative gifts inside you and place yourself in alignment with the synchronicity of your true destiny.

Pearl Sequence — The third and final sequence comprising the Golden Path, the Pearl Sequence is the primary genetic sequence for opening up our mental awareness to operate on a higher plane. Constructed from the positions of Mars, Jupiter and the Sun at the time of your birth, the Pearl Sequence is a contemplative journey using the Gene Keys whose purpose is to open your mind to a transcendent view of the universe. Such a view allows one to see the inherent simplicity of life and move one's energies and resources into alignment with it. Your individual Pearl Sequence is made up of four specific Gene Keys which have a direct bearing on your ability to be efficient and prosperous in your life. Each of these four Gene Keys shows you a shadow pattern that prevents you from living a prosperous and liberating life. As your awareness enters into these patterns and unlocks their hidden gifts, you will discover a resource of untapped genius and creativity. The other great secret of the Pearl Sequence is the power of philanthropy as a worldview. The Pearl allows us to find our closest allies and work together in service to our community and a higher goal.

Prime Gifts — Calculated by the time, date and place of your birth, the Prime Gifts are a series of four Gene Keys that relate profoundly to your overall purpose in life. Known as your Life's Work, your Evolution, your Radiance and your Purpose, the Prime Gifts represent the living field of your genius that was imprinted in your DNA at the point of conception. By understanding and embracing the Shadow aspects of the Gene Keys that correspond to your four Prime Gifts you will activate their higher frequencies and catalyse a mutation to take place deep within your DNA. This process is known as the Activation Sequence. By sustained contemplation on the highest frequencies of your four Prime Gifts, you will witness a complete transformation in your life as you unlock the true genius inside you.

Programming Partners — Two Gene Keys that are holographically bonded together through opposition – in other words, they are exact mirror opposites. There are 32 such programming partners within the genetic matrix, and each creates a biofeedback loop that reinforces the themes of those Gene Keys at every level of frequency. At the Shadow frequency, the programming partners create physical, emotional and mental patterns and complexes that mutually reinforce each other. As awareness penetrates these patterns and transforms them, they release waves of creative energy at the Gift frequency, which in turn are mutually reinforced, leading to a continual raising of one's evolutionary frequency. At the Siddhic frequency, the programming partners no longer oppose each other but dawn as pure consciousness, creating a harmonic so pure that it deletes their difference.

Sanskaras — Biogenetic memories that are passed down all ancestral lines. Your sanskaras are the wound opportunities that you have inherited during your lifetime. Such memories are more than simple memories held in the mind but are charges of kinetic energy that give rise to your behavioural patterns, beliefs and general outlook. Sanskaras are not in any way personal, nor are they the result of actions in past lives. Rather, your sanskaras determine the specific themes of the great challenges you will face during your life. Once you realise that such patterns are not caused consciously by you but are your greatest opportunities for transformation and evolution, the challenges in your life become much easier to bear. The Venus Sequence provides a systematic means of tracking and transforming the specific sanskaras that you carry during your lifetime.

Seven Seals — The Seven Seals describe the specific pattern of awakening for humanity as the higher currents of involution and grace move, over time, through our species. Outlined in allegorical form in the Revelation of St John, the unfolding of the Seven Seals can be understood as a sequential predestined awakening code hardwired into all human DNA.

As each of the Seven Seals is mystically opened, all aspects of the human wound, both individual and collective, will one day be healed. The teachings of the Seven Seals are contained within the transmission of the 22nd Gene Key, whose highest aspect is Grace. In the teaching of the Venus Sequence, you learn the precise science and underlying patterning of your suffering. As your awareness moves deeper into these shadow patterns, you may become aware of the functioning of the Seven Seals as they affect your own individual awakening. At even deeper levels, you may become aware of how the Seven Seals are gradually opening up within the body of humanity. Such insights will lead to a great welling up of compassion and peace inside you.

Shadow (Frequency) — The frequency band relating to all human suffering. The shadow frequency band emerges from an ancient wiring in the human brain. Such wiring is based on individual survival and is linked directly to fear. The unconscious presence of fear in our system continues to enhance our belief that we are separate from the world around us. This deep-seated belief propagates a victim mentality, since the moment we believe we are separate, we feel vulnerable and at the mercy of outside forces. When we live at the shadow frequency band, then we live within a culture of blame and shame. We blame those forces and people that we believe are outside us, and we feel shame when we believe we alone are responsible for our lives. Once you begin to understand how the shadow frequency controls the majority of people in the world, including yourself, you realise how simple it is to move out of its grasp. Simply by shifting your attitude, you release the creative currents hidden within the shadow frequencies, and your life takes on a higher purpose. Your very suffering becomes the source of your salvation. Thus begins your journey away from those internal patterns and traits that keep you believing you are a victim and towards the inherent genius and love that is your true nature.

Shakti — The divine rain that accompanies spontaneous manifestation of grace. Shakti is a subtle fluidic emanation of the most refined frequencies imaginable. It moves from the highest subtle bodies, such as the atmic and buddhic, down into their lower counterparts. When one has lived a life of service and surrender, or if one happens to be the recipient of grace, then shakti will pour through the subtle vehicles, allowing you to experience higher states of absorption and embodiment.

Siddhi (Frequency) — The frequency band relating to full embodiment and spiritual realisation. The very concept of frequencies and levels paradoxically dissolves when the truth is realised as a Siddhi. The word siddhi is from the Sanskrit meaning divine gift. The siddhic state only comes about when all vestiges of the shadow, particularly at a collective level, have been transformed into light. As you enter the state of absorption, this alchemical transformation begins to accelerate until, finally, all falls silent, and you enter the state of embodiment at the Sixth Initiation. There are 64 Siddhis and each one refers to a different expression of divine realisation. Even though the realisation is the same in each case, its expression will differ and can even appear contradictory. The Siddhis spoken of in the Gene Keys Synthesis are not to be confused with the way they are understood in certain other mystical traditions. The 64 Siddhis are not obstacles on the path to realisation but are the very expression and fruition of realisation.

The Six Lines — Relating to the hexagram structure of the I Ching, the six lines describe further nuances of each of the 64 Gene Keys. If we see each Gene Key as a pre-designed archetypal picture, then each line is like the colour of that picture. Once you can see the colour, the whole picture comes alive. Knowing the six lines and their keynotes is an essential skill to master, as it enables you to interpret the many elements of the Hologenetic Profile in a simple and accessible way. There are many, many layers of keynotes for each of the six lines and they are fun to learn and illuminating to apply. The deeper you can feel the resonance of the six lines inside your being, the easier it is to understand your own sequences and share that resonance with others.

Solar Plexus — The area of the body below the diaphragm down to and including the perineum. In the understanding of the Corpus Christi – the science of the rainbow body – the solar plexus is the alchemical laboratory where the higher Christ consciousness is gestating. A great transmutation is occurring in this area throughout the human gene pool, and it will in time bring a new kind of human, Homo Sanctus, into our world. It is through the complex ganglia of the solar plexus that our individual, racial and collective memories (sanskaras) are being transformed. The genetic mutations in our gut will dramatically change our physiology – gradually human beings are evolving the means to draw and digest higher frequency currents into our bodies.

This will, in time, allow us to live on subtler kinds of nutrition. The likelihood is that the human beings of the future will therefore eat less dense foods. They will stop eating meat and eventually even stop eating plants. As our diet changes, so will our solar plexus evolve its new function – to be the equipment for our higher awareness. It is through the belly that human beings will one day realise their oneness with all creation.

Synarchy — The universal principle through which collective intelligence naturally aligns itself in perfect harmony with all that is. Synarchy is the underlying nature of humanity that can only be known once it has emerged from the shadow frequencies. As the new collective consciousness dawns all across our planet, humanity will self-organise its creative genius and manifest the true higher purpose hidden in its DNA – to bring about the New Eden. Whereas the Shadow consciousness manifests on the material plane through the principle of hierarchy, and the Gift consciousness through the principle of heterarchy, the Siddhic consciousness manifests through the principle of synarchy.

Synthesis — The universal principle through which collective intelligence sees and understands the holographic nature of reality. As humanity begins increasingly to understand the world around us as a perfectly interwoven fractal pattern, with us as the eyes of that pattern, we will enter into the Age of Synthesis. In order to know what synthesis means, you must first unlock the latent genius within your DNA. The very manifestation of that genius is to see beyond any single discipline. At the level of genius, one can see and directly know the interconnectedness of all patterns and disciplines. As humanity begins to move more as a synarchy, we will begin to externalise the truth of synthesis.

Syntropy — The universal principle through which collective intelligence transforms itself physically in order to become a synarchy. The law of syntropy holds that all energy in the universe is intricately ordered, even at its most chaotic. The essence of the law of syntropy is contained in the phrase 'Give and you shall receive'. In a syntropic world, everything is entangled with everything else, and everything is responsible for everything else. In human terms, the collective embodiment of syntropy would mean the end of selfishness. When you give for the sake of the whole, you activate the currents of grace and move into synchronicity with the whole.

As human genius awakens, we will organise our living systems to mirror nature and the law of syntropy. A world based upon this law would look utterly alien to us today. It will give us the means to harness free energy anywhere and anytime, and it will demand that we eventually eradicate money from our world. It is money that is the great collective symbol of conditional giving.

Synchronicity — The universal principle of good fortune that is activated whenever you think, act and speak from a higher frequency – that is from a place of unconditional giving. Synchronicity is a universal law at all levels of frequency. When you understand the perfection of the cosmos in every moment, then you can only surrender and trust life. Synchronicity is summed up by the phrase 'Nothing is by chance'. Every event that occurs in your life is an invitation to evolve and unlock the hidden higher gifts inside you. As your higher functioning allows you to see the truth of synarchy, synthesis and syntropy, you will see that the three laws are one, and they are bound together by synchronicity.

Transmission — A higher field of consciousness whose sole purpose is to penetrate and awaken those aspects of itself that still remain unaware of their greater reality. Most transmissions assume the form of a teaching or set of teachings that enter the world exactly when the world is ready to receive them. All transmissions follow natural fractal lines as they spread throughout humanity, bringing with them a higher order of consciousness. Even though transmissions may take the form of words and practices, their true nature lies shrouded in mystery. The Gene Keys transmission is a part of the wave of awakening generated as the Great Change is felt in the world.

Transmutation — The process of dynamic and permanent change that comes about as you surrender and accept mutation. At the Shadow frequency, mutation is something that is greatly feared as it always challenges an established pattern, rhythm or routine. Unless mutations (periods of natural upheaval) are embraced and fully accepted in your life, transmutation cannot occur. Transmutation involves a complete shift from one state or plane to another. After a transmutation, nothing is ever the same again. Transmutation only begins to occur at the Gift frequency band as your deepest cellular victim patterns are transformed through awareness. Periods of intense mutation in your life are always a great opportunity for transmutation.

So long as your attitude is open-hearted and accepting and you embrace and take responsibility for your own state, transmutation will occur in your life. With transmutation come great clarity, freedom and creativity. It is the process through which your genius emerges into the world.

Trinity — The underlying nature of all manifest form. As form emerges from the undivided state of the formless, it spontaneously moves from one to three. To the human mind and our current perception, it appears that all around us is a binary. Because of the limitations of our current awareness, the trinity at the heart of creation is not easily apparent. The trinity is the reflection of the infinite, whereas the binary is the reflection of the finite. As you raise the frequency of your DNA and begin to see through a higher kind of vision, one of the first patterns you will recognise is the trinity. As the fundamental building block of the holographic universe, the pattern of the trinity allows life to keep on transcending and evolving. It ensures that nothing can ever be consistent and fixed, even though things may appear so. As you begin to attune your mind and heart to the secret nature of the trinity, you will begin to relax more and more deeply into the truth of the inherent self-ordering perfection of the universe.

Venus Sequence — The primary genetic sequence for unlocking deep core emotional patterns in your life. As the central axis of the Golden Path, the Venus Sequence is an inner contemplative journey into the dynamics of your specific emotional wounding patterns, inherited through your ancestral DNA. Constructed from the positions of the Earth, Moon, Mars and Venus at the time of your birth, the Venus Sequence outlines a natural pathway of six Gene Keys which govern all your emotional patterns in this life. As your awareness begins to understand and observe the shadow aspects of these six Gene Keys, particularly in your relationships, you will begin to transform the low-frequency patterns into higher-frequency gifts. In this way, your astral body (your emotional nature) moves through a process of transmutation leading to the permanent opening of your heart. During our current phase of evolution known as the Great Change, the Venus Sequence has particular relevance, as its underlying purpose is to open up a new centre of awareness in the solar plexus.

BIBLIOGRAPHY

INTRODUCTION
p. xxvii:

"In the beginner's mind there are infinite possibilities. In the expert's there are few."

Suzuki, Shunriyu (2006). *Zen Mind, Beginner's Mind*. Shambhala.

1ST GENE KEY
p. 2:

"A measure of the disorder or unavailability of energy within a closed system. More entropy means less energy available for doing work."

http://www.pbs.org/faithandreason/physgloss/entropy-body.html

16TH GENE KEY
p. 111:

Quote by holocaust survivor Elie Wiesel

US News & World Report (October 27, 1986).

34TH GENE KEY
p. 272:

"The very softest thing of all
can ride like a galloping horse
through the hardest of things.

Like water, like water penetrating rock.
And so the invisible enters in.

That is why I know it is wise
to act by doing nothing.
And how few, how very few understand this!"

Kwok, Man-Ho; Palmer, Martin; Ramsay, Jay (2003). *The Illustrated Tao Te Ching*. Vega.

36TH GENE KEY
p. 286:

The Beatitudes.

Matthew: 5. *The Bible*. New International Version.

BIBLIOGRAPHY

40ᵗʰ Gene Key

p. 315:

"The individual who senses his aloneness, and only he, is like a thing subject to the deep laws, the cosmic laws. If a person goes out into the dawn or gazes out into the e vening filled with happenings, if he senses what happens there, then all situations fall away from him as from someone dead, even though he stands in the midst of life."

Rilke, Rainer Maria; Burnham, Joan M., translator (2000). *Letters to a Young Poet*. New World Library.

53ʳᵈ Gene Key

p. 414:

The Gospel of Buddha. Compiled from ancient records by Paul Carus (1894). Web publication, Mountain Man Graphics, Australia.

56ᵗʰ Gene Key

p. 457:

"And in the Time to Come,
The Holy One will make a banquet for the Righteous
from the flesh of the leviathan,
and its skin will be used to cover the tent
where the banquet will take place."

The Talmud. Baba Bathra, 74b.

p. 458:

"As it draws you to itself
What pleasure your suffering becomes.
Its fires are like water
Do not tense your face.
To be present in the soul is its work,
And to break your vows.
By its complex art
These atoms are trembling in their hearts."

Rumi, Jelaluddin; Helminski, Kabir Edmund, translator (2005). *The Rumi Collection*. Shambhala.

THE 64 GENE KEYS
SPECTRUM OF CONSCIOUSNESS

	SIDDHI	GIFT	SHADOW		SIDDHI	GIFT	SHADOW
1	Beauty	Freshness	Entropy	33	Revelation	Mindfulness	Forgetting
2	Unity	Orientation	Dislocation	34	Majesty	Strength	Force
3	Innocence	Innovation	Chaos	35	Boundlessness	Adventure	Hunger
4	Forgiveness	Understanding	Intolerance	36	Compassion	Humanity	Turbulence
5	Timelessness	Patience	Impatience	37	Tenderness	Equality	Weakness
6	Peace	Diplomacy	Conflict	38	Honour	Perseverance	Struggle
7	Virtue	Guidance	Division	39	Liberation	Dynamism	Provocation
8	Exquisiteness	Style	Mediocrity	40	Divine Will	Resolve	Exhaustion
9	Invincibility	Determination	Inertia	41	Emanation	Anticipation	Fantasy
10	Being	Naturalness	Self-Obsession	42	Celebration	Detachment	Expectation
11	Light	Idealism	Obscurity	43	Epiphany	Insight	Deafness
12	Purity	Discrimination	Vanity	44	Synarchy	Teamwork	Interference
13	Empathy	Discernment	Discord	45	Communion	Synergy	Dominance
14	Bounteousness	Competence	Compromise	46	Ecstasy	Delight	Seriousness
15	Florescence	Magnetism	Dullness	47	Transfiguration	Transmutation	Oppression
16	Mastery	Versatility	Indifference	48	Wisdom	Resourcefulness	Inadequacy
17	Omniscience	Far-Sightedness	Opinion	49	Rebirth	Revolution	Reaction
18	Perfection	Integrity	Judgement	50	Harmony	Equilibrium	Corruption
19	Sacrifice	Sensitivity	Co-Dependence	51	Awakening	Initiative	Agitation
20	Presence	Self-Assurance	Superficiality	52	Stillness	Restraint	Stress
21	Valour	Authority	Control	53	Superabundance	Expansion	Immaturity
22	Grace	Graciousness	Dishonour	54	Ascension	Aspiration	Greed
23	Quintessence	Simplicity	Complexity	55	Freedom	Freedom	Victimisation
24	Silence	Invention	Addiction	56	Intoxication	Enrichment	Distraction
25	Universal Love	Acceptance	Constriction	57	Clarity	Intuition	Unease
26	Invisibility	Artfulness	Pride	58	Bliss	Vitality	Dissatisfaction
27	Selflessness	Altruism	Selfishness	59	Transparency	Intimacy	Dishonesty
28	Immortality	Totality	Purposelessness	60	Justice	Realism	Limitation
29	Devotion	Commitment	Half-Heartedness	61	Sanctity	Inspiration	Psychosis
30	Rapture	Lightness	Desire	62	Impeccability	Precision	Intellect
31	Humility	Leadership	Arrogance	63	Truth	Inquiry	Doubt
32	Veneration	Preservation	Failure	64	Illumination	Imagination	Confusion

Embrace the Shadow, release the Gift, embody the Siddhi.

WATKINS
1893

The story of Watkins began in 1893, when scholar of esotericism John Watkins founded our bookshop, inspired by the lament of his friend and teacher Madame Blavatsky that there was nowhere in London to buy books on mysticism, occultism or metaphysics. That moment marked the birth of Watkins, soon to become the publisher of many of the leading lights of spiritual literature, including Carl Jung, Rudolf Steiner, Alice Bailey and Chögyam Trungpa.

Today, the passion at Watkins Publishing for vigorous questioning is still resolute. Our stimulating and groundbreaking list ranges from ancient traditions and complementary medicine to the latest ideas about personal development, holistic wellbeing and consciousness exploration. We remain at the cutting edge, committed to publishing books that change lives.

DISCOVER MORE AT:
www.watkinspublishing.com

| Read our blog | Watch and listen to our authors in action | Sign up to our mailing list |

We celebrate conscious, passionate, wise and happy living.
Be part of that community by visiting

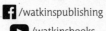 /watkinspublishing   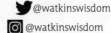 @watkinswisdom

 /watkinsbooks   @watkinswisdom